## CLIMATE CHANGE, PUBLIC HEALTH, AND THE LAW

*Climate Change, Public Health, and the Law* provides the first comprehensive explication of the dynamic interactions between climate change, public health law, and environmental law, both in the United States and internationally. Responding to climate change and achieving public health protections each require the coordination of the decisions and behavior of large numbers of people. However, they also involve interventions that risk compromising individual rights. The challenges involved in coordinating large-scale responses to public health threats and protecting against the invasion of rights make the law indispensable to both of these agendas. Written for the benefit of public health and environmental law professionals and policymakers in the United States and in the international public health sector, this volume focuses on the legal components of pursuing public health goals in the midst of a changing climate. It will help facilitate efforts to develop, improve, and carry out policy responses at the international, federal, state, and local levels.

Michael Burger is the executive director of the Sabin Center for Climate Change Law at Columbia Law School.

Justin Gundlach is a staff attorney at the Sabin Center for Climate Change Law at Columbia Law School.

# Climate Change, Public Health, and the Law

*Edited by*

MICHAEL BURGER

Columbia University, New York

JUSTIN GUNDLACH

Columbia University, New York

CAMBRIDGE
UNIVERSITY PRESS

**CAMBRIDGE**
**UNIVERSITY PRESS**

University Printing House, Cambridge CB2 8BS, United Kingdom

One Liberty Plaza, 20th Floor, New York, NY 10006, USA

477 Williamstown Road, Port Melbourne, VIC 3207, Australia

314–321, 3rd Floor, Plot 3, Splendor Forum, Jasola District Centre, New Delhi – 110025, India

79 Anson Road, #06–04/06, Singapore 079906

Cambridge University Press is part of the University of Cambridge.

It furthers the University's mission by disseminating knowledge in the pursuit of education, learning, and research at the highest international levels of excellence.

www.cambridge.org
Information on this title: www.cambridge.org/9781108417624
DOI: 10.1017/9781108278010



First published 2018

Printed in the United States of America by Sheridan Books, Inc.

*A catalogue record for this publication is available from the British Library.*

ISBN 978-1-108-41762-4 Hardback

# Contents

# About the Contributors

## EDITORS AND AUTHORS

- *Michael Burger* is the executive director of the Sabin Center for Climate Change Law and a research scholar and lecturer in Law at Columbia Law School.
- *Justin Gundlach* is a staff attorney at the Sabin Center for Climate Change Law and an associate research scholar at Columbia Law School.

## AUTHORS

- *Colleen Healy Boufides* is a staff attorney at the Network for Public Health Law for the Mid-States Region.
- *Maxine Burkett* is a professor of law at the University of Hawai'i's William S. Richardson School of Law and a global fellow at the Woodrow Wilson International Center for Scholars.
- *Robin Kundis Craig* holds the James I. Farr Presidential Endowed Chair in Law at the University of Utah's S. J. Quinney College of Law. She also is an affiliate faculty member at the Wallace Stegner Center for Land, Resources, and Environment and at the Global Change & Sustainability Center, as well as serving on the executive board of the Water Center at the University of Utah.
- *Chandra Ganesh* is an assistant professor for health sciences in the Department of Nursing and Health Science at California State University East Bay.
- *Sara Hoverter* is a staff attorney at the Harrison Institute for Public Law and an adjunct professor at Georgetown University Law Center.
- *Jennifer Klein* is an associate attorney in Kramer Levin Naftalis & Frankel LLP's New York office.

- *Jill Krueger* is a director of the Northern Region Office of the Network for Public Health Law.
- *William Onzivu* is a technical officer with the World Health Organization's Regional Office for Africa, and a lecturer in law at the Bradford University School of Law, West Yorkshire, United Kingdom.
- *Margot Pollans* is an associate professor at the Elisabeth Haub School of Law at Pace University, and the faculty director of the Pace-NRDC Food Law Initiative.
- *Jason Smith* is an assistant professor and associate chair of the Health Sciences Program in the Department of Nursing and Health Sciences at California State University East Bay.
- *David C. Vladeck* holds the A. B. Chettle Chair in Civil Procedure at the Georgetown University Law Center, is faculty director of the Law Center's Center on Privacy and Technology, and was the director of the Federal Trade Commission's Bureau of Consumer Protection from 2009 to 2012.
- *Jessica Wentz* is a staff attorney at the Sabin Center for Climate Change Law and an associate research scholar at Columbia Law School.
- *Lindsay F. Wiley* is a professor of law and the director of the Health Law and Policy Program at American University's Washington College of Law.

# Foreword

## Gina McCarthy

For more than thirty-five years I have worked for environmental agencies at every level of government, for both Democratic and Republican administrations, most recently as the Administrator of the US Environmental Protection Agency under President Barack Obama. Over those years, I have come to realize that most people are unaware that EPA is a public health agency or that human health concerns are at the core of our nation's environmental laws. EPA is not the birds, bunnies, or polar bears agency – although we all love birds, bunnies, and polar bears. EPA's efforts to deliver clean air, water, and land, as well as a stable climate, are all focused on environmental protections that are essential to the health and well-being of American families. Saving lives and avoiding impacts and illnesses is EPA's stock in trade.

Too often of late, the importance of environmental protections for public health has been lost in the political rhetoric that inevitably accompanies the prospect of new or more stringent environmental rules, especially when it comes to climate change action. This is more than just a little worrisome, since climate change is the single greatest threat to human health and our children's future. We need quick, broad, and deep actions in this country and internationally to reduce the carbon pollution that fuels climate change and to adapt to the impacts we can no longer deny or avoid. Given the enormity of the climate change challenge, this effort will require a level of public support and engagement that can only come from a common understanding of the threat that climate change poses, and the tremendous public health benefits that carbon reductions provide.

This book, *Climate Change, Public Health, and the Law*, drives home the connections between environmental protection and public health, and it highlights a path forward in the fight against climate change, in two new and important ways. First, by focusing on the ways climate change impacts health, the book makes climate change personal, which should inspire greater support for climate action. Second, by surveying the complex relationships between climate change and the responsibilities of the public health sector it highlights not just the need, but also numerous and specific opportunities to take action to protect public health and well-being.

The book examines the real costs of climate change – not how much it would cost to address the problem, but the ongoing and inevitable human toll climate changes exact from the health of our children, our elderly, the poor, and all those most vulnerable to the impacts of pollution, illness, and disease. It uses both a legal lens and a public health framework to articulate the human health threats posed by climate change, explore the challenges associated with government actions, and make the case for the kind of cross-sector, multidisciplinary response that climate change demands. At a time when the US government has taken a lunch break from its job of protecting public health and welfare, this book should help focus attention on what climate risks really are, and what we should actually be doing about them.

A recent report released by the Yale Program on Climate Change Communication and the George Mason University Center for Climate Change Communication, based on data gathered from a nationally representative survey in the United States back in 2016, revealed that most American's (70 percent) understand that climate change is real and most (59 percent) know that humans are causing it, but that only one in three Americans (35 percent) think global warming is affecting our weather right now, and most see it as a distant threat to future generations (70 percent) or to people in developing countries or the world's poor (62 percent).[1] This suggests that while Americans have accepted the reality of a changing climate, they don't view it as a direct threat to American families today.

This widespread misunderstanding should serve as a wake-up call. Until Americans acknowledge that climate change is an ongoing personal risk to them and their families, we cannot hope to garner the support we need to make the next big leap forward. And, of course, climate change is not about some far off challenge in distant countries, or something that will only be felt by people in some far off time. It's already posing direct and indirect threats to public health in the United States today. As this book makes clear, climate change is, right now, robbing too many Americans of their ability to pursue health, happiness, and prosperity. Climate change doesn't just threaten those who live along our coasts or in low-lying areas prone to flooding; climate

[1] Anthony Leiserowitz et al., *Climate Change in the American Mind* (New Haven, CT: Yale Program on Climate Change Communication, 2017).

impacts aren't limited to hurricanes, forest fires, and droughts – heat waves, infectious diseases, severe storms, and aeroallergens will affect us all. The chapters collected in this volume offer a range of law and policy responses to deal with these threats, across the board.

Using language that is even in tone and clinical in perspective, *Climate Change, Public Health, and the Law* describes how climate change poses broad and ongoing threats to each one of us and to all of us collectively. This clear message is enormously important for the people of this country to hear right now – and that includes people working in the areas of environment, public health, and the law. As citizens of the strongest and most prosperous nation on Earth, we cannot fail to embrace the challenge of addressing climate change with the passion, fervor, innovation, ingenuity, and investment that we have brought to bear in the past when our lives and livelihoods have been at risk. And we cannot allow our country's leaders to stick our collective heads in the sand when it is abundantly clear to the vast majority of us that climate action is needed to keep us healthy and safe and to keep our country's economy strong.

We have made significant progress domestically and internationally over the past two decades in our fight against climate change, and I am proud of the work I performed at EPA to help shape and deliver on President Obama's Climate Action Plan (CAP). EPA set limits on carbon pollution from cars, trucks, landfills, and power plants. We made progress in setting methane limits for the oil and gas production sector and implemented legislation requiring reporting of greenhouse gas emissions from numerous industrial sectors. We considered climate change in programs, permits, and grants, and worked with other federal agencies, as well as states, cities, and tribes to help them adapt to a changing climate. We supported State Department efforts to secure the 2015 Paris Agreement. And as EPA Administrator, I led the US delegation, with the strong support of Secretary of State John Kerry, in the successful negotiation of an amendment to the Montreal Protocol to establish a timeline to phase out the production and use of hydrofluorocarbons, pollutants with a high global warming impact.

Admittedly, it is discouraging to see many of these climate actions facing potential rollback in the Trump Administration. But I am confident that EPA based these final standards on science and the law, and I know each one benefitted from an extraordinary outreach and engagement effort. Ultimately, courts will decide on the legality of any rollbacks, but it is heartening to know that the Trump Administration's denial of climate science and its early efforts to undo federal public health safeguards have reawakened over a dozen states, hundreds of cities, and more than a thousand businesses who have embraced the goals of the 2015 Paris Agreement and are working harder than ever to provide needed US leadership. These actions are helping chart new courses and continue progress in our fight against climate change, especially with the increasing engagement of the private sector and the growth of clean energy – the fastest growing economic sector in the country. In fact, no matter what actions are rolled or not rolled back, the climate change actions taken during the Obama Administration have substantially paid off for our pocketbooks, not just our health. The clean energy train has left the station and there is no turning back. The energy transition – away from fossil fuels to renewable sources and increasing efficiency – will deliver direct benefits to public health, including thousands of fewer premature deaths, asthma attacks, heart attacks, missed school and work days, and so on.

While progress in the fight against climate change will continue, we are nowhere near the carbon pollution trajectory that science demands. We need deeper and faster reductions, and we need more focus on how to address the inevitable impacts climate change is having, and will have, on us as individuals and on the communities we live in.

My hope is that this book will spark a broader conversation among the public health sector and health caregivers, legal scholars, lawyers, and the American public, about how to respond to the direct and indirect threats that climate change poses to public health. The book provides a unique and useful framework to more effectively understand, communicate, and grapple with the human health impacts of climate change we are experiencing now and can expect to face in the near future.

If my thirty-five years in public service has taught me anything, it has taught me that people take the direct health threat to their children personally. While many of us care about nature, any parent (myself included) would tell you that we live and die for our children. If this book succeeds in raising the visibility of the enormous public health impacts that climate change is already imposing on our children, then maybe we can begin to generate the kind of can-do, must-do, and will-do attitude that I have come to expect in this country.

I know no better way to make climate change front and center than to talk clearly about the threat it poses to our kids' health and future – right here in America, as well as in the developing world, and what to do to meet that threat head on. This book makes the case, in the same way we have made the case, that traditional pollutants in our air, water, and land pose threats to public health. Pollution – including carbon pollution – hurts those most vulnerable – our kids, the elderly, and others – and it denies them their right to live healthy lives. These arguments have been the basis for federal, state, and local laws, regulations, and policies over the past four decades that have successfully reduced traditional pollutants.

As this book rightly points out, we need the active engagement of all levels of government, and with the American public, to develop and implement multi-sector, interdisciplinary strategies to reduce carbon emissions and adapt to our changing climate. I could not agree more. Climate change is not solely an environmental issue; it is at once a public health, environmental, economic, and national security challenge. That is why President Obama's Climate Action Plan demanded action across all federal agencies, created a climate cabinet to cut through the silos and ensure that multiple agencies were providing input into the design and implementation of each action, and incorporated extensive public outreach and engagement efforts at each step of the way.

This book also examines a number of specific tools and strategies – some that have already been utilized and others to be considered – to reduce risks to human health through carbon emission reductions and investments in adaptation efforts that will protect people and properties from anticipated climate impacts. It uses a public health framing to examine the task of preparing for the health impacts of climate change: how we build the capacity we need to effectively assess the threats, develop policies to reduce those threats, and develop a plan to promote public health, and finally provide assurance that the conditions that allow people to be healthy are restored. It identifies research and governance gaps that must be filled in order to deliver the cross-sector, multidisciplinary approach necessary to address climate change. It also examines ways in which our laws and our courts have been both providing opportunities for climate action and struggling with the unique legal challenges that climate change poses.

Clearly, the time is right for this book and for the larger conversation and work ahead. *Climate Change, Public Health, and the Law* has all the ingredients needed to motivate people to demand action in our country, and it provides a useful approach to engage developing nations in the fight against climate change. For example, supporting actions in the developing world to clean up the air pollution that is choking their cities, causing untold health impacts on those most vulnerable, and limiting their economic opportunities will also inevitably result in carbon pollution reductions that help protect our collective future.

If we work together at home and abroad to amplify the connection between public health and climate change, we will spur cities, states, tribes, and businesses to work together to examine and implement new tools and strategies – informed by science and the law – that can drive human health and climate benefits. And we will enable continued US international leadership by opening up productive avenues to clean up significant sources of pollution while advancing a clean energy future in the developing world.

Since leaving Washington, DC, I have become a full-time professor of the Practice of Public Health at the T. H. Chan School of Public Health at Harvard University, and taken on the role of director of the Center for Health and the Global Environment with the goal of focusing increased attention on the intersection of health and climate change. I for one will do what I can to expand the conversation that this book has begun – for the sake of my children's health and our collective future.

# PART I

# THE CONTEXT

# 1

# Introduction

Michael Burger and Justin Gundlach

In September 2016 the Conservation Law Foundation (CLF), a regional environmental advocacy group in New England, filed a cutting edge climate change lawsuit in federal district court in Boston against ExxonMobil. CLF's complaint claimed that the fossil fuel giant had failed to adequately prepare its Everett Terminal facility for climate change impacts, including sea level rise and increasingly intense storms, putting the Mystic River at risk for pollution and nearby communities at risk of toxic exposure. One year later, Judge Mark L. Wolf ruled that CLF had standing to sue – meaning CLF's claimed injury was one considered by the law to be a violation of its rights – but only for injuries in the present or "near future." The group did not have standing to sue "for injuries that allegedly will result from rises in sea level, or increases in the severity and frequency of storms and flooding, that will occur in the far future, such as in 2050 or 2100."[1] The decision's implications are clear in one respect: Although CLF could not sue over climate risks projected to manifest later this century, it could sue over pollution that is evidently exceeding the facility's permitted thresholds now or exceedances that are foreseeable in the immediate future. But less clear is whether the court recognized that climate-driven pollution risks will certainly occur in both time frames, or instead wrongly views climate change as a phenomenon that will not manifest in legally significant ways until "the far future."

Judge Wolf was understandably reticent to make his courtroom the place where consideration for climate change could be added to the requirements of existing environmental protection laws – he said he wanted to avoid the case becoming "the Scopes Monkey Trial of the 21st Century."[2] But climate change

[1] Order, *Conservation L. Foundation, Inc.* v. *ExxonMobil Corp.*, C.A. No. 16-11950-MLW, at 2 (D. Mass. Sept. 13, 2017).

[2] Adrianne Appel, *Tone down climate change in suit against Exxon: Judge, Bloomberg* BNA, Sept. 12, 2017, https://perma.cc/8824-5253

is not a problem for the future. As the 2017 Atlantic hurricane season made plain, it is a problem now. Just two weeks before Judge Wolf issued his order, Hurricane Harvey struck Houston and its surrounding Harris County, dumping so much rain so quickly that the National Weather Service had to modify the color-coding on its maps to convey the volume of precipitation.[3] Chemical plants and refineries in the area were caught unprepared for the deluge and attendant flooding, and toxic substances were released into the air and flowed from their grounds into the surrounding area and out into the Gulf of Mexico as a result.[4] This storm was the third so-called 500-year storm to strike Houston and Harris County in just three years. It was not caused by climate change per se – coastal storms would strike Houston even if humans were not adding greenhouse gases to the atmosphere – but there is little doubt that climate change amplified it by warming and swelling the waters of the Gulf and warming the air that carried all that rain.[5] The same can be said of Hurricane Irma, which caused catastrophic damage on a number of Caribbean islands, and Hurricane Maria, which devastated Puerto Rico, all within weeks of Harvey. For those engaged in public health protection and management, these scenarios are the new normal.

The extent of the risks climate change poses to public health and the immediacy of the demand for effective policy responses have been well documented by highly visible, prominent bodies, including the Intergovernmental Panel on Climate Change, established as a joint project of the World Meteorological Organization and the United Nations Environment Program and serving as the entity primarily responsible for developing and issuing scientific reports that inform international climate action; the United States Global Change Research Program, a federal program mandated under the Global Change Research Act of 1990 and tasked with developing the National Climate Assessment every five years; and the *Lancet* Commission on Health and Climate Change, an international

3 National Weather Service, Twitter, Aug. 28, 2017, 7:21 AM, https://perma.cc/7XW4-PZCR ("#Harvey in perspective. So much rain has fallen, we've had to update the color charts on our graphics in order to effectively map it.")

4 Michael Biesecker and Frank Bajak, *Evidence of Spills at Toxic Site During Floods*, AP, Sept. 17, 2017, https://perma.cc/9GGK-UMA5; *Globs of Mercury Wash up After Harvey*, E&E News Greenwire, Sept. 7, 2017, https://www.eenews.net/greenwire/2017/09/07/stories/1060059977; see also Dena Adler, *Hurricanes' Contaminated Floodwaters Might Crest Next Wave of Climate Change Litigation*, Climate Law Blog, Sept. 19th, 2017, https://perma.cc/Z29V-UYLE (describing other examples of toxic releases).

5 Kerry Emanuel, Assessing the present and future probability of Hurricane Harvey's rainfall, *PNAS Early Edition*, Oct. 4, 2017, https://perma.cc/XPV3-8VG2; Lisa Friedman and John Schwartz, How Hurricane Harvey became so destructive, N.Y. *Times*, Aug. 28, 2017, https://perma.cc/7WQC-V5TU.

specialist group of academic scientists formed to map climate impacts on public health and policy responses. The *Lancet* Commission's 2015 report concluded that "tackling climate change could be the greatest global health opportunity of this century."[6] It also recommended that governments invest in research to better understand adaptation needs and the public health co-benefits of climate change mitigation; invest in creating public health systems that are themselves resilient in the face of climate change impacts; direct the transition away from coal as a fossil fuel relied on for energy around the world; and develop sustainable and resilient cities.[7]

As the chapters in this book illustrate, these recommendations are valid and important but merely point in the direction of a necessary set of concrete actions. Others have taken the important intermediate step of describing some of the policy goals and tools that should inform those actions,[8] but those further prescriptions for public health-oriented climate action do not say enough about the legal and institutional contexts in which action is to be taken. And those contexts will, of course, shape and limit whatever might be done. This book, whose authors and editors are expert in the law as well as the policy areas considered, spans that remaining gap. It is the first to do so.

This book can be useful in more than one way. Climate change affects public health in ways that are direct (such as through heat and severe storms), indirect (such as through diseases), and mediated through human institutions (such as through food insecurity). In many instances these effects will exacerbate health problems that already disproportionately impact vulnerable populations. In all instances, achieving public health goals in relationship to climate change effects will mean somehow persuading decision makers of their present relevance. In the broader context of public debates over regulations and spending, the fact of dramatic public health impacts arising from these effects can itself be useful. Several researchers, observing the politicization of climate science and climate policy discussions over the past thirty years, have explored the utility of presenting climate-related issues within different frames, such as public health or national security, as a means of overcoming the filters that prevent the simple presentation of valid data or analysis from leading to greater public understanding, much less inspiring

6 N. Watts et al., Health and climate change: Policy responses to protect public health, 386 *Lancet* 1861–1914, at 1861 (2015); World Health Organization, Climate Change and Human Health: Publications, www.who.int/globalchange/publications/en/ (accessed May 9, 2017).

7 Ibid.

8 See, e.g., Barry Levy and Jonathan Patz, *Climate Change and Public Health* (2015); Howard Frumkin et al., *Climate Change: The Public Health Response*, 98 Am. J. Pub. Health. 435 (Mar. 2008).

decisive action.[9] These researchers' results suggest that framing climate change as a public health problem is no panacea for intransigence on climate action,[10] but that it can potentially matter a great deal.[11]

While public buy-in would be valuable and maybe even necessary, it would not be sufficient to prompt effective public health-oriented climate action – the full scope of the public health sector's response to climate change and the implications of climate change for public health law must be more fully articulated as well. There is an emerging consensus, reflected in the chapters collected here, that there are a number of key goals that ought to define public health law and policy in the Anthropocene era. These include fortifying public health infrastructure against climate change impacts, making public health systems more resilient, coupling public health considerations to any and all mainstreaming of climate change adaptation planning, as well as using the public health frame to inform the appropriate degree of mitigation ambition – the acceptable level of climate risk we, as a global community, are willing to endure.

Thus *Climate Change, Public Health, and the Law* serves at least three important purposes: (1) to provide a foundation for further research into and a more engaged discourse on the dynamic interplay between climate change, public health, and the law, at multiple scales of governance and across the spectrum of climate action, from reducing greenhouse gas emissions to preparing for climate change impacts; (2) to inform stakeholders and decision makers seeking to more effectively address the public health impacts arising from climate change about overarching themes, cross-cutting issues, interdependencies, strategies, and techniques at the intersection of climate law and public health policy; and (3) to provide those actors with information that can help frame key arguments about and deliver actual implementation of important climate-oriented measures or policy changes that hold promise for public health outcomes.

9 See, e.g., Edward W. Maibach et al., Reframing climate change as a public health issue: An exploratory study of public reactions, 10 *BMC Pub. Health* 299 (2010).

10 Compare S. P. Singh and M. Swanson, How issue frames shape beliefs about the importance of climate change policy across ideological and partisan groups, 12 *PLoS ONE* e0181401 (2017) ("Rather than being unaffected by frames, those on the right and Republicans tend to rate climate change policy as *less* important than their counterparts in the control group when the issue is framed."), with Teresa A. Myers et al., A public health frame arouses hopeful emotions about climate change: A Letter, 113 *Climatic Change* 1105 (2012).

11 Solange Gould and Linda Rudolph, Challenges and opportunities for advancing work on climate change and public health, 12 *Int'l J. Envtl. Research Pub. Health* 15649 (2015) (quoting description of public health officials speaking of climate-driven risks during a political campaign as "a messenger beyond reproach, with a message beyond reproach").

So much for its uses. As for this book's contents: It offers a kaleidoscopic view of how the law makes sense of and responds to the impacts of climate change on public health. Its thirteen chapters are arranged in four parts. The first part sets out information about legal rights and public health impacts that is basic to the rest. Coeditor Michael Burger sets the legal stage for the chapters that follow with an examination of how climate change's ubiquitous impacts on public health resonate in rights – including the right to health and human rights – and how those rights for individuals implicate duties for governments to take action to address climate change. From the panoply of rights implicated by climate change impacts, Burger distills five types of governmental obligation: procedural, relating to mitigation, relating to adaptation, international cooperation, and the safeguarding of human rights. In addition to these governmental duties, Burger notes that several sources and authorities have also identified private sector obligations to mitigate and adapt to climate change. Having identified sources and types of rights and duties, Burger then discusses a number of lawsuits, like the CLF lawsuit that began this chapter, in which plaintiffs have sought to vindicate one or more rights and to hold governments or corporate actors accountable for the duties arguably assigned to them by international or domestic law. He gives particular attention to the US context, where questions of rights and duties relating to climate change impacts have been the subject of little legislation but heavy doses of litigation. The status of climate-related jurisprudence rooted in constitutional, common law, and statutory theories in the United States is mixed: Partial success at grounding demands for action by the federal government in the Clean Air Act is counterbalanced by struggles to gain purchase on other legal footing. Insofar as these struggles have resulted from the challenge of persuading a court to worry about *future* harms, the steady accumulation of *current* impacts traceable to climate change holds some promise for turning the legal tide.

Jill Krueger and Colleen Healy provide a technical and institutional overview by mapping the connections between climate change and public health, exploring the challenges these connections pose to the public health sector's ability to carry out its core mission, and proposing several cross-cutting law, policy, and practice approaches that should help inform any public health agency's responses to climate change. Their chapter offers a summary of vulnerabilities and exposures to public health-related climate risks, including heat, air pollution and allergens, extreme weather events, infectious disease, food insecurity, and mental health. Subsequent chapters drill down deeper into each of these areas, but here Krueger and Healy offer a comprehensive overview that sets the stage for their discussion of gaps in, strengths of, and next steps for the public health sector's response. Critically, the authors identify three ways public health practitioners can immediately engage with climate change: incorporating climate change into their existing work; accounting for climate change in long-term public health planning; and bringing their public health expertise, along with the principles of health equity, to bear on specific climate change mitigation and adaptation efforts. Krueger and Healy also offer a multiplicity of analytic and decision-making frameworks that can help shape the integration of climate change response into public health praxis.

The book's second part considers cross-cutting issues that arise both from the public health impacts of climate change and responses to those impacts in key contexts: legal, institutional, and physical. David Vladeck, a former director of the Federal Trade Commission's Consumer Protection Bureau, launches this part by asking the question: *To what extent can governments respond to the public health impacts of climate change by acquiring, using, disseminating, and/or controlling information in a manner intended to influence educate consumers and the public or else to shape public policy?* Vladeck argues that the government has a great deal of discretion when issuing particular public health messages or allocating funding to promote them, though the First Amendment does impose some limits on what the government can require of prospective contractors. He then turns to the issues of information disclosure (and the state of the law on First Amendment limits on government-mandated disclosure) and the role of the government in policing the marketplace for greenwashing and other false or misleading speech. These are issues that, both in relation to climate change and in their own right, are somewhat unsettled, with clear room for further development. To help advance the cause, Vladeck proposes four government initiatives: emissions disclosure requirements for gas-powered consumer products; climate risk disclosure requirements for publicly traded companies; codification of climate disclosures in environmental reviews under the National Environmental Policy Act; and disclosure of government subsidies for fossil fuel companies. The proposals include important caveats, and taken together they cannot fill the entirety of the climate change information gap, but they represent an important set of steps that can be taken now.

The risk of increased exposure to existing and new infectious diseases due to climate change raises a different information quandary than Vladeck's: how to adapt existing institutions and processes to surveil the population and manage the information obtained in ways that effectively address the potential for contagion. Jason Smith and Chandra Ganesh provide a unique introduction to this question, describing the history and scope of public health surveillance in the United States and identifying legal and policy challenges that the public health community must face to stave off the threats posed by climate change. As a practical matter, there are data issues and silo effects to deal with: climate change projections are often developed by one set of actors, who are not well-coordinated with public health departments. As a consequence, the data may be of a type that is not geared toward public health practitioners, may not be immediately available to them, and may be managed by agencies with different goals or located in entirely different jurisdictions (or at different scales of government). Interestingly, Smith and Ganesh challenge the notion that preparedness is the most appropriate model with which to grapple with climate change and public health problems, and posit that practitioners may need to develop a more dynamic set of solutions that can endure the long haul between the present and whenever it is that society gets a grip on climate change.

Climate change triggers the need to produce, gather, disclose, and manage a variety of information types, including direct and indirect risks that greenhouse gas emissions and climate impacts present to businesses, communities, and individuals. It also produces the need to respond to that information. In Chapter 6, co-editor Justin Gundlach and Jennifer Klein describe how climate change ripples through the built environment to impact public health, and how governments and other actors go about the process of adapting to those impacts as they become aware of them. The process and substance of adaptation may seem both familiar and relatively intuitive to those already steeped in the practice, but to those new to the space it can appear overwhelming. This chapter offers an on-ramp, with clear-eyed descriptions of how and what happens in adaptation planning and a universe of resources in the notes for those wanting or needing to go deeper. The chapter also offers an important legal lens, exploring the legal claims that could be brought against governments and private actors both for enacting adaptation measures later found to have deleterious effects and failing to adapt to known risks. The authors detail the universe of takings and negligence claims and defenses in a way that will clarify the situation for the experienced attorney but that can also be understood by the non-lawyers among us.

The third part of the book focuses on particular impacts and sectors where those impacts are felt, and also considers the responses those impacts prompt – or ought to prompt. It begins with heat, which, as Sara Hoverter points out, is typically the first deleterious effect that comes to mind when thinking of climate change, and is the cause of severe direct and indirect impacts on public health. Despite being both obvious and harmful, heat is not an easy problem to address. Hoverter's chapter maps the sprawling and diverse array of laws, institutions, programs, assets, communities, and individuals somehow touched by the public health impacts of heat waves and hotter average temperatures. In sketching that map, Hoverter highlights its fragmentary nature and notes the problems this fragmentation creates for coordinating responses among government agencies, the private sector, community organizations, and individuals in ways that can achieve public health goals. She also points out a number of duties rooted in state constitutions that seem likely to make state governments not only responsible for addressing the risks posed by heat to public health but also potentially liable legally for harms resulting from a failure to address those risks to an adequate degree. It is uncertain whether and how these duties might motivate efforts to better integrate consideration of and responses to heat, but pressure to do so is sure to increase as temperatures and the intensity, frequency, and duration of heat waves rise.

If heat is the first thing that comes to mind when thinking of climate change (the state of affairs formerly known as "global warming"), sea level rise and extreme storms are a close second. Robin Kundis Craig performs a feat similar to Hoverter in her explication of the interrelated public health effects of climate change impacts in the coastal zone. The submersion of populated coastal areas, storm-related flooding and disasters, and the bleaching of coral reefs are high visibility, high impact consequences of climate change; but Craig layers into her survey other dangers, such as the risk of toxic contamination, the loss of water supply from saltwater intrusion, and disease. Taken together, Craig's portrait presents a situation of extraordinary public health risk to coastal areas and demands an innovative set of responses. For her part, Craig recommends a cross-sectoral, "multivalent" approach, one that infuses preparedness to projected impacts not only into disaster risk reduction strategies and emergency preparedness but also into construction standards and siting requirements for critical public infrastructure and industrial facilities, building codes, and remediation standards for contaminated sites. Her call for coordination among laws – disease control and prevention, natural resources, land use, pollution control, water, food, and fisheries management – as well as among actors, resonates with much of the material collected in this book, and highlights the recurrent theme that these new challenges require thinking and acting laterally across sectors and issues.

Climate change is already altering ecosystems, and with them the pathways through which infectious diseases find their way to human hosts. As these changes become still more pronounced, so too will their consequences for public health. Lindsay Wiley describes these changes and consequences, but observes also that the resulting need for integrated responses across policy areas is made especially difficult by key features of the public health field. These include its increasing orientation to health care, as well as the fragmentation of responsibility for basic prerequisites to public health among agencies specializing in environmental protection, sanitation, agriculture, food and drug safety, and workplace health and safety. Having pointed out these barriers to an effective response to novel infectious disease outbreaks like Zika virus, Wiley describes efforts to overcome them, such as the One Health and Health in All Policies approaches. Both of these support transcending narrow policy areas in service to public health goals. They also both align well with a social-ecological public health model, which recognizes that individual choices are constrained by physical, economic, and social circumstances, and which is better suited to the task of climate change adaptation than older and more individualistic agent-based and behavioral public health models. Having identified these approaches as versions of the path forward for addressing climate-driven infectious disease impacts, Wiley acknowledges that this path runs headlong into tensions between individual and collective rights and duties that have long featured in the public health field.

As food is basic to public health, the impacts of climate change on the food system will give rise to public health impacts. Margot Pollans observes that climate change will have significant adverse impacts on the food system, which standard public analyses tend to evaluate using the four criteria that together make up the "food security" rubric: availability, utilization, access, and stability. But, as Pollans explains, that rubric can work against environmental protection efforts and provides little or no insight into the implications of governance, ownership, or control of the means of food production. She argues, therefore, that relying solely on that rubric to guide action responsive to climate-driven challenges would almost certainly fall short of addressing key public health risks. Instead, based on her survey of climate impacts on the food system, current adaptation efforts, and the state of the law as it relates to food production in the United States, she suggests grounding responses in the alternative approaches supported by the "food sovereignty" rubric. In contrast to food security, food sovereignty concerns itself with issues of agroecological production practices, the significance of localizing food production, equity, the preservation of traditional knowledge about food production, and self-determination among producers and consumers. Although seemingly less aligned to public health concerns, this rubric holds greater transformative potential for the food system, and so Pollans proposes adding its elements to the toolkit of adaptive responses to climate change's impacts.

Climate change-induced displacement, migration, and relocation, whether internal or across national borders, might be thought of, in and of themselves, as public health impacts of climate change. Maxine Burkett, however, goes far beyond this superficial point to elucidate the myriad ways in which population movement – both during the course of movement and in the short- and long-term aftermath of temporary or permanent resettlement – can more specifically impact both individuals' and communities' health. Physical impacts, mental health impacts, and new vulnerabilities are all associated with migration, whether it be forced displacement following an extreme event, planned relocation under government supervision, or a more ad hoc form of migration over time. Public health law and existing immigration and migration laws are, Burkett argues, ill prepared to deal with these scenarios. But there are important efforts underway to identify the salient gaps and create appropriate legal and policy reforms. Here, Burkett discusses how identification of at-risk populations and their current conditions – a kind of vulnerability mapping; new regimes to ensure the rights and dignity of climate migrants or those displaced by climate-related disasters, especially vulnerable populations such as women and children; and effective planning to build resilience for individuals and communities, whether they return and remain in place or undertake a planned relocation, can help reduce the risk of adverse effects. Some strategies are prudent in the short-term, and reflect a "no regrets" approach, such as reducing emissions to reduce overall climate change, providing aid and technology transfer to assist the developing countries most vulnerable to climate change, establishing standards for host communities, and increasing coordination. Other strategies, Burkett suggests, may require further research, further thinking, and more time.

The final part of the book turns to the dynamic relationship between climate change, public health, and environmental law and regulation – first in the international context, then in the domestic one. Dr. William Onzivu observes that developing countries can expect to see some of the most severe effects of climate change and to experience the most dramatic impacts on public health, yet they are often less well equipped than the developed world to adapt. He then considers how international organizations and legal frameworks are coming to recognize these points, and to address them. He traces the recent advent of more explicit and direct recognition of the climate-health nexus in instruments adopted by parties to the United Nations Framework Convention on Climate Change, most notably the Paris Agreement. In parallel to the increasingly clear articulation of health as a priority in international climate law, the World Health Organization (WHO) has led international efforts to act on the adverse consequences of climate change for public health. WHO's contributions chiefly include facilitation of cooperation across and within national boundaries, and the collection and provision of useful information. In addition to observing these legal and institutional steps at the international level, Dr. Onzivu examines how they inform implementation efforts within national boundaries, such as in new climate-oriented legislation or through the reformulation of existing laws to take climate change or health into account to a greater degree. Challenges abound, however, and Dr. Onzivu notes that consideration for public outcomes has not been thoroughly integrated into international law or programmatic goals, and further that national governments have struggled to translate new legislation and policy goals into action. He proposes steps to integrate a greater degree of health-oriented adaptive governance at both levels.

In Chapter 13 Justin Gundlach demonstrates that public health has long been central to pollution control laws in the United States, such as the Clean Air Act and the Resource Conservation and Recovery Act. However, these laws are premised, to a greater or lesser extent, on a notion of "stationarity," in which future environmental conditions roughly replicate existing (now former) baselines. Gundlach proposes that the changes to air quality, and to exposure pathways, brought about by changing climatic conditions challenge these existing frameworks, and require direct engagement from regulators and legislators alike. In addition, public health has come to play a central role in the ongoing debates over how and to what extent greenhouse gas emissions should be regulated, as the so-called co-benefits of greenhouse gas regulation – especially the benefits to public health from reduced air pollution – provide a prominent legal and policy rationale, as well as a powerful communications frame, for climate action. At the same time, climate action itself can pose public health risks, or so-called co-harms. Navigating this tricky terrain, Gundlach proposes several ambitious reforms that could help provide greater clarity and certainty: adoption of new legislation that provides for inclusion of climate change factors in environmental permitting decisions, development of analytic methodologies for assessing co-benefits and co-harms that can withstand judicial scrutiny, and insurance and bonding measures to provide greater financial coverage.

In addition to the public health impacts of greenhouse gas emissions felt via the mechanisms of climate change, other impacts also arise from efforts to *mitigate* those emissions and to *adapt* to the changing climate. Jessica Wentz's survey of tools for assessing these impacts begins by pointing out not just the ubiquity of such impacts but their importance – positive or negative. That importance argues for their consistent and thorough consideration using a Health Impact Assessment (HIA) of some sort as part of climate-oriented policy development and implementation. However, Wentz finds that the legal frameworks supporting such consideration, whether a formal HIA or otherwise, are diverse, fragmentary, and generally partial in their coverage. At the federal level, Regulatory Impact Analysis and Environmental Impact

Assessments (EIAs) get at some of these considerations some of the time. At the state level, practice is highly varied, with some states requiring thorough reviews not just of health impacts incidental to environmental impacts, but full formal HIAs. At the local level and with respect to individual facilities, review is similarly mixed – sometimes thorough, more often partial or wholly absent. Building on the map she draws of legal frameworks that support incorporation of HIA (or something like it) into government actions and decisions, Wentz identifies ways that existing approaches sometimes unduly constrain HIA review, and, of course, the susceptibility of HIAs to resource constraints that effectively limit their scope, depth, and quality. Her recommendations for dealing with these gaps and limitations call for building on and improving existing frameworks and authority, not just by making HIA part of decision-making but also by streamlining review processes to make that additional aspect of regulatory processes practically feasible.

* * *

This book charts the relationships between climate change, public health, and the law in 2018, with particular focus on the United States. These relationships are tangled, complex, and rife with gaps as well as content. The authors and editors aim to bring the nature and arrangement of these relationships into fuller view, and also to provide a foundation for further research and action – as both are sorely needed. Perhaps the most important pieces of that foundation are a set of four problems that recur across the contexts discussed herein:

*Policy silos.* Those responsible for formulating or implementing policy of one sort often do so in ignorance of how their decisions will affect – or could be improved by – information gathered and analyses conducted by those responsible for an officially unrelated policy area. This is especially obvious in relation to the built environment (Chapter 6) and food and agriculture (Chapter 10), and can be addressed in part by public health practitioners becoming more directly engaged in climate change mitigation and adaptation planning efforts (Chapter 3) and by employing health impact assessments in different contexts (Chapter 14).

*Governance gaps.* Law tends to emerge in response to the experience of a problem or conflict. Thus novel problems – or old problems on an unprecedented scale – often go unaddressed legally, or are addressed only indirectly or incidentally by existing laws. Clear examples of this relate to climate change itself (Chapter 2), the United States' surprisingly informal approach to disease surveillance (Chapter 5), the heat impacts climate change generates (Chapter 7), and its effects on coasts through sea level rise and storms (Chapter 8), among others.

*Lacking institutional capacity*. Similarly, because existing institutions are generally built to deal with past circumstances rather than present ones, problems of an unprecedented nature or scale can exceed institutional capacities and are rarely if ever met with an adequate response that was scaled-up in anticipation or out of precaution. Infectious disease (Chapter 9) and migration (Chapter 11), as well as adaptation efforts in the developing world (Chapter 12), each provide examples of this phenomenon.

*Costly information and underinformed policy choices*. The costs of thoroughly examining policy options are often more readily apparent and immediately felt than the benefits of such examination. This can be true even when the benefits are likely to include revelations so significant that they could sway the policy choice. In consequence, as the examples of regulatory co-benefits (Chapter 13) and health impact assessment (Chapter 14) illustrate, policy makers can miss significant opportunities as a result of this simple bias against investments in discovering more information about their options. On the other hand, as discussed in relation to governmental or compulsory speech (Chapter 4), remedies are available to address this sort of bias among individual citizens and institutional actors.

It is difficult to draw broad conclusions across this diverse array of climate risks forcing new and various responses in the multiplicity of law and policy settings covered in this book. Yet, it is obvious that the problems climate change presents for the community of lawyers, policy professionals, medical professionals, academics, researchers, and others engaged in public health practice demand creative law and policy responses. This book provides not only the foundation but also a set of practical recommendations for carrying this work forward.

# 2

# The Duty to Protect Public Health from Climate Change Impacts

Michael Burger

*Under existing international and domestic law, governments and private actors have obligations to individuals and communities affected by climate change impacts on public health. This chapter discusses such obligations and the legal authorities – international, constitutional, statutory, and common law – under which they arise.*

## 2.1 INTRODUCTION

International law has long recognized that environmental quality is essential to individuals' and communities' full enjoyment of fundamental human rights, including the right to health and the right to life. The extraordinary impacts of climate change have laid bare the depth and extent of the interdependence between environmental protection, a stable climate system, and this panoply of human rights. The World Health Organization, for example, has estimated that climate change could cause some 250,000 additional deaths per year from malnutrition, malaria, diarrhea, and heat stress between 2030 and 2050,[1] and the Intergovernmental Panel on Climate Change Fifth Assessment Report, published in 2014, details how climate change will adversely affect billions of people all around the world, along with the ecosystems, natural resources, and social and physical infrastructure with and within which we live.[2] These impacts will result both from extreme events that pose a direct threat to human lives and safety, such as severe storms and heat waves, and slow-onset changes that can undermine access to clean air and clean

[1] World Health Organization, *Quantitative Risk Assessment of the Effects of Climate Change on Selected Causes of Death, 2030s and 2050s* (2014); World Health Organization, *Fact sheet no. 266, Climate Change and Health*, www.who.int/mediacentre/factsheets/fs266/en/

[2] IPCC, *Climate Change 2014: Impacts, Adaptation, and Vulnerability, Contribution of the Working Group II to the Fifth Assessment Report of the Intergovernmental Panel on Climate Change* (Cambridge: Cambridge University Press, 2014).

water, detract from food security, and threaten the habitability of human settlements. This vision of a world at risk is not science fiction, futurist speculation, or a theoretical construct. It reflects impacts that are happening now, including increased temperatures, altered hydrologic conditions, degraded ecosystem functioning, decreased agricultural productivity, and the immediate prospect of displacement for some communities at risk from sea level rise, desertification, and disasters brought on by extreme events such as hurricanes and wildfires.

This chapter surveys government and private actor obligations to take action to protect communities and individuals from the adverse impacts of climate change on public health, both through reducing greenhouse gas (GHG) emissions and adapting to climate change impacts. The chapter operates from the assumption that the science of climate change attribution is well enough settled to satisfy any call for proof of causation, triggering the need to take action even in the face of uncertainty about the efficacy of any individual action. Section 2.2 reviews some of the ways in which climate impacts infringe on the right to health and other, related rights, both in the United States and abroad. Section 2.3 then examines the obligation to tackle the climate change–public health nexus under a rights-based approach to international law. Section 2.4 does the same in regard to US law. Section 2.5 concludes with reference to the ethical imperative to satisfy the duty to take action on climate change.

## 2.2 CLIMATE IMPACTS ON THE RIGHT TO HEALTH AND OTHER FUNDAMENTAL RIGHTS

The Intergovernmental Panel on Climate Change (IPCC)'s Fifth Assessment Report (AR5) provides a detailed picture of how observed and predicted climatic changes will adversely affect global populations and the ecosystems, natural resources, and physical infrastructure upon which they depend.[3] The AR5 also provides an of-the-moment summary of climate change impacts on human health.[4] Similarly, the US Global Change Research Program

3 See ibid.

4 K. R. Smith et al., Human health: impacts, adaptation, and co-benefits. In: IPCC, *Climate Change 2014: Impacts, Adaptation, and Vulnerability. Part A: Global and Sectoral Aspects. Contribution of Working Group II to the Fifth Assessment Report of the Intergovernmental Panel on Climate Change* (Cambridge: Cambridge University Press, 2014, 709–754).

(USGCRP)'s Third National Climate Assessment describes how climate change will disrupt ecosystems, infrastructure, and human health within the United States,[5] providing a detailed look at specific climate impacts on human health.[6] The nature and scope of these impacts are summarized in Chapter 3 of this volume and restated in various permutations and combinations in the chapters that follow, and I will not reiterate them all here. However, it bears noting that climate change touches on almost if not every aspect of human civilization: our freshwater resources, terrestrial ecosystems and species, coastal systems and low-lying areas, oceans, and food security; our key economic systems and infrastructure, and our urban and rural places; our livelihoods, our health, and our security.[7]

The IPCC's AR5 and the USGCRP's Third Assessment Report reflect different foci, but they are in accord on at least three critical points in regards to the impacts of climate change on human health. First and foremost, both recognize that climate change threatens human health and well-being in a great variety of different ways, including through illnesses transmitted by food, water, and disease-carriers; direct health impacts from increased extreme weather events, wildfire, and decreased air quality; and the threats to mental health caused by extreme events such as storms and heatwaves, as well as temporary and permanent displacement. Importantly, both reports also recognize that some of these public health impacts are already occurring. Second, climate change not only creates new threats but also amplifies existing ones, particularly for vulnerable populations, such as children, the elderly, the sick, the poor, some communities of color, and some workers. That is, people already at increased risk of asthma or heat-related illness will see those risks increase in many instances as temperatures rise, air quality worsens, and extreme heat events become more common. Third, all hope is not lost. Appropriate responses from the public health community, including increasing preparedness and prevention as forms of climate change adaptation, can have a significant impact in their own right, and lessen adverse health outcomes associated with climate change.

A less well-appreciated facet of the climate problem is that government and private sector responses to climate change can also affect individuals' and

5 US Global Change Research Program, National Climate Assessment (2014).

6 George Luber et al., Ch. 9: Human health, In: J.M. Melillo et al. *Climate Change Impacts in the United States: The Third National Climate Assessment*, 220–256. (U.S. Global Change Research Program, 2014) doi: 10.7930/J0PN93H5.

7 For a more detailed description of how climate change impacts enumerated human rights, see United Nations Environment Programme, *Climate Change and Human Rights*, 2–10, (December 2015) http://columbiaclimatelaw.com/files/2016/06/Burger-and-Wentz-2015-12-Climate-Change-and-Human-Rights.pdf.

communities' enjoyment of the right to health, along with other fundamental rights.[8] This is true both for mitigation actions and projects undertaken to adapt to the impacts of climate change. For example, hydroelectric projects financed and developed through international climate finance mechanisms can lead to displacement of local people and the destruction of the ecosystems upon which they depend; they can also reduce river flows, harming the health and livelihoods of those living downstream.[9] Similarly, biofuels policies and projects can contribute to food shortages and price shocks, water stress and scarcity, deforestation, and land acquisitions that displace indigenous peoples and small-scale farmers.[10] Meanwhile, adaptation efforts might result in secondary impacts on other communities, such as when a sea-wall forces wave action onto a neighboring community and increased erosion or increased storm risk. Given these realities, the obligation to protect against adverse impacts to public health plainly pertains not only to the imperative to take action on climate change, but to the actions taken, as well.

## 2.3 INTERNATIONAL HUMAN RIGHTS LAW

The obligation under international human rights law to take action on climate change to protect human health is best understood in relationship to the long-running discourse on the right to a healthy, clean, and/or functioning environment, and governments' commitments to protect the right to health. The idea of such a rights-based obligation is primarily based on the rights enumerated in the Universal Declaration of Human Rights (UDHR), the

8 For a broad and insightful exploration of these issues, see Damilola Olawuyi, *The Human Rights Based Approach to Carbon Finance* (Cambridge: Cambridge University Press, 2016). For a more targeted analysis about how to address these issues in implementation of the Paris Agreement, see Sebastien Duyck et al., *Delivering on the Paris Promises: Combating Climate Change while Protecting Rights Recommendations for the Negotiations of the Paris Rule Book* (Center for International Environmental Law, 2017).

9 See, e.g., IFC Compliance Advisor Ombudsman, Complaint, Guatemala / Real LRIF-01/ Coban (Oct. 29, 2014) (complaint submitted to International Finance Corporation alleging social and environmental harms from Santa Rita dam); Sébastien Jodoin and Katherine Lofts, *Economic, Social & Cultural Rights and Climate Change: A Legal Reference Guide* 68–69 (Centre for International Sustainable Development Law, GEM Initiative, and Academics Stand Against Poverty 2013); Naomi Roht-Arriaza, *"First, Do No Harm": Human Rights and Efforts to Combat Climate Change*, 38 *Ga. J. Int'l. & Comp. L.* 593, 597–598 (2010).

10 See, e.g., Food and Agriculture Organization, *Climate Change Adaptation and Mitigation: Challenges and Opportunities in the Food Sector*, 7 (2012); International Council on Human Rights, *Climate Change and Human Rights: A Rough Guide*, 33–36 (2008); United Nations Development Programme, *Human Development Report 2007/2008, Fighting Climate Change: Human Solidarity in a Divided World*, 143 (2007).

International Covenant on Civil and Political Rights (ICCPR), and the International Covenant on Economic, Social and Cultural Rights (ICESCR), but commitments under the United Nations Framework Convention on Climate Change (UNFCCC) and under the auspices of the World Health Organization (WHO), as well as other agreements, also carry weight. This section reviews the legal foundations of the right to environment and the right to health, before turning to the linkages between climate change, public health, and legal obligations under international law.

### 2.3.1 *The Right to a Clean Environment*

Despite numerous academic and civil society efforts over the years to develop a basis for agreement on a new treaty, convention, or agreement,[11] international law does not yet recognize a freestanding right to a clean, healthy, and/or functioning environment. However, governments, including the US government, have frequently recognized that environmental harms, including climate change, can undermine the effective enjoyment of other enumerated rights, such as the rights to health, life, food, and water.[12] Some UN human rights treaties make the link explicit. For instance, the ICESCR advances "improvement of all aspects of environmental and industrial hygiene" in order to achieve full enjoyment of the right to health,[13] and the Convention on the Rights of the Child directs states to account for the "dangers and risks of environmental pollution" to ensure full implementation of the right to health for children.[14] What is more, various declarations, reports, and other soft law instruments support recognition of the linkage, going all the way back to the 1972 Stockholm Declaration.[15] The Office of the UN High Commissioner for Human Rights (OHCHR) has interpreted the Stockholm Declaration to reflect a "general recognition of the

[11] This movement has been centered most recently on France, where two different groups are advancing versions of a global treaty. The Centre International de Droit Comparé de l'Environment has put forward the *Draft of the International Covenant on the Human Right to Environment*, available at https://cidce.org/wp-content/uploads/2016/08/Draft-of-the-International-Covenant-on-the-Human-Right-to-the-Environment_15.II_.2017_EN.pdf. Le Club des Juristes has convened a drafting committee and is working with high-level politicians to advance the *Global Pact for the Environment*. See www.leclubdesjuristes.com/toward-a-global-pact-for-the-environment-action-for-the-planet-action-through-law/.

[12] United States Department of State, *Observations by the United States of America on the Relationship Between Climate Change and Human Rights*, para. 14, www.state.gov/documents/organization/138855.pdf.

[13] International Covenant on Economic, Social and Cultural Rights, art. 12(2)(b).

[14] Convention on the Rights of the Child, art. 24(2)(c).

[15] Declaration of the UN Conference on the Human Environment ("Stockholm Declaration"), Principle 1 (June 16, 1972).

interdependence and interrelatedness of human rights and environment."[16] Offering further support, the UN Independent Expert on Human Rights and the Environment concluded that "[h]uman rights law includes obligations relating to the environment."[17] Finally, the right to a clean environment has either been expressly included in or else interpreted as a component of regional human rights agreements and national and sub-national constitutions.[18]

### 2.3.2 *The Right to Health*

The right to the enjoyment of the highest attainable standard of physical and mental health, or the right to health, was first articulated in the World Health Organization's Constitution, which states in the preamble that "the enjoyment of the highest attainable standard of health is one of the fundamental rights of every human being without distinction of race, religion, political belief, economic or social condition."[19] Article 25 of the Universal Declaration of Human Rights includes health as part of the right to an adequate standard of living.[20] Article 12 of the ICESCR enshrines the right to health as one of the enumerated human rights recognized by international law.[21]

Achieving the right to health is closely related to the vindication of other human rights, including the right to food, housing, work, education, nondiscrimination, access to information, and participation.[22] But the right to health also has a direct relationship to the prospective right to a clean environment – or at the very least to environmental quality. Article 12 of the ICESCR provides that the steps to be taken by states to achieve the full realization of that right "shall include those necessary for ... the improvement of all aspects of environmental and industrial hygiene."[23] The Convention on the Rights of

[16] OHCHR further explains: "the United Nations human rights treaty bodies all recognize the intrinsic link between the environment and the realization of a range of human rights ... " Office of the United Nations High Commissioner for Human Rights, *Report of the OHCHR on the Relationship between Climate Change and Human Rights*, paras. 17-18 (Jan. 15, 2009) UN Doc. A/HRC/10/61.

[17] John H. Knox, *Report of the Independent Expert on the Issue of Human Rights Obligations Relating to the Enjoyment of a Safe, Clean, Healthy and Sustainable Environment*, paras. 79–81, (United Nations Human Rights Council Dec. 30, 2013) UN Doc. A/HRC/25/53.

[18] See, e.g., David Boyd, *The Environmental Rights Revolution: A Global Study of Constitutions, Human Rights, and the Environment* (2012).

[19] Constitution of the World Health Organization (1946).

[20] Universal Declaration of Human Rights, art. 25 (1948).

[21] International Covenant on Economic, Social and Cultural Rights, art. 12 (1966)

[22] For a more robust analysis of the right to health, see Committee on Economic, Social, and Cultural Rights, *General Comment No 14 on the Right to Health* (2000).

[23] International Covenant on Economic, Social and Cultural Rights, art. 12, para. 2(b).

the Child directs states to account for the "dangers and risks of environmental pollution" to ensure full implementation of the right to health for children.[24]

Moreover, as the Committee on Economic, Social and Cultural Rights has acknowledged, the right to health extends to the "underlying determinants of health," including, among other things, safe drinking water, adequate sanitation, safe food, adequate nutrition, adequate housing, and healthy working and environmental conditions.[25] The Committee has also stated that individuals should be given full and equal access to information concerning water and the environment, and in its responses to country reports, it has urged States to assess the impacts of actions that may have adverse environmental effects on the right to health and other rights within its purview.[26] Similarly, the Committee on the Rights of the Child, in its concluding observations on country reports, regularly recognizes environmental hazards as barriers to the realization of the right to health and other rights.[27] The Special Rapporteur on Hazardous Substances and Wastes noted the dangers that exposure to mercury through artisanal mining poses to women's enjoyment of their right to health.[28]

Climate change poses obvious challenges to satisfying the criteria necessary to protect the right to health, as it threatens availability of water and sanitation; impacts occupational health and safety; disrupts housing through severe storms and other disaster impacts; and threatens food security. The relationship between climate change and human rights – including the right to health – is explored more fully in the next section.

### 2.3.3 *Climate Change and Human Rights*

The discourse on human rights and climate change was arguably launched in December 2005, when the Chair of the Inuit Circumpolar Conference submitted a petition to the Inter-American Commission on Human Rights (IACHR) requesting relief for human rights violations resulting from the impacts of global warming on the Inuit.[29] The petition specifically alleged

24 Convention on the Rights of the Child, art. 24(2)(c).

25 Office of the United Nations High Commissioner for Human Rights, *Fact Sheet no. 31, The Right to Health* (2008); ICESCR, general comment No. 14 (2000).

26 International Covenant on Economic, Social and Cultural Rights, *Report*, sect. III(A)(1)

27 Convention on the Rights of the Child, Report, sect. II.

28 Calin Georgescu, *Report of the Special Rapporteur on the Human Rights Obligations Related to Environmentally Sound Management and Disposal of Hazardous Substances and Waste*, paras. 32,33 (United Nations Human Rights Council July 2, 2012) UN Doc. A/HRC/21/48.

29 Inuit Circumpolar Council Canada, *Inuit Petition Inter-American Commission On Human Rights To Oppose Climate Change Caused By The United States Of America* (2005), www.inuitcircumpolar.com/inuit-petition-inter-american-commission-on-human-rights-to-oppose-climate-change-caused-by-the-united-states-of-america.html.

that the United States had violated the Inuit's human rights by failing to adopt adequate GHG emissions controls. The IACHR never issued a decision on the merits, but the petition initiated an extensive dialogue about the human rights implications of climate change that continues to this day.[30]

In November 2007, the Small Island Developing States (SIDS) adopted the Male' Declaration on the Human Dimension of Global Climate Change.[31] The Male' Declaration explicitly recognized that "climate change has clear and immediate implications for the full enjoyment of human rights," and called upon the Conference of the Parties (COP) to the UNFCCC and the UN human rights bodies to assess the human rights implications of climate change. That same month, OHCHR publicly acknowledged that "climate change can adversely affect the fundamental human rights of present and future generations" and stated that governments have both moral and legal obligations to protect and promote basic human rights when tackling climate change.[32] A few months later the Human Rights Council issued a resolution, expressing concern that climate change "poses an immediate and far-reaching threat to people and communities around the world and has implications for the full enjoyment of human rights," and directing OHCHR to conduct a "detailed analytical study of the relationship between climate change and human rights" in consultation with states and other relevant international organizations and intergovernmental bodies.[33] Since that time, HRC and OHCHR have repeatedly reaffirmed the core connection between climate change and human rights, including the right to health.[34] Notably, in May 2016 OHCHR published a detailed report titled, *Analytical study on the relationship between climate change and the human right of everyone to the enjoyment of the highest attainable standard of physical and mental health*.[35]

30 Hari M. Osofsky, The Inuit petition as a bridge? Beyond dialectics of climate change and indigenous peoples' rights, 31 *Am. Indian L. Rev.* 675 (2007).

31 Malé Declaration on the Human Dimension of Global Climate Change (Nov. 14, 2007).

32 Office of the United Nations High Commissioner for Human Rights, *The Human Rights Impact of Climate Change, UN Joint Press Kit for Bali Climate Change Conference*, 3–14 (Nov. 2007).

33 United Nations Human Rights Council *Res.* 7/23, *Human rights and climate change* (March 28, 2008) UN Doc. A/HRC/Res/7/23.

34 See United Nations Human Rights Council, *Res.* 35/20, *Human Rights and Climate Change*, UN Doc. A/HRC/Res/35/20; United Nations Human Rights Council, *Res.* 32/33, *Human Rights and Climate Change*, UN Doc. A/HRC/Res/32/33; United Nations Human Rights Council, *Res.* 29/15, *Human Rights and Climate Change*, UN Doc. A/HRC/Res/29/15; United Nations Human Rights Council, *Res.* 26/27, *Human Rights and Climate Change*, UN Doc. A/HRC/Res/26/27; United Nations Human Rights Council, *Res.* 18/22, *Human Rights and Climate Change*, UN Doc. A/HRC/Res/18/22; United Nations Human Rights Council, *Res.* 10/4, *Human Rights and Climate Change*, UN Doc. A/HRC/Res/10/4. See also Office of the United Nations high Commissioner for Human Rights, *Key Messages on Human Rights and Climate Change*, www.ohchr.org/Documents/Issues/ClimateChange/KeyMessages_on_HR_CC.pdf

35 Office of the United Nations High Commissioner for Human Rights, *Analytical Study on the Relationship Between Climate Change and the Human Right of Everyone to the Enjoyment of*

At the same time, the public health impacts of climate change have long been central to the climate change conversation, and are even embedded in the UNFCCC framework. Article 1 of the UNFCCC defines the "adverse effects of climate change" as "changes in the physical environment or biota resulting from climate change which have significant deleterious effects on . . . human health and welfare."[36] Article 3 then calls upon Parties to the Convention to "take precautionary measures to anticipate, prevent or minimize the causes of climate change and mitigate its adverse effects,"[37] which as defined include health. Article 4 requires Parties to minimize the public health impacts of mitigation and adaptation projects and measures they undertake.[38] The Cancun Agreements, reached in 2010, explicitly incorporated reference to human rights.[39] The Paris Agreement, signed in 2015 and entered into force in 2016, in its preambular recitations specifically stated that "Parties should, when taking action to address climate change, respect, promote and consider their respective obligations on human rights, the right to health, the rights of indigenous peoples, local communities, migrants, children, persons with disabilities and people in vulnerable situations and the right to development . . ."[40]

### 2.3.4 *Obligations to Protect the Right to Health in the Age of Climate Change*

Governments are generally understood to be obligated to respect, protect, and fulfill human rights, meaning they must refrain from taking action that violates rights, protect rights against violations by third parties, and take measures to promote the realization of rights for all persons.[41] Numerous entities have recognized that governments have a general duty to "adopt legal and

*the Highest Attainable Standard of Physical and Mental Health* (May 6, 2016) UN Doc. A/HRC/32/23.

36 United Nations Framework Convention on Climate Change, art. 1(1).

37 United Nations Framework Convention on Climate Change, art. 3(3).

38 United Nations Framework Convention on Climate Change, art. 4(1)(f).

39 United Nations Framework Convention on Climate Change Conference of the Parties, *Report of the Conference of the Parties on its Sixteenth Session, Held in Cancun from 29 November to10 December 2010, Addendum, Part Two: Action Taken by the Conference of the Parties at its Sixteenth Session,* UN Doc. FCCC/CP/2010/7/Add.1.

40 United nations Framework Convention on Climate Change Conference of the Parties, *Adoption of the Paris Agreement* (Dec. 12, 2015) U.N. Doc. FCCC/CP/2015/L.9./Rev.1.

41 Some documents also recognize a related but distinct obligation to "promote universal respect for, and observance and protection of, all human rights." Vienna Declaration and Programme of Action, art. 1 (June 25, 1993); United Nations General Assembly, *Declaration on the Right and Responsibility of Individuals, Groups and Organs of Society to Promote and Protect Universally Recognized Human Rights and Fundamental Freedoms,* art. 1 (December 9, 1998) UN Doc. A/RES/53/144.

institutional frameworks that protect against, and respond to, environmental harm that may or does interfere with the enjoyment of human rights."[42] States have the discretion to balance environmental protection against other goals but, according to the Special Rapporteur on Human Rights and the Environment, "the balance struck cannot be unreasonable or result in unjustified, foreseeable infringements of human rights."[43]

The HRC, OHCHR, human rights treaty bodies, special procedures mechanisms, human rights tribunals, and scholars have all started to elucidate how these general human rights obligations translate in the climate change context.[44] Taken together, one can abstract five distinct obligations for governments that pertain to climate change and the right to health.[45] First, governments possess *procedural obligations* to ensure their citizens have access to relevant information, are involved in decision-making, and have access to justice. Second, governments possess *mitigation obligations* to reduce and control greenhouse gas emissions. Third, governments possess *adaptation obligations* to protect people from climate change impacts. Fourth, governments possess *international cooperation obligations*. Finally, governments possess *obligations to safeguard human rights* in the context of climate change policies and response measures. In addition to those five, sources reviewed also recognize *private sector obligations* to mitigate and adapt to climate change.

These obligations for governments and private actors, as filtered through domestic constitutions, have been tested in a number of lawsuits in foreign courts over the years, though precious little had been definitively resolved by the time of this writing.[46] For instance, in *Gbemre* v. *Shell Petroleum Development Company of Nigeria Ltd.* the Federal High Court of Nigeria held that gas flaring, which contributes to global warming as well as local

42 See also Office of the United Nations High Commissioner for Human Rights, *Mapping Human Rights Obligations Relating to the Enjoyment of a Safe, Clean, Healthy and Sustainable Environment: Focus Report on Human Rights and Climate Change*, 47 (June 2014).

43 United Nations Human Rights Council, *Report of the Special Rapporteur on the Issue of Human Rights Obligations Relating to the Enjoyment of a Safe, Clean, Healthy and Sustainable Environment*, para. 67 (Feb. 1, 2016) UN Doc. A/HRC/31/52.

44 See, e.g., United Nations Environment Programme, *Climate Change and Human Rights*, *supra* note 7+.

45 These obligations are discussed in greater detail in Michael Burger and Jessica Wentz, Climate change and human rights, In: James May and Erin Daly, *Human Rights and the Environment: Indivisibility, Dignity, and Legality*, (Cheltenham: Edward Elgar, forthcoming).

46 See generally Michael Burger and Justin Gundlach, *The Status of Climate Change Litigation: A Global Review* (United Nations Environment Programme, 2017).

air pollution, is an unconstitutional practice in breach of the fundamental human right to a clean and healthy environment, as protected under the Nigerian Constitution and the African Charter on Human and People's Rights.[47] In that case, however, the climate impacts of gas flaring were wholly secondary to more immediate impacts. More recently, international human rights law has been both the subject of and an element within climate change litigation. In *Greenpeace Southeast Asia et al.*, environmentalists and Filipino citizens petitioned the Philippine Commission on Human Rights to investigate whether a group of fossil fuel companies, the so-called Carbon Majors, violated Filipinos' human rights under domestic and international law.[48] In *Urgenda* v. *Kingdom of the Netherlands*, an environmental organization alleged the Dutch government's backsliding on its previous GHG emissions commitment violated, among other things, human rights protected under international law. Although the Hague District Court did not adopt that argument, it did refer to international human rights law in reaching its determination that there had been a violation of the government's duty of care.[49] (At the time of writing, that case is on appeal.) In *Leghari* v. *Pakistan* the Lahore High Court Green Bench reached a similar conclusion to the district court in *Urgenda*.[50] Plaintiffs in similar suits brought in Norway, Belgium, and Switzerland, among other places, also alluded to international human rights and obligations.[51] Finally, in May 2018 ten families from across Europe, Kenya and Fiji brought an action in the European Union General Court, alleging that the EU's failure to adequately address climate change violated European human rights law.[52]

[47] *Gbemre* v. *Shell Petroleum Dev. Co. Nigeria Ltd.*, EHC/B/CS/53/05 FHCNLR (Nigeria 2005), www.climatelaw.org/cases/case-documents/nigeria/ni-shell-nov05-judgment.pdf.

[48] Greenpeace Southeast Asia and Philippine Rural Reconstruction Movement, *Petition To the Commission on Human Rights of the Philippines Requesting for Investigation of the Responsibility of the Carbon Majors for Human Rights Violations or Threats of Violations Resulting from the Impacts of Climate Change* (hereinafter Petition of Greenpeace Southeast Asia), https://perma.cc/2S8RTTKN.

[49] *Urgenda Foundation* v. *Kingdom of the Netherlands*, HAZA C/09/00456689, appeal filed (Netherlands 2015).

[50] *Ashgar Leghari* v. *Federation of Pakistan*, W.P. No. 25501/2015 (Lahore High Court Green Bench, September 4, 2015).

[51] See Burger and Gundlach, *The Status of Climate Change Litigation*, *supra* note 46. For the most up-to-date compilation of such cases, see Sabin Center non-US Climate Change Litigation database: http://climatecasechart.com/non-us-climate-change-litigation/.

[52] Armando Ferrão Carvalho and Others v. The European Parliament and the Council, GC Case no. T-##/18.

The outcome of these and other cases will give shape to governments' and private actors' duties to address the public health effects of climate change in the years to come. But there is little question that a strong foundation exists in international law, and that this foundation can influence interpretations of domestic rights in different countries. The issue is, unfortunately, somewhat less clear the United States.

## 2.4 US LAW

The obligation under domestic law to take action on climate change to protect human health is a subject of no small amount of controversy. Some have sought to inscribe this obligation as a constitutional right, based on the Fifth Amendment's protections of life, liberty, and property. Others have turned to the common law, seeking redress through public nuisance and other torts. Still others have relied on federal statutes to force government and private action. Without making a pretense to comprehensiveness, this section reviews some of the potential constitutional, common law, and statutory bases for the right of the public – collectively and as individuals – to have health protected from the adverse impacts of climate change. (Most importantly, this chapter does not, in general, address legal rights that might relate to a failure to adapt to climate impacts. Those issues are addressed in Chapter 13.)

### 2.4.1 *Constitutional Rights*

The general view has long been that the United States does not recognize a constitutional right to health that imposes an affirmative obligation on the federal government to provide any particular level of health or environmental quality to citizens or residents. In *DeShaney* v. *Winnebago Cnty. Dep't of Social Servs.*, the Supreme Court held that "[a]s a general matter ... a State's failure to protect an individual against private violence simply does not constitute a violation of the Due Process Clause,"[53] explaining that while the Fifth Amendment's Due Process Clause bars the state from directly depriving "individuals of life, liberty, or property without due process of law ... its language cannot fairly be extended to impose an affirmative obligation on the State to ensure that those interests do not come to harm through other means."[54] This rule is subject to two exceptions: the "special relationship" exception and the "danger creation" exception."[55] The "special

53 *DeShaney* v. *Winnebago Cnty. Dep't of Social Servs.*, 489 U.S. 189, 197 (1989).
54 Ibid. at 195. 55 See, e.g., *L. W.* v. *Grubbs*, 974 F.2d 119, at 121 (9th Cir. 1992).

relationship" exception applies when the government takes an individual into custody against his or her will. The "danger creation" exception permits a substantive due process claim when the government's "deliberate indifference to their safety" places a person in jeopardy. To the extent that the federal government has recognized a right to health, it has made clear that any such theoretical right is narrowly tailored to avoid resembling a judicially enforceable one.[56]

Nonetheless, the specific applicability of the US Constitution to climate change and its impacts has been the subject of litigation seeking to force the federal government to take action. Two cases bear brief mention, one a substantive due process claim, one a takings claim.

In *Juliana* v. *United States*, twenty-one youth plaintiffs filed a lawsuit in federal district court against the US government, asking the court to compel the government to take action to reduce $CO_2$ emissions so that atmospheric $CO_2$ concentrations will, consistent with the global commitment to keep global warming below 2 degrees Celsius, be no greater than 350 parts per million by 2100, claiming that the nation's climate system was critical to their constitutional rights to life, liberty, and property, and that the federal defendants had violated their substantive due process rights by allowing fossil fuel production, consumption, and combustion at dangerous levels.[57] The plaintiffs also raised the issue of intergenerational equity, claiming that the government's failure to control $CO_2$ emissions constituted a violation of their constitutional right to equal protection before the law, as they were being denied the fundamental rights afforded to prior and present generations.[58] At the time of this writing, the case was still being litigated. A decision upholding the existence of any such constitutional rights would be a transformational moment in the development of climate law, and an order directing the federal government to take action in accordance with the best

56 See *Observations by the United States of America on "The Right to Health", Fact Sheet No. 31*, www.state.gov/documents/organization/138850.pdf.

57 See Complaint for Declaratory and Injunctive Relief, Kelsey Cascadia Rose Juliana, *Xiuhtezcatl Tonatiuh M. Et Al.* v. *United States.*, No. 6:15-cv-01517-TC (D. Or. Aug. 12, 2015).

58 Ibid. at para. 145. The complaint and other recent cases in the United States also allege that federal and state governments have violated their public trust obligation by failing to adequately mitigate the GHG emissions that contribute to climate change. See, e.g., Our Children's Trust, *State Legal Actions*, www.ourchildrenstrust.org/state-legal-actions/. These cases do not involve express claims to a right to health, but there is a clear relationship between governments' public trust obligations – which require the maintenance and preservation of common environmental resources for the benefit of current and future generations – and any such obligations. See David Takacs, *The Public Trust Doctrine, Environmental Human Rights, and the Future of Private Property*, 16 NYU Envtl. L.J. 711 (2008).

available climate science would mark a major victory in securing climate action. While this would be a welcome evolution, it would also represent a radical departure from domestic courts' longstanding reluctance to recognize and enforce environmental or health-based rights.

The case of *St. Bernard Parish Government* v. *United States* raised a different Fifth Amendment issue, namely, a takings claim.[59] The case focused on the role of the Mississippi River Gulf Outlet (MRGO) shipping channel in worsening the impact of Hurricane Katrina on the city of New Orleans. Since the Army Corps of Engineers finished excavating MRGO in 1968, natural wave action, storms, and the wakes of large ships had caused it to widen from 500 feet to nearly 2000 feet, and by 2005 its banks sat close to the levees built to protect New Orleans neighborhoods from flooding. The Federal Court of Claims found that the Corps could be held liable for damages for an uncompensated temporary taking due to the storm surge's impacts, because the Corps' management of the channel – and implicitly its failure to account for the changing conditions in part caused by climate change – may have contributed to the impacts. In April 2018 the Federal Circuit Court of Appeals reversed the Court of Claims, finding that that the government cannot be liable for inaction on a takings theory and that the construction and operation of MRGO had not been shown to be the cause of the flooding.[60] It seems unlikely that a similar claim more directly related to health impacts would succeed.

### 2.4.2 *Common Law*

The first wave of common law climate cases came in the mid-2000s, most visibly in *Connecticut* v. *American Electric Power*, where state, city, and environmental plaintiffs sued five power companies, alleging that their contribution to climate change constituted a public nuisance under federal common law. The trajectory of the case illustrates the range of issues common law approaches have continued to face since then: First, a federal district court judge in New York concluded that climate change was a "patently political" and "transcendently legislative" issue, and that the political question doctrine barred the court from hearing the case.[61] Next, the Second Circuit Court of Appeals concluded the opposite, that regulation of climate change-causing emissions was not a political question, that plaintiff states had standing to sue, and that a suit against GHG emissions sources could sound in federal

59 *St. Bernard Parish Gov't v. U.S.*, 121 Fed. Cl. 687 (2015).

60 St. Bernard Parish Gov't v. U.S., 887 F.3d 1354 (Fed. Cir. 2018). For a further discussion of this case, see Chapter 6.

61 *Connecticut* v. *Am. Elec. Power*, 406 F. Supp. 2d 265, 267 (S.D.N.Y. 2005).

common law public nuisance.[62] Finally, the US Supreme Court unanimously held that Congress, in enacting the federal Clean Air Act and authorizing the US Environmental Protection Agency to address climate change, had "displaced" the judiciary's authority to provide a remedy for such public nuisance claims.[63]

In a similar suit brought under state common law, *Comer* v. *Murphy Oil USA*, property owners in Louisiana alleged that certain power and chemical companies' GHG emissions contributed to climate change, which in turn exacerbated the harmful effects of Hurricane Katrina, constituting a private nuisance, as well as a public nuisance, trespass, negligence, unjust enrichment, fraudulent misrepresentation, and civil conspiracy. The significance of the case is unclear, as it involved a convoluted procedural history with inconsistent decisions between the district and appellate courts, ultimately culminating in the refusal of the court to hear the case on the merits.[64] However, there were some notable opinions along the way, including rejection of the idea that climate change is a political question barred from courts and the grant of standing to plaintiffs to make their claims. Importantly, one reviewing court found that a diversity suit brought under state common law for damages was materially distinguishable from public nuisance claims brought under federal common law,[65] leaving the door open for future state law-based suits.

In 2017 and 2018, a number of state and local governments across the U.S. filed lawsuits charging fossil fuel companies with liability for a range of torts, including public nuisance, failure to warn, design defect, private nuisance, negligence, and trespass, as well as statutory violations.[66] The lawsuits present a recurrent, and compelling, storyline: The fossil fuel industry knew that climate change was happening, that fossil fuel production and use was causing it, and that continued fossil fuel production and use would make it worse, and yet the industry hid what it knew, lied about it, and paid other people to lie about it for them, all the while

62 *Connecticut* v. *Am. Elec. Power*, 582 F.3d 309 (2d Cir. 2009).

63 *Am. Elec. Power* v. *Connecticut*, 564 U.S. 410, 415 (2011). See also *Native Village of Kivalina* v. *ExxonMobil Corp.*, 696 F.3d 849, at 856 (9th Cir. 2012), cert. denied, 133 S. Ct. 2390 (2013).

64 See *Comer* v. *Murphy Oil USA, Inc.*, 839 F. Supp. 2d 849, 855–68 (S.D. Miss. 2012) (dismissing the re-filed complaint on preemption, political question, standing, res judicata and collateral estoppel grounds), *aff'd*, 718 F.3d 460 (5th Cir. 2013).

65 Ibid. at 879.

66 See e.g., *Cnty. of San Mateo* v. *Chevron Corp.*, 17CIV03222 (Cal. Super. Ct. 2017); *Cnty. of Marin* v. *Chevron Corp.*, CIV1702586 (Cal. Super. Ct. 2017); *City of Imperial Beach* v. *Chevron Corp.*, C17-01227 (Cal. Super. Ct. 2017); *People of State of California* v. *BP p.l.c. (San Francisco)*, CGC-17-561379 (Cal. Super. Ct. 2017); *People of State of California* v. *BP p.l. c. (Oakland)*, RG17875889 (Cal. Super. Ct. 2017).

profiting from their ongoing activities, and plotting to profit more in the future. These actions have contributed significantly to global warming and sea level rise, and thereby caused harm, are continuing to cause harm, and are contributing to future harm to the plaintiff governments and their residents. The lawsuits seek compensation for the costs the local governments are incurring to adapt to sea level rise and for the companies' own willful, deceptive, and malicious behavior. These lawsuits face significant legal obstacles – including, though not limited to, questions of standing, political question doctrine, preemption, and causation.[67] Yet their outcome is uncertain. The common law can certainly accommodate these claims, just as the US Constitution can accommodate the claims in Juliana. The question that remains is whether judges will extend well-established principles of tort law to the new and complex context supplied by climate change.

### 2.4.3 *Statutes*

There is, at this point, no serious question that the federal Clean Air Act secures the people of the United States a right to have climate change addressed through direct regulation of GHG emissions. In *Massachusetts v. U.S. Environmental Protection Agency*, the US Supreme Court held that carbon dioxide falls within the Clean Air Act's definition of "air pollutant," and therefore the EPA had the authority to regulate it, and a corresponding obligation to determine whether regulation was necessary to protect public health and welfare.[68] In 2009, EPA issued an endangerment finding concluding, based on an extensive record of evidence linking the human-induced increase in atmospheric greenhouse gases to global climate changes and to specific impacts such as sea level rise, that GHG emissions from motor vehicles do endanger public health and welfare.[69] That decision was challenged by an industry group and upheld in the D.C. Circuit Court of Appeals.[70] In reaching its decision the court explained that EPA had addressed

67 Michael Burger, *Local Governments in California File Common Law Lawsuits Against Largest Fossil Fuel Companies*, Climate Law Blog, July 18, 2017, http://blogs.law.columbia.edu/climatechange/2017/07/18/local-governments-in-california-file-common-law-claims-against-largest-fossil-fuel-companies/.

68 *Massachusetts v. EPA*, 549 U.S. 497 (2007).

69 *See Endangerment and Cause or Contribute Findings for Greenhouse Gases Under Section 202 (a) of the Clean Air Act*, 74 Fed. Reg. 66, 496 (Dec. 15, 2009).

70 *Coal. for Responsible Regulation, Inc.* v. *E.P.A.*, 684 F.3d 102, at 121 (D.C. Cir. 2012), aff'd in part, rev'd in part sub nom. *Util. Air Regulatory Grp. v. E.P.A.*, 134 S. Ct. 2427, 189 L. Ed. 2d 372 (2014), and amended sub nom. *Coal. for Responsible Regulation, Inc.* v. *EPA*, 606 F. App'x 6 (D.C. Cir. 2015).

each link in the causal chain connecting anthropogenic greenhouse gas emissions to harmful impacts on public health and welfare and that EPA had provided ample evidence to support the finding.[71]

There are, however, limits on EPA's authority to utilize the Clean Air Act to address all sources of GHG emissions. In *Utility Air Regulatory Group* v. *Environmental Protection Agency*, the Supreme Court rejected EPA's interpretation that Section 169 of the Clean Air Act compelled the agency to require Prevention of Significant Deterioration and/or Title V permits from stationary sources based solely on the sources' GHG emissions.[72] What's more there have been a variety of lawsuits challenging the Clean Air Act regulations that EPA has promulgated respecting the control of GHG emissions.[73]

The prospect of immediate harms from climate-related extreme events may also result in statutory violations, and the associated statutory right to protection from them. For example, it was widely reported that the 2017 hurricanes Harvey and Irma inundated industrial plants, wastewater treatment plants, and Superfund sites, causing toxic chemicals and sewage to leak into floodwaters and releasing toxic pollutants into the air.[74] These contaminated floodwaters and toxic air emissions could indicate potential statutory violations under the Clean Water Act, the Oil Pollution Act of 1990, the Resource Conservation & Recovery Act, the Clean Air Act, the Emergency Planning and Community Right-to-Know Act, and the Comprehensive Emergency Response, Compensation, and Liability Act.[75] Litigation seeking to enforce these provisions might not even have to engage in the particular nuances of climate change attribution: Facilities and municipalities are liable for failing

[71] Ibid. at 121. [72] *Util. Air Regulatory Grp. v. EPA*, 134 S. Ct. 2427 (2014).

[73] See *West Virginia v. EPA*, No. 15-1306 (D.C. Cir.); *North Dakota v. EPA*, No. 15-1381 (D.C. Cir.); *Americans for Clean Energy v. EPA*, No. 16-1005 (D.C. Cir). Summaries and pleadings for each of these cases are available from the Sabin Center-Arnold & Porter US Climate Change Litigation database: http://climatecasechart.com/

[74] See, e.g., Troy Griggs, et al., *More Than 40 Sites Released Hazardous Pollutants Because of Hurricane Harvey*, N. Y. Times, Sept. 8, 2017, www.nytimes.com/interactive/2017/09/08/us/houston-hurricane-harvey-harzardous-chemicals.html?mcubz=0&_r=0; Center for Biological Diversity, *Center for Biological Diversity Analysis of Air Emissions Data from Texas Petrochemical Facilities, Provided by the Texas Commission on Environmental Quality as of August 31, 2017*, www.biologicaldiversity.org/programs/environmental_health/pdfs/Air_pollutant_releases_from_Texas_petrochemical_facilities_as_of_Aug_31_2017.pdf; Union of Concerned Scientists, *Hurricane Harvey's Impact on Energy and Industrial Facilities*, https://ucsusa.maps.arcgis.com/apps/MapJournal/index.html?appid=1e958eff5c3e45a983e52ad523c2ffdd; Emily Atkin, *Florida's poop nightmare has come true*, New Republic, Sept. 14, 2017, https://newrepublic.com/article/144798/floridas-poop-nightmare-come-true.

[75] See Dena Adler, *Hurricanes' Contaminated Floodwaters Might Crest Next Wave of Climate Change Litigation*, *supra* note 4.

to undertake adequate preparatory, safety, and control measures without needing to prove a link between global climate change, the particular storm, and actual harms suffered.[76] Thus, one way to protect against immediate risks from climate change impacts is to set aside future climate risk and focus on the risks climate change has already produced.[77]

## 2.5 CONCLUSION

As we enter deeper into the Anthropocene era we find ourselves at a critical juncture in human history. We have already locked in anthropogenic forcing of climate change that is now and will continue in the future to cause serious harm to human health and well-being, in the United States and all around the world. The most vulnerable among us will bear the largest share of the impacts, suffer the greatest number of harms, and endure the most extreme disruptions. The moral imperative is clear: We (the international community, national governments, subnational governments, corporate actors, local communities, and individuals) simply must take action to bend the GHG emissions trajectory toward zero, and eventually to net negative levels, in order to secure the conditions for flourishing in this and future generations. The law – under the international climate regime and its interpenetration with human rights, and under the US Constitution, common law, and federal statutes – demands that we endeavor to accomplish this goal. Failure, as they say, is not an option.

76 Ibid.

77 *Order, Conservation Law Foundation, Inc.* v. *ExxonMobil Corp.*, C.A. No. 16-11950-MLW, at 2 (D. Mass. Sept. 13, 2017).

# 3

# The Public Health Sector's Challenges and Responses

Jill Krueger and Colleen Healy Boufides

*Climate change poses an extraordinary and unprecedented threat to human health, not only by producing direct health consequences, but also by exacerbating disparities in social determinants of health and by adversely affecting ecosystems, food systems, and public health infrastructure. To date, United States public health responses to climate-related health effects have been shaped largely by the Centers for Disease Control & Prevention's Building Resilience against Climate Effects (BRACE) framework. Public health activities have primarily focused on assessing potential health impacts and developing responsive interventions. The Five Essential Public Health Law Services framework, which provides a series of iterative steps based upon past successes in public health, may help to build the field's capacity to communicate, implement, and evaluate legal interventions. When crafted through cross-sectoral collaboration and grounded in principles of health equity, the law may serve as an effective tool for building climate resilience and ameliorating the adverse health effects of climate change.*

## 3.1 INTRODUCTION

As atmospheric changes lead to rising temperatures, weather patterns influence activity of disease vectors, and harmful algal blooms proliferate in drinking water, it is unsurprising yet alarming to observe the enormous and unprecedented impacts of climate change on human health. The health threats posed by climate change are many and varied; they are unevenly distributed among populations and geographic regions; and many threats remain poorly understood or altogether unrecognized. Health impacts are compounding and cumulative, and they are inextricably connected with the health of plants, animals, and ecosystems also affected by climate change. Although there is no universal solution, adaptive management suggests a process of ongoing learning as communities plan and implement solutions to complex problems.[1]

This chapter begins in Section 3.2 by offering a survey of climate change impacts on public health. Section 3.3 then describes the public health community's responses to climate change, including the pathways for practitioners' engagement with climate change; the Centers for Disease Control and Prevention's Climate Ready Cities and States Initiative; success stories from the field; and next steps for the public health response. This section also discusses the Five Essential Public Health Law Services as a framework to ensure that law- and policy-related activities support public health efforts to address the negative health impacts of climate change. Section 3.4 offers three recommendations for continuing to promote public health and enhance resilience in the face of climate change.

## 3.2 MAPPING THE CONNECTION BETWEEN CLIMATE CHANGE AND PUBLIC HEALTH

Climate change affects human health directly and indirectly, through numerous interconnected and sometimes unexpected pathways. Health effects may be realized through increasing severity or frequency of familiar threats as well as through introduction of new threats.[2] This section discusses vulnerability to climate-related health effects, and provides a broad overview of what those effects are. As one of the most recent and exhaustive resources on the topic, the 2016 US Global Change Research Program Report, *The Impacts of Climate Change on Human Health in the United States*,[3] was a key resource in the development of this chapter.[4] Many of the health effects reviewed in this

1 In keeping with this spirit of ongoing learning, the authors gratefully acknowledge Lorraine Cameron, Rosemary Chaudry, Aaron Ferguson, Jerry King, Donna Levin, Jeff Phillips, and Joan Trendell for valuable feedback on an earlier version of this chapter.

2 John Balbus et al., Introduction: Climate change and human health, In: Alison Crimmins et al., *The Impacts of Climate Change on Human Health in the United States: A Scientific Assessment* 25, 31 (U.S. Global Change Research Program, 2016).

3 Allison Crimmins et al., *The Impacts of Climate Change on Human Health in the United States: A Scientific Assessment* (U.S. Global Change Research Program, 2016).

4 This chapter does not attempt to independently review, synthesize, and analyze the vast body of existing climatology and public health research, but rather draws heavily on the work of scientists who have applied their expertise to identify correlations and make predictions. Readers are encouraged to consult the resources cited within this chapter to gain a better understanding of the level of confidence and likelihood associated with the connections and predictions described. See, e.g., U.S. Global Change Research Program, *The Impacts of Climate Change on Human Health in the United States: A Scientific Assessment*, at xi.

chapter, along with the public health sector's legal efforts to address them, are explored in greater detail in subsequent chapters of this book.

### 3.2.1 *Vulnerability to Climate-Related Health Impacts*

Some of the most obvious pathways through which climate change affects health include rising temperatures, more frequent extreme weather events such as floods, and increasing air pollution. Each of these climate changes is linked not only to direct adverse health outcomes, but also to other climate changes with additional health consequences of their own. For example, higher average and extreme temperatures are linked not only to increased heat-related morbidity and mortality, but also to increased frequency and duration of wildfires (and associated degradation of air quality) and to changing biological processes of vectors of disease (and increased exposure to vector-borne illness).[5]

On a broader scale, climate change harms the health of a vast array of plants, animals, and insects and changes the balance of whole ecosystems.[6] Though not always immediately evident, eventual impacts on human health are inevitable, as all species within an ecosystem are interrelated. For example, plants', animals', and insects' seasonal behavior patterns are dependent on one another; thus, when species within an ecosystem respond differently to climate change, there is a "desynchronization" in the ecosystem as each species adapts.[7] Ultimately, desynchronization can lead to new combinations of species living together, potentially enabling emergence of novel diseases or new exposure pathways for humans.[8] Climate-related trophic cascades (cascading changes to a food chain resulting from removal of a predator), destruction of habitat, and loss of biodiversity may have similar consequences for ecosystem and human health. Given this complexity, an interdisciplinary "One Health" approach that combines public health, medical, and veterinary perspectives may be useful to address emerging health effects of climate change.[9]

The burden of climate change is experienced differently between and within communities. Geography is a key factor influencing exposure to climate-related risks because baseline climate characteristics (e.g. equatorial,

5 See generally Marcus C. Sarofim et al., US Global Change Research Program, Temperature related death and illness, In: *The Impacts of Climate Change on Human Health in the United States: A Scientific Assessment*, 1, 4 (2016).

6 Jonathan A. Patz and Micah B. Hahn, *Climate change and human health: A one health approach*, 366 Current Topics in Microbiology and Immunology 141, 141–166 (2012).

7 Ibid. at 161.

8 Ibid. at 161–166.

9 Ibid. at 142, 166.

coastal, or arid) as well as rural versus urban landscapes influence the nature and severity of climate impacts.[10] Individual and community risk factors further mediate the effects of climate change, increasing the vulnerability of certain populations to adverse health impacts. Differentiating between unavoidable risk factors, such as those associated with physiological differences like age, and avoidable risk factors, including the social determinants of health, is essential to understanding, predicting, preventing, and responding to climate-related health effects.[11]

A useful framework for assessing vulnerability to climate change includes three components: exposure, sensitivity, and resilience.[12] Exposure refers to contact between an individual or community and a climate stressor; sensitivity refers to physiological, environmental, social, or other conditions that affect the degree to which an individual or community is affected by the climate stressor; and resilience refers to the capacity of the individual or community to respond or adapt to and recover from climate stressors.[13]

In general, vulnerability tends to be high for children due to their immature immune systems and physiology, age-related behaviors and contact with the environment, and high ratio of air, food, and water intake relative to body mass.[14] Older adults tend to be vulnerable because they are more likely to have preexisting health conditions, reduced capacity for thermoregulation, or mobility challenges.[15] Climate-related exposures are associated with adverse pregnancy and newborn outcomes, rendering pregnant women uniquely vulnerable to climate stressors.[16] Chronic illness, immunosuppression, and disabilities affecting communication, mobility, or cognitive functioning may also increase vulnerability to climate impacts.[17] Individuals in the United States with limited English proficiency may face increased risks arising from

[10] See Kirk. R. Smith et al., Human health: Impacts, adaptation, and co-benefits, In: C.B. Field et al., *Climate Change 2014: Impacts, Adaptation, and Vulnerability. Part A: Global and Sectoral Aspects. Contribution of Working Group II to the Fifth Assessment Report of the Intergovernmental Panel on Climate Change* 709, 717 (2014).

[11] See Carla Stanke et al., *Health Effects of Drought: A Systematic Review of the Evidence,* Pub. Lib. Sci. Currents: Disasters 1, 13 (2013), 10.1371/currents.dis.7a2cee9e980f91ad7697b570bcc4b004.

[12] B. L. Turner II et al., *A Framework for Vulnerability Analysis in Sustainability Science*, 100 Proceedings Nat'l Academy Sci. 8074 (2003), 10.1073/pnas.1231335100. See also Janet L. Gamble et al., U.S. Global Change Research Program, Populations of concern, In: Alison Crimmins et al., *The Impacts of Climate Change on Human Health in the United States: A Scientific Assessment* 247, 249.

[13] Turner et al., *A Framework for Vulnerability Analysis in Sustainability Science, supra* note 12; U.S. Global Change Research Program, *Populations of Concern, supra* note 12.

[14] U.S. Global Change Research Program, *Populations of Concern, supra* note 12, at 255.

[15] Ibid. at 257. [16] Ibid. at 256–257. [17] Ibid. at 256–261.

communication barriers.[18] Outdoor workers and military personnel may be at higher risk during extreme temperature and weather events and may face increased exposure to vector-borne diseases.[19]

Each component of vulnerability described in this section further interacts with the social determinants of health (the societal, economic, political, and structural conditions that shape daily life and health outcomes[20]) to place certain groups at heightened risk of adverse health outcomes. The result is that low-income communities, communities of color, and indigenous groups shoulder a disproportionate burden of adverse effects associated with climate change.[21] The reasons for unevenly distributed harms are cumulative and rooted in a long history of discrimination; they include greater exposure to pollution, such as that associated with the burning of fossil fuels; poor housing conditions; underlying health inequities; aging community infrastructure; and fewer resources for adapting or responding to climate change.[22]

Recognition of the disparate impacts of climate change is at the heart of the ongoing climate justice movement, which is grounded in principles of environmental and social justice.[23] Climate justice calls not only for equitable distribution of burdens and benefits, but also for inclusion, participation, procedural justice, and deconstruction of the societal conditions constructs that perpetuate inequity.[24]

### 3.2.2 *Heat-Related Health Impacts*

Climate change is associated with increasing average temperatures as well as with increasing variability in temperature.[25] Rising average temperatures, extreme temperatures, and increasingly frequent, intense, and prolonged heatwaves are expected to lead to an increase in heat-related death and illness,

18 Ibid. at 252. 19 Ibid. at 258–259.

20 See generally World Health Organization Commission on Social Determinants of Health, Closing the gap in a generation: health equity through action on the social determinants of health, In: *Final Report of the Commission on Social Determinants of Health* (2008).

21 David Schlosberg and Lisette B. Collins, *From Environmental to Climate Justice: Climate Change and the Discourse on Environmental Justice*, 5 WIREs Climate Change 359, 359–364 (2014). See also U.S. Global Change Research Program, *Populations of Concern supra* note 12, at 252–253.

22 David Schlosberg and Lisette B. Collins, *From Environmental to Climate Justice: Climate Change and the Discourse on Environmental Justice*, at 359–364; U.S. Global Change Research Program, *Populations of Concern* at 252–253.

23 Ibid. at 362–364. 24 Ibid. at 361.

25 Simon N. Gosling et al., *Climate change and heat-related mortality in six cities*, Part 2, 53 Int. J. Biometeorology 31 (2008).

especially among vulnerable populations.[26] Though extreme cold weather events are projected to decline as global warming continues, associated reductions in mortality are not expected to offset the increased mortality associated with heat events.[27] Quantification of heat-related mortality can be difficult (and is likely underestimated) because heat is often a contributing rather than primary cause of death; nevertheless, heatwaves have been identified as the primary weather-related cause of death in the United States.[28] Moreover, heat-related premature deaths in the United States are expected to increase by thousands to tens of thousands each year as a result of climate change.[29]

By hindering thermoregulation (the body's regulation of internal temperature), extreme external temperatures may lead to illnesses including heat exhaustion and heat stroke.[30] Heat exhaustion may impair physical and cognitive functioning, while heat stroke may lead to loss of consciousness; permanent organ damage resulting in severe functional impairments; or death.[31] Heat exposure may also exacerbate preexisting chronic diseases, and individuals with cardiovascular, respiratory, cerebrovascular, or renal diseases, or with depression or diabetes, are at higher risk during extreme heat.[32]

Extreme high temperatures are especially problematic to health when they are accompanied by high humidity, persist for a long duration, or are punctuated by relatively high minimum temperatures; in such cases, the body has little opportunity for relief before again facing the stress of heat exposure.[33] Because physiological acclimatization may increase individual tolerance to heat over time, health effects of a given temperature may be less severe at the end of the summer compared to the beginning, or for populations living in generally warmer climates compared to those living in cooler climates.[34]

26 Sarofim et al., *Temperature-Related Death and Illness*, *supra* note 5, at 43, 57, 58.

27 Ibid. at 57.

28 George Luber and Michael McGeehin, *Climate Change and Extreme Heat Events*, 35 American J. Preventive Med. 429, 430 (2008).

29 Sarofim et al., *Temperature-Related Death and Illness*, *supra* note 5, at 57.

30 Ibid. at 46. See also Smith et al., *Human Health: Impacts, Adaptation, and Co-Benefits*, *supra* note 10, 720; R. Sari Kovats and Shakoor Hajat, *Heat Stress and Public Health: A Critical Review*, 29 Annual Review of Public Health 9.1, 9.2 (2008).

31 Kovats and Hajat, *Heat Stress and Public Health*, *supra* note 30, at 9.2; Smith et al., *Human Health: Impacts, Adaptation, and Co-Benefits*, *supra* note 10, 720; Luber and McGeehin, *Climate Change and Extreme Heat Events*, *supra* note 28, 429.

32 Kovats and Hajat, *Heat Stress and Public Health*, *supra* note 30, at 9.7; Luber and McGeehin, *Climate Change and Extreme Heat Events*, *supra* note 28, 430.

33 Luber and McGeehin, *Climate Change and Extreme Heat Events*, *supra* note 28, 430.

34 Sarofim et al., *Temperature-Related Death and Illness*, *supra* note 5, at 49.

Adverse health outcomes may be ameliorated by improvements in infrastructure (e.g. increased access to air conditioning and incorporation of cooling centers, green space, and cool roofs in "urban heat islands") and societal responses (e.g. improved preparation services and awareness systems).[35] However, the use of air conditioning may itself increase emissions that contribute to climate change, and power outages due to increased electricity usage during heatwaves may limit advances made in urban areas as well as increasing migration into these areas.[36] Moreover, many low-income populations continue to lack access to these adaptive measures.[37]

Socioeconomic status is a key factor in determining vulnerability to extreme temperatures, in part because low-income individuals are more likely to live in poorer quality housing, lack access to air conditioning, and/or suffer from chronic diseases that are exacerbated by heat.[38] In addition, less efficient or effective thermoregulation elevates heat-related health risks for young children, older adults, and individuals taking common medications that incidentally inhibit thermoregulation.[39] Preexisting chronic diseases, social isolation, or limited mobility may also place older adults at increased risk during periods of high heat.[40] High temperatures are associated with negative health effects for pregnant women, including preterm birth, and with poor birth outcomes, such as low birth weight and infant mortality.[41] Outdoor workers and homeless persons may be subject to increased risk due to prolonged exposure to unregulated outdoor temperatures.[42] Finally, race and ethnicity are significant and compounding factors effecting vulnerability, with non-Hispanic Black persons facing a higher risk of heat-related health risks as compared to other racial or ethnic groups in the United States.[43]

Increasing temperatures may trigger or exacerbate other environmental changes and adverse health effects. Higher temperatures are associated with increased concentrations of air pollutants and increased risk and severity of wildfires, especially during drought.[44] Over time, temperature changes may alter population levels, geographic boundaries, and biological processes of vectors of infectious disease, potentially leading to increased rates of infection

35 Ibid. at 58; Kovats and Hajat, *Heat Stress and Public Health*, *supra* note 30, at 9.8; Luber and McGeehin, *Climate Change and Extreme Heat Events*, *supra* note 28, at 432–433.

36 Sarofim et al., *Temperature-Related Death and Illness*, *supra* note 5, at 51, 54.

37 Luber and McGeehin, *Climate Change and Extreme Heat Events*, *supra* note 28, at 431.

38 Sarofim et al., *Temperature-Related Death and Illness*, *supra* note 5, at 54–55; Kovats and Hajat, *Heat Stress and Public Health*, *supra* note 30, at 9.8.

39 Ibid. at 9.7. 40 Sarofim et al., *Temperature-Related Death and Illness*, *supra* note 5, at 54.

41 Ibid. 42 Ibid. at 54–55. 43 Ibid. 44 Ibid. at 54-55.

and to emergence or re-emergence of infectious diseases in new geographic areas.[45]

### 3.2.3 *Air Pollution and Aeroallergen Exposure*

Climate change is a product of and a contributor to air pollution. Air pollutants such as ozone and particulate matter are associated with a variety of direct adverse respiratory and other health effects. In addition, air pollution and other climate changes increase exposure and susceptibility to aeroallergens (airborne allergens such as pollen and mold), yielding additional widespread adverse health impacts.[46]

Concentrations of ground-level ozone ($O_3$), a component of smog, play a significant role in outdoor air quality.[47] Ozone is harmful to the respiratory system and is associated with premature mortality as well as respiratory impacts including shortness of breath, coughing, lower respiratory tract infection, decreased lung function, injury to or inflammation of airways, and systemic oxidative stress.[48] Even minimal increases in ground-level ozone concentration are associated with increased risk of premature death.[49] The quantity of ground-level ozone at a given time depends on the amount of ozone precursors (pollutants of human or natural origin that form ozone) emitted into the atmosphere as well as weather-related conditions.[50] Warm temperatures, sunlight, and stagnant air masses facilitate ozone formation and are becoming increasingly common due to climate change; thus, ozone concentrations are expected to increase even if emissions are reduced in coming years.[51]

Particulate matter (PM), comprising solid or liquid airborne particles such as dust and aerosols, is another component of outdoor air pollution that is particularly detrimental to health.[52] Exposure to $PM_{2.5}$ (particles with

[45] See generally Charles B. Beard et al., Ch. 5: Vector-borne diseases, In: Alison Crimmins et al., *The Impacts of Climate Change on Human Health in the United States: A Scientific Assessment*, 129–156 (U.S. Global Change Research Program, 2016).

[46] Neal Fann et al., Ch. 3: Air quality impacts, In: Alison Crimmins et al. *The Impacts of Climate Change on Human Health in the United States: A Scientific Assessment*, 69, 84–86 (U.S. Global Change Research Program, 2016); Gennaro D'Amato, et al., *Climate Change and Respiratory Diseases*, 23 European Respiratory Review 161, 163–165 (2014); Smith et al., *Human Health: Impacts, Adaptation, and Co-Benefits, supra* note 10, at 727–730.

[47] Fann et al., *Ch. 3: Air Quality Impacts, supra* note 46, at 72–73.

[48] D'Amato et al., *Climate Change and Respiratory Diseases, supra* note 46 at 164; Fann et al., *Ch. 3: Air Quality Impacts, supra* note 46, at 72.

[49] Smith et al., *Human Health: Impacts, Adaptation, and Co-Benefits, supra* note 10, at 729.

[50] Fann et al., *Ch. 3: Air Quality Impacts, supra* note 46, at 72.

[51] Ibid. at 72–73.

[52] Ibid. at 76.

diameter less than 2.5 microns) is especially harmful, as these small particles are linked to premature mortality even at low levels of exposure.[53] In addition, $PM_{2.5}$ exposure is linked to development and exacerbation of asthma, as well as to respiratory symptoms, chronic bronchitis, decreased lung function and lung cancer, chronic obstructive pulmonary disease (COPD), and cardiovascular disease.[54] Concentrations of particulate matter are likely to increase as wildfires and droughts increase in frequency and severity, releasing dust and smoke into the air.[55]

The burden of asthma and allergic diseases – already sources of substantial health, economic, and quality-of-life impacts in the United States and throughout the world – is expected to increase with climate change.[56] Exposure to aeroallergens may cause or exacerbate asthma episodes and may lead to allergic illnesses such as hay fever, sinusitis, conjunctivitis, hives, eczema, and anaphylaxis, which can be life-threatening.[57] The burden of asthma and allergic illness is projected to increase as changes in temperature, precipitation, and extreme weather events alter the quantity, allergen content, geographic distribution, and timing of aeroallergens.[58] Moreover, pre-exposure to air pollution is likely to increase sensitivity to and exacerbate the effects of exposure to aeroallergens.[59]

Indoor air quality can be particularly problematic to health because so much time is spent indoors.[60] In addition to the harmful effects of outdoor air and aeroallergens that inevitably enter buildings, climate change may lead to warmer and damper conditions that increase growth of mold, bacteria, and other contaminants, exacerbating asthma and respiratory infections.[61] Warm, damp conditions may increase incidences of Legionnaire's disease as well, a potentially fatal type of pneumonia caused by aerosolized *Legionella* bacteria.[62]

53 D'Amato et al., *Climate Change and Respiratory Diseases, supra* note 46, at 164.

54 Ibid. at 164–165; Fann et al., *Ch. 3: Air Quality Impacts, supra* note 46 at 76.

55 Fann et al., *Ch. 3: Air Quality Impacts, supra* note 46, at 77.

56 Colleen E. Reid and Janet L. Gamble, *Aeroallergens, Allergic Disease, and Climate Change: Impacts and Adaptation*, 6 EcoHealth 458, 458–467 (2009).

57 Fann et al., *Ch. 3: Air Quality Impacts, supra* note 46, at 77.

58 Reid and Gamble, *Aeroallergens, Allergic Disease, and Climate Change, supra* note 56, at 460–463.

59 D'Amato et al., *Climate Change and Respiratory Diseases, supra* note 46, at 165; Fann et al., *Ch. 3: Air quality impacts, supra* note 46, at 79.

60 Fann et al., *Ch. 3: Air Quality Impacts, supra* note 46, at 79.

61 D'Amato et al., *Climate Change and Respiratory Diseases, supra* note 46, at 164; Fann et al., *Ch. 3: Air Quality Impacts, supra* note 46, at 81.

62 Fann et al., *Ch. 3: Air Quality Impacts, supra* note 46, at 81.

Children are particularly vulnerable to the effects of aeroallergen exposure because their respiratory and immune systems are still developing.[63] Adverse health impacts of air pollution are also amplified for individuals with preexisting conditions or compromised immune systems.[64] Racial and ethnic minorities and low-income groups may be disproportionately exposed to indoor and outdoor air pollutants associated with poor neighborhood and housing conditions.[65]

### 3.2.4 *Health Impacts of Extreme Weather*

Climate change is associated with increases in the frequency, severity, and duration of precipitation, storms, flooding, droughts, and wildfires.[66] Serious adverse health effects including illness or death may result directly from extreme weather events, while damage to infrastructure and housing and loss of livelihood may have long-lasting effects on physical and mental health. Adverse health effects may be compounded if multiple extreme events occur in quick succession in the same location.[67]

#### 3.2.4.1 Precipitation, Storms, and Floods

Climate change is projected to increase the frequency and severity of extreme precipitation events, hurricanes, and flooding.[68] High temperatures and increased humidity may increase the likelihood and violence of storms such as tornadoes. In addition, storm surges combined with rising sea levels and high tides exacerbate the risk of flooding in coastal areas.[69] Flooding is already a significant cause of death in the United States and throughout the world.[70] Drowning and acute trauma are among the most common causes of immediate fatalities during floods, with the most significant loss of life associated with flash floods and coastal floods.[71] Flooding may lead to injuries from falling and

63 Ibid. at 82. 64 Ibid.

65 See generally Gary Adamkiewicz et al., *Moving Environmental Justice Indoors: Understanding Structural Influences on Residential Exposure Patterns in Low-Income Communities*, 101 Amer. J. Pub. Health S238 (Supp. 1, 2011).

66 Bell et al., Ch. 4: Impacts of extreme events on human health, In: Alison Crimmins et al, *The Impacts of Climate Change on Human Health in the United States: A Scientific Assessment*, 99, 101 (U.S. Global Change Research Program, 2016).

67 Ibid. at 101. 68 Ibid. at 102. 69 Ibid.

70 Katarzyna Alderman et al., *Floods and Human Health: A Systematic Review*, 47 Env't Int' 37, 38 (2012).

71 Ibid. at 70.

fast-moving debris or objects, animal bites or stings, electrocution, and motor vehicle accidents.[72]

Additional health implications are associated with the runoff that often follows a heavy rainfall, as runoff may increase the concentrations of pathogens and nutrients in fresh and marine water.[73] Nutrient loading may in turn result in proliferation of naturally occurring pathogens as well as algal blooms that produce toxins harmful to human health.[74] Exposure to waterborne pathogens or toxins may occur via drinking water, recreational water, or consumption of fish and shellfish.[75] Toxins from harmful algal blooms may yield a variety of adverse health impacts, ranging from gastrointestinal illness, dermatologic, respiratory, or neurologic symptoms, acute respiratory illness, and eye irritation following recreational exposure,[76] to paralytic or neurotoxic shellfish poisoning.[77] Health impacts of waterborne pathogens are discussed in further detail in Section 3.2.5.2.

Infrastructure damage resulting from extreme precipitation poses additional health challenges.[78] Damage to roads may impede evacuations and emergency response, especially in rural areas and for individuals with disabilities if alternate routes are not available.[79] Power outages may impair access to electrically powered medical equipment such as ventilators; contribute to food spoilage and associated foodborne illness; and increase risk of carbon monoxide poisoning associated with improper use of gas-powered energy sources.[80] Overwhelmed stormwater drainage and wastewater treatment systems may contaminate floodwaters or drinking water, leading to infections and gastrointestinal illnesses.[81] Finally, infrastructure failures may impede communication and transportation during and after extreme weather events and may affect hospitals and other providers of health care services.[82]

[72] Bell et al., Ch. 4: *Impacts of Extreme Events on Human Health, supra* note 66, at 107; Alderman et al., *Floods and Human Health, supra* note 70, at 39; David G. C. McCann et al, *The Water/Health Nexus in Disaster Medicine: I. Drought Versus Flood*, 3 Current Opinion in Envtl. Sustainability 480, 482 (2011).

[73] Juli Trtanj et al., Ch. 6: Climate impacts on water-related illness, In: *The Impacts of Climate Change on Human Health in the United States: A Scientific Assessment* 157, 173 (2016).

[74] Ibid. at 161, 173. [75] Ibid. at 162. [76] Ibid. 165.

[77] Ibid. at 170. See also Smith et al., *Human Health: Impacts, Adaptation, and Co-Benefits, supra* note 10, at 726.

[78] Bell et al., *Ch. 4: Impacts of Extreme Events on Human Health, supra* note 66, at 105.

[79] Ibid. at 105. [80] Ibid.

[81] Trtanj et al., *Ch. 6: Climate Impacts on Water-Related Illness, supra* note 73, at 161.

[82] Bell et al., *Ch. 4: Impacts of Extreme Events on Human Health, supra* note 66, at 105.

#### 3.2.4.2 Drought and Wildfire

The nature and severity of health risks associated with drought depend substantially on local infrastructure and on socioeconomic factors influencing community resiliency.[83] Thus, drought-related health risks vary widely across the globe. In some regions, food shortages, micronutrient deficiency, and contamination of food products are chief impacts of drought, leading to malnutrition and increased morbidity and mortality.[84] Elsewhere, health effects may relate primarily to diminished water quality as low water flow allows pollutants, chemicals, or pathogens to accumulate before ultimately washing into water sources when precipitation occurs.[85] Alternatively, water quality issues may relate to increased animal and human use of a single water source, accelerating transmission of water- and vector-borne infectious diseases.[86] Additional drought-related health impacts arise from proliferation of airborne dust and pathogens (such as Coccidioides, the fungus that causes Valley Fever)[87] and increased transmission of vector-borne diseases associated with domestic water storage or increases in vector populations following drought.[88]

Drought may also contribute to conditions favorable to wildfires.[89] Exposure to wildfire smoke is associated with low infant birth weight, asthma, COPD, and premature death, as well as increased incidence of acute respiratory illness and exacerbation of respiratory symptoms.[90] Wildfires can degrade air and water quality many miles downwind and downstream from the location of the fire.[91]

### 3.2.5 *Exposure to Infectious Disease*

Warming water and air temperatures and extreme precipitation events are likely to increase opportunities for transmission of vector-borne, zoonotic, waterborne, and foodborne infectious diseases. Immunosuppression and old

83 Stanke et al., *Health effects of drought,* at 4. 84 Ibid. at 12–13.

85 Ibid. at 13–14; Bell et al., *Ch. 4: Impacts of Extreme Events on Human Health, supra* note 66, at 108.

86 Stanke et al., *Health Effects of Drought, supra* note 11, at 13.

87 Ibid. at 14; Bell et al., Ch. 4: Impacts of extreme events on human health, at 109–110.

88 Stanke et al., *Health Effects of Drought, supra* note 11, at 14–15. See also Bell et al., *Ch. 4: Impacts of Extreme Events on Human Health,* note 66, at 108.

89 Stanke et al., *Health Effects of Drought, supra* note 11, 17–18.

90 Bell et al., *Ch. 4: Impacts of Extreme Events on Human Health, supra* note 66, at 110.

91 Ibid.

or young age contribute to individual vulnerability to infectious diseases.[92] The risk of exposure and infection, however, may be drastically reduced by strengthening existing public health interventions as well as investing in local and regional infrastructure and capacity to respond to new and emerging disease threats.[93]

### 3.2.5.1 Vector-Borne and Zoonotic Diseases

Vector-borne diseases are caused by pathogens transmitted from arthropods to humans (e.g. malaria, transmitted by mosquitoes), while zoonotic diseases are transmitted by vertebrate animals (e.g. hantavirus, transmitted by mice and rats).[94] There is frequent overlap between these types of diseases, as some diseases are both vector-borne and zoonotic (e.g. Lyme disease, transmitted by ticks from nonhuman reservoirs, such as mice, to humans).[95] Climate change is likely to increase the health burden associated with both vector-borne and zoonotic diseases.[96]

Increased transmission of disease is expected to accompany the climate-related expansion of geographic and seasonal distributions of vectors and animals, leading to increased and prolonged opportunities for contact among vectors, animals, and humans.[97] Additionally, changes in precipitation, drought, temperature, and humidity patterns may increase the "carrying capacity" of a given ecosystem, enabling increased vector, reservoir, or host population density.[98] For example, heavy rainfall may incubate breeding sites for mosquitoes, while elevated winter temperatures may increase the number of a species that survive the winter.[99] Climate changes may also alter the biological processes (e.g. rates of development and reproduction, biting, or viral replication) of vectors, animals, or pathogens, leading to increased opportunities for human exposure to disease.[100] Finally, weather disasters may indirectly contribute to the spread of infectious disease by disrupting public health intervention programs, delaying treatment of

92 Beard et al., *Ch. 5: Vector-borne diseases, supra* note 45, at 142.

93 See Ulisses E. C. Confalonieri et al., *Climate Change and Adaptation of the Health Sector: The Case of Infectious Diseases*, 6 Virulence 554, 555–556 (2015).

94 James N. Mills, *Potential Influence of Climate Change on Vector-Borne and Zoonotic Diseases: A Review and Proposed Research Plan*, 118 Envtl. Health Persp. 1507, (2010).

95 Ibid. at 1507. 96 Ibid. at 1513.

97 Ibid. at 1508; Beard et al., Ch. 5: Vector-Borne Diseases, *supra* note 45, at 144–146.

98 Ibid. at 1509. 99 Ibid.

100 Ibid. at 1509–1510; Beard et al., Ch. 5: Vector-borne diseases, at 140.

acute infections, and redistributing diseases through population displacement.[101]

### 3.2.5.2 Waterborne and Foodborne Infectious Disease

Exposure to water- and foodborne infectious disease is also projected to increase in connection with rising temperatures and extreme precipitation. Water contamination is especially common following floods, as runoff may increase concentrations of pathogens and nutrients in nearby bodies of water.[102] Nutrient loading, in turn, may amplify growth of naturally occurring pathogens.[103] Furthermore, warming water temperatures may expand geographic ranges and seasonal growth windows for waterborne pathogens, yielding still more opportunities for human exposure.[104] Key pathways for exposure to waterborne pathogens include ingestion of contaminated drinking water, direct contact with recreational water, and consumption of fish and shellfish.[105]

Warming air temperatures and humidity are associated with growth and spread of foodborne pathogens.[106] Flooding or other extreme weather may interrupt electricity (and therefore refrigeration) as well as the food supply chain, contributing further to spoilage and proliferation of foodborne pathogens.[107]

Health outcomes associated with contaminated drinking water include gastrointestinal illnesses (including cholera and other diarrheal diseases), respiratory tract and skin infections, and Hepatitis A and E infection.[108] Infectious diseases transmitted through direct skin contact, such as Leptospirosis or infections associated with *Vibrio* bacteria species, may lead to diarrheal illness or eye, ear, and wound infections.[109] Shellfish poisoning linked to *Vibrio* bacteria may cause gastroenteritis or bloodstream infection.[110] Finally, foodborne pathogens such as *Salmonella* and *Escherichia coli* (*E. coli*) are most commonly associated with diarrhea, but in severe cases may cause long-lasting health impacts or death.[111]

101 Confalonieri et al., *Climate Change and Adaptation of the Health Sector, supra* note 93, at 554.

102 Trtanj et al., *Ch. 6: Climate Impacts On Water-Related Illness, supra* note 73, at 173.

103 Ibid. 104 Ibid. at 166. 105 Ibid. at 162.

106 Lewis Ziska et al., Ch. 7: Food safety, nutrition, and distribution, In: Alison Crimmins et al., *The Impacts of Climate Change on Human Health in the United States: A Scientific Assessment*, 195 (U.S. Global Change Research Program, 2016).

107 Ziska et al., *Ch. 7: Food Safety, Nutrition, and Distribution, supra* note 106, at 195–200; M.E. Brown et al., *Climate Change, Global Food Security, and the U.S. Food System* 10(2015), www.usda.gov/oce/climate_change/FoodSecurity2015Assessment/FullAssessment.pdf.

108 Alderman et al., *Floods and human health, supra* note 70, at 42–43. 109 Ibid.

110 Trtanj et al., *Ch. 6: Climate Impacts on Water-Related Illness, supra* note 73, at 160; Ziska et al., *Ch. 7: Food Safety, Nutrition, and Distribution, supra* note 106, at 194.

111 See Ziska et al., *Ch. 7: Food Safety, Nutrition, and Distribution, supra* note 106, at 194.

### 3.2.6 *Food Insecurity and Nutrition*

Food security exists "when all people at all times have physical, social, and economic access to sufficient, safe, and nutritious food to meet their dietary needs and food preferences for an active and healthy life."[112] Thus, food security depends not only on adequate production of food products, but also on accessibility, utilization and nutrient content, and stability.[113] Indeed, hunger and malnutrition persist globally even though enough food is produced to satisfy the existing level of need.[114]

The effects of climate change on food production and distribution are complex and influenced by geography. Rising temperatures may lead to decreased production at lower latitudes but increased production at higher latitudes.[115] The food crops grown in a given region may shift with changing precipitation patterns, and drought conditions may eliminate certain crops or all agriculture from affected areas. Climatic changes affecting marine and freshwater ecosystems disproportionately affect many tribal and coastal communities that are heavily dependent on fish and seafood for nutrition.[116] More generally, extreme weather events, increased variability in temperature and rainfall, and seasonal and geographic shifts in plant, insect, pest, and pathogen patterns may yield a wide range of interconnected impacts on agricultural and livestock production and on the stability of local and global food systems.[117]

Food prices are expected to rise in connection with climate change, likely causing or exacerbating food insecurity in low-income households.[118] Elevated levels of atmospheric $CO_2$ may reduce the nutritional content of many major food crops by decreasing protein concentration and essential minerals (e.g. calcium, iron, zinc).[119] Extreme weather

112 John R. Porter et al., Food security and food production systems, In: C. B. Field et al., *Climate Change 2014: Impacts, Adaptation, and Vulnerability. Part A: Global and Sectoral Aspects. Contribution of Working Group II to the Fifth Assessment Report of the Intergovernmental Panel on Climate Change* 485, 490 (citing Food and Agricultural Organization of the United Nations (FAO) World Summit, *Rome Declaration and World Food Summit Plan of Action* 13 (1996), www.fao.org/docrep/003/w3613e/w3613e00.HTM. See also Brown et al., *Climate Change, Global Food Security, and the U.S. Food System, supra* note 107, at ix.

113 Porter et al., *Food Security and Food Production Systems, supra* note 112, at 490; Brown et al., *Climate Change, Global Food Security, and the U.S. Food System, supra* note 107, at 1.

114 Porter et al., *Food Security and Food Production Systems, supra* note 112, at 490.

115 Ibid. at 491.

116 See Trtanj et al., *Ch. 6: Climate Impacts on Water-Related Illness, supra* note 73, at 171.

117 Porter et al., *Food Security and Food Production Systems, supra* note 112, 497–503.

118 Ibid. Brown et al., *Climate Change, Global Food Security, and the U.S. Food System, supra* note 107, at 9.

119 Ziska et al., *Ch. 7: Food Safety, Nutrition, and Distribution, supra* note 106, at 198–199; Porter et al., *Food Security and Food Production Systems, supra* 112, at 501.

events may also increase prices of and decrease access to nutritious foods by disrupting transportation and other activities at different points along the supply chain; in addition, flooding and humidity may increase mold growth or food spoilage during storage, transport, or handling.[120] Certain types of mold that grow on crops may produce carcinogenic toxins that can impair child development, suppress the immune system, or ultimately result in death.[121]

The effects of climate change on food security, nutrition, and related health outcomes are linked predictably to socioeconomic conditions and adaptive capacity of communities; thus low-income communities will disproportionately suffer from both food insecurity and malnutrition.[122] Infants, children, pregnant women, the elderly, and immuno-compromised individuals are also vulnerable to malnutrition, with infants and children potentially experiencing lifelong impacts associated with stunted growth.[123]

### 3.2.7 *Mental Health Impacts*

The negative impact of major natural disasters upon mental health is well-documented.[124] Symptoms of depression, anxiety, anger, despair, and post-traumatic stress may persist long after the physical evidence of a natural disaster has been cleaned up. There is mounting evidence that even the threat of climate change can take a toll by triggering stress responses that have a cumulative impact on mental health, especially when exacerbated by media coverage that induces feelings of helplessness and hopelessness.[125]

The public health challenges posed by climate changes have different pathways. For example, many psychoactive medications reduce the

[120] Ziska et al., *Ch. 7: Food Safety, Nutrition, and Distribution, supra* note 106, at 195–197, 199–200; Brown et al., *Climate Change, Global Food Security, and the U.S. Food System, supra* note 107, at 10.

[121] Ziska et al., *Ch. 7: Food Safety, Nutrition, and Distribution, supra* note 106, at 196.

[122] Porter et al., *Food Security and Food Production Systems, supra* note 112, at 502–503.

[123] Ziska et al., *Ch. 7: Food Safety, Nutrition, and Distribution, supra* note 106, at 200.

[124] S. Clayton et al., *Beyond Storms & Droughts: The Psychological Impacts of Climate Change*, 51 (American Psychological Association and ecoAmerica, Washington, DC, 2014) http://ecoamerica.org/wp-content/uploads/2014/06/eA_Beyond_Storms_and_Droughts_Psych_Impacts_of_Climate_Change.pdf.

[125] D. Dodgen et al., Ch. 8: Mental health and well-being, In: Alison Crimmins et al., *The Impacts of Climate Change on Human Health in the United States: A Scientific Assessment*, 217–246 (U.S. Global Change Research Program, Washington, DC, 2016) http://dx.doi.org/10.7930/J0TX3C9H.

ability to regulate body temperature, making persons who take these medications especially vulnerable to extreme heat.[126] Indeed, their risk of death during heat waves is three times that of the general population.[127]

Groups at higher risk for mental health impacts include those with preexisting mental illness as well as children and older adults, pregnant and post-partum women, people with low incomes, and emergency workers.[128] Those who depend upon the natural environment for their livelihoods, such as farmers and those working in tourism, may experience economic stress related to actual and potential climate impacts.

Stress related to extreme heat, flooding, drought, limited water supplies, and food shortages can lead to increased aggression and violence, including domestic violence. Displacement and involuntary relocation may also have adverse mental health impacts.[129]

## 3.3 PUBLIC HEALTH RESPONSES TO CLIMATE CHANGE

As the foregoing discussion demonstrates, changing weather patterns pose substantial challenges to current public health institutions in the United States by exacerbating existing public health threats and by creating new ones that differ in degree, location, and/or kind. In many respects, the public health sector is in a good position to address the human health impacts of climate change through its focus on gathering health-related data, engaging the community to understand concerns and priorities, designing evidence-based solutions, and evaluating the effectiveness of the solutions in order to make needed improvements. In other respects, the public health response to climate change may be hampered by a lack of knowledge and training in climate science among public health staff. State and local health departments may have neither specific legal authority to address climate change nor dedicated funds for climate-related work. This section turns now to consider strengths of the public health response to climate change and offer suggestions to bolster that response.

[126] Dodgen et al., *Mental Health and Well-Being*, *supra* note 125, at 226. [127] Ibid. at 222.

[128] Ibid. at 223–227.

[129] Clayton et al., *Mental Health and Our Changing Climate: Impacts, Implications, and Guidance*, 25 (American Psychological Association, Climate for Health, and ecoAmerica, 2017).

### 3.3.1 Pathways for Public Health Practitioners' Engagement

One important framework for conceptualizing public health responses to climate change is the Ten Essential Services of Public Health.[130] The Essential Services include:

(1) Monitor health status to identify and solve community health problems.
(2) Diagnose and investigate health problems and health hazards in the community.
(3) Inform, educate, and empower people about health issues.
(4) Mobilize community partnerships and action to identify and solve health problems.
(5) Develop policies and plans that support individual and community health efforts.
(6) Enforce laws and regulations that protect health and ensure safety.
(7) Link people to needed personal health services and ensure the provision of health care when otherwise unavailable.
(8) Ensure competent public and personal health care workforce.
(9) Evaluate effectiveness, accessibility, and quality of personal and population-based health services.
(10) Research for new insights and innovative solutions to health problems.[131]

FIGURE 3.1 Ten essential services of public health[132]

All of these are critical to the public health response to climate change.

Using this familiar framework can emphasize the ways in which building climate resilience is an extension of, rather than a major break from, public health practice. Indeed, many climate intervention strategies, such as promoting a plant-rich diet, reducing food waste, and developing walkable cities, have clear health benefits and may fit easily into existing public health programs.[133] Other, more interdisciplinary strategies for mitigating or adapting to climate change, such as increasing availability of mass transit and electric vehicles and supporting solar farms and refrigerant management, may require public health practitioners to learn about and work in new subject areas with new

[130] H. Frumkin et al., *Climate change: The public health response*, Am. J. of Pub. Health, 98(3) 435–445 (2008).

[131] Ibid. at 435–445.

[132] Ibid.

[133] See Paul Hawken, *Drawdown: The Most Comprehensive Plan Ever Proposed to Reverse Global Warming* (2017).

partners to improve health outcomes.[134] Thus, public health practitioners may engage with climate change along three primary pathways:

#### 3.3.1.1 They May Incorporate Aspects of Climate Change into Their Existing Work

For example, public health practitioners focused on emergency preparedness and health equity may use heat vulnerability maps to locate cooling centers in low-income communities and near concentrations of senior citizens. Environmental public health and infectious disease specialists may address emerging diseases with interventions ranging from enhanced epidemiological surveillance systems and vector control practices[135] to stormwater management and water treatment systems[136] to food safety regulations.[137] In addition, as trusted and credible sources of health information in their communities, public health nurses and community health workers are uniquely situated to educate patients, families, policymakers, and others about local health impacts of climate change and to advise at-risk populations and inform policymakers on locally appropriate adaptation strategies.[138]

#### 3.3.1.2 They May Account for Climate Change in Long-term Public Health and Community Health Improvement Planning

For example, community stakeholders may identify recurring flooding as an unresolved health threat during the community health improvement process. In such cases, public health practitioners may partner with climatologists, planners, and environmentalists to identify natural defenses which may be restored and strengthened in order to improve floodplain management and

[134] See Hawken, *Drawdown, supra* note 133 (2017).

[135] See Confalonieri et al., *Climate Change and Adaptation of the Health Sector, supra* note 93, 554–556.

[136] Trtanj et al., *Ch. 6: Climate Impacts on Water-Related Illness, supra* note 73, at 163.

[137] Ziska et al., *Ch. 7: Food Safety, Nutrition, and Distribution, supra* note 106, at 189, 197.

[138] See, e.g., Lucia Sayre, et.al., *Climate Change and Human Health: The Role of Nurses in Confronting the Issue*, 34 Nurs. Admin. Q. 334–342 (2010), at 337–341; Jennifer Miller and Courtney Gonzales, Working with community clinics to link climate and health in San Diego, In: *Act on Climate: California Communities Addressing Climate Change and Health* 12–13 (Center for Climate Change and Health, Public Health Institute, 2015), www.phi.org/uploads/application/files/9ki2v8jlwg7bzjb26o2shd7qs6duoeikqzm48mfn9zryuqz22j.pdf.

protect community health.[139] Similarly, climate change may be incorporated into all-hazards mitigation planning, as was done in Baltimore, Maryland.[140] This approach leverages assessments of risk and vulnerability that must be completed for all-hazards mitigation planning to support climate planning efforts as well. Other long-term planning processes, such as master planning, watershed management planning, long-range transportation planning, and capital improvement planning, may be vehicles for increasing climate resilience at the local level.[141]

#### 3.3.1.3 They May Engage in Specific Climate Change Mitigation and Adaptation Planning Efforts

For example, epidemiologists may identify environmental public health indicators related to climate change, and collect data related to these indicators. Public health practitioners may participate in development of cross-sector climate adaptation plans for their jurisdiction, thus ensuring that public health is involved and considered in decisions pertaining to everything from building codes to transit plans. Finally, public health practitioners may develop climate and health action plans with the input of community stakeholders. Climate and health action plans may include established programs and activities, as well as enhanced or new programs and activities.[142]

### 3.3.2 *Adaptive Management and the CDC's Climate Ready Cities and States Initiative*

The Centers for Disease Control and Prevention (CDC) plays a principal role in mobilizing the public health community's response to climate change. Climate change poses unique challenges for public health, in part due to the

139 Stacy Small-Lorenz et al., *Natural Defenses in Action: Harnessing Nature to Protect our Communities* (National Wildlife Federation, 2016).

140 Jason Vogel et al., *Climate Adaptation: The State of Practice in U.S. Communities*, 61 (Abt Associates and the Kresge Foundation, 2016).

141 See, e.g., Land Information Access Association and Beckett & Raeder, Inc., *Planning for Community Resilience in Michigan: A Comprehensive Handbook* (2017), www.resilientmichigan.org/downloads/michigan_resiliency_handbook_web.pdf.

142 G. Marinucci et al., *Building Resilience Against Climate Effects – A Novel Framework to Facilitate Climate Readiness in Public Health Agencies*, 11(6) Int. J. Envtl. Res. Public Health. 6433, 6447 (2014).

limited understanding of its complexity, scope, and future course. To address these challenges, the CDC and many others in the public health community have adopted an adaptive management approach, the elements of which impart both flexibility and rigor to interventions in complex systems. Adaptive management has six elements:

(1) Management objectives that are regularly revisited and revised;
(2) A model of the system(s) being managed;
(3) A range of management choices;
(4) Monitoring and evaluation of outcomes;
(5) Mechanisms for incorporating learning into future decisions; and
(6) A collaborative structure for stakeholder participation and learning.[143]

Adaptive management is, in sum, an "iterative process that acknowledge[s] complexity and uncertainty, emphasize[s] ongoing learning, and allow[s] for continuous stakeholder input."[144]

The CDC builds capacity through its Climate Ready States and Cities Initiative. The initiative involves cooperative agreements between the CDC and participating states and cities to implement an adaptive management approach tailored to public health departments addressing the health impacts of climate change. The initiative is implemented by grantees in accordance with the Building Resilience against Climate Effects (BRACE) framework.

The BRACE framework shapes the public health response to climate change in several important ways. It guides health departments to identify the specific public health threats of greatest concern in the area; it relies upon assessments and modeling to project the potential disease burden; it emphasizes a practice of regularly assessing interventions in light of emerging information; it calls for development and implementation of comprehensive climate and health adaptation plans; and it prescribes ongoing evaluation and quality improvement.[145] The BRACE framework incorporates public health practices and values by drawing on the literature of social determinants of health for vulnerability assessments and emphasizing evidence-based evaluation of possible public health interventions.[146]

143 Marinucci et al., *Building Resilience*, *supra* note 142, at 6435, (internal citation omitted).
144 Jeremy J. Hess et al., *Integrating Climate Change Adaptation into Public Health Practice: Using Adaptive Management to Increase Adaptive Capacity and Build Resilience*, Envtl. Health Persp 171 (Feb. 2012).
145 Marinucci et al., *Building Resilience*, *supra* note 142, at 6435.
146 Ibid.

The BRACE framework includes five steps, set forth in Figure 3.2.

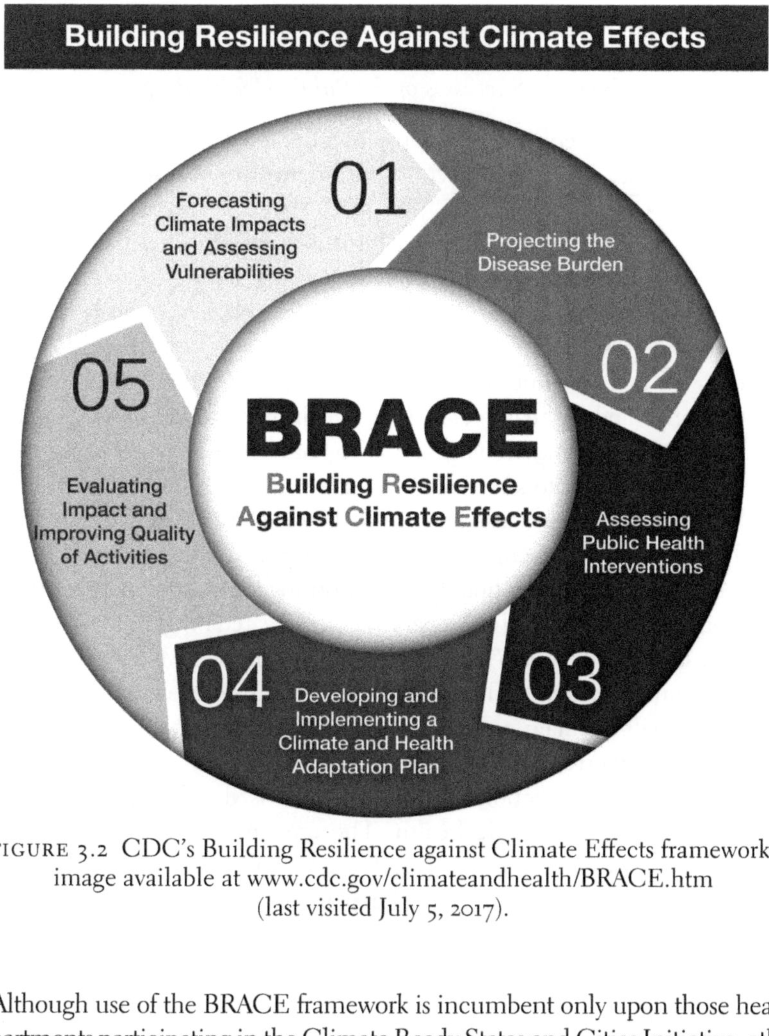


FIGURE 3.2 CDC's Building Resilience against Climate Effects framework, image available at www.cdc.gov/climateandhealth/BRACE.htm (last visited July 5, 2017).

Although use of the BRACE framework is incumbent only upon those health departments participating in the Climate Ready States and Cities Initiative, other health departments are free to adopt the framework as well. BRACE has been described as a tool to enable "systematic prioritization" of climate adaptation activities for public health departments with limited resources.[147] The BRACE framework assists participating health departments in identifying the climate threats of greatest concern in their jurisdiction, prioritizing issues for action,

[147] Marinucci et al., *Building Resilience, supra* note 142, at 6437.

choosing the most effective and appropriate public health interventions to address the issues identified in the local context, and crafting an overall plan.[148]

### 3.3.3 *Notable Successes of the Public Health Response*

Public health has three core functions: assessment, policy development, and assurance.[149] It has made notable contributions to climate resilience in each of these areas.

The first core function of public health is assessment of health threats. This includes identifying, investigating, and monitoring threats to human health. The 2016 US Global Change Research Program Report, *The Impacts of Climate Change on Human Health in the United States*, is an essential tool for public health. It will help to educate, inform, and mobilize communities to address climate change and its health-related impacts in particular.[150] The Council of State and Territorial Epidemiologists (CSTE) has developed a set of state environmental health indicators related to climate change.[151] These state indicators provide a model for more granular assessments at the local level.[152] The indicators should help facilitate movement toward consensus about the nature of the forecasting, assessing, and projecting that occur during steps 1 and 2 of the BRACE framework. Taken together, the Global Change Research Program Report and the indicators developed by CSTE provide the means to develop a baseline of consistent climate-related health information. Measuring the same factors across jurisdictions will facilitate comparison of the effectiveness of different public health laws and interventions.

The second core function of public health is development of policies and plans to promote public health. The creation of a wealth of climate and health action plans, as well as tools to collect and disseminate the plans, is a major contribution.[153] The plans position the states and cities

148 Marinucci et al., *Building Resilience*, note 142, at 6437, 6451.

149 Institute of Medicine, *The Future of Public Health*, 7–8 (1988).

150 Alison Crimmins et al., *The Impacts of Climate Change on Human Health in the United States: A Scientific Assessment*, *supra* note 3.

151 P. B. English et.al., *Environmental Health Indicators of Climate Change for the United States: Findings from the State Environmental Health Indicator Collaborative*, 117 Envtl. Health Persp. 1673–1681 (2009) http://dx.doi.org/10.1289/ehp.0900708; See also www.cste.org/?page=ehindicatorsclimate.

152 Council of State and Territorial Epidemiologists, *Developing Climate Change Environmental Public Health Indicators: Guidance for Local Health Departments* (September 2013).

153 Georgetown Climate Center, *Adaptation Clearinghouse*, www.adaptationclearinghouse.org/.

that have developed them to implement the most promising interventions to address the most critical challenges in their communities. Disseminating the plans as well as the lessons learned during their development will accelerate the ability of other states and cities to engage in their own planning processes.[154]

Some states and cities have also responded to local or regional health threats by crafting responsive laws or policies. For example, occupational health specialists in California addressed the heightened vulnerability of outdoor workers to extreme heat with 2015 revisions to state regulations to require a heat illness prevention plan for specified categories of outdoor workers.[155] The public health response to recent outbreaks of Legionellosis associated with climate change also demonstrates that policy development must be attuned to local conditions. The city of New York promulgated amendments to its building code pertinent to cooling towers, while the city of Milwaukee engaged in education and outreach to the HVAC industry.[156]

The third core function of public health is assurance. One essential public health service included in this core function is assuring a competent public health and healthcare provider workforce. Associations for public health professionals, including the American Public Health Association (APHA), the Council of State and Territorial Epidemiologists (mentioned above), and the National Association of County and City Health Officials (NACCHO), have each taken steps to increase the competence and confidence of their members with respect to climate change. The theme of the American Public Health Association annual meeting in 2017 was "Climate Changes Health," signifying a commitment to supporting professional development with respect to climate resilience for the public health workforce. The National Association of County and City Health Officials published a useful checklist of twelve concrete tasks to operationalize climate change work in local health departments.[157] Many of these tasks will already be familiar to public health practitioners, even if their application to climate change is new. The Los Angeles County Department of Public Health worked with partners to

[154] Centers for Disease Control and Prevention, *Adaptation in Action: Grantee Success Stories from CDC's Climate and Health Program*, 2015, www.cdc.gov/climateandhealth/pubs/adaptation-in-action.pdf.

[155] Cal. Code of Regs. tit. 8, § 3395.

[156] Compare New York City Department of Health and Mental Hygiene, *Cooling Tower Requirements: What Building Owners Should Know* (2015), www1.nyc.gov/assets/doh/downloads/pdf/cd/cooling-tower-FAQs.pdf with Paul Biedrzycki et al., *Notes from the Field: Increase in Reported Legionellosis – Milwaukee, Wisconsin, June-September 2013*, Morbidity and Mortality Weekly Report (January 24, 2014).

[157] National Association of County and City Health Officials, *12 Steps to Operationalize Climate Change in a Local Health Department* (December 2014).

develop a series of in-depth workshops on climate and health for staff with the goal of developing department-wide strategies for addressing climate change.[158] Additionally, the public health sector has worked to educate healthcare providers about the current health impacts of climate change.[159] A climate-competent workforce is a necessary precursor to other assurance activities.

### 3.3.4 *Next Steps in the Public Health Response*

The public health response demonstrates notable successes in the assessment and planning steps of the BRACE framework. Next steps in the public health response include more widespread, systematic, and strategic communication, implementation, and evaluation. Whereas implementation and evaluation are identified as steps in the BRACE framework, communication is not.[160] All three response activities would benefit from elevation at this stage of the national public health response, which may gather speed more rapidly if it is able to incorporate lessons learned from previous campaigns on topics such as tobacco control and emergency preparedness, as will be discussed further in this section.

#### 3.3.4.1 Communication

Messages about human health impacts of climate change are important. They help people relate to climate change as a personal and immediate challenge, and if framed appropriately, they can help people to feel greater confidence in individual and societal capacity to reduce adverse health impacts and increase climate resilience.[161] The BRACE framework acknowledges that health departments may wish to consider their "political, cultural, and logistical environment" when identifying locally suitable strategies to address climate change.[162] The Government Accountability Office concluded in 2015 that the Department of Health and Human Services could do more to address climate change, particularly by taking further steps to use communication to enhance the public's understanding of public health

158 Linda Rudolph et al., *Climate Change, Health, and Equity: Opportunities for Action* 29 (Public Health Institute, 2015).

159 See, e.g., *Climate Change and Health: From Science to Practice, CDC Grand Rounds* (December 16, 2014) (archived webinar).

160 Communication is addressed as an important factor in public health work to address climate change in the literature.

161 E. Maibach et al., *Conveying the Human Implications of Climate Change: A Climate Change Communication Primer for Public Health Professionals* (Fairfax, VA: George Mason University Center for Climate Change Communication, 2011).

162 Marinucci et al., *Building Resilience, supra* note 142, at 6446.

risks.[163] Communication and community engagement are part of the CDC's conception of the BRACE framework, yet they are not listed explicitly among the five steps. More strategic communication would strengthen public health responses to climate change.

#### 3.3.4.2 Implementation

Perhaps the most pointed critique of local and state government responses to climate change is not limited to health departments, but it does implicate them. The critique is that there has been too much study and not enough action. Local government officials have noted that many resources related to climate adaptation are focused upon the early assessment and planning phases of the adaptation process, and fewer are focused on designing, implementing, enforcing, and assessing effective interventions.[164] To some extent, this is consistent with best practices being identified in the emerging field of implementation science, which suggests that there are fourteen steps in implementation of evidence-based interventions in a variety of sectors, and that completing the first ten steps prior to actual implementation of the intervention may optimize outcomes.[165]

Nonetheless, local government officials have called specifically for model adaptation ordinances and regulations that would accelerate implementation in their own communities.[166] In light of the urgency and severity of the public health threat posed by climate change, this concern must be taken seriously. One can acknowledge public health's commitment to evidence-based solutions, and balance that with a commitment to implement innovative solutions. Implementing interventions based upon the best evidence currently available, and then evaluating the effectiveness of those interventions, is entirely consistent with the principles of adaptive management and the BRACE Framework.

#### 3.3.4.3 Evaluation

Recently, a group of BRACE grantees released a technical report setting forth their findings regarding the effectiveness of over fifty interventions to reduce

[163] U.S. Government Accountability Office, *Climate Change: HHS Could Take Further Steps to Enhance Understanding of Public Health Risks* (October 2015) www.gao.gov/assets/680/672913.pdf.

[164] John Nordgren al., *Supporting Local Climate Change Adaptation: Where We Are and Where We Need to Go*, 66 Envtl. Sci. Policy 344–352 (2016).

[165] Duncan C. Meyers et al., *The Quality Implementation Framework: A Synthesis of Critical Steps in the Implementation Process*, Am. J. of Community Psychol. (2012).

[166] John Nordgren et al., *Supporting Local Climate Change Adaptation*, *supra* note 164.

the negative health effects of climate change.[167] A small minority of interventions met their standards to be considered scientifically supported, but a larger proportion of the interventions studied had at least some evidence. Some strides are being made toward filling the gaps with respect to intervention effectiveness, especially in the broader climate adaptation and mitigation literature.[168] Significant additional evaluation research is needed.[169] Ideally, evaluation will not hinder or delay implementation, but will occur on a contemporaneous and expedited basis with evaluation of interventions and their implementation in particular communities.

A brief illustration of the problem may be helpful. According to the recent BRACE grantee evaluation, the intervention related to wildfires with greatest scientific support in the literature is evacuation.[170] Evacuation is certainly a powerful intervention after a wildfire has occurred, but it does not address prevention. Although not in the context of the CDC's Climate Ready States and Cities Initiative, communities in the Southwest Crown region of Montana were concerned about the economic, safety, and health impacts of frequent wildfires, and decided to engage in forest restoration.[171] They resolved to prioritize evaluation in order to inform their own efforts and to contribute to the evidence. Thus, rather than waiting for an accumulation of evidence before responding to an urgent threat, this region chose to contribute to the development of that body of evidence.

With consistent execution of steps 3 ("assessing public health interventions"), 4 ("developing and implementing a climate and health action plan"), and 5 ("evaluating impact and improving quality of activities") of the BRACE framework, more findings about the effectiveness of climate interventions in promoting public health will be generated, and the tension between evidence-based practice and the urgency to act may diminish.

167 H. Anderson et al., BRACE Midwest and Southeast Community of Practice, *Climate and Health Intervention Assessment: Evidence on Public Health Interventions to Prevent the Negative Health Effects of Climate Change. Climate and Health Technical Report Series,* (2017) www.cdc.gov/climateandhealth/docs/ClimateAndHealthInterventionAssessment_508.pdf.

168 Vogel et al., *Climate Adaptation: The State of Practice in U.S. Communities, supra* note 140, at 37, 211 (noting that evaluation is often still anecdotal, but some projects have prioritized contemporaneous formal evaluation).

169 Jeremy J. Hess et al., *An Evidence-Based Public Health Approach to Climate Change Adaptation,* 122 Envtl. Health Persp. 1177, at 1184 (Nov. 2014).

170 Centers for Disease Control and Prevention, *Adaptation in Action: Grantee Success Stories from CDC's Climate and Health Program* (2015) www.cdc.gov/climateandhealth/pubs/adaptation-in-action.pdf.

171 Vogel et al., *Climate Adaptation: The State of Practice in U.S. Communities, supra* note 140, at 211.

### 3.3.5 *The Five Essential Public Health Law Services Framework*

The Five Essential Public Health Law Services framework offers one model to support and strengthen the public health response to climate change.[172] Drawing upon the example of tobacco control, as well as other successful movements in public health, the framework calls for a strategic partnership among public health practitioners, lawyers, health advocates, and researchers to improve population health and health equity through the coordinated use of law and policy.[173] The framework is rooted in the effort to identify policies with a reasonable likelihood of success along several dimensions – political, legal, and epidemiological – from the outset.[174] It recognizes that "building understanding and support is often the most difficult phase of the policy project."[175]

Like the BRACE framework, the Five Essential Public Health Law Services framework sets forth an iterative process and bears a kinship to the ten essential services of public health. Specifically, the Five Essential Public Health Law Services framework elucidates the functions necessary for implementing the policy elements of the ten essential services (including, but not limited to, "develop[ing] policies and plans that support individual and community health efforts" and "enforcing laws and regulations that protect health and ensure safety"). Moreover, the Public Health Accreditation Board (PHAB) standards, which are used to measure and document health departments' performance of the ten essential public health services), establish an expectation that health departments will review existing laws and work with governing entities and officials to update laws as needed. The associated PHAB commentary acknowledged the need to assess laws for their consistency with evidence-based public health and with newly emerging public health issues and information.[176] The Five Essential Public Health Law Services provide a framework for doing just that.

The Five Essential Public Health Law Services are:

(1) Access to Evidence and Expertise
(2) Expertise in Designing Legal Solutions
(3) Help Engaging Communities and Building Political Will

172 Scott Burris et al., *Better Health Faster: The Five Essential Public Health Law Services*, 131(6) Public Health Reports 747–753 (October 13, 2016).

173 For an approach even more explicitly focused on promoting health equity, see Rudolph et al., *Climate Change, Health, and Equity, supra* note 158.

174 Burris et al., *Better Health Faster, supra* note 172, at 748.

175 Ibid. at 749.

176 Public Health Accreditation Board, *Standards & Measures, Domain 6: Version 1.5* (December 2013), *available at* www.phaboard.org/wp-content/uploads/SM-Version-1.5-Board-adopted-FINAL-01-24-2014.docx.pdf (last visited July 5, 2017).

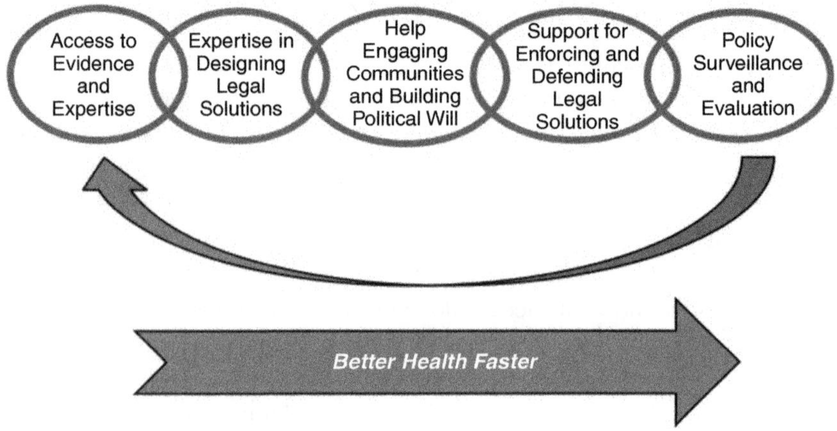


FIGURE 3.3 The Five Essential Public Health Law Services, image from Scott Burris et al., Better Health Faster: The Five Essential Public Health Law Services, 131(6) Public Health Reports 747, 748 (October 13, 2016). (reprinted with permission.)

(4) Support for Enforcing and Defending Legal Solutions
(5) Policy Surveillance and Evaluation[177]

Whereas implementation of the BRACE framework has so far involved greater emphasis on preparatory steps such as monitoring and assessment, the Five Essential Public Health Law Services framework is aimed squarely at creating and spreading effective solutions. Indeed, the originators of the Five Essential Public Health Law Services framework state that the mission of public health law is to "support the design, implementation, monitoring, evaluation, and scale-up of legal measures that improve public health and health systems."[178] By focusing on ensuring efficacy through the adoption and implementation of laws and policies, the Five Essential Public Health Law Services framework offers a complementary approach to the BRACE framework already in wide use.

### 3.3.6 *Harmonizing the BRACE and Five Essential Public Health Law Services Frameworks*

The BRACE framework is intended to build capacity in the public health sector to reduce the adverse health impacts of climate change. The Five

[177] Burris et al., *Better Health Faster, supra* note 172, at 747–753. [178] Ibid. at 748.

Essential Public Health Law Services are intended to support cross-sector collaborative efforts, assess gaps in efforts to change social norms, and produce "better health faster." State, local, and tribal health departments often convene climate and health planning processes. When inviting cross-sector partners, they may want to consider attorneys. Including lawyers on the climate resilience team may result in a salutary early focus on crafting, implementing, and enforcing legal solutions in the planning process. More broadly, harmonizing the BRACE and Five Essential Public Health Law Services perspectives may bring dividends in communication, implementation, and evaluation.

#### 3.3.6.1 Communication

Communication is not one of the steps in the BRACE framework, though it is acknowledged as a beneficial contributing factor in the literature. Engaging communities and building political will are recognized as one of the Five Essential Public Health Law Services. Purposeful, two-way communication adds value to the process of any effort to change social norms, because it melds awareness of local conditions with strategic efforts to shape local attitudes for greater understanding and support of the desired change, in this case, adoption of climate adaptation and mitigation practices. This is not to say that all communication strategies must be identical.

A recent overview of the state of adaptation practice in the United States reported a wide variety of communication strategies across communities.[179] In light of the politically divisive nature of the topic, some communities embrace climate change framing, others eschew any framing related to climate change, and still other communities fall somewhere in between.

Research from the fields of political science and moral psychology offer helpful lessons for communicating about climate change across political boundaries. For example, Moral Foundations Theory provides insight into the moral values that underlie American conservative and liberal viewpoints. Research suggests that framing climate change to encompass a broader range of moral foundations, including values such as patriotism, respect for authority, and defense of the purity and sanctity of nature – as compared to traditionally liberal messages concerning injustice and harm – may have greater

[179] Vogel et al., *Climate Adaptation: The State of Practice in U.S. Communities*, *supra* note 140, at 30–31, 47.

appeal in more conservative political environments.[180] Likewise, recent political science research indicates that while the majority of Climate Ready States and Cities initiative grantees are clustered along the coasts of the United States, substantial adaptation and mitigation activities are also being carried out by small and medium-sized local governments in the middle of the country.[181] However, survey responses from local government leaders in 11 Great Plains states indicate that due to cultural and political skepticism about whether climate change is caused by human activity, adaptation and mitigation activities are often framed around economic development and other more culturally resonant goals and values.[182]

The Great Plains research provides an example of strategic framing relative to developing walkable cities, identified as one of the top 100 strategies to reduce global warming.[183] This research shows that nearly 60 percent of communities surveyed had community-wide trails for hiking and biking.[184] When public health practitioners participate in development of a hiking or walking path with the stated goal of promoting physical activity and preventing obesity, reduced carbon emissions and increased climate resilience may be an incidental bonus, or a co-benefit.[185] The public health activities may be informed to some extent by climate science, but they do not generally take place in the context of an overall plan to address climate change.

Harmonizing the BRACE and Five Essential Public Health Law Services approaches might result in a greater emphasis on strategic efforts to engage the community in culturally meaningful terms, paired with an understanding of the procedural and substantive benefits of dedicated climate change planning and prioritization. Rather than stopping at what is politically and culturally acceptable in the moment, the combined frameworks could assist in strategic efforts to expand the field of acceptable interventions and goals. Highlighting improvements in economic outcomes, social connectedness, quality of life, and health outcomes as a result of climate mitigation and adaptation

[180] Christopher Wolsko et al., *Red, White, and Blue Enough to be Green: Effects of Moral Framing on Climate Change Attitudes and Conservation Behaviors*, 65 J. of Experimental Soc. Psychol., 7–19 (2016).

[181] Robert S. Wood et al., *An Examination of Local Climate Change Policies in the Great Plains*, 31(6) *Rev. of Pol'y Res.* 529, 540–546 (2014).

[182] Ibid. at 529, 540–546. [183] Paul Hawken, *Drawdown, supra* note 133, at 86.

[184] Wood et al., *An Examination of Local Climate Change Policies in the Great Plains, supra* note 181, at 529, 540–546.

[185] Walkable cities also enhance social connectedness, contributing to mental health and well-being.

interventions while simultaneously pointing out their climate benefits may diminish any perceived conflict between climate resilience and other goals.[186]

### 3.3.6.2 Implementation

As noted in the previous section, the implementation component of the BRACE framework, which is paired with climate and health planning in step 4, has not yet been fully realized. The Five Essential Public Health Law Services framework brings a legal lens to implementation by specifying actionable steps to achieve effective legal solutions. It encourages strategic decision-making, including which legal tools to use, how to design a policy in order to withstand legal challenge, and how to simplify administration and enforcement.[187]

There is keen interest in standard legal tools such as ordinances, bonds, memoranda of understanding, taxes, fees, incentives, and cost-sharing to accelerate implementation of climate interventions.[188] In addition, attorneys are crucial partners when evaluating the legal authority of a particular jurisdiction, anticipating and avoiding preemption-based challenges, and assuring that key pieces of a proposed law – such as enforcement mechanisms – are not lost as the proposal moves forward.[189] More collaboration between public health practitioners and attorneys to create generally applicable templates as well as to discuss considerations for tailoring legal interventions to a particular jurisdiction could serve to advance implementation more quickly and systematically.

### 3.3.6.3 Evaluation

There is no single, immediate, or easy solution for addressing the adverse human health effects of climate change. There are a multitude of potential interventions, some of which are more effective than others. The BRACE Framework guides public health practitioners to engage in a careful assessment of the local community, prioritize the greatest threats, identify the most effective interventions, and then evaluate their implementation. This

186 Michael Bloomberg and Carl Pope, *Climate of Hope: How Cities, Businesses, and Citizens can Save the Planet* (2017).

187 Burris et al., Better Health Faster, *supra* note 172, at 749–750.

188 Rudolph ed., *Climate Change, Health, and Equity*, *supra* note 158, at 13; Vogel et al., *Climate Adaptation: The State of Practice in U.S. Communities*, *supra* note 140, at 45; Nordgren et al., *Supporting Local Climate Change Adaptation*, *supra* note 164.

189 Burris et al., Better Health Faster, *supra* note 172, at 749–750.

approach has many strengths, and if we faced a threat with more time for a richer and more robust scientific literature to emerge, it might be enough. However, we should be wary of Martin Luther King, Jr.'s "paralysis of analysis," and recall the principles of adaptive management, which recognize the need to act despite incomplete and imperfect information, and then to adjust the course of action as new information emerges.

The Five Essential Public Health Law Services framework adds two helpful refinements to the BRACE framework. First, it offers the insight that the decision to conduct the most critical research need not be left to chance, but researchers, practitioners, advocates, and policymakers can develop "relationship networks" to facilitate prioritization of research with the highest potential impact.[190] To be clear, the intent is not to dictate the conclusions of the research, but merely to have input on the selection and sequence of research topics. Second, it offers the concept of "policy surveillance," or the use of scientific methods and rigor to evaluate and compare the effectiveness of various legal approaches to public health challenges.[191] To conduct policy surveillance, interdisciplinary teams of researchers analyze and compare the elements of statutes, regulations, or ordinances across jurisdictions, and then compare health outcomes in the various jurisdictions over time. This is a tremendously powerful tool in promoting the spread of the most effective interventions, in both public health and legal terms.

One instance where policy surveillance might add substantial value is in the use of zoning and building regulations requiring infrastructure to reduce ambient and indoor heat in residential and commercial development plans – cool roofs, green roofs, thicker insulation, shade tree planting, energy audits, climate risk assessments, and the like.[192] BRACE grantees have identified this as an area where there is insufficient evidence of the effectiveness of the intervention.[193] Of course, insufficient evidence does not mean no evidence, and climate professionals may seek to enact zoning and building regulations based on existing evidence.[194] At the same time, evaluation of the effectiveness of the legal interventions, undertaken by teams of attorneys and public health practitioners in accordance with policy surveillance methods, would be invaluable in comparing policy and health outcomes and identifying policies most worthy of emulation. Policy surveillance permits consideration not just of the practices, such as installation of cool roofs, but also of the underlying legal

[190] Ibid. at 748–749. [191] Ibid. at 751.

[192] Bloomberg and Pope, *Climate of Hope*, *supra* note 186, at 30, 110–116, 226, 238, and throughout. Chapter 6 discusses interventions like this in greater detail.

[193] Anderson et al., *Climate and Health Intervention Assessment*, *supra* note 167, at 18.

[194] Vogel et al., *Climate Adaptation: The State of Practice in U.S. Communities*, *supra* note 140.

mechanisms to support implementation. Thus, it facilitates analysis of legal authority related to mitigation and adaptation activities, including mechanisms to encourage practices or enforce requirements such as incentive structures, regulatory requirements, provision of funding, dedicated staff positions, and training and technical assistance for staff.

## 3.4 MOVING THE PUBLIC HEALTH RESPONSE FORWARD

Addressing health-related impacts of climate change through adaptation and mitigation is increasingly a function of public health. Public health efforts may occur as part of an iterative process of long-term planning (as under the BRACE framework), through implementation of specific adaptation or mitigation practices, or as part of a coordinated legal and policy strategy. Following are three recommendations for how the public health community could build upon and strengthen its efforts to promote health and increase climate resilience.

### 3.4.1 *Engage in Cross-Sector Collaboration to Inform Public Health Responses*

Collaboration across sectors is essential to effective work to address climate change. For public health practitioners, helpful partners may include climatologists, urban and regional planners, architects, transportation specialists, regenerative agriculture and forestry experts, researchers, lawyers, and many others. Public health practitioners may benefit from seeking out "boundary organizations," which assist in translating specialized technical information, such as climate science, for a more general audience.[195] The growth in literature related to climate change may itself create a need for such boundary organizations to assist with evaluating the literature for quality and relevance.[196] Based on surveys of local government officials engaged in climate adaptation activities, some experts are calling for a centralized web platform to facilitate access to resources specifically designed to support community-based climate adaptation, as well as a centralized climate adaptation "extension service," "service bureau," or "help desk."[197] In addition, membership in or engagement with relevant professional associations may

195 D. H. Guston, *Boundary Organizations in Environmental Policy and Science: An Introduction*, 26(4) Sci., Tech., and Hum. Values 399–408 (October 1, 2001).

196 Stults et al., Assessing the climate adaptation resource and service landscape. In: Bullock et al., *Living with Climate Change: How Communities are Surviving and Thriving in a Changing Climate* (Boca Raton, FL: CRC Press 2015).

197 Nordgren et al., *Supporting Local Climate Change Adaptation*, *supra* note 164.

help address gaps in the knowledge, skills, and resources of public health practitioners and public health lawyers who seek to address the adverse health impacts of climate change.[198] Engaging with these professional associations is likely to benefit the public health field as a whole, and not just individual practitioners.

The One Health approach described earlier in this chapter provides an important example of the cross-sector and interdisciplinary learning and collaboration that may facilitate development of strategic and comprehensive responses to systemic causes and consequences of climate change. The One Health approach has been defined as "a coordinated, collaborative, interdisciplinary and cross-sectoral approach to address potential or existing risks that originate at the animal–human–ecosystems interface."[199] One Health scientists have recognized infectious disease, food security, and food safety as interconnected issues occurring at this interface, and have observed that these issues are not only affected by the changing climate, but also by the same "fundamental drivers" that have contributed to climate change, including "population growth, consumption patterns and economic models that are tied to infinite growth."[200] Given the scale of the threat posed by climate change, One Health scientists have called for the inclusion of a still broader range of disciplines in their work, specifically including policymakers and others capable of supporting social, economic, and political changes informed by science.[201]

The Health in All Policies (HiAP) framework provides another strategy for engaging across sectors to address health impacts of climate change. The HiAP approach seeks to incorporate health and equity considerations into all aspects of policymaking by formalizing and institutionalizing a process through which every sector of government applies a public health lens to decision-making.[202] The goal of this approach is to ensure that decision-makers are

198 Relevant associations and resources include the American Planning Association, American Psychological Association, American Society of Adaptation Professionals, Association of Climate Change Officers, Association of State Floodplain Managers, City Energy Project, Health Care Without Harm, ICLEI – Local Governments for Sustainability, International Transformational Resilience Coalition, National Environmental Health Association, National Sustainable Agriculture Coalition, Network for Emerging Leaders in Sustainability, Risky Business Project, Smart Growth America, Sustainability Accounting Standards Board Alliance, Sustainable Communities Leadership Academy, US Climate and Health Alliance, US Conference of Mayors Climate Protection Agreement, US Green Building Council, and We Are Still In.

199 Peter F. Black and Colin D. Butler, *One Health in a World with Climate Change*, 33 Rev. Sci. Tech. Off. Int. Epiz. 465, 466 (2014).

200 Ibid. at 465, 469. 201 *Ibid.*

202 Linda Rudolph et al., *Health in All Policies: Improving Health through Intersectoral Collaboration*, 1 (2013).

aware of and accountable for the potential health harms and/or benefits of different policy options and the manner in which law and policy may shape social determinants of health.[203] A defining strength of the framework is that it provides a pathway for engaging diverse and nontraditional governmental and community-based partners to address multicausal health challenges.[204] Moreover, by fostering collaboration and innovation, HiAP maximizes often scarce public health resources and reduces duplicative efforts.[205] HiAP has been identified as a key strategy for addressing complex public health issues with cross-sector origins and impacts, including climate change.[206]

### 3.4.2 *Incorporate Public Health Law into Public Health Responses*

Among the many strategies public health practitioners may employ to improve health, changes to the environmental context for individual decision-making (so that the "default" decision is healthy) and interventions addressing social determinants of health are recognized as having the greatest population health impact while requiring the least amount of individual effort.[207] By nature, these types of interventions often require widespread societal change and/or collective action, often prompted, required, or incentivized by changes in the law.[208] Thus, law has been a major component of many of the great achievements of public health, including smoking reductions, immunizations, motor vehicle safety, community water fluoridation, and food safety.

The Five Essential Public Health Law Services framework distills some of the best practices from these earlier achievements to provide guidance for spreading effective, public health legal interventions. The Five Essential Public Health Law Services framework complements the BRACE framework by setting forth the steps necessary for implementation of legal interventions in particular. Harmonizing the two approaches, especially in the areas of communication, implementation, and evaluation, could strengthen the public health response to climate change considerably by facilitating the types of contextual and societal changes that have the greatest potential for improving population health.

203 Ibid. 204 Ibid. 7. 205 Ibid.

206 *See* Kathryn J. Bowen and Kristie L. Ebi, *Governing the Health Risks of Climate Change: Towards Multi-Sector Responses*, 12 Current Opinion in Environmental Sustainability 80, 83 (2015).

207 Thomas R. Frieden, *A Framework for Public Health Action: The Health Impact Pyramid*, 100 Am. J. of Pub. Health 590, 591–592 (2010).

208 Ibid. at 594.

### 3.4.3 *Incorporate Health Equity Principles into Public Health Responses*

In almost every aspect of climate change, the health impacts among vulnerable populations are worse than among the general population. As noted in this chapter, individuals or communities may face heightened vulnerability to climate change due to increased exposure and sensitivity to climate stressors as well as decreased resilience to climate impacts.[209] Social determinants of poor health exacerbate vulnerability to climate change while also independently degrading personal and public health outcomes. Moreover, in addition to the inequitable distribution of health harms directly linked to climate change, some climate response strategies may also produce adverse effects on health (referred to as "co-harms") that disproportionately affect vulnerable communities.[210] For example, increased production of biofuel from traditional food crops such as corn may increase demand for these crops, leading to increased food prices that disproportionately burden low income households.[211]

The Office of Minority Health within the United States Department of Health and Human Services has defined health equity as the "attainment of the highest level of health for all people."[212] Health equity has also been described as the absence of "health differences that are socially produced, systematic in their distribution across the population, and unfair."[213] Given the interrelationship between exposure, sensitivity, and resilience to climate change and the social determinants of health, climate change promises to exacerbate existing health disparities unless health equity is prioritized in developing public health responses.[214] Thus, when developing community engagement strategies, as well as when leading climate and health action planning and designing legal solutions, reaching the highest level of health

[209] Turner et al., *A Framework for Vulnerability Analysis in Sustainability Science*, *supra* note 12; U.S. Global Change Research Program, *Populations of Concern*, *supra* note 12.

[210] Rudolph et al., *Climate Change, Health, and Equity*, *supra* note 158, at 18.

[211] Ibid.

[212] National Partnership for Action to End Health Disparities, Office of Minority Health, United States Dept. of Health & Human Servs., *National Stakeholder Strategy for Achieving Health Equity* 9 (2011), www.minorityhealth.hhs.gov/npa/templates/content.aspx?lvl=1&lvlid=33&ID=286. The definition further provides that "[a]chieving health equity requires valuing everyone equally with focused and ongoing societal efforts to address avoidable inequalities, historical and contemporary injustices, and the elimination of health and healthcare disparities."

[213] See O. Solar and A. Irwin, World Health Organization, A *Conceptual Framework for Action on the Social Determinants of Health*, 12 (2010).

[214] Rudolph et al., *Climate Change, Health, and Equity*, *supra* note 158, at 19. See also World Health Organization Commission on Social Determinants of Health, *Closing the Gap in a Generation*, *supra* note 20, at 1, 196.

for all people should be the goal. Moreover, as championed by climate justice advocates, a just public health response to climate change requires not only an equitable distribution of health benefits and harms, but also of power and participation.[215]

## 3.5 CONCLUSION

The adverse health impacts of climate change are both occurring now and likely to grow more severe. Public health practitioners and public health lawyers have access to frameworks and logic models to support prompt and strategic action. There is no one-size-fits-all model, but adaptive management, the BRACE Framework, the Five Essential Public Health Law Services, Health in All Policies, and principles of health equity provide tools to aid in the development of locally appropriate, legally sound, and highly effective public health responses.

[215] Schlosberg and Collins, *From Environmental to Climate Justice, supra* note 21, at 361. See also James K. Boyce et al., *Power distribution, the environment, and public health: A state-level analysis*, 29 Ecological Economics 127, 138 (1999) (providing empirical support for the notion that unequal distribution of power is associated with increased overall environmental degradation and unequal distribution of public health costs and benefits).

# PART II

# CROSS-CUTTING ISSUES

# 4

# Government Speech and the First Amendment

David C. Vladeck

*Information is a powerful tool that government can use to inform the public about the risks of climate change, how to mitigate those risks, and how to reduce emissions of the greenhouse gases responsible for climate change. This chapter addresses the role the First Amendment plays when the government uses information to promote public policies on climate change. The chapter explains that the government has vast leeway to collect and disseminate information and the First Amendment presents no barrier to the government's use of information tools when the government itself is the speaker. But the First Amendment does come into play when the government enlists or compels others to carry the government's message. The First Amendment may, in some cases, limit the government's power to compel speakers to engage in speech on matters of opinion or on subjects that might be considered controversial. The First Amendment generally limits government's ability to force speakers to adhere to the government's views. And the First Amendment may constrain the government's ability to rid the marketplace of advertising claims that may over-promise environmental benefits, but are not demonstrably false or misleading. These constraints, however, do not pose a serious obstacle to the government's ability to use information tools to help in the fight against climate change.*

## 4.1 INTRODUCTION

Reye's syndrome is a rare but often fatal disease. In the early 1980s, there were as many as 500 cases of Reye's syndrome – a disease that causes death and brain damage in children – each year. In 1981, the federal Centers for Disease Control identified a link between Reye's syndrome and the use of aspirin by children with flu and chicken pox. No one suggested that aspirin be taken off the market, but the Food and Drug Administration drafted a proposed regulation that would have required a warning label. The aspirin industry pressured the Reagan Administration, and the FDA was forced to table its proposal. But a federal appeals court questioned the FDA's decision, and in February 1986 the FDA at last issued a final regulation requiring a warning label on aspirin products. Since then, Reye's syndrome has virtually disappeared.[1]

The eradication of Reye's syndrome is a testament to the power of government information to fight public health threats. Climate change, in all of its many manifestations, poses many threats to public health that are just as urgent and far more widespread than was the threat of Reye's syndrome decades ago.

The fundamental question that this chapter aims to answer is to what extent the government may respond to the public health threats engendered by climate change by acquiring, using, and disseminating information to educate the public and to shape regulatory and legislative policy? The short answer, elaborated on in this chapter, is that the government has vast leeway when it collects information and uses it to inform, educate, and warn the public about public health risks. But when the government enlists others to speak on its behalf, through incentives or compulsion, the First Amendment does, to some extent, place restraints on the government.

For this reason, the government's power to speak on matters of public health is at its zenith when the government itself is the speaker. And make no mistake, the government constantly engages in speech on public health matters – speech that often is critical to an informed public debate. Federal agencies have long issued reports and published scientific papers, statistical analyses, technical documents, and other information on climate change and related fields to inform the public and to assist in regulatory decision-making.

Much of the information disseminated by the government is generated by the government itself. The government obtains information from countless sources: from weather satellites and ocean probes deployed by science agencies like the National Oceanographic and Atmospheric Administration, to meteorological data compiled by the National Weather Service, to data about wildfires compiled by the Forest Service, to information about energy consumption from the Department of Energy, and to air quality data compiled by the Environmental Protection Agency. This information is crucial to an informed public debate on environmental matters, and is central to regulatory decision-making. After all, regulators need a wealth of reliable information in order to make sound regulatory decisions. The government's ability to gather and disseminate the facts that are the raw material for policy development and public education is not limited by the First Amendment; it is bounded only by budgetary and political constraints.

[1] See *Pub. Citizen Health Res. Group v. Comm'r, FDA*, 740 F.2d 21 (D.C. Cir. 1984); see also Center for Disease Control, *CDC Study Shows Sharp Decline in Reye's Syndrome Among U.S. Children* (May 6,1999) www.cdc.gov/media/pressrel/r990506.htm.

But not all of the information agencies use and make public comes from the government's own endeavors. Much of it is provided by companies under statutes that compel them to submit data to the government. For instance, data about emissions from coal-fired power plants are critical to environmental policy-setting, but they are not produced by the government. Rather, power plant operators are required to submit that data to the EPA under the Clean Air Act. And there are many other environmental statutes, such as the Clean Water Act and the Emergency Planning and Community Right-to-Know Act, that similarly impose a statutory mandate on companies to submit data to the EPA or other regulatory agencies and place a duty on the federal agencies to make that information available to the public, either on the agency's own initiative or in response to public requests. That information also provides a basis for government speech on public health matters and aids in the formulation of policy. Once again, the First Amendment rarely limits the ability of government to demand information from companies subject to federal regulation.

Government speech, however, cannot reach everyone and often cannot provide information to the public when that information is most needed. Consider Reye's syndrome. Warnings from the Centers for Disease Control about the likely link between Reye's syndrome and aspirin certainly reached some parents. But it was the warnings on aspirin packages and bottles that alerted parents just when they needed to be told about the serious risks of giving aspirin to children with the flu or chickenpox.

For this reason, our regulatory system depends on companies complying with laws that require disclosures to convey important information directly to consumers at the right time. After all, the government is not omnipresent: Information-forcing regulations provide the disclosure of information needed to make educated decisions in the marketplace. Some of these disclosures are intended to shield consumers from false or misleading information. But many advance other governmental interests. Food labels provide consumers with the information they need to have confidence in what they're purchasing. Nutritional labels help consumers understand the nutritional content of the foods they consume. Drug labels warn patients about the risks of the drugs they take, and the possible interactions the drug might have with other medications or foods.

This is where the First Amendment comes into play. When it comes to the fight against climate change, the First Amendment does not limit government speech. But the First Amendment does, at times, constrain how the government may interact with nongovernmental entities the government either compels to speak or pays to disseminate the government's message. Courts have ruled that the First Amendment, in some cases, limits the government's power to compel companies to engage in speech on matters of opinion or on subjects that might be considered controversial, such as putting graphic warning labels on cigarette packages, or alerting buyers that products might contain "conflict minerals." The First Amendment also plays a role when the government provides incentives to enlist others to engage in speech to further the government's objectives. The government has broad latitude in crafting the message it wants to convey. But the First Amendment may limit the government's ability to choose speakers based on the speaker's viewpoint. And finally, the First Amendment may constrain the government's ability to rid the marketplace of claims that may over-promise environmental benefits, but are not demonstrably false or misleading, leaving marketers free to claim that products are environmentally friendly, even when the claimed environmental "benefits" may be overstated or illusory.

This chapter proceeds in two sections. Section 4.2 discusses the interplay between the First Amendment and government speech, mainly to drive home that the First Amendment is not a significant obstacle to the government's ability to communicate climate change-related health threats to the public. But it makes clear as well that the First Amendment generally forbids the government from forcing anyone, even government contractors, to adhere to a specific viewpoint or belief, and limits the government's ability to compel private actors to engage in speech they deem to be nonfactual, controversial, or antithetical to their interests. Section 4.2 also addresses the extent to which the First Amendment may limit the government's ability to police the marketplace for unproven claims of environmental benefit. Section 4.3 then discusses the kinds of messages the government might convey to the public to safeguard the public's health and to provide guidance about policy choices relating to climate change. Section 4.3 also addresses the practical limits on government speech relating to climate change.

## 4.2 GOVERNMENT SPEECH AND THE FIRST AMENDMENT

The government participates in today's vast information market in three ways that raise issues under the First Amendment: when the government funds speech on public health; when it compels companies to disseminate the government's public health messages that the company does not want to disseminate or may disagree with; and when it polices the marketplace to ensure that consumers are not led astray by false or misleading commercial messages.

### 4.2.1 Government Funding of Speech

The government spends billions of dollars every year to promote important public health messages, both domestically and abroad. In so doing, the government has broad discretion to determine the message that it wants to deliver and the type of entity it wants to convey that message. But there are limits on the conditions the government may impose on those the government enlists to disseminate its message.

One case that illustrates those limits is *Agency for International Development (AID)* v. *Alliance for Open Society International*,[2] which was decided by the Supreme Court in 2013. The *AID* case grew out of a federal initiative to combat the spread of AIDS/HIV worldwide. Congress made substantial funding available to nongovernmental organizations to assist in the cause of fighting HIV. But Congress limited eligibility for funding to only those organizations that agreed to two conditions: One was that no funds made available under the program "may be used to promote or advocate the legalization or practice of prostitution or sex trafficking," and the second prohibited funding "to any group or organization that does not have a policy explicitly opposing prostitution or sex trafficking."

The constitutional question did not relate to the first condition; no one disputed that the government was free to decide not to spend money on advocating "the legalization or practice of prostitution or sex trafficking," even though some organizations maintained that legalization would help stem the AIDS epidemic. It was the second condition – the one requiring funding recipients to adopt a policy "explicitly opposing prostitution or sex trafficking" – that was alleged to violate the First Amendment.

The Supreme Court found that the requirement ran afoul of the First Amendment. The Court said that as a general rule "freedom of speech prohibits the government from telling people what they must say." The question, as the Court framed it, was whether the government could nonetheless impose that requirement as a condition of federal funding. After all, some conditions on funding *define* the scope of federal program, as did the first condition barring program participants from using federal funds to promote the legalization of prostitution or sex trafficking. But the Court found that the second condition went well beyond defining the program's scope – it compelled a grant recipient to adopt publicly a particular belief as a condition of funding.[3] In so doing, the Court ruled that the requirement violated the

[2] *Agency for Int'l Dev. v. Alliance for Open Soc'y Int'l, Inc.*, 570 U.S. 205 (2013).

[3] The line between conditions that define a program and conditions that require adherence to a particular point of view is, at times, hazy. For instance, in *U.S. v. Am. Library Ass'n*, 539 U.S.

First Amendment for two related reasons. One was the general rule that the First Amendment forbids the government from compelling adherence to a particular belief.[4] The second was that the government's efforts to compel recipients to adopt a policy imposed an "unconstitutional condition" on prospective grantees. The "unconstitutional conditions" doctrine forbids the government from using the leverage of a grant or the threat of denial of funding or other benefit to suppress a viewpoint the government finds offensive, or, as in the *AID* case, to force adherence to a viewpoint an individual or entity might disagree with or find offensive.[5]

How might these doctrines impact government programs that seek to enlist nongovernmental organizations to spread the word about the public health risks presented by climate change? The answer is that the effect of these doctrines should be minimal, so long as the government is careful to define the objectives of an information campaign in ways that focus tightly on the messages that the government wants to convey. Consider this example. Suppose the government wants to ensure that coastal communities take measures to mitigate the risks of climate change by adapting their environments to cope with rising seawater levels. And suppose that the government decides that, given the large number of coastal communities in need, it has to enlist nongovernmental organizations to carry its message. The government could, for example, set as conditions of eligibility that only organizations that have demonstrated expertise in the field and a proven track record in effectively communicating these measures to local governments will be considered for funding. Defining the project's mission in that way would inevitably exclude groups that believe that climate change is a hoax or lack expertise in mitigation strategies, and that exclusion would be perfectly acceptable under the Constitution.[6] Or suppose that the government wanted to enlist local organizations to monitor emission levels of mercury or other toxic pollutants

194, at 212 (2003) (the Court rejected the Library Association's argument that conditioning funds for Internet access on the libraries installing filtering systems to screen out pornography violated their First Amendment rights because "[t]o the extent that the libraries wish to offer unfiltered access, they are free to do so without federal assistance.")

4 See, e.g. *Rumsfeld* v. *Forum for Academic and Institutional Rights, Inc.*, 547 U.S. 47, 61 (2006); *Board of County Com'rs, Wabaunsee County, Kan.* v. *Umbehr*, 518 U.S. 668 (1996); *Wooley* v. *Maynard*, 430 U.S. 705, 717 (1977).

5 See, e.g. *United States* v. *United Foods, Inc.*, 533 U.S. 405 (2001).

6 The government might even be able to forbid grantees from advocating against climate change mitigation measures if it were so inclined. In *Rust* v. *Sullivan*, 500 U.S. 173, 180–181 (1991), the Court upheld a regulation barring organizations that received federal grants to nonprofit healthcare providers for family planning projects from using those funds in programs where abortion is a method of family planning. The Court made clear that the condition barring any advocacy of abortion passed constitutional muster.

associated with coal combustion in their communities. The government could certainly condition funding in ways that would ensure that only locally based organizations competent to monitor mercury emissions accurately were eligible for funding, and again, using program-specific eligibility criteria would raise no constitutional issue.

As should be evident, the government has broad authority to use its vast spending power to enlist speakers to deliver the government's message. So long as the government carefully defines what its messages and objectives are, and makes no attempt to compel adherence to a particular belief, the First Amendment poses no obstacle to the fulfillment of the government's goals.[7]

### 4.2.2 *Government-Compelled Speech*

#### 4.2.2.1 Background

Contemporary regulation of the marketplace depends on disclosures to prevent deception and unfairness, to protect public health and safety, to ward off environmental harm, and to ensure that markets are transparent so consumers are able to make educated choices. It is hard to go to any store or to buy any product that does not bear at least one government-required disclosure or label. Every store has at least one prominent "Exit" sign. All foods, drugs, cosmetics, home furnishings, toys, clothes, mattresses, and virtually anything else one buys or consumes bear a government-required statement. Many of these labels simply tell consumers that the T-shirt they're buying is made of cotton, or that hamburger meat contains 15 percent fat, weighs nineteen ounces, and may not be sold after a certain date.[8]

For the most part, companies do not object to complying with disclosure laws because they recognize that truthful disclosures foster consumer confidence. They also understand that any objection would likely be futile. As the Supreme Court has recognized time and again, government may compel disclosures when they serve important purposes: to lessen the risk of deception,

[7] The First Amendment has no particular political valence so this broad authority can be a double-edged sword. It is not far-fetched to think that the Trump Administration might use its power to disseminate information to promote policies that undercut efforts to stem the use of fossil fuels or other efforts to mitigate the effects of climate change.

[8] These kinds of requirements are longstanding and noncontroversial. See, e.g., 21 C.F.R. § 201.57 (requirements mandating warnings on drug labels, including prominent "black box" warnings that emphasize a particular hazard); 21 U.S.C. § 343(i)(1) (requiring most food product labels to bear an ingredient list); 21 CFR § 101.9 (requiring nutrition labels on food packages and identifying the information that the labels must address, such as calories, dietary fiber, and certain vitamins and minerals).

promote public health and safety, mitigate environmental harms, and enable the public to make informed choices in the marketplace.[9]

#### 4.2.2.2 The "*Zauderer* Test"

First Amendment theorists maintain that the First Amendment was crafted to serve two reinforcing values – safeguarding democratic discourse and insuring the free flow of information to the public. Most importantly, the First Amendment is the "guardian of our democracy" because it "guarantees the right of all persons to participate in the formation of public opinion" on matters of self-governance. The First Amendment also serves the "informational function" of "circulating knowledge to those who wish to participate in public discourse."[10]

The First Amendment accords varying levels of protection based on the kind of speech the government seeks to regulate. Core speech, such as political speech, receives full-bore constitutional protection because it serves both First Amendment values. Government restraints on core speech are subject to strict scrutiny by the courts, and may be upheld only where the restraints serve a compelling government interest and do so in the narrowest way possible.

On the other hand, the Supreme Court has always drawn "the commonsense distinction between speech proposing a commercial transaction . . . and other varieties of speech," because commercial speech has only an informational value and the government has a keen consumer protection interest in regulating commercial transactions. For that reason, commercial speech was not accorded First Amendment protection until 1976. In *Virginia State Board of Pharmacy* v. *Virginia Citizens Consumer Council*[11], the Court held that commercial speech merits some First Amendment protection, but less than core speech. Since then, the Court has made clear that restraints on commercial speech may be upheld so long as they serve a substantial governmental interest, directly advance the interest asserted, and are not more extensive than is necessary to serve that interest.[12]

This background helps explain why courts generally uphold government efforts to compel companies to provide information relating to commercial transactions to the public. After all, providing more information to consumers

[9] See generally Robert Pitofsky, *Beyond Nader: Consumer Protection and the Regulation of Advertising*, 90 Harv. L. Rev. 661 (1977); Cass Sunstein, *Informing America: Risk, Disclosure and the First Amendment*, 29 Fla. St. U. L. Rev. 653 (1993).

[10] Robert Post, *Compelled Commercial speech*, 117 W. Va. Univ. L. Rev. 867, 874–875 (2015). The framing of the First Amendment values in the text is largely drawn from this article.

[11] *Va. State Bd. of Pharmacy v. Va. Citizens Consumer Council, Inc.*, 425 U.S. 748 (1976).

[12] *Cent. Hudson Gas & Elec. Corp. v. Pub. Serv. Comm'n*, 447 U.S. 557 (1980).

fosters the "informational function" of the First Amendment. There are times, however, when companies object to disseminating the government's messages, and invariably they argue that forced disclosure violates their First Amendment rights.

The key Supreme Court case addressing this issue is *Zauderer* v. *Office of Disciplinary Counsel*.[13] One issue in *Zauderer* was whether a lawyer could be disciplined for failing to comply with a state rule requiring that his newspaper advertisements disclose information about the fees he charged in contingency cases. In upholding the discipline, the Court drove home that mandatory disclosures prevent consumer deception, but they also further the broader goal of fostering more informed decision-making. A core value of the First Amendment is encouraging the circulation of accurate and useful information. Mandatory disclosures do just that. For that reason, the Court in *Zauderer* held that a company's "constitutionally protected interest in *not* providing any particular factual information in advertising is minimal" and that "the First Amendment interests implicated by disclosure requirements are substantially weaker than those at stake when speech is actually suppressed." So long as the disclosure mandates are "reasonably related" to the public interests served and are not unduly burdensome, under *Zauderer* they almost invariably pass constitutional muster.[14]

As a result, most First Amendment attacks on mandatory disclosure requirements have been rebuffed by the courts. For instance, courts have upheld:

* New York City rules requiring that some chain restaurants list the caloric content of the foods and beverages they serve;[15]
* EPA rules requiring cities with storm sewers that discharge into national waters to educate the public about the impacts of stormwater discharge on water bodies and the hazards of improper waste disposal;[16] and
* New York State rules requiring manufacturers of mercury-containing light bulbs to include information about their safe disposal on packaging.[17]

[13] *Zauderer v. Office of Disciplinary Counsel of Supreme Court of Ohio*, 471 U.S. 626 (1985). The author was co-counsel for Philip Zauderer in the Supreme Court and helped draft Mr. Zauderer's submissions to the Supreme Court.

[14] *Zauderer*, 471 U.S. at 650–654. See also *Milavetz, Gallop & Milavetz, P.A.* v. *United States*, 559 U.S. 229 (2010) (applying the *Zauderer* "reasonably related" standard to disclosure requirements "intended to combat the problem of inherently misleading commercial advertisements").

[15] *N.Y. State Restaurant Ass'n v. N.Y.C. Bd. of Health*, 556 F. 3d 114 (2d Cir. 2009).

[16] *Environmental Defense Center* v. *EPA*, 344 F.3d 832, 849 (9th Cir. 2003).

[17] *Nat'l Elec. Mfr. Ass'n v. Sorrell*, 272 F. 3d 104 (2d Cir. 2001).

These cases, and many more like them, reflect the general rule that most compelled-disclosure laws easily survive constitutional review. But not all do. There are a number of cases where courts have ruled that mandatory disclosures violate the First Amendment. More often than not, these are cases where companies argue that the required disclosure places a company's product in a negative light, addresses "controversial" matters, or expresses opinions, not fact. In these cases, companies argue that courts should apply a far more stringent test, the *Central Hudson* test that is used to assess whether restraints on commercial speech offend the First Amendment. In *Central Hudson*, the Court said that restraints on commercial speech that concern lawful activity must be supported by a substantial governmental interest, must directly advance the interest asserted, and may not be more extensive than is necessary to serve that interest.[18]

#### 4.2.2.3 The Battle over *Zauderer*'s Application

To illustrate this conflict over *Zauderer*'s scope, let's focus on the United States Court of Appeals for the District of Columbia Circuit, commonly known as the D.C. Circuit, which had been divided on whether *Zauderer* applies in cases in which the mandated disclosure serves interests apart from guarding against deception. In two cases, the court ruled that *Zauderer* did not apply, and applying *Central Hudson*, struck down the Food and Drug Administration's regulation requiring graphic warning labels on cigarette packs and the Security and Exchange Commission's regulation requiring companies that sell "conflict minerals" to disclose that fact. In contrast, applying *Zauderer*, the court upheld a Department of Agriculture rule requiring meat and poultry products to list their country of origin.[19]

Recently, the full D.C. Circuit reheard the country of origin labeling case *American Meat Institute (AMI)* v. *Department of Agriculture* to resolve the differing approaches the court has taken in compelled commercial speech cases. The core question before the court was "whether the test set forth in *Zauderer* applies to government interests beyond consumer deception." *AMI* was not an easy case for the government. After all, the government required country of origin labeling on meat and poultry products, not to prevent

[18] *Cent. Hudson Gas & Elec. Corp. v. Pub. Serv. Comm'n,*, 447 U.S. 557 (1980).

[19] See *R.J. Reynolds Tobacco Co.* v. *FDA*, 696 F.3d 1205 (D.C. Cir. 2012); *Nat'l Ass'n of Mfr. v. SEC*, 748 F.3d 359 (D.C. Cir. 2014); and *American Meat Inst. v. Dep't of Agriculture*, 746 F.3d 1065 (D.C. Cir. 2014). The full D.C. Circuit overruled the First Amendment analysis of each of these decisions in *Am. Meat Inst. v. Dep't of Agriculture*, 760 F.3d 18 (D.C. Cir. 2015) (en banc).

consumer deception or to protect the public health, but because consumers had shown a keen interest in that information and often preferred American-made products. The question in *AMI* thus distilled down to whether informing consumers is a sufficient governmental interest to justify compelling a company to disclose information. As discussed in more detail in Sections 4.2.3 and 4.3, this is a particularly important question for information-programs designed to educate the public about the risks of climate change.

Unfortunately, the full court's opinion leaves many questions unanswered. On the positive side, the court found the country of origin label constitutional. And the court did make clear that "the test in *Zauderer* applies to government interests beyond consumer deception" because "*Zauderer*'s characterization of the speaker's interest in opposing forced disclosure of . . . as 'minimal' seems inherently applicable beyond the problem of deception."[20] But then, in defining the *Zauderer* test, the court appeared to conflate the two standards, suggesting that "one could think of *Zauderer* largely as 'an application of *Central Hudson*, where several of *Central Hudson*'s elements have already been established.'" Applying this hybrid test, the court upheld the disclosure requirement because it clearly served the government's asserted interest of providing consumers information they would deem relevant to their purchasing decisions.

As scholars have pointed out, the test that seems to have emerged from the D.C. Circuit's opinion is hard to square with *Zauderer*, which explicitly distinguished between restraints on lawyer advertising, which had to be justified under *Central Hudson*, and compelled disclosures, which "are adequately protected as long as disclosure requirements are reasonably related to the State's interest in preventing deception of consumers."[21] That distinction makes sense. Restraints run counter to the basic pro-information thrust of the First Amendment. Restraints dampen information flow and may keep people in the dark about information they would like to receive. On the other hand,

[20] In this respect, the D.C. Circuit's opinion aligns with those from other Circuits, which have rejected the idea that *Zauderer*'s rational relationship test applies only where the state mandates a disclosure to prevent deception. See, e.g. *CTIA-The Wireless Ass'n v. City of Berkeley*, 854 F.3d 1105 (9th 2017); *Disc. Tobacco City & Lottery, Inc. v. United States*, 674 F.3d 509, 556 (6th Cir. 2012); N.Y. *State Rest. Ass'n v.* N.Y. *City Bd. of Health*, 556 F.3d 114, 133 (2d Cir. 2009); *Pharm. Care Mgmt. Ass'n* v. *Rowe*, 429 F.3d 294, 310 (1st Cir. 2005) (Torruella, J.); *Ibid* at 316 (Boudin, C.J. & Dyk, J.); Ibid. at 297–298 (per curiam) (explaining that the opinion of Chief Judge Boudin and Judge Dyk is controlling on the First Amendment issue); *Environmental Defense Center, Inc. v. EPA*, 344 F.3d 832, 849 (9th Cir. 2003); *Nat'l Elec. Mfrs. Ass'n v. Sorrell*, 272 F.3d 104, 113–15 (2d Cir. 2001); see also *Beeman v. Anthem Prescription Management, LLC.*, 315 P.3d 71, 89 (Cal. 2013).

[21] See, e.g., Post, *Compelled Commercial Speech*, *supra* note 10, 867.

mandated disclosures have the opposite effect; they foster the flow of information and leave the company substantial freedom to frame appropriate statements. The differences between the two standards are considerable. Under *Central Hudson*, the government must shoulder the weighty burden of showing that the restraint directly furthers the asserted state interest, and does so in a narrow and tailored way. But under *Zauderer*, the burden is on the speaker to show that the requirements are not reasonably related to the government's interest. This difference is not just one of semantics; it often dictates outcomes.

### 4.2.2.4 *Zauderer*'s Uncertain Future

The D.C. Circuit's ruling in *AMI* raises a number of questions that will be addressed in future cases. The good news is that *AMI* ruled that disclosure requirements beyond those aimed at avoiding consumer deception presumptively pass muster under the First Amendment. But the law is less clear in a discrete and limited class of cases; namely, cases in which a company claims that a government-mandated statement disparages its product or strays from the purely factual to opinion, conjecture, or controversy. Three examples mark this class of cases.

First, before the full court's ruling in *AMI*, a divided panel of the D.C. Circuit held that the FDA's proposed graphic warnings on cigarette packages violated the First Amendment because they compelled the use of images that "do not impart purely factual, accurate, or uncontroversial information to consumers."[22] The dissenting judge took issue with that characterization, contending that the disclosures were fact-based and truthful, and that the government had a powerful interest in effectively conveying the risks of smoking to consumers. Although the full court in *AMI* vacated the portion of the graphic warning label opinion that suggested that *Zauderer* applied only to compelled disclosures aimed at preventing deception, it is far from clear that the D.C. Circuit would reach a different conclusion under the full court's decision in *AMI*.

That was the result in the SEC conflicts mineral case. Although the full court's opinion in *AMI* vacated the First Amendment analysis used in an earlier ruling to strike down the SEC "conflict minerals" regulation, a divided panel of the court nonetheless found that the regulation violated the First Amendment.[23] The regulation required companies to disclose whether

[22] *R.J. Reynolds Tobacco Co.*, *supra* note 19, 10 696 F.3d at 1217.

[23] "Conflict minerals" are certain valuable minerals that can be sold by armed groups in the Congo and neighboring countries to finance their war operations.

certain minerals were from the Democratic Republic of the Congo (DRC) and thus have "not been found to be 'DRC conflict free.'" Congress thought that the disclosure would discourage companies from purchasing conflict minerals or minerals whose provenance was uncertain, thereby stemming the humanitarian crisis in the DRC. The panel majority, however, said that the required disclosure was not "factual and non-ideological," but was instead "a metaphor that conveys moral responsibility for the Congo war. It requires an issuer to tell consumers that its products are ethically tainted, even if they only indirectly finance armed groups." Applying *Central Hudson*, the majority invalidated the regulations. The dissent strongly objected to the majority's use of the *Central Hudson* test, emphasizing that the prescribed statements were entirely factual, unlikely to be misunderstood, and served the goal of informing prospective consumers of the minerals' origins, thus allowing consumers to make an informed choice.

The D.C. Circuit's decisions do not stand alone. Years earlier, a different circuit court struck down a Vermont statute that required any dairy products derived from herds treated with rBST (a growth hormone) to carry a label disclosing that fact. Vermont law permitted store owners to place signs acknowledging that the Food and Drug Administration had "determined that there is no significant difference between milk from treated and untreated cows." Vermont claimed that the label's purpose was to "help consumers make informed shopping decisions" and to respond to "strong consumer interest and the public's right to know" about the use of rBST. But the court found these interests insufficient to support a disclosure mandate that did not advance the public health, but instead carried an implication that milk from rBST-treated herds was somehow inferior to other milk products.[24]

The upshot of these cases is they cast a shadow over disclosures in cases in which the mandated disclosure could be argued to be controversial or to disparage a product. On the other side of the ledger, however, there are cases that have upheld disclosure mandates notwithstanding the charge that the disclosure was controversial. For instance, a Vermont statute requiring that light bulbs containing mercury set out on the labels instructions for the bulbs' safe disposal was upheld by the court of appeals despite claims that the instructions would deter consumers from purchasing the bulbs. And more recently, the United States Court of Appeals for the Ninth Circuit rejected a request by industry to enjoin enforcement of a city ordinance requiring stores

[24] *Int'l Dairy Foods Ass'n v. Amestoy*, 92 F.3d 67 (2d Cir. 1995).

selling cell phones to disclose the risk of radio-frequency emissions from carrying a cell phone close to the body.[25]

#### 4.2.2.5 *Zauderer* and Climate Change

How might the issue over compelled statements that could be seen as "controversial" play out with regard to climate change? Consider a couple of hypotheticals: Suppose the government wanted to inform the public about the risks of rising seas levels. Could it require sellers to inform prospective purchasers that "This building has not been assessed for vulnerability to floods using projections that take into account sea level rise"? Or suppose the government wanted the public to understand the environmental risks of generating electricity from coal-fired power plants. Could the government require consumer products to bear a label that said: "This Product is Made with Electricity from a Coal-Fired Power Plant: Coal-Fired Plants Emit Pollutants that Contribute to Global Warming"? Would a court uphold these statements on the ground that they provide consumers information that helps them make better-educated choices in the marketplace?

There are no clear-cut answers to these questions. Given the weight of authority favoring application of *Zauderer* to fact-based disclosures, it is likely that many courts would uphold these statements. After all, facts cannot be one-sided, so long as there is no reasonable controversy about their truth. And yes, facts can be persuasive and at times provocative, but they are still facts. So it is possible that courts will come to a common understanding that compelled factual disclosures reasonably related to the government's interest are reviewed under *Zauderer*, even if the statement may be viewed by some as controversial or provocative.[26]

But it is also possible that some courts would find statements like the hypotheticals to be "controversial" and strike them down under *Central Hudson*. Courts have not yet come to a consensus that demonstrably truthful, factual statements are, by definition, noncontroversial. As we have seen, even

[25] CTIA-The Wireless Ass'n v. City of Berkeley, 854 F.3d 1105 (9th 2017). However, a panel of the the same court set aside a regulation issued by San Francisco requiring soft drinks to bear a warning label that "[d]rinking beverages with added sugars contribute to obesity, diabetes, and tooth decay" on the ground that the message was "one-sided" and "misleading." *Am. Beverage Ass'n v. City and Cnty. of S.F.*, 871 F.3d 884 (9th Cir. 2017). That decision, however, was vacated and is awaiting reconsideration by the *en banc* court. 880 F.3d 1019 (9th Cir. 2018) (vacating panel opinion and ordering the case to be reheard *en banc*).

[26] One Circuit, the Sixth, has made clear that "whether a disclosure is scrutinized under *Zauderer* turns on whether the disclosure conveys factual information or an opinion, not on whether the disclosure emotionally affects its audience or incites controversy." *Discount Tobacco City & Lottery, Inc. v. U.S.*, 674 F.3d 509, at 569 (6th Cir. 2012) (Stranch, J., for the majority).

purely factual statements have spawned disagreements within courts about whether the statements are too "controversial" to be evaluated under *Zauderer*. As long as this issue remains unresolved, First Amendment challenges to our hypotheticals could end up being reviewed under *Central Hudson* and upheld only if the government could make out a compelling case that the mandates fulfilled a substantial government interest and did so in a narrow and tailored way.

It is important, however, to emphasize again that this small category of cases constitutes the exception rather than the rule. Every Circuit to address the issue has now concluded that *Zauderer*'s "reasonably related" test for government-mandated disclosures goes beyond those that seek to prevent consumer deception.[27] For that reason, the mine run of government-compelled disclosures will be upheld by courts. As addressed in Part 4.3, the government's power to require disclosures aimed at educating the public on issues relating to climate change gives the government a tremendous tool to amplify its voice and reach a broad audience. For that reason, the evolution of the law applying *Zauderer* matters in assessing the tools that the government has at its disposal and how these tools should be deployed.

### 4.2.3 *The Government's Role in Policing the Marketplace*

#### 4.2.3.1 For False or Misleading Speech

One critical role that the government plays is policing the marketplace to rid it of false and misleading claims about goods or services. The Supreme Court has often said that the First Amendment does not inhibit the government from "insuring that the stream of commercial information flow[s] freely as well as cleanly."[28] False or misleading statements about goods or services pollute the "stream" of commercial information and distort the marketplace. They deprive consumers of the benefit of the bargain they strike, and they tilt the playing field in favor of market actors willing to deceive consumers in order to

[27] See, e.g. *CTIA-The Wireless Ass'n* v. *City of Berkeley*, 854 F.3d 1105 (9th 2017); *Disc. Tobacco City & Lottery, Inc. v. U.S.*, 674 F.3d 509, at 556 (6th Cir. 2012); *N.Y. State Rest. Ass'n v. N.Y. City Bd. of Health*, 556 F.3d 114, at 133 (2d Cir. 2009); *Pharm. Care Mgmt. Ass'n* v. *Rowe*, 429 F.3d 294, 310 (1st Cir. 2005) (Torruella, J.); Ibid. at 316 (Boudin, C.J. & Dyk, J.); Ibid. at 297–298 (per curiam) (explaining that the opinion of Chief Judge Boudin and Judge Dyk is controlling on the First Amendment issue); *Envtl. Defense Center, Inc. v. EPA*, 344 F.3d 832, at 849 (9th Cir. 2003); *Nat'l Elec. Mfrs. Ass'n* v. *Sorrell*, 272 F.3d 104, at 113–115 (2d Cir. 2001); see also *Beeman v. Anthem Prescription Mgmt., LLC.*, 315 P.3d 71, at 89 (Cal. 2013).

[28] *Va. State Bd. of Pharmacy v. Va. Citizens Consumer Council, Inc.*, 425 U.S. 748, at 772 (1976); accord *Edenfield* v. *Fane*, 507 U.S. 761, at 768 (1993).

get a competitive advantage. And the government, at all levels – federal, state, and local – has long fought against false and deceptive acts and practices in the marketplace.

The government's power to take action against bogus claims made in connection with the sale or promotion of a product is considerable. In a typical deceptive advertising case, the government first determines what messages a reasonable consumer would construe a given claim to convey. Next, government turns to whether the claims are deceptive. As a general matter, a claim is considered deceptive if the advertiser lacks a "reasonable basis" to support it; that is, there is in fact insufficient evidentiary substantiation under the circumstances to back up the claim.[29] If the evidence fails to substantiate the claim, then the claim is deceptive as a matter of law and the government can seek a broad range of remedies, including a court order barring the seller from repeating the claim, forcing the seller to return ill-gotten gains to consumers, and, where appropriate, compelling the seller to correct the claim through advertising or to publicly acknowledge its false claims.[30]

Perhaps the most celebrated compelled corrections case is the government's racketeering case against the tobacco industry, *United States* v. *Philip Morris*, which found that tobacco companies had engaged in a decades-long conspiracy to deceive the American people about the risks of smoking. Among the statements the tobacco companies were compelled to make is the following:

> *A Federal Court has ordered Altria, R.J. Reynolds Tobacco, Lorillard, and Philip Morris USA to make this statement about the health effects of smoking.*
> *Here is the truth:*
> * *Smoking kills, on average, 1,200 Americans. Every day.*
> * *More people die every year from smoking than from murder, AIDS, suicide, drugs, car crashes, and alcohol, combined.*
> * *Smoking causes heart disease, emphysema, acute myeloid leukemia, and cancer of the mouth, esophagus, larynx, lung, stomach, kidney, bladder, and pancreas.*
> * *Smoking also causes reduced fertility, low birth weight in newborns, and cancer of the cervix.*[31]

[29] Zauderer makes clear that "[w]hen the possibility of deception is . . . self-evident," no further evidence is required. *Zauderer, supra* note 13, 471 U.S. at 652–653 (quoting *FTC v. Colgate–Palmolive Co.*, 380 U.S. 374, at 391–392 (1965)).

[30] See, e.g. *Warner-Lambert Co.* v. *FTC*, 562 F.2d 749, 757–58 (D.C. Cir. 1977); *Novartis* v. *FTC*, 223 F.3d 783 (D.C. Cir. 2012).

[31] *U.S. v. Philip Morris*, 164 F. Supp. 3d 121, at 124–25 (D.D.C. 2016). Similarly worded statements are required for the companies to acknowledge (a) the addictiveness of smoking and nicotine;

The theory behind these compelled statements was developed in cases brought by the Federal Trade Commission seeking corrective advertising from companies whose earlier misleading advertisements created or reinforced in the public's mind a false belief about the product and where the belief might linger even after the false advertising ceases. For instance, for years the Warner-Lambert Company falsely claimed that Listerine mouthwash prevents, cures, or alleviates the common cold and sore throats. The FTC ordered that in future advertising the company would have to affirmatively and prominently disclose that "Listerine will not help prevent colds or sore throats or lessen their severity."

The possibility of compelled corrective statements as a remedy for deceptive claims has particular relevance in the fight over climate change. There is deep concern that, like the tobacco industry, the fossil fuel extraction industry has long known about the adverse effects fossil fuels have on climate. Nonetheless, the industry has contested the risks of climate change and the role that fossil fuels play in it, has funded proxies to challenge the scientific basis for efforts to develop responsible policies to mitigate climate change, and has misled financial regulators about the risks climate change poses to these companies' future profitability.[32] A number of state Attorneys General have initiated investigations into these issues. Although it is impossible to forecast whether any action will be taken as a result of these investigations, should any of these investigations bear fruit, the government could insist on corrective statements of the kind required in the *Philip Morris* or *Listerine* cases. Corrective statements, however, are just one of many remedies the government can deploy when it demonstrates that a claim is false or misleading.

It is not just the fossil fuel industry that misled consumers about the environment. Today's marketplace is rife with claims of environmental benefits that may not deliver on their promises. Increasingly, consumers are gravitating to products or services that claim to be environment-friendly, such as claims that companies provide carbon offsets or make products with recycled materials, or with renewable materials or energy. Some of these claims may exaggerate the product's characteristics or environmental benefits, a practice that now is known as "greenwashing."

The First Amendment constrains the ability of government to force marketers to stop making problematic claims that are not demonstrably false or

(b) the manipulation of cigarette design and composition to ensure optimum nicotine delivery; and (c) the adverse health effects of exposure to secondhand smoke.

[32] See, e.g. Naomi Oreskes and Erik M. Conway, *Merchants of Doubt* (2011); Katie Jennings et al., *What Exxon knew about the Earth's melting Arctic*, L.A. Times, Oct. 9, 2015, http://graphics.latimes.com/exxon-arctic/.

misleading. And this is a particular problem with environmental benefit claims, which are relatively new. For many of these claims, there is no consensus about what they mean, which hinders government enforcement. For instance, what is meant by claims that a product is made with "renewable material"? Is any product made from wood renewable, even if it is made from old growth, hardwood trees that take decades or even centuries to regenerate? Or what is meant by "ozone friendly" claims? Does the presence of even trace amounts of any ozone-depleting substance make the claim deceptive? Or even more perplexing, what does a claim that a product is "sustainable" mean? Does it mean that the product is made from renewable material, or made in a special way, or simply that it is durable?

When are claims like these deceptive? To prove deception, the government must demonstrate that "a significant minority of reasonable consumers" would "likely" interpret the claim to provide a benefit that, in fact, the product does not deliver.[33] Absent some baseline of consumer expectation, the government can be hamstrung in protecting consumers from claims that over-promise benefits.

To address the lack of commonly shared definitions for many environmental claims, the Federal Trade Commission has published the "Green Guides" to give guidance to industry on how to ensure that environmental claims are not deceptive.[34] The guides address a broad range of common claims, including claims that a product is made from renewable materials or recycled content, that a company will use carbon offsets in connection with certain activities, such as automobile use or air travel, or that a product is "carbon neutral" in that the company purchases offsets to counter emissions associated with the production of its products.

The Guides discourage claims that are not adequately substantiated. For example, the FTC urges marketers to refrain from making general claims of environmental benefit – such as "eco-friendly" or "green" – because they have no common meaning. The same is true with "sustainable," which also has no single meaning for most consumers. The FTC warns marketers who want to use "sustainable" claims to test those claims in the context of their advertisements to ensure they can substantiate them.

[33] See *POM Wonderful, LLC* v. *FTC*, 777 F.3d 478, 489–90 (D.C. Cir. 2015) (citing *In re Telebrands Corp.*, 140 F.T.C. 278, 291 (2005), *aff'd*, 457F.3d 354 (4th Cir. 2006).

[34] The most recent version of the guides is set forth at 15 C.F.R. Part 260 (2013); 77 Fed. Reg. 62122 (Oct. 11, 2012). The author was Director of the FTC's Bureau of Consumer Protection at the time the 2012 guides were published and played a role in their development. At the same time the guides were published, the FTC released a separate document explaining the guides' basis and purpose, which is available here: www.ftc.gov/sites/default/files/attachments/press-releases/ftc-issues-revised-green-guides/greenguidesstatement.pdf.

For other claims that have a clearer meaning, like claims that products are "made from renewable materials," the FTC pushes marketers to provide specifics about the composition of the materials and, in cases where the product is not made entirely from renewables, clearly disclose the nonrenewable content. For instance, if a company claims that its products are made with renewable wood materials, it should make clear that the wood comes from tree stocks that regenerate quickly, and specify precisely what the company means by renewable. The FTC provided an example of a well-qualified renewable claim: "Our flooring is made from 100 percent bamboo, which grows at the same rate, or faster, than we use it." And for carbon offsets, the FTC advises marketers to disclose when and how the offsets would be used, and how the company would track offsets to ensure that there is no double-counting.

Growing concern over climate change has started to prompt governments and responsible companies alike to seek to inform consumers about products' carbon footprints. This is an important step forward. Product labeling reminds consumers that their purchasing decisions have an impact on carbon emissions. Unless consumers understand the carbon emissions associated with the products they purchase, they cannot make informed decisions based on environmental considerations. Carbon footprint labeling can fill that gap. Berkeley, California, now requires gas pumps to bear a label that says:

> ***GLOBAL WARNING ALERT: BURNING GASOLINE EMITS $CO_2$***
> *The City of Berkeley Cares About Global Warming*
> *The State of California has Determined that Global Warming Caused by $CO_2$ Emissions Poses a Serious Threat to the Economic Well-Being, Public Health, Natural Resources, and the Environment of California.*[35]

Other cities, including Seattle and San Francisco, are considering similar measures.[36] New York and California now require carbon emission labeling for new motor vehicles.[37] And the British government has created the "Carbon Trust," an independent company charged with moving the economy away

[35] See, e.g., CBS SF Bay Area, *Berkeley Approves Plan For Global Warming Labels On Gas Pumps* (Nov. 19, 2014), http://sanfrancisco.cbslocal.com/2014/11/19/berkeley-approves-plan-for-global-warming-labels-on-gas-pumps-climate-change/.

[36] See, e.g., Daniel Beekman, *Pump gas in Seattle, get warning you're harming the Earth?*, The Seattle Times (Nov. 8, 2015), www.seattletimes.com/seattle-news/politics/seattle-gas-pumps-may-sport-climate-warning/.

[37] See Jason J. Czarnezki, The *Future of Food Eco-Labeling: Organic, Carbon Footprint and Environmental Life-Cycle Analysis*, 30 Stanford Envtl. L. J. 3, 19 and n. 66 (2011).

from carbon-based energy by developing uniform standards for carbon labeling.[38]

Companies are also beginning to take climate change seriously. For instance, Timberland, the shoe and clothing company, rates each of its products with its "green index" that gives consumers information about the product's "climate impact" (measured by $CO_2$ emissions), use of recycled material, and the use of chemicals linked to environmental harm.[39] And considerable efforts to devise ways of tracking carbon emissions through a product's lifecycle are ongoing. For instance, the highly regarded International Standards Organization recently adopted a standard, ISO 14067, which specifies principles, requirements, and guidelines for the quantification and communication of the carbon footprint of a product, based on lifecycle assessment for quantification and on environmental labels and declarations for communication.[40] To be sure, these and other approaches are still being developed. No consensus has yet emerged on how best to conduct lifecycle emissions analysis. Gas pump labeling has failed to gain political traction, or to reduce consumers' gasoline use. But these efforts to enlist consumers in the fight against climate change are important and hold considerable promise, notwithstanding the hurdles that must be overcome.

At the same time as efforts are underway to convey *accurate* information about climate risks to consumers, *misleading* environmental marketing claims are undermining consumer confidence. The FTC has started bringing enforcement cases challenging misleading environmental claims. The FTC has sued manufacturers who claim that their products are made entirely from recycled plastic when, in fact, a significant portion of the plastic was not recycled.[41] The agency has brought cases against sellers deceptively claiming that its products are biodegradable, compostable, save energy, or are free of toxic chemicals.[42] And, recently, the FTC sued Volkswagen for deceiving car-

38 Ibid. at 20 and nn. 68–69.

39 Timberland's green index is here: http://greenindex.timberland.com/browse/.

40 See International Standards Organization, *ISO/TS 14067:2013: Greenhouse gases – Carbon footprint of products – Requirements and guidelines for quantification and communication*, https://asq.org/quality-press/display-item?item=T940E; see also Tiantian Liu, Qunwei Wang, and Bin Su, *Review of Carbon Labeling: Standards, Implementation, and Impact*, 53 Renewable & Sustainable Energy Reviews 68 (2016).

41 See, e.g., *In the Matter of Engineered Plastic Systems, LLC*, FTC No. 132 3204 (Sept. 11, 2014); *In the Matter of American Plastic Lumber, Inc.*, FTC No. 132 3200 (July 18, 2104); *In the Matter of N.E.W. Plastics Corp.*, FTC No. 132 3126 (Feb. 28, 2104).

42 See, e.g. *In the Matter of Down to Earth Designs, Inc. d/b/a gDiapers*, FTC No. 122 3268 (March 19, 2014); see also FTC Press Release, *FTC Cracks Down on Misleading and Unsubstantiated Environmental Marketing Claims; Actions Challenge Deceptive Biodegradable Plastics Claims for the First Time* (Oct. 29, 2013) (announcing six enforcement

buyers by rigging 500,000 of their diesel-powered cars to pass emission tests even though they emitted harmful, smog-producing NOx (nitrogen oxides) at levels up to forty times EPA-compliant levels during normal on-road driving conditions. The settlement also requires VW to invest $2 billion toward improving infrastructure, access, and education to support and advance zero emission vehicles. The investments will be made over ten years, with $1.2 billion directed toward a national EPA-approved investment plan and $800 million directed toward a California-specific investment plan that will be approved by the California Air Resources Board.[43]

More work remains to be done in order for the public to have confidence in claims of environmental benefit. Definitional ambiguities still inhibit trustworthy advertising and hold back the growth of market-based solutions to environmental problems. Commercial speech merits First Amendment protection because it helps consumers make educated choices in the marketplace. An accepted vocabulary of environmental terms would facilitate the development and marketing of products that actually deliver environmental benefits. Consumers could then vote with their wallets in favor of those products. In a well-functioning market, the role of the government is to ensure that sellers that tell the truth are not crowded out by sellers who tout, but do not deliver on their promises. And as a consensus emerges about the meaning of environmental claims, the flow of reliable information about products' environmental benefits will play an important role in society's response to climate change.

## 4.3 GOVERNMENT INFORMATION PROGRAMS THAT CAN HELP MITIGATE THE RISKS FROM CLIMATE CHANGE

As discussed in the previous sections, the government has broad authority to collect information, to compel regulated parties to provide information, and to disseminate information in a countless variety of forms. The question facing policymakers is how to harness the government's informational powers to address most effectively the public health risks posed by climate change. It is

actions against companies deceptively claiming that is products were biodegradable); *FTC Press Release, FTC Approves Final Orders Settling Charges Against The Sherwin-Williams Co. and PPG Architectural Finishes, Inc.; Issues Enforcement Policy Statement on "Zero VOC" Paint Claims* (March 6, 2013).

43 FTC Press Release, *Volkswagen to Spend up to $14.7 Billion to Settle Allegations of Cheating Emissions Tests and Deceiving Customers on 2.0 Liter Diesel Vehicles* (June 28, 2016), www.ftc.gov/news-events/press-releases/2016/06/volkswagen-spend-147-billion-settle-allegations-cheating.

worth reflecting on just how substantial those risks are; indeed, it is daunting simply to catalogue those risks. According to the EPA and leading environmental groups,[44] the most urgent public health threats include:

* Armed conflict, like the one in Syria, resulting from water shortages and famines;[45]
* Heat-related spread of vector-borne diseases (diseases and illnesses that are transmitted by mosquitoes, ticks, and fleas) like West Nile Virus, Zika, and Lyme Disease;[46]
* Disruption of global food supplies triggered by extreme weather events;[47]
* Devastation and property loss caused by the severe and destructive flooding of coastal communities and 1000-year floods in riverine floodplains;[48]
* Massive deforestation resulting from wildfires and tree-destroying bark beetles;
* Acidification of oceans caused by greenhouse gas pollution, wiping out half of Australia's Great Barrier Reef;[49]
* Threats to drinking water supply, with one-third of the world's major lakes and rivers drying up;[50]
* Increases in heat-related deaths, like the heatwave that took 70,000 lives in Europe in 2003;[51] and
* Rapid rise in respiratory diseases and deaths.[52]

44 U.S. Global Change Research Program, *Climate and Health Assessment* (2016), https://health2016.globalchange.gov/; International Panel on Climate Change, *AR5, Chapter 11, Near-term Climate Change: Projections and Predictability* (2013), www.ipcc.ch/report/ar5/wg1/.

45 Department of Defense, *National Security Implications of Climate-Related Risks and a Changing Climate* (July 3, 2015), http://archive.defense.gov/pubs/150724-congressional-report-on-national-implications-of-climate-change.pdf?source=govdelivery.

46 See, e.g., CDC, Climate and Health, www.cdc.gov/climateandhealth/.

47 United Nations, *The World's Food Supply is Made Insecure by Climate Change* (Feb. 15, 2016), https://academicimpact.un.org/content/world%E2%80%99s-food-supply-made-insecure-climate-change.

48 Berkeley Labs, Insurance in a Climate of Change (2017), available at http://insurance.lbl.gov/availability-affordability.html.

49 ABC News, *Great Barrier Reef: Only 7 Per Cent Not Bleached, Survey Finds* (April 28, 2106), www.abc.net.au/news/2016-04-20/great-barrier-reef-bleaching/7340342.

50 U.S. Global Change Research Program, *Climate and Health Assessment* (2016), https://health2016.globalchange.gov/.

51 Union of Concerned Scientists, *The Changing Northeast Climate: Our Choices, Our Legacy* (2015), www.ucsusa.org/sites/default/files/legacy/assets/documents/global_warming/The-Changing-Northeast-Climate.pdf; see also The Lancet Commissions, *Managing the health effects of climate change*, 373 The Lancet 1693, 170 (May 16, 2009).

52 EPA, *Climate Change, Public Health, and Environmental Justice* (May 2016), www.epa.gov/sites/production/files/2016-10/documents/ej-health-climate-change.pdf.

Information-based policies alone are obviously not going to solve these problems. But information is the oxygen that informs policy-setting, educates the public and can be the catalyst for important changes. What follows are four examples that illustrate the possibility of the government using information-forcing and information-disclosing reforms to focus attention on climate and effect change on the part of individuals and institutions. With these threats in mind, consider the proposals set out in the following subsections.

### 4.3.1 *Proposed Government Information-Forcing Initiatives*

#### 4.3.1.1 Proposal One: Use Information-Forcing Measures to Induce Reductions of Emissions That Contribute to Climate Change That Come from Sources Other than Stationary Sources and Vehicles

The Clean Air Act regulates emissions by industrial polluters. And fuel economy standards for vehicles similarly force reductions in emissions. These programs are mainstays in the fight against climate change and pollution.

But other significant sources of emissions are not subject to disclosure requirements aimed at influencing consumers' purchasing behavior. For instance, there are few existing requirements to test the emissions of consumer products and disclose the results to consumers. Consider lawnmowers. Gas-powered lawnmowers – the kind that are used by millions of Americans every weekend in suburban America – account for 5 percent of the annual emissions of toxic substances in the United States. According to the EPA, a new gas-powered lawnmower produces volatile organic compounds (VOCs) and nitrogen oxides emissions in one hour of operation as eleven new cars being driven for one hour. More than 5 million new mowers are sold in the United States each year.[53] Other consumer products powered by gasoline engines (e.g. leaf blowers, chain saws, tillers, and snow blowers) are also responsible for substantial additional emissions. In contrast, electric-powered equipment uses a fraction of the energy, emits none of these harmful pollutants, and works just as well as their gas-guzzling counterparts.

No doubt some consumers realize that gas-powered lawn mowers pollute. But consumers are not made aware of the environmental harm they cause when that information is most relevant – namely, when they are looking to

[53] See EPA, *Cleaner Air: Gas Mower Pollution Facts*, www.peoplepoweredmachines.com/faq-environment.htm; EPA, *National Emissions from Lawn and Garden Equipment* (2015), www.epa.gov/sites/production/files/2015-09/documents/banks.pdf.

purchase a replacement for their old mower or snow blower. The case of Reye's syndrome shows how important it is to get information in consumers' hands when it is most needed.

To enable consumers to make informed choices, Congress should enact legislation requiring manufacturers to test the emissions of all gas-powered consumer products and submit the results to the EPA for verification. Manufacturers should also be required to label these products to (a) disclose that gas-powered consumer products contribute to climate change; (b) disclose the amount of their product's emissions; and (c) compare their product's emissions to electric-powered mowers, comparable gas mowers, and automobiles.[54]

#### 4.3.1.2 Proposal Two: Require Publicly Traded Companies to Disclose the Financial Risks They Face from Climate Change and Their Global Greenhouse Gas (GHG) Emissions[55]

The risks climate change poses to major corporations and financial institutions are considerable and growing. Just look at the economic peril facing coal companies in the United States.[56] There is little evidence that these risks are factored into companies' economic forecasts, let alone forecasts disclosed to investors and the public. U.S. securities laws require disclosure of material risks, and the Securities and Exchange Commission (SEC) in 2010 warned

[54] There is also no uniform federal standard limiting the emissions of the products used in home construction and maintenance, or disclosing the amount of emissions to consumers. These products also emit high levels of VOCs that contribute to climate change, especially adhesives, sealants, caulking compounds, paints, and coatings. The EPA and several states now require the manufacturers of these products to test their emissions, and set limits on emissions, but this patchwork is far from uniform, and information about the emissions of products is not communicated effectively to the public. Few of these products bear labels alerting consumers to these emissions. A consumer who seeks to find low-VOC paints, for example, is hard pressed to do so. The FTC has taken action against major paint brands for claiming that their paints were VOC-free, when in fact, the pigments in the paints had high levels of VOCs. See FTC Press Release, *FTC Approves Final Orders Settling Charges Against The Sherwin-Williams Co. and PPG Architectural Finishes, Inc.; Issues Enforcement Policy Statement on "Zero VOC" Paint Claims* (March 6, 2013). These cases make clear that there is strong consumer demand for nonpolluting home construction materials; labeling requirements could be the foundation of a competitive market where less polluting products would have a market advantage.

[55] This proposal is drawn from a report by The Global Commission on the Economy and Climate entitled, *Better Growth, Better Climate, the New Climate Economy Report*, at 51 (Sept. 2014).

[56] Economic forecasts predict significant declines in the coal industry, even though the Trump Administration is seeking to bolster coal production. See Cecilia Jamasmie, *US coal industry to decline even further in 2017 – IEEFA*, InfoMine (Jan. 19, 2017), www.mining.com/us-coal-industry-decline-even-2017-ieefa/.

companies that they must disclose material financial information about their exposure to climate risks, whether physical, regulatory, or marketplace.[57] But few companies have taken on this responsibility, and the SEC has been unwilling to force the issue.

Energy companies, for example, have avoided making disclosures by asserting uncertainties in the marketplace. For instance, the annual securities filings by Arch Coal and Alpha Natural Resources at the end of 2014 provided little guidance about their serious exposure to climate risk. Alpha maintained that it could not estimate the financial impact of impending regulation of GHG emissions. Arch Coal reassured investors that coal was likely to remain the dominant fuel for power generation and that, though subject to regulatory risks and competition from gas and renewables, coal would be competitively priced. The following year both companies filed for protection in bankruptcy court.[58] And Exxon's 2015 SEC disclosure statement briefly mentions the possibility of impacts to its business due to climate change and evolving regulatory and market conditions, but its statement is so vague that it would be of no use to regulators, shareholders, or investors.[59] Nor, at present, is there any specific, express obligation under state or federal law in the United States that businesses disclose the extent to which their operations

[57] SEC, *Commission Guidance Regarding Disclosure Related to Climate Change* (Feb. 2, 2010), www.sec.gov/rules/interp/2010/33-9106.pdf.

[58] John W. Miller and Peg Brickley, *Arch Coal Files for Bankruptcy: Company Seeks to Cut $4.5 Billion in Debt*, *Wall Street Journal* (Jan. 11, 2016); Alpha Natural Resources, *About Our Company*, www.alphanr.com/about/pages/history.aspx www.alphanr.com/about/Pages/default.aspx. After the court approved a reorganization plan, Alpha emerged from bankruptcy in 2016. See Ibid.

[59] Exxon's 2015 10-K filing with the SEC says only this about climate change: "Climate change and greenhouse gas restrictions. Due to concern over the risk of climate change, a number of countries have adopted, or are considering the adoption of, regulatory frameworks to reduce greenhouse gas emissions. These include adoption of cap and trade regimes, carbon taxes, restrictive permitting, increased efficiency standards, and incentives or mandates for renewable energy. These requirements could make our products more expensive, lengthen project implementation times, and reduce demand for hydrocarbons, as well as shift hydrocarbon demand toward relatively lower-carbon sources such as natural gas. Current and pending greenhouse gas regulations may also increase our compliance costs, such as for monitoring or sequestering emissions." Although the filing recounts all of Exxon's reserves and refining capacity, it makes no mention of the risk that any of these assets might be "stranded" as a consequence of reduced demands for fossil fuels or increased regulation. The 10-K is available at www.sec.gov/Archives/edgar/data/34088/000003408816000065/xom10k2015.htm. Two lawsuits have already been brought against Exxon, challenging the adequacy of the company's disclosures about the risks climate change poses to the company. http://climatecasechart.com/case/fentress-v-exxon-mobil-corp/ and http://climatecasechart.com/case/ramirez-v-exxon-mobil-corp/.

depend on or generate GHGs, or the potential risk of stranded fossil fuel assets, either as part of their SEC submissions or other regulatory filings.

Investment and innovation in key low-carbon and climate-resilient technologies are essential ingredients in any effort to mitigate the effects of climate change. But current reporting requirements mask the risks to fossil fuel-dependent companies and thus may skew investment decisions. Energy companies must be required to come clean about the challenges climate change poses to their company's financial health and should disclose their global GHG emissions. Manufacturers dependent on fossil fuels also must be compelled to address forthrightly risks posed by climate change as well as disclose their GHG emissions. Business reporting on GHG emissions and climate risk should be mandatory, integrated with financial reports, and standardized. After all, companies dependent on energy that emits GHGs to make or ship their products are especially vulnerable to new regulatory requirements, shifting energy costs, the possibility of stranded assets, and potential liability.[60] Perhaps most importantly, the investment firms, banks, and capital investment companies that invest in energy companies should be required to conduct climate (and wider environmental) risk assessments of their portfolios as part of their fiduciary duty, and make those assessments available to regulators and the public.[61]

#### 4.3.1.3 Proposal Three: Restore the Requirement that the Government Assess Environmental Impacts in Ways That Explicitly Account for Direct and Indirect Climate Impacts

The National Environmental Policy Act (NEPA) requires federal agencies to conduct environmental impact assessments before taking or authorizing any

60 The liability risks for fossil fuel companies are not insignificant. For instance, ExxonMobil recently lost a case involving excess emissions in one of the Texas facilities and was fined over $20 million. Diana Wray, *Exxon ordered to pay $20 million for air pollution violations at Baytown complex*, Houston Press (April 27, 2017).

61 The United States would not be the first nation to institute this kind of reporting obligation. Article 173 of the French Energy Transition Law came into force on January 1, 2016. It requires mandatory reporting for listed companies and introduced carbon reporting requirements for institutional investors. A description of the law is available here: www.sseinitiative.org/wp-content/uploads/2016/04/PRI-French-Energy-Transition-Law-ENGLISH.pdf. In addition to the French legislation, a global industry group chair by Michael Bloomberg, the Task Force on Climate-Related Financial Disclosures, is working to come up with standardized, voluntary disclosures for companies worldwide. See Phase I Report (March 31, 2016) at www.fsb-tcfd.org/.

significant action.[62] The range of actions covered by NEPA is broad and includes, among other things, making decisions on permit applications, adopting federal land management actions, and constructing highways and other publicly owned facilities. NEPA also created the Council on Environmental Quality ("CEQ"), an office with the Executive Office of the President, to coordinate the government's reviews under NEPA.

On August 1, 2016, CEQ issued a directive entitled "Final Guidance for all Federal Departments and Agencies on Consideration of Greenhouse Gas Emissions and the Effects of Climate Change in National Environmental Policy Act Reviews."[63] CEQ's guidance advised agencies that the statute requires them to engage in a thorough consideration of the impact of proposed federal actions on climate, either directly or indirectly.[64] Given the sweeping scope of federal programs and the centrality of the federal government in building and maintaining the nation's infrastructure, CEQ's guidance was intended to ensure that questions about climate impacts do not go unasked, and that the government fixes threats to climate rather than unwittingly exacerbating them. The guidance was an important step forward in the battle against climate change.

Unfortunately, on April 15, 2017, CEQ withdrew that guidance, as it was required to do by President Trump's Executive Order 13783, which seeks to roll back virtually all of President Obama's initiatives to mitigate climate change.[65] To be sure, existing law still requires agencies to evaluate environmental impacts, including those from climate change. But CEQ's guidance required agencies to place a premium on a detailed analysis of climate change's effects, and urged agencies, where possible, to quantify those effects. The withdrawal of CEQ's guidance document suggests to agencies that they may turn a blind eye to federal actions that might contribute to climate change – without bothering to explore alternatives or mitigation measures. Congress should amend NEPA to require agencies to give careful consideration to possible direct and indirect impacts on climate, and to disclose their analysis to the public.

62 42 U.S.C. § 4321 *et seq.*

63 CEQ's 2016 Guidance Document is available at https://ceq.doe.gov/docs/ceq-regulations-and-guidance/nepa_final_ghg_guidance.pdf

64 See, e.g., Michael Burger and Jessica Wentz, *Downstream and Upstream Greenhouse Gas Emissions: The Proper Scope of Review*, 41 Harv. Env. L. Rev. 109 (2017).

65 82 Fed. Reg. 16576 (April 15, 2017). E.O. 13783 is published at 82 Fed. Reg. 16093 (March 31, 2017).

#### 4.3.1.4 Proposal Four: Force Disclosure of Government Subsidies for Companies That Extract and Use Fossil Fuels[66]

Unfortunately, there is no coherent national industrial carbon policy in the United States. As a result, it appears that the federal government is providing subsidies for the development of both fossil and renewable fuels, perhaps with the balance tilting toward fossil fuels. Estimates of government subsidies for fossil fuel production vary widely – from a low estimate of $5 billion to a high estimate of nearly $40 billion or more. Whatever the amount, these subsidies raise important policy questions. After all, what justification is there for using taxpayer money to subsidize a fossil fuel industry that is enormously profitable?[67] And why shouldn't society support the development of renewable and low carbon alternatives, not just for the sake of the climate, but to end dependency on a fuel that, at some point, will be depleted, or, like coal, will be rendered economically unviable?

Until the curtain is pulled back and there is a clear accounting of where taxpayer dollars are going, forging a political consensus for change will be difficult. It is hard to win an argument over facts where the facts are unknown or easily contested. Congress should demand from the White House a clear and comprehensive annual accounting of the energy subsidies provided by the US government. The accounting should carefully identify every subsidy provided to support fossil fuels, renewables, and nuclear, and set forth whether the subsidy took the form of a direct transfer of funds (e.g. a grant, loan, or infusion of equity); a potential transfer of funds or liabilities (e.g. a loan guarantee); or a foregone government revenue (e.g. a tax credit). If it turns out that we are still spending disproportionate amounts to prop up the fossil fuel industry, then it may be possible to galvanize support for change. But until

66 The World Trade Organization defines a subsidy as "any financial contribution by a government, or agent of a government, that is recipient-specific and confers a benefit on its recipients in comparison to other market participants." Subsidies include "direct transfer of funds (e.g. grants, loans, and equity infusion), potential direct transfers of funds or liabilities (e.g. loan guarantees); government revenue that is otherwise due is foregone or not collected (e.g. fiscal incentives such as tax credits); government provision of goods or services other than general infrastructure, or purchase of goods, below market-value income or price support." See WTO, *Agreement on Subsidies and Countervailing Measures ("SCM Agreement")*, www.wto.org/english/tratop_e/scm_e/subs_e.htm.

67 A recent report for the Council on Foreign Relations found that ending tax subsidies for fossil fuel producers would have little or no impact on their profitability. See Gilbert E. Metcalf, *The Impact of Removing Tax Subsidies for US Gas and Oil Production* (Aug. 2016), www.cfr.org/energy-policy/impact-removing-tax-preferences-us-oil-gas-production/p38150?cid=otr-marketing_use-TaxPreferencesPaper/.

there is a shared factual predicate for the argument, it will be difficult to make progress on forging a coherent national economic policy on carbon.

* * *

It may be easy to dismiss these proposals because their scope seems modest compared with the threats. But disclosure regimes empower informed individual choice and enable collective action by facilitating the ability of citizens to come together to put pressure on private companies to improve their behavior or to force improvements through the political process. Information-forcing mandates have brought about important change. For instance, in 1986 Congress enacted the Emergency Planning and Community Right to Know Act ("EPCRA"), which requires companies to report on the quantities of hazardous chemicals they release into the environment. Based on the data, the EPA compiles the "toxic release inventory" or TRI, which identifies the top polluters in every state and reports on each polluter's emissions.[68] Two things happened when EPCRA went into effect. As economist Paul Portney observed, "when firms are required to make public their emissions, they feel pressure to reduce those emissions even when they are perfectly legal." Many firms did just that. But many firms did not. Media and environmental and community groups transformed the TRI from stodgy EPA filings to "top ten" polluter lists and a steady stream of news stories to make the case that stricter emission controls were needed, leading to the passage of the landmark 1990 Clean Air Amendments.[69] There is no reason why the information programs described in this section cannot provide a similar impetus for change.

### 4.3.2 *Practical Considerations*

There are, however, important caveats that must come with any call for government-mandated disclosure programs. After all, information is worth little without political will or institutional capacity to use it. But it is equally true that without information, political will or institutional capacity will achieve little. The lesson of the TRI program is that the right information, in the right hands, can result in dramatic advances that have a real impact on climate policy.

The first caveat is that successful information regimes must be carefully targeted to an audience or audiences. The goal of the TRI program was not just to inform regulators; it was to provide evidence to the public about the high volume of emissions that pollutes our air and water and to identify the

68 The TRI website is here: www.epa.gov/toxics-release-inventory-tri-program

69 See Cass Sunstein, *Informing America: Risk, Disclosure and the First Amendment*, 20 Fla. St. U.L. Rev. 653, 662–663 (1993).

companies that bear responsibility for that pollution. The TRI data still plays that role today. Another successful information-forcing program is California's Proposition 65, which requires labeling and disclosures of products containing listed carcinogens and reproductive toxins. That program has been credited with bringing about wholesale reformulation of products and enhanced consumer understanding of public health risks.[70]

The second caveat is that information-forcing initiatives are almost inevitably opposed by powerful industry groups and political forces aligned with them. That is not surprising. Disclosing information about public health risks rarely benefits the disclosing firm. Instead, information is costly to generate, and disclosure often confirms that the company is seeking to externalize its costs onto the public. Disclosures can also lead to direct regulation or to litigation and liability findings. The fight over disclosure initiatives has been particularly bitter on climate change issues. Consider the fate of President Obama's initiative to force oil and gas producers to measure the amount of methane that leaks from their pipelines and production and processing facilities. According to the EPA, "[m]ethane, the key constituent of natural gas, is a potent greenhouse gas (GHG) with a global warming potential more than 25 times greater than that of carbon dioxide."[71] But the EPA's efforts to require companies to monitor methane emissions prompted a fierce response from the industry, and it now appears that the industry may prevail.[72]

The third caveat is that regulation is not a substitute for information, and information is not a substitute for regulation. The classic example of a regulatory program that lacks an adequate information-forcing mandate is the Toxic Substances Control Act (TSCA). Except for chemicals produced in

[70] California's Proposition 65 has been intensively studied by scholars who maintain that it has had many beneficial effects. See, e.g., Clifford Rechtschaffen, *How to Reduce Lead Exposures With One Simple Statute: The Experience of Proposition* 65, 29 Envtl. L. Rep. 10,581, 10,582–10,591 (1999) (stating that manufacturers removed lead from faucets, ceramics, crystal glassware, foil caps on wine bottles, and other consumer products to avoid Proposition 65 warning requirements); John S. Applegate, *The Perils of Unreasonable Risk: Information, Regulatory Policy, and Toxic Substances Control*, 91 Colum. L. Rev. 261, 309–310 and n.263 (1991); David Roe, *Toxic Chemical Control Policy: Three Unabsorbed Facts*, 32 Envtl. L. Rep. 10,232 (2002). Randolph B. Smith, *California Spurs Reformulated Products*, Wall Street Journal, Nov. 1, 1990, at B1 (describing Proposition 65's national impact as manufacturers reformulate products to avoid its labeling requirements).

[71] EPA News Release, *EPA Releases First-Ever Standards to Cut Methane Emissions from the Oil and Gas Sector* (May 12, 2016), www.epa.gov/newsreleases/epa-releases-first-ever-standards-cut-methane-emissions-oil-and-gas-sector

[72] EPA News Release, *EPA Withdraws Information Request for the Oil and Gas Industry* (March 2, 2017), www.epa.gov/newsreleases/epa-withdraws-information-request-oil-and-gas-industry

high volumes and posing a substantial risk of exposure, TSCA provides authority to impose testing requirements on new chemicals only if the EPA can demonstrate that the existing data are "insufficient" to assess the chemical and the EPA has reason to suspect that the new chemical "may present" a risk or hazard – a burden the EPA cannot meet without testing. Because Congress failed to give the EPA broader authority to require that new chemicals be tested before they enter the marketplace, little is known about the effects of many chemicals used in commerce.[73]

The final caveat is that we face what some call a "post-disclosure age," where the sheer volume of disclosures results in sensory overload, so much so that many disclosures go unread and warnings unheeded.[74] That consideration is important, but it points to the need for better and more targeted disclosure. Whatever force that claim may have with regard to the fine print disclosures in online contracts, it does not apply to the examples set out in this chapter. Nor does it apply to the proposals set out in this section. Proposals Two, Three, and Four are addressed to specialized audiences that have a keen interest in the information that the proposals aim to generate. And Proposal One is aimed at consumers at the moment they are making purchasing decisions, when they are most receptive to relevant information. The success of disclosure mandates like TRI and California's Proposition 65 bear this out, as do the FTC's Green Guides, which are based on significant empirical evidence that sellers and consumers care deeply about environmental claims.

## 4.4 CONCLUSION

Information is a powerful tool. The battle over how the government should respond to the threats of climate change depends heavily on information. The government needs information to understand the threats we face, to educate the public about those threats, and to devise policies that best meets those threats. The First Amendment gives the government broad leeway to collect and use information to advance these goals, and the modest constraints it imposes should not impair the government's ability to use information-based programs as a centerpiece in the fight against climate threats.

[73] See 15 U.S.C. § 2603. Although TSCA was substantially amended in 2016, the amendments did not change this provision of the law. For a critique on TSCA's lack of information-forcing provisions, see Wendy Wagner, *Commons Ignorance: The Failure of Environmental Law to Produce Needed Information and Health and the Environment*, 53 Duke L. J. 1619, 1665–1668 and nn.157 and 158 (2004).

[74] See, e.g., Omri Ben-Sharar and Carl Schneider, *More Than You Wanted to Know: The Failure of Mandated Disclosure* (Princeton, NJ: Princeton University Press, 2014).

# 5

# Disease Surveillance

Jason A. Smith and Chandrakala Ganesh

*Climate change will pose a serious and intractable threat to the health of populations. The ability of the public health system to address this problem and to preserve and protect the health of the public depends on a robust surveillance system. While public health has a long history of surveillance, the current system is not prepared to respond to the health effects of climate change. In order to develop a robust surveillance system, the public health infrastructure must address several issues, including coordination problems, conflicting legal authorities, jurisdictional issues, an overemphasis on preparedness, and complex issues of management and long-term planning.*

## 5.1 INTRODUCTION

Climate change poses serious challenges for the public health system. The disruptions related to climate change have significant long-term impacts on health. These include extreme heat, increased frequency of extreme weather events, increases in vector-borne diseases, waterborne and foodborne diseases, food insecurity, mental health impacts, and violence related to civic disruption and migration.[1] These health disruptions are multi-sectorial with disruptions in one area amplifying disruptions in another. They also exacerbate existing health disparities and health inequities.[2] Surveillance is a key part of addressing this issue:

> [S]urveillance in real time is necessary to inform the models, track vectors and pathogens in time and space, measure risk, and inform decision-support

[1] Barry S. Levy and Jonathan Patz. *Climate Change and Public Health* (Oxford; New York: Oxford University Press, 2015).

[2] S. Friel et al., *Health Inequities and the Added Pressure of Climate Change: An Action-Oriented Research Agenda*, 88(5) J. of Urban Health 886 (2011).

> systems that direct intervention. Because outbreaks occur intermittently in time and space, surveillance must be performed indefinitely and preparedness for implemented intervention measures must be maintained so that they can be implemented rapidly to prevent or curtail outbreaks.[3]

This chapter describes the need to address legal and policy issues to build a robust public health surveillance system that is critical for public health protection in the face of the health impacts and disruptions that are associated with climate change.

## 5.2 PUBLIC HEALTH SURVEILLANCE

There are two concepts that are critical to understanding public health surveillance. First, the main characteristic of surveillance is its goal of protecting and promoting public health.[4] Second, surveillance is praxis, not a particular tool or theory.[5]

### 5.2.1 *Protecting and Promoting Public Health*

Public health activities by governments seek to assure "the conditions in which people can be healthy" through three core functions: assessment, policy development, and assurance.[6] These three functions form the core of public health management and operate in a cyclical form. Assessment, policy development, and assurance are essential for population health. The first function, the assessment and monitoring of the health of communities, is the systematic collection of information, investigation of issues, and the analysis and determination of specific health needs of the population.[7] The development of policies to address identified health issues includes advocating for community health issues, prioritizing health needs, and planning policies that are specific to local health needs.[8] Finally, the public health authority assures that the policies implemented address the initial health need through resource

3 Levy and Patz, *Climate Change and Public Health*, 146.

4 H. Irene Hall, Adolfo Correa, Paula W. Yoon, Christopher R. Braden, and Centers for Disease Control and Prevention. *Lexicon, Definitions, and Conceptual Framework for Public Health Surveillance*, MMWR Surveill Summ 61 10–14 no. Suppl (2012).

5 Ibid.

6 Institute of Medicine (US) Committee for the Study of the Future of Public Health, *The Future of Public Health*. Vol. 88, no. 2. National Academy Press, 1988.

7 S. Declich and A. O. Carter, *Public Health Surveillance: Historical Origins, Methods and Evaluation*, 72(2) Bulletin of the World Health Organization 285 (1994).

8 Ibid.

management and program evaluation.[9] This cyclical process allows for periodic evaluation and adjustment of policies as necessary.

The ten essential public health services provide a more granular expression of the three core functions (see Figure 5.1), outlining shared assumptions about the functions of public health authorities and providing a tool for conceptualizing them.[10] Community health assessment and the assurance of policy effectiveness both rely heavily on the collection and analysis of information to support public health – generally referred to as *surveillance*. Public health surveillance, or simply, *surveillance*, is "the systematic, ongoing collection, management, analysis, and interpretation of data followed by the dissemination of those data to public health programs to stimulate public health action."[11]

The Ten Essential Public Health Services are:

(1) Monitor health status to identify and solve community health problems.
(2) Diagnose and investigate health problems and health hazards in the community.
(3) Inform, educate, and empower people about health issues.
(4) Mobilize community partnerships and action to identify and solve health problems.
(5) Develop policies and plans that support individual and community health efforts.
(6) Enforce laws and regulations that protect health and ensure safety.
(7) Link people to needed personal health services and assure the provision of health care when otherwise unavailable.
(8) Assure competent public and personal health care workforce.
(9) Evaluate effectiveness, accessibility, and quality of personal and population-based health services.
(10) Research for new insights and innovative solutions to health problems.

FIGURE 5.1

[9] Ibid.

[10] Ibid. See also, Centers for Disease Control and Prevention, *The Public Health System and the 10 Essential Public Health Services*, www.cdc.gov/nphpsp/essentialservices.html

[11] S. B. Thacker et al., Control Centers for Disease, and Prevention, *Public Health Surveillance in the United States: Evolution and Challenges*, 61(3) MMWR Suppl 3–9 (July 27, 2012).

### 5.2.2 *Ongoing Implementation of Processes*

Surveillance is praxis. Rather than one particular tool or system, surveillance is an approach to receiving data, interpreting that data, and using it to support decision making. This practice includes, "1) planning and system design, 2) data collection, 3) data analysis, 4) interpretation of results of analysis . . ., 5) dissemination and communication of information, and 6) application of information to public health programs and processes."[12] Rather than building a one-time system or tool, surveillance requires the development of a comprehensive ongoing practice with careful attention to the timing and frequency of measurements.[13] Given the complex set of issues, health conditions, and management problems involved in the health effects of climate change, any attempt to address the surveillance issues must also grapple with complex issues of funding, analysis, and data collection.[14] Providing an administrative structure to facilitate the process of surveillance is a complex political, legal and regulatory challenge, and is perhaps the primary legal issue in the age of climate change.

## 5.3 HISTORY OF SURVEILLANCE

### 5.3.1 *Surveillance, Public Health, and Governance*

The collection and analysis of population health data has been an integral part of the development of the modern state,[15] with the development of stable governments being key to the establishment of surveillance systems.[16] For examples, the use of population statistics as tools to govern in England,

[12] Hall et al., *Lexicon, Definitions, and Conceptual Framework, supra* note 4, 11.

[13] Ibid. at 10–14.

[14] See, e.g., J. del Corral et al., *Climate Information for Public Health: The Role of the IRI Climate Data Library in an Integrated Knowledge System*, 6(3) Geospat Health S15–24 (Sep 2012); M. Pascal et al., *How Can a Climate Change Perspective Be Integrated Into Public Health Surveillance?* 126(8) Public Health 660–667 (2012); Howard S. Burkom, *Evolution of Public Health Surveillance: Status and Recommendations*, 107(6) Am. J. of Pub. Health 848–850 (2017).

[15] Dorothy Porter, *Health, Civilization, and the State: A History of Public Health from Ancient to Modern Times* (London; New York: Routledge, 1999).

[16] Stephen B. Thacker and Ruth L. Berkelman, History of public health surveillance. In: William Halperin and Edward L. Baker, Jr., *Public Health Surveillance*, 1–2 (New York: Wan Nostrand Reinhold, 1992). Thacker and Berkelman also note the importance of a classification system for disease as critical along with the establishment of statistical methods. See, I. M. Moriyama et al., *History of the Statistical Classification of Diseases and Causes of Death* (Hyattsville, MD: National Center for Health Statistics. 2011) for a discussion of the development of disease classification systems. See, D. F. Stroup and R. L. Berkelman, History

Sweden, and Germany led to the development of the fields of statistics and epidemiology.[17] In the United States, Article I of the US Constitution incorporates the measurement of population in its requirement for a census.[18] Public health surveillance continues to play a critical role in effective governance and represents a modern expression of these historic governance practices.

In the United States, primary authority for public health lies with the states and is a function of their police power.[19] Accordingly, states conduct most surveillance in the United States pursuant to their police powers. The federal government has limited authority to enact public health laws and is bounded by the enumerated powers of the federal government.[20] National surveillance initiatives generally rely on practice, voluntary sharing, and cooperative agreements. This authority-sharing between state and federal agencies raises complex coordination issues for surveillance in the United States. Climate change and the anticipated disruptions to civil society demand that societies identify and resolve these coordination issues. A survey of historic, traditional, and emerging types of surveillance illustrates some of these challenges.

### 5.3.2 *Disease Reporting*

One of the most salient uses of public health surveillance is in the detection of epidemics, with an emphasis on "identifying and controlling individuals with communicable disease."[21] This function of public health surveillance has roots in the earliest days of the nation, going back to the colonial era. Requirements for national disease reporting were intermittent through the mid-nineteenth century. However, in 1850, a seminal report by Lemmuel

of statistical methods in public health. In: D. F. Stroup S. M. Teutsch, *Statistics in Public Health: Quantitative Approaches to Public Health Problems* (New York: Oxford University Press, 1998, pp. 1–18) for a discussion of the development of statistical methods and epidemiology.

17 Porter, *Health, Civilization and the State, supra* note 15, at 48–50; 65–78.

18 U.S. Const. Article I, sec. 2.

19 See, e.g., *Gibbons* v. *Ogden*, 22 U.S. 1, 203, 6 L. Ed. 23 (1824); *Gonzales* v. *Oregon*, 546 U.S. 243, at 270, 126 S. Ct. 904, at 923, 163 L. Ed. 2d 748 (2006), Erwin Chemerinsky, *Constitutional Law: Principles and Policies*, 5th edn. §5.1 (New York: Wolters Kluwer, 2015). See, generally, Lawrence O. Gostin and Lindsay F. Wiley, *Public Health Law: Power, Duty, Restraint*, 2016, 73–112 (discussion of the police power and modern public health authority); Ernst Freund, *The Police Power: Public Policy and Constitutional Rights* (1904) (discussion of traditional understanding of police power and limits).

20 Chemerinsky, *Constitutional Law*, §3.1. See also, Wendy E. Parmet. *Public Health and Constitutional Law: Recognizing the Relationship*, 10 J. of Health Care L. and Pol'y 13 (2007).

21 Gostin and Wiley, *Public Health Law: Power, Duty, Restraint, supra* note 19, at 308.

Shattuck of the Massachusetts Sanitary Commission underlined the relationship between communicable diseases and living conditions and the importance of a standardized system of reporting and collection of disease and death.[22] This report helped spur disease and sanitary conditions reporting efforts.[23] In addition to Shattuck's work in Massachusetts, the federal government's interest in controlling the recurring epidemics of yellow fever and cholera in the South through the nineteenth century also contributed to the development of a national disease reporting system.[24] The federal government began collection of mortality data in 1850, followed by morbidity data from periodic epidemics of smallpox, cholera, and yellow fever.[25] The modern public health surveillance system is an ongoing expression of these efforts to control these epidemics.

Today, the sharing of disease-related health information on a local, state, territorial, federal, and international level is through the National Notifiable Diseases Surveillance System (NNDSS). A patchwork of existing systems in the mid-twentieth century, after experiences with the 1916 poliomyelitis epidemic and the 1918–1919 influenza epidemics, led to the establishment of this system. This system remains collaborative and voluntary; only state-level authorities mandate reporting.[26] The system has continued to evolve and has been instrumental in addressing poliomyelitis, smallpox, malaria, HIV, hantavirus, SARS, and H1N1.[27]

### 5.3.3 *Vital Statistics*

States and localities have compiled vital statistics since the colonial period.[28] As with disease reporting, the primary responsibility for the collection and maintenance of vital statistics lies with the states, pursuant to their police power. The only federal law mandating the collection of vital statistics in the United States is the U.S. Constitution's census requirement. The early

22 John Duffy, *The Sanitarians: A History of American Public Health*, 98 (Urbana: University of Illinois Press, 1990).

23 Ibid.

24 Duffy, *The Sanitarians supra* note 22 Margaret Humphreys, *Yellow Fever and the South* [in English] (Baltimore; London: Johns Hopkins University Press, 1999).

25 Gostin and Wiley, *Public Health Law: Power, Duty, Restraint supra* note 19, at 308.

26 Ibid.

27 Gostin and Wiley, *Public Health Law: Power, Duty, Restraint, supra* note 19, at 308; Thacker et al., *Public Health Surveillance in the United States, supra* note 11, at 3–9.

28 U.S. Department of Health and Human Services, Centers for Disease Control and Prevention, National Center for Health Statistics, *U.S. Vital Statistics System: Major Activities and Developments*, 1950 – 95, by Alice M. Hetzel, Report, 1997, 44–45.

reporting of vital statistics focused on key events in the lives of citizens for the purposes of securing individual rights, such as property, paternity and inheritance.[29] Only after the interest in disease reporting grew, thanks to efforts such as the Shattuck report, was there an increased interest in vital statistics as a tool for measuring population health status.[30] The interest in vital statistics and their expansion increased with World War II and the creation of the modern welfare state.[31] Just like disease reporting, the modern vital statistics system relies on cooperative agreements, state authority, and the ongoing interest of the federal government in compiling and reporting vital statistics.[32] With regard to climate change, vital statistics records would be an invaluable tool, as they include geographic location of subjects and provide important information about births and causes of death. They would also help to identify vulnerable populations and target resources appropriately.

### 5.3.4 *Environmental Reporting*

The environment has a significant impact on our health and is a key variable for effective surveillance of climate-related health impacts. Reporting on environmental threats to health has not been part of traditional surveillance in the United States. In 2000, the Pew Environmental Health Commission released a report concluding that the environmental public health system was "fragmented, neglected, and ineffective."[33] It stated that there was an "environmental gap" due to lack of information to document the links between environmental triggers and chronic disease. Without adequate monitoring and data reporting, it would not be possible to ascertain the links between chronic disease and environmental hazards.[34] Following this report, the U.S. Centers for Disease Control and Prevention (CDC) created the National Environmental Public Health Tracking Program (EPHT), with an appropriation of funds from Congress.[35] The tracking program has three distinct components: identifying environmental hazards, determining human exposure to environmental hazards, and identifying the health impacts of human exposure to such hazards.[36] Health and environmental agencies at the local, state, and federal levels, as well as nongovernmental

[29] Ibid. [30] Ibid. at 47. [31] Ibid. at 1–4. [32] Ibid.

[33] Pew Environmental Health Commission, *America's Environmental Health Gap: Why the Country Needs a Nationwide Health Tracking Network* (2000).

[34] *Ibid.*

[35] A. E. Charleston et al., *Environmental Public Health Tracking: Driving Environmental Health Information*, [In english], 21(Suppl 2) J. Pub. Health Manag. Pract. S4–11 (Mar–Apr 2015).

[36] Ibid. at 3.

agencies, assist EPHT with collection, integration, analysis, interpretation, and dissemination of environmental health data. The CDC currently funds local and state tracking programs that work to collect data that impact environmental health and allow for swift decision making by policymakers.[37] While this program has made progress in coordinating data on the environment and health, much work is pending. This program is also dependent on ongoing funding, voluntary cooperation, and political will to continue it. Unlike other traditional surveillance areas, this type of environmental monitoring has not been a part of traditional state public health efforts.

### 5.3.5 *BioSurveillance*

A recent phenomenon in public health surveillance was the creation of a biosurveillance component or perspective, which, in many ways, is duplicative of extant efforts in public health surveillance. Biosurveillance is surveillance applied to human, animal, plant, food, or environmental data to detect and counter biological threats. Its emphasis is on national security. The terrorist attacks of Sept. 11, 2001, and the subsequent anthrax attacks led to considerable investment in and development of biosurveillance systems in the United States. Another important development post-9/11 included the drafting of the Model State Emergency Health Powers Act (MSEHPA or the "Model Act") by the Center for Law and Public Health at the request of the CDC.[38] The Model Act contains several provisions relating to surveillance, reporting, and tracking.[39] The Model Act was extremely controversial.[40] Since its development, twenty-eight states have

37 Ibid.; G. D. Kearney et al., A *Decade of Environmental Public Health Tracking (2002–2012): Progress and Challenges*, [In eng], 21(Suppl 2) J. Public Health Manag. Pract. S23–35 (Mar–Apr 2015).

38 Model State Emergency Health Powers Act (MESHPA) (2001).

39 Model State Emergency Health Powers Act §301 et seq. (2001).

40 See, e.g., L. O. Gostin. *The Model State Emergency Health Powers Act: Public Health and Civil Liberties in a Time of Terrorism,* 13(1) Health Matrix Clevel. 3–32 (Winter 2003) (defending the Model Act and discussing its history and development); See generally, John M. Colmers and Daniel M. Fox. *The Politics of Emergency Health Powers and the Isolation of Public Health,* 93(3) Am. J. of Pub. Health 397–399 (2003) (discussing controversy of Model Act); Annas, G. J. Bioterrorism, *Public Health, and Civil Liberties,* 346(17) N. Engl. J. Med. 1337–1342 (Apr 25, 2002) (critique of Model Act and civil liberties implications); James G. Hodge, The Evolution of Law in Biopreparedness, In: TV Inglesby and DA Henderson, *Biosecurity and Bioterrorism: Biodefense Strategy, Practice and Science* 10(1) (November 2012) (discussing history of MSEHPA evolution of public health law preparedness in the context of biopreparedness).

adopted the provisions relating to surveillance, in whole or partially.[41] In 2003, the Department of Homeland Security created the BioWatch program to detect, monitor, and analyze any airborne biological agents. Subsequently, the 9/11 Commission Recommendations Act of 2007 formally authorized the establishment of the National Biosurveillance Integration Center (NBIC) to "rapidly identify, characterize, localize, and track a biological event of national concern; integrate and analyze data relating to human health, animal, plant, food, water; and disseminate alerts and pertinent information."[42] It is unclear to what extent the biosurveillance program, which focuses on national security, overlaps with public health surveillance more generally. Syndromic surveillance is another tool developed to better identify disease outbreaks and bioterrorism. Syndromic surveillance is different from the traditional surveillance systems in that it seeks to identify clusters of disease or other symptoms before an outbreak.[43] Rather than reports of disease diagnoses, syndromic surveillance relies on near real-time data, reporting clusters of symptoms before diagnosis or on nonclinical data such as over-the-counter medication purchases, absenteeism at work and school, and other nontraditional data sources.[44] With regard to climate change, syndromic surveillance would be useful in identifying infectious disease outbreaks or water contamination related to climate change. This form of surveillance would be less useful in addressing slower changes in population health related to climate change.

## 5.4 LEGAL AND POLICY CHALLENGES IN INTEGRATING PUBLIC HEALTH INFORMATION, MANAGEMENT, AND CLIMATE CHANGE

Integrating data related to climate change into the public health surveillance system presents multiple complex law and policy issues.[45] The focus of this

[41] Network for Public Health Law, *The Model State Emergency Health Powers Act Summary Matrix*, last updated June 2012, www.networkforphl.org/_asset/80p3y7/MSEHPA-States-Table-022812.pdf.

[42] Office of Health Affairs, *National Biosurveillance Integration Center Fact Sheet, Department of Homeland Security*, www.dhs.gov/sites/default/files/publications/2015-Factsheet-NBIC_0.pdf.

[43] R. Katz, et al., *Redefining Syndromic Surveillance*, 1(1) J. Epidemiol. Glob. Health 21–31 (Dec 2011).

[44] *Ibid.* at 26.

[45] See generally, M. Pascal, et al., *How can a climate change perspective be integrated into public health surveillance?* 126(8) Pub. Health 660–667 (2012) (discussing framework to integrate climate into public health surveillance); S. J. Connor et al., Health and Climate – Needs, 1 Procedia Envtl. Sci. 27–36 (2010) (discussing information needs for health action); French Institute for Public Health Surveillance, *Workshop on Public Health Surveillance and Climate*

section is to highlight how the development of a comprehensive surveillance framework for climate change and public health presents several areas of concern: First, coordinating climate data with public health surveillance may, as a practical matter, be a complex endeavor. Second, differing regulatory goals and jurisdictional issues may prove challenging. Third, analyzing data within any framework presents a unique priority-setting challenge. Fourth, resisting the tendency to address climate change through a preparedness model is critical. Finally, climate change and its health-related effects require a long view, which is not currently present in the public health system, as explained in this section. It is clear that for the legal and policy system to grapple effectively with the serious issues of climate change and its impacts on health, the surveillance system must be sufficiently robust to support any legal regime applied to address those impacts. Ultimately, if the public health surveillance system is to include the information and data necessary to allow actors to make regulatory and policy decisions amid a changing climate, a clear effort to alter the norms of the field may be necessary as well as concrete steps to adapt public health and healthcare systems themselves to climate-related disruptions.[46]

### 5.4.1 *Coordination*

The coordination issues in incorporating climate-related data into public health surveillance are complex. As the discussion at the start of this section outlined, public health surveillance has developed historically in an ad hoc fashion and remains relatively fractured today. The CDC has created the Building Resilience against Climate Effects (BRACE) framework to begin to address this complex set of public health challenges by helping communities and policy makers prepare for the health impacts of climate change.[47] The five sequential processes in the BRACE framework are: (i) Forecasting climate

*Change* (conference proceedings, Saint-Maurice, France, 2010) (discussing surveillance needs and climate change generally).

46 *See*, Solange Gould and Linda Rudolph. *Challenges and Opportunities for Advancing Work on Climate Change and Public Health*, 12 (12) Int'l J. of Envtl. Research and Pub. Health 15649–15672 (2015); See, World Health Organization, *Operational Framework for Building Climate Resilient Health Systems*, 2015, www.who.int/globalchange/publications/building-climate-resilient-health-systems/en/

47 Gino Marinucci et al., *Building Resilience Against Climate Effects – A Novel Framework to Facilitate Climate Readiness in Public Health Agencies*, [In english]. 11(6) Int'l J. of Envtl. Research and Pub. Health 6433–6458 (2014).

impacts and assessing vulnerabilities, (ii) Projecting the disease burden, (iii) Assessing public health interventions, (iv) Developing and implementing a climate and health adaptation plan, and (v) Evaluating impact and improving quality of activities.[48] Adaptive management tools guide this five-step approach. While there has been criticism of adaptive management and its application, this continues to be the dominant model. Currently, the CDC has funded eighteen cities and state health departments or agencies to begin implementing climate adaptation strategies. There are, to-date, no studies of the effectiveness of the CDC program or of the funded programs in each health department. However, following the work of Solange and Rudolph, we could assume that without ongoing funding and support of these projects at CDC, it would be difficult to maintain the institutional support required to continue these efforts.[49]

### 5.4.2 *Differing Regulatory and Administrative Goals, Legal Authority*

Many of the agencies producing and maintaining climate data have markedly differing administrative goals. Environmental protection agencies aim to protect and preserve the environment. Efforts to achieve this goal often yield concomitant improvements for human health, but human health outcomes are often not the basic measure of how well such agencies are performing. On the other hand, the goal of public health agencies is the control and prevention of disease and the maintenance of public health. While environmental conditions and quality have a significant impact on this goal, environmental quality itself is not the core competency of public health officials or staff. The silo effect remains a problem, and the restriction of environmental and public health agencies and programs to their funded mandates can be an institutional barrier to effective protection.[50] An example from California is illustrative.

In 2012 California passed SB 535, the California Global Warming Solutions Act of 2006: Greenhouse Gas Reduction Fund, designed to ensure that funding for climate mitigation in California were directed toward projects that benefited disadvantaged communities.[51] CalEnvrioScreen, a tool that

48 *Ibid.*

49 Solange and Rudolph, *Challenges and Opportunities for Advancing Work on Climate Change and Public Health*, *supra* note 46.

50 Ibid.

51 Vien Truong, *Addressing Poverty and Pollution: California's SB 535 Greenhouse Gas Reduction Fund*, 49 Harv. CR-CLL Rev. 493 2014.

combined measures of income and exposure to environment-related health impacts of climate change, helped identify these disadvantaged communities.[52] The goal of SB 535 was to address ongoing issues of environmental justice and to provide financial resources to communities that had experienced poor health outcomes related to air pollution, climate change, and other environmental impacts.[53] The law requires that mitigation projects not only benefit disadvantaged communities but also require the spending of a percentage of the monies in those disadvantaged communities.[54] Since the law was not coordinated with larger adaptation efforts and priorities, it is likely that some projects may have exacerbated health disparities.[55] SB 535 was an environmental law and considered only environmental issues. Because of lack of coordination and differing administrative goals, the planning process did not incorporate broader health issues and disparities. For example, one mitigation project increased the amount of affordable housing in one city under the theory that it reduces the number of drivers commuting on the road. However, the project overlooked the fact that it involved building affordable housing units in a city with very high particulate matter (PM2.5) concentrations and poor air quality.[56]

### 5.4.3 *Jurisdictional Issues*

Jurisdictional issues also complicate the coordination of climate-related data with public health surveillance. The most significant challenge will likely be the relationship of the federal government and the states. As discussed in this section, the public health surveillance system in the United States has developed in an ad hoc fashion with the authority for surveillance remaining in the states and federal efforts relying on cooperative agreements and federally funded projects. As the election of November 2016 demonstrated, the policies and priorities of the federal government can change rapidly. To the extent that the current system relies on federal leadership to facilitate cooperation rather than legal mandates per se, the system is vulnerable. It is possible that smaller subnational groupings in the United States and regional agreements will rise in importance as the federal commitment shrinks. There are no current regional surveillance networks. Instead, surveillance remains coordinated at the state level with federal coordination of larger networks such as the

[52] California Office of Environmental Health Hazard Assessment, *CalEnvrioScreen, version* 3.0.

[53] Ibid. [54] Truong, *Addressing Poverty, supra* note 51. [55] Ibid.

[56] Chandrakala Ganesh and Jason Smith, *Climate Change, Public Health and Policy: A California Case Study* (publication forthcoming), August 7, 2017.

Environmental Health Tracking Program.[57] Surveillance and public health efforts can consider regional agreements and activities similar to the Western Climate Initiative, which was formed to coordinate the cap and trade market policies between California and Canadian provinces in the area of mitigation.[58]

### 5.4.4 *Priority Setting*

Given the complexity of climate change, its long-term nature, and the scarcity of resources available to address it, it is necessary to set priorities for adaptation policies. Thus, for instance, governments must decide whether to adapt particular systems and infrastructure to the risk of climate disruptions or to retreat and abandon those systems. Public health surveillance should be a key part of the analysis that informs this prioritization. To support such priority setting, surveillance systems must develop and employ social and health indicators so as to enable governments to identify not only the health impacts of climate change but also to identify populations that are vulnerable to those impacts – both now and over a multi-generational timescale. Priority setting raises complex legal and policy value problems. Communities must determine what metrics to use to evaluate competing responses to climate change and must then make difficult choices based on what those metrics indicate. This process implicates foundational values and principles of governance.

These issues are not legal per se. Rather, the complexity of these coordination issues around information and surveillance may require the development of new administrative structures. The considerable theoretical work in this area has identified two coordination issues as being especially important. First, there must be a combination of both mitigation and adaptation planning in any coordination and analysis.[59] These efforts are interrelated in complex ways that require adaptive management techniques. The two concepts are tightly related and to treat one without the other is to invite failure. For example, mitigation efforts that attempt to reduce greenhouse gas emissions by reducing urban sprawl and emphasizing public transit can increase urban density. This increase in urban density can produce heat islands that exacerbate heat-related health problems and can increase pressure on power grids with increased use of air conditioning. Second includes development of new

57 See text accompanying notes 30ff.

58 See, *Western Climate Initiative*, www.wci-inc.org/index.php.

59 Jason A. Smith et al., *Climate Change and Public Health Policy*, 45(1) The J. of L., Medicine & Ethics, 82–85 (2017).

administrative approaches to address public health issues, which are primarily adaptation issues, through a "mitigation-adaptation lens."[60] Parker-Flynn proposes one such approach: a *mitigation-adaptation research agency or commission* (MARA). This sort of dedicated agency, ideally located at the federal level, would perform four functions: information gathering and multi-scalar, multi-agency coordination to support research on effective strategies; analyzing proposed mitigation or adaptation strategies; providing funding and model approaches; and conducting monitoring and assessment.[61] These functions are similar to the core functions of public health, namely, assessment, policy development, and assurance.

### 5.4.5 *Rejecting Preparedness as the Only Model*

One legal framework often used to understand climate and health impacts is the preparedness model. The preparedness model, which focuses on responding adequately to emergencies and disease outbreaks, is ubiquitous in the public health community but is a mismatch for many of the impacts of climate change.[62] The concept of preparedness focuses on responding to emergencies, infectious disease outbreaks, and emergent infectious disease.[63] The preparedness approach should not be the *only* framework for understanding climate change and health. While it is important to design the public health surveillance system to alert the public health authorities of emergent events, applying this framework alone will lead authorities to ignore much else that is also important. That is, a focus on preparedness alone tends to focus policy makers on large-scale and sudden events: hurricanes and sudden storms, disease outbreaks, heat waves, etc. But to effectively address climate change impacts through surveillance, practitioners must focus on the larger, perhaps more significant, impacts: sea level rise, agricultural disruptions, migration, and so forth. If anything, climate change will represent a slowly evolving set of challenges that requires a more dynamic and responsive system, one not focused on preparedness alone.[64]

60 Ibid.; James E. Parker-Flynn. *The Intersection of Mitigation and Adaptation in Climate Law and Policy*, 1 Environs Envtl. L. and Pol'y J. 30 (2014).

61 *Ibid.* at 40.

62 A. D. Moulton et al., *What is Public Health Legal Preparedness?* 31(4) J. of L., Medicine and Ethics 672–683 (2003).

63 Ibid.

64 Lindsay F. Wiley, *Adaptation to the Health Consequences of Climate Change as a Potential Influence on Public Health Law and Policy: From Preparedness to Resilience*, 15 Widener L. Rev. 483–519 (2010).

### 5.4.6 *Long View*

Climate change, unlike other problems that might affect public health, is unique in regards to both its timeline and scope. Instead of operating on one-, five-, and ten-year scales, surveillance systems must be developed or adapted to operate on multi-decade time frames and to incorporate into public health perspectives the interests of future generations.[65] Indeed, the health issues related to climate are severe, pervasive, and, to some extent, unavoidable. They touch all aspects of society. Moreover, based on models of climate disruption, even if all proposed mitigation actions were taken right now to arrest the introduction of greenhouse gases into the environment, the planet would still be locked into a multi-decade period of climate disruption and warming. The long-term nature of this problem requires that approaches to coordination and surveillance be adapted to take a long view for data collection and analysis. This reframing of public health action and management is challenging and will require significant attention to reforming many of the public health approaches and systems so that public health practitioners can effectively act not only for populations now but also for population health fifty years from now. There is not, to date, a clear mechanism to reframe public health systems to include this long-term forecasting and management. The history of public health in the United States, particularly in the twentieth century, has tended to make public health practitioners responsive rather than proactive with a narrow area of focus that remains underfunded compared to other public programs and endeavors. Successful public health efforts that have been controversial have been successful when linked to larger social and cultural movements. For example, public health advocates were able to take concrete steps in addressing HIV infection only under pressure from, and with the support of, AIDS advocacy and political organizations. The ability of public health to address climate change and its effects would be helped by a larger cultural and advocacy movement that demanded these efforts.

## 5.5 CONCLUSION

Surveillance will be a key component of effectively managing the health impacts of climate change. The current surveillance system in the United States developed haphazardly in response to historic events and the political structure of the United States. This surveillance system requires cooperation,

[65] H. Frumkin and A. J. McMichael. *Climate change and Public Health: Thinking, Communicating, Acting*, 35(5) Am. J. Prev. Med. 403–410 (2008), doi: 10.1016/j.amepre.2008.08.019.

custom, and the efforts of state governments. Given the complexity of managing the effects of climate change, public health practitioners must reconsider the organization of the surveillance systems at a fundamental level. It is critical to make climate change an area of emphasis right now, not only to address climate-driven issues occurring currently but also to address issues that are likely to occur in the future.

# 6

# The Built Environment

Justin Gundlach and Jennifer Klein

*The built environment, which includes not only buildings but infrastructure, mediates several important climate impacts on public health and is also subject to diverse legal requirements. It is a subject of particular focus for policy efforts aimed at promoting adaptive responses to climate change on the part of institutions and individuals. This chapter presents key examples of public health impacts that arise from climate change but are mediated – possibly mitigated, possibly exacerbated – by elements of the built environment. It also describes the process and substance of adaptive responses to those impacts. Having presented these physical and policy contexts in its first section, this chapter's second section considers the role the law could play as individuals, organizations, and localities react to climate-driven harms and seek to adapt.*

## 6.1 INTRODUCTION

The built environment – meaning buildings and the infrastructure systems on which they and their occupants rely[1] – arises from layered human decisions.[2] In 2008, a team of public health researchers put it this way:

> Distinct from the natural environment, the built environment is comprised [sic] of manmade components of people's surroundings, from small-scale settings (e.g. offices, houses, hospitals, shopping malls, and schools) to large-scale settings (e.g. neighborhoods, communities, and cities), as well as roads, sidewalks, green spaces, and connecting transit

[1] National Institute of Standards and Technology, *Special Publication 1190, Community Resilience Planning Guide for Buildings and Infrastructure Systems vol. I*, 13 (May, 2016), https://perma.cc/ZVH7-GEJ2.

[2] William Cronon, *Nature's Metropolis* 62 (1991) (contrasting the natural environment, or "first nature," with "structures of the human economy," or "second nature").

> systems. The development of the built environment involves many sectors, including urban planning, architecture, engineering, local and regional governments, transportation design, environmental psychology, and land conservation.[3]

They also noted, of course, that "[t]he built environment influences human choices, which in turn affect health," specifically, "physical activity, respiratory and cardiac health, injury risk, chronic disease risk, social connectedness, and mental health ... "[4] However, because *all* social and economic institutions rely to some degree on the built environment,[5] as climate change redraws shorelines and modifies seasons, temperatures, and weather patterns, the responsive changes to the built environment will implicate a host of interests, public health just one among them.

The chapter is divided into two sections. The first discusses key climate impacts on public health that are mediated in some way by the built environment. It also discusses the process and substance of adaptive responses to those impacts. The second section considers the role the law could play as individuals, organizations, and localities react to climate-driven harms and seek to adapt.

Before proceeding to those discussions, it is important to first define some key terms. Climate change is the basic cause of a large number of immediate and intermediate effects – that is, effects that cause further effects. The public health and disaster management communities have developed terminology that is useful to climate change adaptation policymakers who must sort through the problem of which effects/causes to address and how to track and coordinate the success of adaptation efforts.[6] In this lexicon, *stressor* and *hazard* both refer to an underlying

[3] Margalit Younger et al., *The Built Environment, Climate Change, and Health: Opportunities for Co-Benefits*, 35(5) Am. J. Preventive Med. 517–526, 517 (2008).

[4] Ibid.

[5] National Institute of Standards and Technology, *Special Publication 1190, Community Resilience Planning Guide for Buildings and Infrastructure Systems vol. II*, 27 tbl. 10–4 (Oct. 2015), https://perma.cc/ZVH7-GEJ2 (providing thorough list of links between institutions and buildings, including direct and indirect impacts arising from damage to buildings).

[6] B. L. Turner II et al., A *Framework for Vulnerability Analysis in Sustainability Science*, 100 Proceedings Nat'l Academy Sci. 8074 (2003), 10.1073/pnas.1231335100; see also Janet L. Gamble et al., Populations of concern, In: Alison Crimmins et al., *The Impacts of Climate Change on Human Health in the United States: A Scientific Assessment*, 247, 249 (U.S. Global Change Research Program, 2016).

cause, such as rising ambient temperatures. Stressor tends to refer to chronic and slow-moving causes;[7] hazard is applicable both to slow-moving causes, such as drought, and sudden-onset causes such as destructive coastal storms. If drought or storms strike in a location far removed from human populations or development, then there is no *exposure* to them. The degree of exposure to a stressor or hazard varies with location and the ability of people or structures in that location to endure it without disruption – thus, someone with central air conditioning who works indoors might not be highly exposed to extreme heat even if her location experiences a heatwave.[8] Among the populations and assets that are exposed and cannot mitigate or avoid that exposure, some are more *sensitive* – that is, susceptible or unable to cope – than others. For instance, as noted in Chapter 3, children, the elderly, and the disabled are generally physically less able to endure hazards such as heat or air quality made worse by climate-driven stressors. Those who are both sensitive and exposed are *vulnerable*. Some populations that are vulnerable are also *resilient*, however, meaning that they are capable of recovering quickly from exposure to a hazard even though they are vulnerable to it.[9]

## 6.2 ADAPTING THE BUILT ENVIRONMENT TO ADDRESS PUBLIC HEALTH IMPACTS OF CLIMATE CHANGE

This subsection begins with a description of key public health impacts of climate change that relate directly to the built environment. It is not meant to be exhaustive and seeks to avoid redundancy with the descriptions of impacts covered by other chapters. After providing that brief survey of impacts, it addresses the task of adapting the built

7 But see Office of Energy Policy and Systems Analysis, U.S. Dep't of Energy, *A Review of Climate Change Vulnerability Assessments: Current Practices and Lessons Learned from DOE's Partnership for Energy Sector Climate Resilience* 10–12 (May 2016), https://perma.cc/G3YB-R546 (referring to sudden and slow-onset climate-driven effects as "stressors").

8 See, e.g., Amiche Alcindor, *In Sweltering South, Climate Change is Now a Workplace Hazard*, N.Y. Times, Aug. 3, 2017, https://perma.cc/4ULU-DXRQ ("They don't know what's going on... because they are in cool houses and in offices," Mr. Guerra said.).

9 UNISDR, 2009 *UNISDR Terminology on Disaster Risk Reduction* (2009); see also UNISDR, *The Sendai Framework for Disaster Risk Reduction 2015–2030* (2015 www.unisdr.org/files/43291_sendaiframeworkfordrren.pdf; *Hyogo Framework for Action 2005–2015: Building the Resilience of Nations and Communities to Disasters, Extract from the final report of the World Conference on Disaster Reduction*, A/CONF.206/6) (2005), www.unisdr.org/files/1037_hyogoframeworkforactionenglish.pdf.

environment in response to those impacts. It begins by noting how various frameworks would organize that task, and then examines key elements of the process and substance of adaptation efforts focused on the built environment.

### 6.2.1 *Key Examples of Climate-Driven Public Health Impacts Transmitted by the Built Environment*

Atop the list of issues that adaptation planning must address in the built environment are public health vulnerabilities traceable to the following hazards: too much water due to sea level rise and/or extreme precipitation; temperatures that are more variable and generally higher; too little water, i.e. drought; and extreme events such as storms and wildfires.[10] Exposure to these hazards takes a variety of forms. Some are direct, such as outdoor workers encountering long periods of hot and humid weather. Others, particularly where the climate-driven stressor interacts with other hazards, are indirect, such as outdoor or indoor air quality made worse by higher temperatures boosting air pollution levels, mold growth promoted by more heat and humidity in buildings, or exposure to toxic waste introduced into the environment by the flooding of a brownfield or a waste storage facility located on a coastline. Still other vulnerabilities that arise from indirect exposure to hazards owe to those hazards' disruption of infrastructure: transit systems, road networks, drinking water provision, wastewater management, and electricity distribution.[11]

Rather than attempting a comprehensive review of how climate change adaptation efforts respond to the diverse climate-driven hazards and vulnerabilities that bear upon and are mediated by the built environment, this chapter focuses on two hazards in particular: hotter ambient temperatures and coastal flooding driven by sea level rise (SLR).[12]

[10] Many of the climate-driven vulnerabilities related directly to the built environment are described at length in Chapters 3, 7 (heat), 8 (storms), 9 (infectious disease), and 10 (food and agriculture) of this volume. See also S. L. Cutter et al., Ch. 11: Urban systems, infrastructure, and vulnerability, In: J. M. Melillo et al., *Climate Change Impacts in the United States: The Third National Climate Assessment* (2014).

[11] See NOAA, *U.S. Climate Resilience Toolkit: Built Environment*, https://toolkit.climate.gov/topics/built-environment.

[12] Riverine flooding driven by severe precipitation and early snowmelts, wildfires, and drought are all also examples of climate change-driven impacts that affect public health via the built environment. This chapter recognizes the importance and relevance of these impacts, see, e.g., Dennis M. Knobloch, *Moving a Community in the Aftermath of the Great 1993 Midwest*

#### 6.2.1.1 Heat

Cities amplify the direct effects of hotter ambient temperatures driven by climate change; this phenomenon is termed the urban heat island effect.[13] In cities, asphalt, concrete, and other artificial surfaces with low albedo (reflectivity) absorb solar radiation and then express it as heat, and the relative (or total) absence of vegetation often means that these surfaces are largely unshaded and are not cooled through the evapotranspiration of moisture. Cities' absorptive capacity for heat is compounded by urban layouts that obstruct cooling breezes, and by the presence of people, vehicles, and other heat sources – including, during warm seasons, air conditioning units.[14]

Within and beyond the bounds of cities, hotter temperatures also impair the functioning of transportation systems and electricity infrastructure.[15] Those systems' vulnerability to heat compounds the heat-related vulnerabilities of people and assets that rely on their smooth operation. Thus, these vulnerable systems can be the source of indirect adverse public health effects.

The other key indirect effects of hotter temperatures result from their contribution to poorer outdoor and indoor air quality – a leading scourge of public health.[16] This relates to the built environment because, as discussed in Chapters 3, 7, and 13, hotter temperatures promote ozone formation in places downwind of roads, power plants, and other sources of nitrogen oxides and volatile organic compounds[17] – a hazard to which

*Flood*, 130 J. Contemporary Water Res. & Edu. 41 (Mar. 2005), https://perma.cc/39T9-VTTV, but space constraints put discussion of them beyond its scope.

13 See Joyce Klein-Rosenthal and Jeffrey Raven, *Urban Heat and Urban Design – An Opportunity To Transform In NYC*, The Sallan Foundation: Snapshot (July 18, 2017) https://perma.cc/X45Q-RDRR (describing effect and its causes).

14 Ibid.

15 Sofia Aivalioti, *Electricity Sector Adaptation to Heat Waves* (Jan. 2015), https://perma.cc/ZML6-3QVQ; Henry G. Schwartz, Michael Meyer et al., Ch. 5: Transportation, In: J. M. Melillo et al., *Climate Change Impacts in the United States: The Third National Climate Assessment* 130, 132–133 (2014) (noting effects of heat on roads, bridges, rails, and aircraft performance).

16 Philip J. Landrigan, *Air Pollution and Health*, 2(1) The Lancet: Public Health e4–e5 (Jan. 2017) ("Air pollution is one of the great killers of our age."); Frank J. Kelly and Julia C. Fussell, *Air pollution and public health: Emerging hazards and improved understanding of risk*, 37(4) Envtl. Geochemistry & Health 631–649 (2015).

17 James N. Galloway and William H. Schlesinger et al., Ch. 15: Biogeochemical cycles, In: J. M. Melillo et al., *Climate Change Impacts in the United States: The Third National Climate Assessment* 350, 357 (2014) ("Rates of ozone formation are accelerated by higher temperatures, creating a reinforcing cycle between rising temperatures and continued human alteration of the nitrogen and carbon cycles. Rising temperatures also work against some of the benefits of air pollution control.").

children and the elderly are especially vulnerable.[18] More frequent or intense precipitation and/or flooding can also promote mold growth indoors, which in turn can visit an array of adverse public health impacts on residents.[19]

### 6.2.1.2 Coastal Flooding

Recent events in the Northeast, Florida, and Louisiana illustrate the vulnerabilities traceable to the coastal flooding that is being made ever more severe and frequent by rising sea levels and intensifying coastal storms.

Superstorm Sandy,[20] which struck the New Jersey and New York coasts in November 2012, drowned dozens of people, destroyed hundreds and damaged thousands of coastal homes and businesses, and wreaked damage and severe disruptions on transit and electricity systems, wastewater treatment plants, and other assets and infrastructure.[21] Acute public health impacts included injury, death, exacerbated illnesses, and "bypass events" (i.e. flows of raw sewage into coastal waters) resulting from the powering down of several wastewater treatment plants. Longer-term public health impacts included mold growth in flooded homes and various adverse mental and physical health effects of residents' displacement from their homes.[22]

18 Radley Horton and Gary Yohe et al., Ch. 16: Northeast, In: J. M. Melillo et al., *Climate Change Impacts in the United States: The Third National Climate Assessment* 371, 377 (2014).

19 George Luber et al., Ch. 9: Human Health, In: J. M. Melillo et al., *Climate Change Impacts in the United States: The Third National Climate Assessment* 220, 222 (U.S. Global Change Research Program, 2014) doi. 10.7930/J0PN93H5.

20 Although the storm's wind speed rated at the low end of Category 1 hurricanes, its pressure rating was comparable to that of a Category 3 hurricane. FEMA, *Mitigation Assessment Team Report Hurricane Sandy in New Jersey and New York*, i (Nov. 2013), https://perma.cc/U5GH-LJ4M.

21 CDC, *Deaths Associated with Hurricane Sandy – October–November 2012, Morbidity and Mortality Weekly Report* 62(20), 393–397 (May 24, 2013) https://perma.cc/RH2M-K8GM; FEMA, 6 Months Report: Superstorm Sandy from Pre-Disaster to Recovery, Apr. 25, 2013, https://perma.cc/SA37-QGN3.

22 John Manuel, *The Long Road to Recovery: Environmental Health Impacts of Hurricane Sandy, Envtl. Health Persp.*, 121: A152–A159 (2013), https://perma.cc/D6VE-VXE9 ("Of the long-term health threats posed by Sandy, the most significant is mold growth in homes that were not properly remediated after flooding."). Recent research highlights that the total costs of recovery from storms like Sandy greatly exceed those classified as disaster aid in scale and duration, and that non-disaster costs arise from healthcare needs. Tatyana Deryugina, *The Fiscal Cost of Hurricanes: Disaster Aid versus Social Insurance*, 9 (3) Am. Econ. J.: Econ. Pol'y 168–198 (2017), https://perma.cc/GLF6-J4TN.

Consistent with the general distributional pattern of disaster impacts, Sandy was disproportionately hard on poorer neighborhoods and people with disabilities.[23]

Miami Beach and Miami are among the cities most notoriously vulnerable to the slow-moving SLR-driven hazards of nuisance flooding and the salt water infiltration of groundwater resources,[24] though they are not alone.[25] Nuisance flooding visits substantial cumulative damage on stormwater and wastewater management systems, and can also facilitate transmission of infectious disease and exposure to toxic chemicals.[26] Infiltration of groundwater by saltwater can compromise drinking water systems relied upon by large populations.[27] Though these results of SLR are generally not a source of dramatic images or news stories, their cumulative impacts on infrastructure are often just as destructive.[28]

Populations in coastal Louisiana – rural and urban alike – face a combination of impacts, some of them climate-driven. Those populations endured the flooding that accompanied hurricanes Katrina and Rita in 2005, followed by the slower-emerging impacts of toxic spills resulting from those storms.[29] In addition, they continue to experience significant land loss due to rapid subsidence,[30] which has lately

23 Furman Center for Real Estate & Urban Policy at NYU School of Law, *Sandy's Effects on Housing in New York City* (Mar. 2013), https://perma.cc/2V4A-E4AV (reporting housing-related impacts of storm surge, broken down by income level); see also Adrien A. Weibgen, *Note: The Right to be Rescued: Disability Justice in an Age of Disaster*, 124 Yale L. J. 2406 (2015) (discussing plight of elderly and disabled during Sandy and in the storm's aftermath).

24 David Smiley, *Mainland Miami Ponders Returning Neighborhoods to Nature in Order to Survive Rising Seas*, Miami Herald, June 9, 2017, https://perma.cc/4A9F-QE6W.

25 William V. Sweet and John J. Marra, NOAA, 2015 *State of U.S. "Nuisance" Tidal Flooding* (June 2016) http://perma.cc/9PHH-2ZJ6 (reporting on prevalence and severity of phenomenon across U.S.).

26 Stephanie Kruel, *The Impacts of Sea-Level Rise on Tidal Flooding in Boston, Massachusetts*, 32 (6) J. Coastal Res. 1302–09, 1308 (Nov. 2016) https://perma.cc/3D2A-45K5 (listing among public health risks of recurrent flooding: transmission of infectious diseases, exposure to toxic chemicals, growth of mold in residences).

27 Robin Kundis Craig, *A Public Health Perspective on Sea-Level Rise: Starting Points for Climate Change Adaptation*, 51 Widener L. Rev. 521 (2010); see also Chapters 8 and 9 of this volume.

28 Hamed R. Moftakhari et al., *Cumulative Hazard: The Case of Nuisance Flooding*, 5 Earth's Future 214–223 (2017), doi:10.1002/2016EF000494.

29 Mark Schleifstein, *Extent of oil spills from Hurricanes Katrina and Rita is still being assessed*, Times-Picayune, Aug. 19, 2010, https://perma.cc/62D8-J6MP.

30 Jaap H. Nienhuis et al., *A New Subsidence Map for Coastal Louisiana*, 27 GSA Today (May 2017) https://perma.cc/6G93-5W9A ("the fundamental culprit is the isolation of the sediment-

been exacerbated by oil spill damage to coastal vegetation – a natural buffer to storms and source of coastal stability.[31] Thus SLR and more intense storms act as stressors on coastal Louisiana's existing vulnerabilities.[32]

### 6.2.2 *Adapting to those Impacts and others*

The Intergovernmental Panel on Climate Change, having examined examples of adaptation efforts around the world, stated that "[t]here is no single approach to adaptation planning because of the complex, diverse, and context-dependent nature of adaptation to climate change."[33] This makes it difficult to distill a general description of adaptation efforts down to a brief summary. However, it is possible – and useful – to take note of the key dimensions of adaptation efforts identified by researchers examining numerous programs funded by the UN's Global Environment Facilities:

* Timing relative to stimulus (anticipatory, concurrent, reactive);
* Intent (autonomous, planned);
* Spatial scope (local, regional, national);
* Form (e.g., technological, behavioral, financial, institutional); and
* Degree of necessary change (incremental, transformational).[34]

delivery system (the Mississippi River) from its delta plain and the adjacent coastal zone due to the construction of flood-protection levees. As a result, the majority of the sediment carried by this system is funneled into the deep waters of the Gulf of Mexico, rather than offsetting the naturally occurring high subsidence rates.").

31 U.S. Geological Survey NASA *Study Finds Widespread Coastal Land Losses from Gulf Oil Spill, Nov.* 17, 2016, https://perma.cc/FU92-8DX8 (identifying oiling of shorelines from oil spill and erosion due to coastal storm as distinct, material causes of shoreline erosion).

32 *Quantifying Vulnerability*, LSU College of Engineering News, Apr. 4, 2013, https://perma.cc/XSC7-QZ7V (discussing use of GIS mapping to identify where land subsidence is likely to compromise access to hurricane evacuation routes).

33 Nobuo Mimura et al., Adaptation planning and implementation, In: C.B. Field et al., *Climate Change 2014: Impacts, Adaptation, and Vulnerability. Part A: Global and Sectoral Aspects. Contribution of Working Group II to the Fifth Assessment Report of the Intergovernmental Panel on Climate Change* 869–898, 871 (2014).

34 See Bonizella Biagini et al., A *Typology of Adaptation Actions: A Global Look at Climate Adaptation Actions Financed Through the Global Environment Facility*, 25 Global Environmental Change 97–108 (Mar. 2014). See also City and Cnty. of Denver, *Climate Adaptation Plan* (2014), https://perma.cc/FA34-H4XK (dividing adaptation activities into short-, medium-, and long-term).

As this list implies, the scope of adaptation efforts focused on the built environment could encompass an enormous array of measures. One such might be a nationwide vulnerability assessment for airports or rail transportation systems conducted in anticipation of coastal flooding and heatwaves that exceed parameters assumed by those systems' current engineering.[35] Another might be a municipality's adoption of a local ordinance that categorizes green roofs as stormwater management infrastructure so that their installation can be funded – like other capital expenses – using municipal bonds.[36] Examples of measures responsive to the impacts highlighted in Sections 6.2.1.1 and 6.2.1.2 are discussed in more detail in Section 6.2.3.

### 6.2.3 *Process*

There is broad consensus on how adaptation efforts should – and do, in practice – proceed;[37] one pair of authors have called the steps involved a "ladder."[38] The highly conventional steps include most or all of the following: identify hazards, assess vulnerabilities, specify objectives, explore options, plan, implement, and evaluate. A large and growing number of jurisdictions have undertaken at least some of these steps with respect to those portions of the built environment for which they are responsible. However, few climb the full ladder: many have identified hazards, fewer have assessed vulnerabilities, fewer still have translated the resulting insights into plans and plans into actions, and very few have evaluated the effectiveness of those actions.[39] Notably, larger cities – which have more resources

[35] See also Schwartz et al., *Ch. 5: Transportation, supra* note 15, at 130–149, 133–134.

[36] See Justin Gundlach, *Putting Green Infrastructure on Private Property in New York City*, 28 Envtl. L. in N.Y. 140, 148 (Sept. 2017) (noting financing challenges arising from accounting treatment of green infrastructure).

[37] See, e.g., NOAA, *U.S. Climate Resilience Toolkit: Steps to Resilience*, https://toolkit.climate.gov/#steps; 2017); Helge Bormann et al., Guiding regional climate adaptation in coastal areas, In: Walter Leal Filho, *Handbook of Climate Change Adaptation* 337, 350–52; P. M. Groffman et al., Ch. 8: Ecosystems, biodiversity, and ecosystem services, In: J. M. Melillo, *Climate Change Impacts in the United States: The Third National Climate Assessment* (2014) 195–219, 202; Timothy Carter et al., Technical guidelines for assessing climate change impacts and adaptations, In: Robert T. Watson, *IPCC, Climate Change 1995 – Impacts, Adaptations and Mitigation of Climate Change: Scientific-Technical Analyses* 823 (1995).

[38] E. M. Hamin and N. Gurran, Climbing the adaptation planning ladder: Barriers and enablers in municipal planning, In: Leal Filho *Handbook of Climate Change Adaptation* (2015).

[39] Several surveys of adaptation efforts by local governments report that many complete the initial steps of identifying hazards but fewer conduct vulnerability assessments, fewer

and can more readily dedicate staff to the task – are more likely than smaller localities to have undertaken a vulnerability assessment and planning effort.[40]

### 6.2.3.1 Identifying Hazards

An effort to adapt necessarily begins by surveying the current and expected future climate in a particular area, and by identifying hazards potentially affecting the population, assets, or infrastructure located there. The process requires interpreting both historical data about the parameters of interest (e.g., temperature, flood risk) and data generated by climate models that predict future climatic circumstances. Data of the latter sort are challenging to derive with high degrees of accuracy for the Earth as a whole, and more so when "climate model downscaling" focuses in on the smaller geographic area involved in any given adaptation effort.[41] Predictions are also consequential to publish because they can inform land use and investment decisions.

Several publicly available tools can help an entity or jurisdiction identify hazards arising from SLR, temperature and humidity changes, or changes in precipitation patterns,[42] though these tools tend to yield coarse projections at

integrate their findings into planning efforts, fewer still implement those plans, and very few evaluate those plans' effectiveness. See John Nordgren et al., *Supporting Local Climate Change Adaptation: Where We Are and Where We Need to Go*, 66 *Envtl. Sci. & Pol'y* 344–352, 347 (Dec. 2016); Linda Shi et al., Global patterns of adaptation planning: results from a global survey, In: Karen C. Seto et al., *The Routledge Handbook of Urbanization and Global Environmental Change* 336, 341–43 (2015) (reporting results from 2011 survey); National Association of Regional Councils, A *Survey of Regional Planning for Climate Adaptation* (2012), https://perma.cc/K3WA-WDDJ.

40 Shi et al., *Global Pattens of Adaptation Planning, supra* note 39, at 341–343.

41 G. Flato et al., Evaluation of climate models, In: T. F. Stocker et al., *Climate Change 2013: The Physical Science Basis. Contribution of Working Group I to the Fifth Assessment Report of the Intergovernmental Panel on Climate Change* 741, 817 (2013) (describing regional climate models as valid but emphasizing their sensitivity to imprecise inputs: "This underlines the importance of both the quality of the boundary conditions and the downscaling method.").

42 See, e.g., NOAA, *Global and Regional Sea Level Rise Scenarios for the United States* (Jan. 2017) (including regional projections), https://perma.cc/N9AG-8Y6S; NOAA Office for Coastal Management, *Sea Level Rise Viewer*, https://coast.noaa.gov/digitalcoast/tools/slr (accessed June 30, 2017); U.S. Climate Resilience Toolkit, *MACA CMIP5 Statistically Downscaled Climate Projections*, https://toolkit.climate.gov/tool/maca-cmip5-statistically-downscaled-climate-projections (last updated Jan. 5, 2017); see also Climate Central, *Surging Seas*, http://sealevel.climatecentral.org/.

smaller scales.[43] New York City's Panel on Climate Change (NPCC) has arguably set the gold standard for developing downscaled projections of SLR, temperature, and precipitation.[44] However, the NPCC is both an expensive endeavor and one that, notwithstanding its relative superiority, generates projections subject to large uncertainties.[45] Whatever data are used, the climate scenarios generated by this step provide parameters useful to, for instance, departments of transportation that need to determine how high to build a bridge, electric utilities that need to design components that can operate even during long durations of high ambient temperatures, floodplain managers that want to guide land use and design decisions for long-lived assets, and municipal authorities that need to match the carrying capacity of a stormwater management system to the precipitation expected during its useful life.

#### 6.2.3.2 Assessing Vulnerabilities

This step builds on the previous one by identifying conflicts between projected climate parameters and the location or operation of existing people, assets, and infrastructure. Whether it is being undertaken by an electricity distribution utility,[46] a city, state, or federal agency,[47] or some

[43] The US Climate Resilience Toolkit homepage cautions that "[c]limate projections are not predictions," that "[t]he increased spatial resolution of statistically downscaled projections available for temperature and precipitation may not be available for all parameters. In addition, increased resolution does not necessarily equate to greater fidelity or reliability." U.S. Climate Resilience Toolkit, *Statistically Downscaled Climate Projections*, *supra* note 42. (last updated Jan. 5, 2017). It also counsels against DIY applications of downscaled modeling data: "For decisions involving the use of climate model projections, you may want to consider seeking expertise." Ibid.

[44] See *Building the Knowledge Base for Climate Resiliency: New York City Panel on Climate Change 2015 Report*, 1336 Annals of the N.Y. Acad. of Sci. (Jan. 2015). The third round of the panel's reporting, NPCC3, included sea level rise projections for three areas in southeastern New York State that have since been adopted by the Department of Environmental Conservation and so must be considered in future planning decisions. See New York Department of Environmental Conservation, *Projected Sea-Level Rise-Express Terms*, 6 NYCRR 490 (2017), www.dec.ny.gov/regulations/103877.html.

[45] See, e.g., New York City Panel on Climate Change [NPCC2], *2015 Report-Executive Summary*, 1336 Annals of N.Y. Acad. Sci. 9-17 (2015) at ES-10, n.e (no probabilities assigned to projected mean temperatures because of multiple uncertain factors), ES-11 (frequency and intensity of coastal storms are "uncertain at local scales"), ES-13 (coastal flooding projections subject to multiple uncertainties).

[46] Pacific Gas and Electric Company, *Climate Change Vulnerability Assessment* (2015), https://perma.cc/N2NR-S79A.

[47] See, e.g., U.S. Federal Highway Administration, *The Gulf Coast Study, Phase 2: Assessing Transportation System Vulnerability to Climate Change: Synthesis of Lessons Learned and Methods Applied* (Oct. 2014) https://perma.cc/N3R9-DEFN.

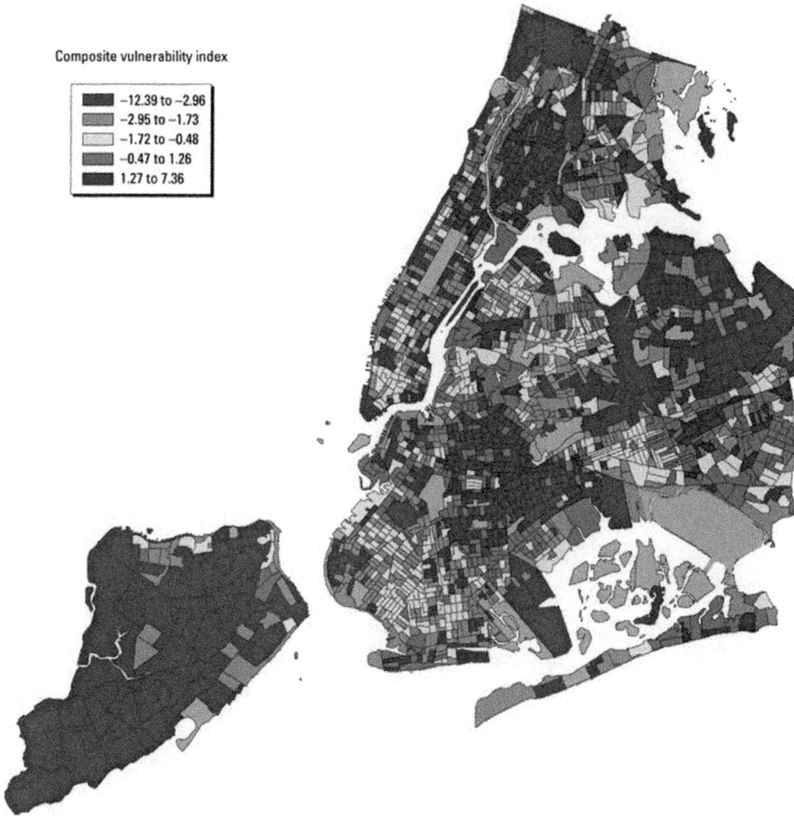


FIGURE 6.1 Heat Vulnerability Index for New York City.[48]

other entity, it tends to involve the rendering in layers of several data sets over a geographic map. Consider the example in Figure 6.2, which excerpts just two of the multiple maps collated by the Minnesota Department of Public Health in the appendices of its *Extreme Heat Toolkit*.

[48] Jaime Madrigano et al., *A Case-Only Study of Vulnerability to Heat Wave–Related Mortality in New York City (2000–2011)*, 123 Envtl. Health Persp. 672, 675 fig.1 (July 2015), https://perma.cc/8D9A-LNT7 ("NYC census tracts according to composite heat vulnerability index. The index is composed of z-scores of the following variables: (+) proportion of homes receiving public assistance, (+) proportion of non-Hispanic black residents, (+) proportion of overall deaths occurring in the home, (+) relative surface temperature, (–) proportion of trees. A higher composite index score indicates a residential area with a higher risk of heat-related mortality."). The New York City Department of Health and Mental Hygiene based its own HVI on this one. New York City, *Cool Neighborhoods NYC A Comprehensive Approach to Keep Communities Safe in Extreme Heat* 9 fig.3 (2017) https://perma.cc/A3NZ-MR8D.

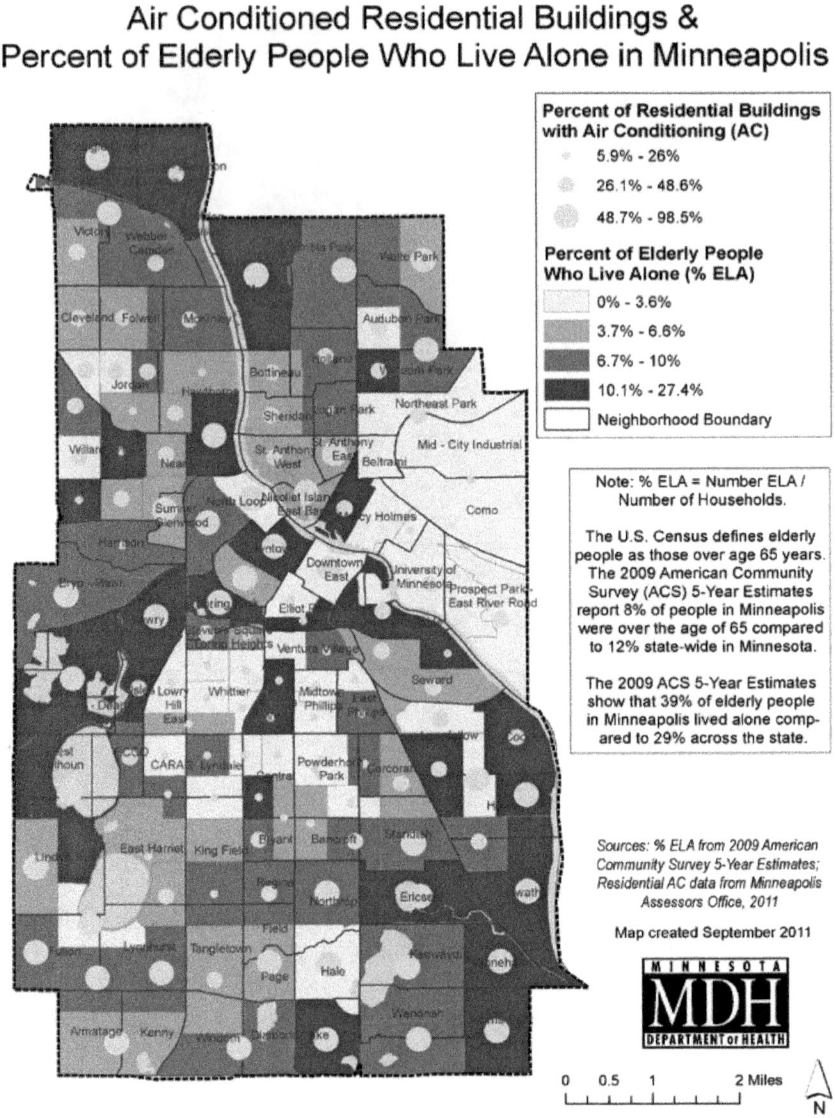


FIGURE 6.2 Map layers showing county-level vulnerability and adaptive capacity to heat.[49]

This pair of maps contains layers of information about the geographic density of a subpopulation that is especially vulnerable to extreme heat

[49] Minnesota Department of Health, *Extreme Heat Toolkit, Appendix H*, H-5 & H-8 (June 2012) https://perma.cc/XM4X-2AQE.

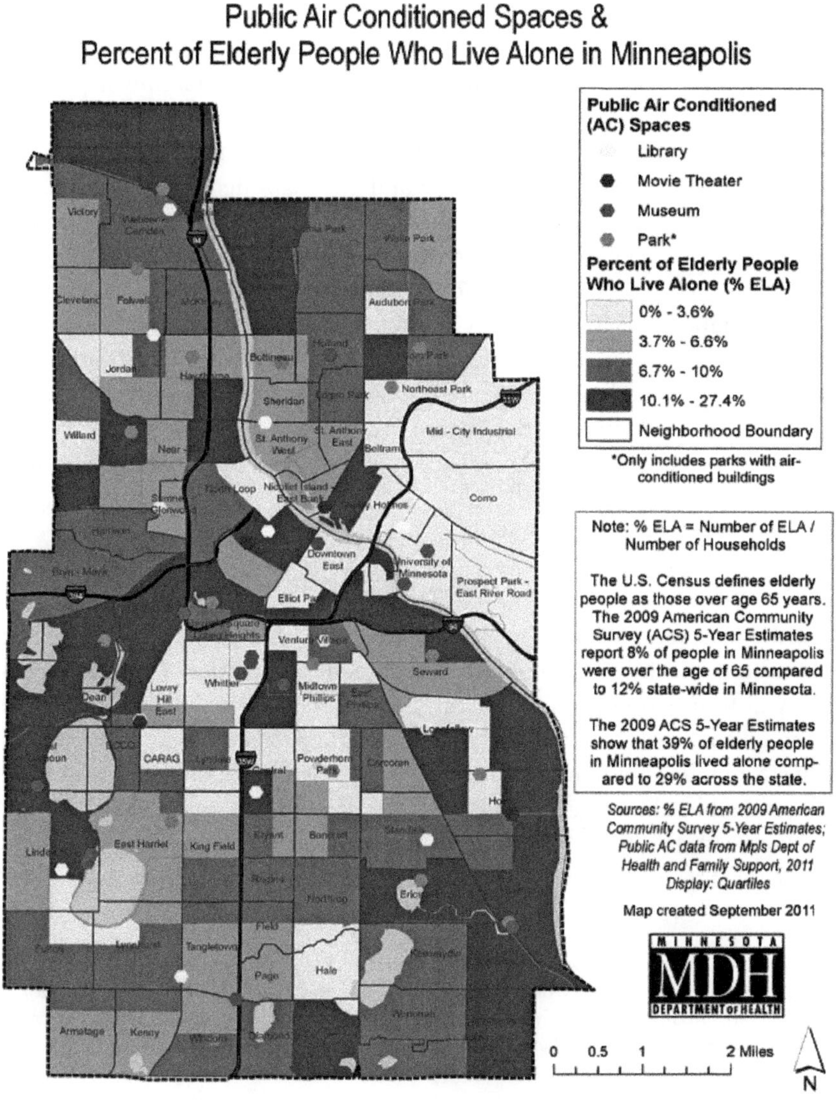


FIGURE 6.2 (cont.)

(elderly who live alone), average access to residential air conditioning across counties, and the location and availability of air conditioned public spaces in those counties. Integrating data like these with data sets that capture relative temperature levels as well as factors correlated positively or negatively with vulnerability (e.g. density of residents receiving public housing assistance and density of tree cover) makes it possible to derive and map a Heat Vulnerability Index for a given jurisdiction. (See Figure 6.2.)

Another factor for possible inclusion in urban heat vulnerability maps is the presence of ventilation corridors – channels through a cityscape that, if present, can allow the wind's passage to diminish ambient heat.[50]

Like the maps shown in Figures 6.1 and 6.2, maps depicting flood-related vulnerabilities compile data on the hazard (i.e. water levels and wave action under different scenarios) and on the area's relevant features, including topography, toxics storage sites, and infrastructure such as bridges or electric grid substations. San Mateo County, California's *SLR Vulnerability Assessment*, for instance, includes "Asset Exposure Maps" that overlay the areas subject to flooding under different SLR scenarios with key components of the built environment.[51] (See Figure 6.3.)

Assets located in projected flood zones are thus readily identified as potentially vulnerable.

### 6.2.3.3 Specifying Objectives

In some instances, it can be relatively easy to specify feasible climate change adaptation objectives for the built environment and public health. For electricity distribution utilities, which are under a statutory obligation to provide a minimum level of service to a defined group of end-users, the obvious objective vis-à-vis public health is simply to meet those obligations in spite of foreseeable climate change-driven hazards and the vulnerabilities – for instance, substations not hardened against flooding[52] – arising from them. Similarly, the basic adaptation objective of a prison might be to provide

50 See X. M. Hu, and M. Xue, *Influence of Synoptic Sea-Breeze Fronts on the Urban Heat Island Intensity in Dallas–Fort Worth, Texas*, 144(4) Monthly Weather Rev. 1487–1507 (2016); M. Roth, Urban heat island, In: *Handbook of Environmental Fluid Dynamics vol. 2*, 2 (2013).

51 Cnty. of San Mateo, *Sea Level Rise Vulnerability Assessment – Draft Report, Appendix B* (Apr. 2017), https://perma.cc/3RUW-H45G (report without appendices), https://perma.cc/6ZXM-G5YW (Appendix B).

52 The experience of Superstorm Sandy, flooding from which led to the explosive destruction of the 13th Street substation in lower Manhattan, led, among other things, to a thorough vulnerability assessment and plans to harden all low-lying substations in New York City against flood risk. See Consolidated Edison, *2015 Capital Work Plan: Storm Hardening–East 13th St 138kV & 345 kV Substation*, https://perma.cc/63UV-D72S (accessed Aug. 13, 2017) ("Based on storm hardening evaluations, this project will raise the existing perimeter flood wall to elevation 18.2" to provide higher storm surge protection, relocate the grade level control room to a higher elevation on the second floor, replace the protection scheme of station equipment with new microprocessor relays and fiber optics communication in between relays, provide capability to raise transformer control cabinets during a flood event . . . ").

FIGURE 6.3 San Mateo County's Built Asset Exposure Map, showing San Francisco International Airport.[53]

53 Ibid. at App. B-4.

adequately for the health and wellbeing of its inmates in spite of the effects of heat waves on indoor temperatures.[54] However, for institutions with a wider array of competing duties, such as municipalities, specifying broadly acceptable and realistic objectives can present challenges.[55] Furthermore, where vulnerabilities are especially large relative to the capacity of the institution charged with their address – for instance, the government of a small coastal town[56] – setting out ambitious objectives can damage credibility.

### 6.2.3.4 Exploring Options and Planning

Whereas the foregoing steps do not necessarily require public engagement, exploring options for how to achieve adaptation objectives for the built environment and then planning how to do so means, inevitably, seeking the approval of a larger group of stakeholders.[57] Put another way, even if the previous steps are done in technocratic fashion, deciding what measures to

54 See *Cole* v. *Collier*, Case 4:14-cv-01698, at 40, 67 (S.D. Tex. July 19, 2017) (ordering prison to maintain indoor temperatures below heat index of 88°F despite testimony about high costs involved in providing air conditioning capacity adequate to the task, and citing Daniel W. E. Holt, Heat in U.S. Prisons and Jails: Corrections and the Challenge of Climate Change, Columbia Law School Sabin Center for Climate Change Law (Aug. 2015)).

55 The Long-Range Transportation Plan (LRTP) developed by Hampton Roads following a grim vulnerability assessment is illustrative. As the Georgetown Climate Center explains: "The LRTP concludes that climate impacts, specifically sea-level rise, might eventually require the relocation or rebuilding of regional roadways. The LRTP explains why it may be difficult to adopt transportation adaptation strategies due to financial constraints, and emphasizes that policy alternatives to adapt transportation infrastructure to the impacts posed by hurricanes and flooding are limited. The LRTP does not provide for specific, concrete measures that should be taken to safeguard infrastructure against rising sea levels." Georgetown Climate Center, *Adaptation Clearinghouse: Hampton Roads Climate Change Adaptation Project*, https://perma.cc/2R7T-X8R7 (last updated July 7, 2016); see also Hampton Roads Transportation Planning Organization, *2034 Long Range Transportation Plan* (June 2015).

56 See, e.g., Elisabeth M. Hamin et al., *Barriers to Municipal Climate Adaptation: Examples from Coastal Massachusetts' Smaller Cities and Towns*, 80 J. Am. Planning Ass'n 110–122 (Sept. 2014) ("The planners interviewed reported that barriers to adaptation actions tend to be interconnected; for example, the strength of private property interests often limits local political leadership on the issue. Without such leadership, it is difficult for planners to allocate time and/or money to adaptation activities.").

57 Several how-to adaptation guides encourage adaptation planners to consider the "STAPLEE" categories of factors and the stakeholders concerned with each of them: social, technical, administrative, political, legal, economic, and environmental. See South Florida Regional Planning Council, *Adaptation Action Areas: A Planning Guidebook for Florida's Local Governments Regional Climate Action Framework: Implementation Guide* 63 (2015), https://perma.cc/2H39-7WUC; NOAA, *Adapting to Climate Change: A Planning Guide for State Coastal Managers* 52–53 (2010), https://perma.cc/E4M2-M6Y7. The STAPLEE list was first developed for the purposes of disaster mitigation planning. *See* FEMA, *Developing the Mitigation Plan:*

take, how much should be spent on them, and how to pay for them involves weighing those measures against the competing priorities of other stakeholders or the public as a whole. This is so chiefly because the tools available to adapt (or to push others to adapt) the built environment take the form of laws and regulations that govern land use planning, infrastructure design, and building design and construction. All of these are highly significant to numerous and diverse stakeholders, and are subject to the purview of state or local government.

These tools lend themselves to "mainstreaming," that is "integrating climate adaptation into existing management plans (for example, hazard mitigation, ecosystem conservation, water management, public health, risk contingency, and energy)."[58] Researchers have found that "[m]ainstreaming prevents adaptation from becoming a solely environmental issue, reduces the risks of agenda sidelining and lowers administrative and implementation costs."[59] The alternative to mainstreaming would be to develop adaptation plans independent of land use or infrastructure planning processes, and then to implement those independent plans by altering the results of conventional planning processes. An example of mainstreaming helps to illustrate its strengths: The Boston Water and Sewer Commission (BWSC)'s three-year capital improvement programs for stormwater and wastewater have, since 2015, incorporated precipitation and sea-level parameters based on not just historical weather patterns but also on projections of more severe future precipitation and sea-level rise.[60] They will thus steer design and procurement decisions in ways that better adapt the city's stormwater and wastewater infrastructure to a changing climate, and will also *avoid* maladaptive investments that could put the city's drinking water or wastewater management systems – and thus public health – at risk. These three-year programs reflect

*Identifying Actions and Implementing Strategies* (2003), https://perma.cc/56PU-K5CS (listing STAPLEE factors in detail).

58 Rosina Bierbaum et al., Ch. 28: Adaptation, In: J. M. Melillo et al., *Climate Change Impacts in the United States: The Third National Climate Assessment* 670–706, 682 (2014); see also Ebinezer R. Florano, Mainstreaming integrated climate change adaptation and disaster risk reduction in local development plans in the Philippines, In: Walter Leal Filho, *Handbook of Climate Change Adaptation* 433, 435 tbl.1 (2015) (noting numerous points of convergence between adaptation and disaster mitigation policies and arguing for "mainstreaming" of adaptation by unifying both policy areas).

59 Linda Shi et al., *Global Patterns of Adaptation Planning, supra* note 39, at 337.

60 Boston Water and Sewer Commission, *Capital Improvement Program 2017–2019*, at 7 (Nov. 2016), https://perma.cc/2JV3-5S3P ("Critical elements of this Plan include … Assessment of the Commission's Service Goals and other factors affecting long-term planning including changing regulatory requirements, **climate change** and financial **conditions**") (emphasis added).

the parameters established by the climate change vulnerability assessment BWSC conducted from 2010 to 2015 as part of its most recent twenty-five-year capital asset plan.[61]

### 6.2.3.5 Implementing and Evaluating

Though implementation can mean various things in relation to different adaptation measures, in all cases it means the allocation of scarce resources to realize plans intended to respond to a climate-related vulnerability. Whether for small-scale programs like the deployment of temporary cooling centers, or larger-scale, transformational measures like the inland relocation of a coastal community, funds must be appropriated, decisions made, and work plans drafted and executed. Taking this step to respond to *future* risks can be difficult – harder, certainly, than assessing vulnerabilities. Unsurprisingly, the authors of the *Adaptation* chapter in the Third National Climate Assessment observed that the key barriers to effective adaptation efforts include a "lack of resources to begin and sustain adaptation efforts; fragmentation of decision-making; institutional constraints; lack of leadership; and divergent risk perceptions/cultures and values."[62] In instances where the scale of planned adaptation would be not just incremental but transformational, high absolute costs and uncertainty about costs and benefits rise to the top of the list of barriers.[63] Researchers and agencies alike advise that, to overcome at least some of these barriers, preference should be given to adaptation measures that are likely to yield co-benefits.[64] Investment in such measures will – the reasoning goes – be easier to justify and more likely to yield manifestly net-positive outcomes regardless of whether their adaptation benefits ever materialize.

Evaluation is no mere procedural formality: Because the future is uncertain, planners cannot safely assume that parameters relevant to their decisions – from

[61] See Charlie Jewell et al., *BWSC Climate Change Risk Assessment, Findings and Mitigation/Adaptation Strategies for Wastewater and Strom Drainage–NEWAE Conference Presentation* (Jan. 28, 2015), https://perma.cc/384G-R8WC.

[62] Bierbaum et al., *Ch. 28: Adaptation*, at 683–686.

[63] Ibid.; see also Klaus Eisenack et al., *Explaining and Overcoming Barriers to Climate Change Adaptation*, 4 Nature Climate Change 867, 868 (Oct. 2014).

[64] William H. Butler et al., *Low-Regrets Incrementalism: Land Use Planning Adaptation to Accelerating Sea Level Rise in Florida's Coastal Communities*, 36 J. of Planning Education and Research 319–332 (May 2016), https://doi.org/10.1177/0739456X16647161 (describing examples of adaptation plans that seek to support co-benefits); June J. Cheng and Peter Berry, *Health Co-Benefits and Risks of Public Health Adaptation Strategies to Climate Change: A Review of Current Literature*, 58 International Journal of Public Health 305–311 (Apr. 2013).

the rate of sea level rise to the range and seasonality of infectious disease vectors – will be consistent with multi-decadal projections. Thus, because effective adaptation efforts are necessarily "dynamic iterative learning processes,"[65] they should assume that more information about climate-driven hazards and vulnerabilities will be revealed over time, and that preliminary decisions should be evaluated as new information corrects earlier assumptions – whether about the environment or the measure's design.[66] Notably, however, while this "adaptive management" approach is widely considered best practice,[67] "[t]here is little literature evaluating the effectiveness of adaptation actions," and most efforts at evaluation "to date, have focused on the creation of process-based rather than outcome-based indicators."[68] Evaluations of hazard mitigation efforts not expressly oriented to climate change are more numerous and can serve as models for evaluating adaptation efforts, which often involve similar measures and goals.[69]

### 6.2.4 *Substance*

Hundreds of cities and smaller localities have engaged in some version of the process described in the previous subsections,[70] in some instances on their own initiative and in others steered or supported by state and/or federal law or policy.[71] Many of the resulting adaptation measures have been demonstrated

65 Mimura et al., *Adaptation Planning and Implementation*, *supra* note 33, at 871.

66 See Sierra C. Woodruff, *Planning for an Unknowable Future: Uncertainty in Climate Change Adaptation Planning*, 139 Climatic Change 445–459 (Dec. 2016), doi:10.1007/s10584-016-1822-y.

67 See, e.g., Elisabeth Hamin and Nicole Gurran, Climbing the adaptation planning ladder: Barriers and enablers in municipal planning, In: Walter Leal Filho, *Handbook of Climate Change Adaptation* 839, 843–44 (2015); Kristi L. Ebi, Overview: Adaptive management for the health risks of climate change, In: James D. Ford and Lea Berrang-Ford, *Climate Change Adaptation, in Developed Nations* 121–131 (2011).

68 Bierbaum et al., *Ch. 28: Adaptation*, at 682.

69 See, e.g., FEMA, *Reducing Losses through Higher Regulatory Standards: Best Practices and Cost-Effective Strategies Report* (2015), https://perma.cc/HT6S-5VJJ.

70 Two of the leading organizations devised to facilitate and coordinate cities' adaptation (and mitigation) efforts are ICLEI-Local Governments for Sustainability and the Compact of Mayors. See ICLEI-Local Governments for Sustainability USA, *Our Members*, https://perma.cc/2D3V-L9DU (listing 191 member localities and noting that ICLEI's international membership includes over 1000 localities in 86 countries); Compact of Mayors, *Cities Committed to the Compact of Mayors* (indicating 142 members located in the United States and 668 in total committed to a program of self-reported goals, vulnerabilities, plans, and actions taken).

71 For examples of state laws that have prompted and guided local adaptation planning, see, e.g., CRRA (NY); Peril of Flood (Fla). But see also Thomas Ruppert and Alexander Stewart, *Summary and Commentary on Sea Level Rise Adaptation Language in Florida Local*

to reduce vulnerability to climate change-driven impacts.[72] Some do so by reducing exposure to hazards, for instance by retreating from a coastline or elevating coastal structures above the level of floodwaters expected during the structures' life. Others do so by reducing the hazards themselves, for instance by installing large amounts of urban green infrastructure to mitigate the urban heat island effect. And still others do so by increasing resilience, for instance by enlarging the capacity of stormwater management systems in anticipation of downpours that would otherwise leave streets temporarily flooded.

To explore these points with respect to the built environment, this subpart returns to the hazards described in Section 6.2.1 and considers examples of adaptation measures – ranging from the immediate to the long-term, the local to the regional, and the incremental to the transformational – that are responsive to them.

#### 6.2.4.1 Heat, Especially in Cities

Adaptive responses to heat range widely and are not reducible to as unified and straightforward a framework as adaptations to coastal flooding. As described in this section, they cover the full ranges of geographic scope, timeframe, degree of change, and other dimensions of adaptation measures listed in Section 6.2.1 of this chapter.

Measures taken to cope directly with individuals' vulnerability over short timeframes include providing warnings on hot days, establishing and publicizing the presence of temporary cooling centers where vulnerable populations live or work, and creating community support networks like the "Be A Buddy" component of New York City's *Cool Neighborhoods* initiative, which links community organizations with vulnerable individuals and facilitates phone or in-person check-ins on hot days.[73] These measures generally do not alter the built environment, but instead identify and correct for instances where permanent features of the built environment fail to reduce a sensitive population's exposure to the hazard of heat – or exacerbate that exposure.[74]

*Government Comprehensive Plans and Ordinances*, 4 (July 2015), http://perma.cc/7VU6-ZGF4 (noting that most ordinances' language is not self-executing and that this "may result in situations in which comprehensive plan language appears more proactive than the tangible actions of a local government in day-to-day operations.")

72 H. Anderson et al., CDC, *Climate and Health Intervention Assessment: Evidence on Public Health Interventions to Prevent the Negative Health Effects of Climate Change* (2017), https://perma.cc/P88P-E62E.

73 New York City, *Cool Neighborhoods*, *supra* note 48, at 23–25.

74 See notes 334–338, and accompanying text, which discusses programs in Minnesota and New York City.

Other measures reduce individuals' vulnerability indirectly by making changes to design specifications and maintenance protocols for electricity infrastructure and transit networks.[75] These changes generally modify the composition and operation of elements of the built environment – a bit like a house that remains adjacent to the shore but is raised on stilts in anticipation of flooding. They can also, however, be more transformative, for instance by relocating system components or reformulating the system's physical and operational features to support different capabilities, like the "islanding" of segments of an electricity distribution grid when transmission lines are inoperable. The geographic scope of these changes generally reflects the jurisdiction of the regulatory authority requiring and/or authorizing them.[76]

Finally, some measures seek to reduce the hazard itself by changing the materials and morphology of a cityscape. These include increasing the albedo of roofs, pavements, and walls;[77] replacing impervious surfaces with green infrastructure to increase both albedo and (under some circumstances) cooling action through evapotranspiration;[78] and preserving or creating wind corridors.[79] "Cool roof" programs are quick, low-cost examples of such measures that alter almost nothing about the built environment but achieve measurable temperature reductions on surfaces and within buildings covered by those surfaces.[80] Replacing impervious surfaces with green infrastructure, such as rain

75 U.S. Department of Energy, *Climate Change and the Electricity Sector: Guide for Climate Change Resilience Planning*, 84 (Sept. 2016), https://perma.cc/63KM-R9GK (listing measures recommended by AVANGRID to improve resilience to heat waves); North Jersey Transportation Planning Authority, Climate Change Risk Assessment of New Jersey's Transportation Infrastructure 87 (2011) (listing heat ratings for rail system components and indicating risks of exceedance).

76 See, e.g., Emerald Coast Utilities Authority, *Wastewater Services*, https://perma.cc/AD8D-VGWN (describing relocation and redesign of water reclamation facility from near coastal flood zone to higher ground in Escambia Cnty., Florida, following Hurricane Ivan).

77 See, e.g., Haley Gilbert et al., *Keeping California cool: Recent cool community developments*, 114 Energy and Buildings 20–26 (Feb. 2016) (describing school districts' and cities' cool pavements and cool roofs pilot programs and preliminary outcomes).

78 Kieron Doick and Tony Hutchings, UK Forestry Commission, *Research Note: Air Temperature Regulation by Urban Trees and Green Infrastructure*, 3 (2013), https://perma.cc/7T93-WQBC (describing mechanisms of cooling, including evaporative cooling and evapostranspiration).

79 See Leyre Echevarría Icaza and Franklin van der Hoeven, *Regionalist Principles to Reduce the Urban Heat Island Effect*, 9 Sustainability 677, 679 (2017) (listing "creation of cool wind corridors" among measures used to mitigate UHI effect); Hu and Xue, Influence of Synoptic Sea-Breeze Fronts,1487–1507.

80 Kevin Krajick, *New York Roofs: Brighter, Whiter, Cooler*, Earth Institute: State of the Planet (Mar. 7, 2012), (quoting physicist Stuart Gaffin regarding the use of white acrylic paint for cool roofing: "It's the lowest hanging fruit. It's very cheap to do; it's a retro-fit. You don't need a skilled labor force. And you don't have to wait for a roof to be retired."); see also Stuart

gardens and extensive (shallow) or intensive (deep) green roofs, is more involved than cool roofing and has a more complicated relationship to adaptation to the heat hazard. First of all, while some tree canopies clearly reduce air temperatures in their vicinity by raising albedo and performing evapotranspiration, small patches of greenery are less certain to do so,[81] which means that a city seeking to abate its urban heat island cannot expect success to follow from merely accumulating fragmented and disparate patches of low-cost green infrastructure. Furthermore, while green infrastructure has several justifications linked to public health, because its primary justification is often stormwater management, its design and placement tends to be oriented to detaining stormwater rather than cooling the ambient air. As for the preservation or creation of urban wind corridors, this is an especially clear example of transforming a cityscape on a permanent basis for the sake of adaptation to heat hazards.

#### 6.2.4.2 Coastal Flooding

Adaptive responses to encroaching seas and more intense coastal storms take one (or a combination) of three forms: protection, accommodation, or retreat.[82] Protection means interposing barriers – whether "hard armoring," such as sea walls, or "soft armoring," such as living shorelines[83] – between rising seas and landward assets and people with the goal of preserving existing patterns of development and activity. Accommodation means staying in the same place – "living with water"[84] – but changing local land uses and building and infrastructure design in ways that reduce vulnerability and improve

R. Gaffin et al., *Bright is the New Black: Multi-Year Performance of High-Albedo Roofs in an Urban Climate*, 7 Environmental Research Letters 014029 (2012).

81 Kieron Doick and Tony Hutchings, UK Forestry Commission, *Research Note: Air Temperature Regulation by Urban Trees and Green Infrastructure*, 4 (2013), https://perma.cc/7T93-WQBC (describing scalar factors important to effectiveness of greenery for cooling).

82 John R. Nolon, *Protecting the Environment Through Land Use Law: Standing Ground*, 221 (2014).

83 Robert Verchick and Joel Scheraga, Protecting the coast, In: Michael B. Gerrard and Katrina Kuh, *The Law of Adaptation to Climate Change: United States and International Aspects* 18–19 (2012). Hard armoring tends to create expensive problems over time; soft armoring is generally favored by scientists, planners, and civil engineers, but is usually feasible only where development (i.e. asphalt, concrete foundations, structures, and infrastructure) can be displaced or has not encroached too close to the water's edge. Gary B. Griggs, The effects of armoring Shorelines – The California experience, In: Hugh Shipman et al., *Puget Sound Shorelines and the Impacts of Armoring – Proceedings of a State of the Science Workshop, May 2009* (2010), https://perma.cc/FN54-7425.

84 C.P. Veerman and M.J. Stive, *Working Together with Water: A Living Land Builds for its Future – Findings of the Deltacommissie 2008*, 61 (2008), https://perma.cc/9TJH-A7Y2 (describing the Dutch Leven met Water research program).

resiliency to flooding. Concretely, this could include compelling real estate sellers to disclose vulnerability to flooding, elevating mechanical or electrical components within buildings, elevating whole structures, or uprating machinery to endure inundation by saltwater.[85] Partial or full retreat, which involves abandoning land and assets made vulnerable by rising seas, is only simple conceptually; its planning and implementation are legally and politically complex.[86] Consider Miami Beach, where seawalls cannot prevent saltwater from flooding city streets by infiltrating via the porous bedrock.[87] Though retreat – managed or otherwise – appears inevitable, the City Engineer recently remarked: "When somebody says, 'How much are you willing to fall back?' I say, 'Not one inch.' We are defending this city at the shoreline. Miami Beach is only one mile wide. If we drop back a mile, we don't have a city."[88]

Each of these approaches can take smaller-scale "incremental" forms that aim to cope with hazards, or larger "transformational" forms that aim to greatly reduce or even eliminate vulnerabilities. Protection: in contrast to a modest seawall or living shoreline seaward of one parcel of property, the State of Louisiana's Comprehensive Master Plan envisions wetland and barrier island restorations on a scale large enough to stem – and ideally offset – the rapid subsidence and erosion of the state's entire coastline.[89] Accommodation: like the elevation of a beach house, only

85 See, e.g., City of New York Department of City Planning, *Coastal Climate Resilience: Designing for Flood Risk*, 16–17 (June 2013) https://perma.cc/7VWS-BLFL.

86 See C. Kousky, *Managing Shoreline Retreat: A US Perspective*, 124 Climatic Change 9, 9 (2014), https://perma.cc/5MY2-NVP3 (discussing institutional factors that impede managed retreat by whole communities in the United States). The Quinault Tribe of Washington State is an exception that proves the rule. Their cohesive and homogeneous community will not repair the seawall that is currently losing its battle to protect Taholah village from a rising Pacific Ocean. Instead, the tribe, using federal funds, will move the 700-person village to higher ground at a cost of about $350,000 per person. NOAA, U.S. Climate Resilience Toolkit, *Case Studies: Quinault Indian Nation Plans for Village Relocation*, https://perma.cc/3PC4-79B3; see also Quinault Indian Nation, *Taholah Village Relocation Master Plan*, https://perma.cc/3LA8-63UZ.

87 Stan Cox and Paul Cox, *A Rising Tide: Miami is Sinking Beneath the Sea – But Not Without a Fight, The New Republic*, Nov. 8, 2015, http://bit.ly/21pwWBD. The city is planning to spend several hundred million dollars on a system of elevated streets and electrically powered pumps, but even these efforts at accommodation struggle to contain and expel seawater faster than it flows in. Joey Flechas et al., *Emily Rain Pounds South Florida and Beach Pumps Fail Without Power*, Miami Herald, Aug. 1, 2017, https://perma.cc/25YH-JNYX.

88 Pam Radtke Russel, *Special Report: How Engineers are Preparing for Sea-Level Rise: From Seattle to Cape Cod, See What's Being Done at 18 Different Locations*, Engineering News Record, Aug. 11, 2017, https://perma.cc/7JXL-X3VU.

89 Ibid. at 94–95, 100–102.

on the scale of citywide infrastructure, Miami Beach has raised many of its streets and sidewalks by several feet, such that floodwaters inundate them less severely and less often.[90] Retreat: this approach is inherently transformational, but it can be undertaken in a disorganized fashion, as happened in the aftermath of Hurricane Katrina,[91] or in coordinated fashion, as has happened with state support and guidance on Staten Island and with federal support in Isle de Jean Charles, Louisiana, and Taholah Village, Washington.[92]

Among the policy tools used to effectuate these categories of coastal adaptation are:

***Land Use Rules and Restrictions***
* Conditional development;
* Conservation easements;
* Floodplain regulations;
* Hard- and soft-armoring permits;
* Land Trusts;
* Real estate disclosures;
* Rebuilding restrictions;
* Setbacks and buffers;
* Transferable development rights;
* Zoning and overlay zones.

Design Prescriptions
* Building codes;
* Infrastructure design parameters.

Fees and Financing
* Coastal land acquisition programs;
* Flood insurance requirements;
* Impact fees;
* Level of service downgrades;
* Special assessments;
* Stormwater utility.

[90] Joey Flechas, *Miami Beach to Begin New $100 Million Flood Prevention Project in Face of Sea Level Rise*, Miami Herald, Jan. 28, 2017, https://perma.cc/H9FB-LLG6.

[91] Elizabeth Maly et al., Experience from the United States: Post-Katrina and Sandy, In: *Land Use Management in Disaster Risk Reduction* 79, 86–89 (Dec. 2016).

[92] Ibid. at 100; State of Louisiana, *Isle de Jean Charles Resettlement Project*, https://perma.cc/EG6T-VLVB; U.S. Climate Resilience Toolkit, *Quinault Indian Nation Plans for Village Relocation*, https://perma.cc/V58H-TP33.

Each of these is described in greater depth elsewhere.[93] For this chapter's purposes, the key point to take from the list is that the tools for adapting to coastal flooding are the same as conventional planning in coastal communities. Importantly, public health seldom factors directly into the use of these tools, which draw instead on inputs relating to land values, engineering specifications, and aesthetic preferences.

## 6.3 LEGAL ISSUES

The preceding section discusses policies and measures responsive to particular climate-driven impacts, but leaves discussion of legal requirements and constraints that those responses are sure to encounter for this section. This section's discussion is not comprehensive with respect to legal issues arising from efforts to adapt – it does not, for instance, discuss disputes among private actors, nor conflicts and entanglements arising from government-led adaptation efforts intersecting with legal obligations imposed by federal laws, such as the Clean Water Act's requirements for stormwater management. Instead, it focuses on situations in which state or local governments run the risk of litigation brought by private actors. One such situation arises when governments' efforts to reduce adverse public health impacts by adapting land uses and the built environment to climate change cause some individuals to arguably be harmed economically by decreased property values or limitations on the use of their property. In such cases, people may attempt to sue the regulating entity to compensate them for their loss, or else to try to stop the project altogether. Another situation arises when a lack of action by governments to adapt to climate change impacts arguably causes injuries or property damage that could have been prevented by improved infrastructure or other measures. Thus governments can expose themselves to legal risk whether they take a passive or active role in adapting the built environment to climate change.

93 See, e.g., South Florida Regional Planning Council, *Adaptation Action Areas: A Planning Guidebook for Florida's Local Governments Regional Climate Action Framework: Implementation Guide* 63 (2015), https://perma.cc/2H39-7WUC; Anne Siders, *Managed Coastal Retreat: A Legal Handbook on Shifting Development Away from Vulnerable Areas* (Oct. 2013), https://perma.cc/Z5A2-ALQB (providing illustrative examples). For an example of how an ordinance can introduce downgrades to the level of service for a branch of networked infrastructure, see Thomas Ruppert et al., *Environmentally Compromised Road Segments – A Model Ordinance*, at para. 1, https://perma.cc/3RLM-DY7K ("any road categorized as 'environmentally compromised' under this ordinance shall be the subject of a requested design/maintenance exception").

This section explores two types of claims litigants may bring against governments either for adapting to climate change or for failing to do so. First, it considers a "takings" claim, also known as an inverse condemnation claim, which may be brought by a property owner against a government where the government's actions diminish the property's value. Such claims can be expected when governments make decisions related to coastal infrastructure or land use in particular.[94] Next, it considers a negligence claim, which alleges that an entity or individual's actions fell below a reasonable standard of care, and therefore harmed the individual bringing the lawsuit. These are not the only legal theories that could be brought by parties seeking to challenge action or inaction related to climate change or its impacts, but they are among the most readily available to parties vying with government defendants. They thus provide a useful illustration of the legal risks governments face and the limits to redress available to plaintiffs in the context of a changing climate.

### 6.3.1 *Takings Claims*

The Takings Clause of the Fifth Amendment of the United States Constitution provides that "private property [may not] be taken for public use, without just compensation."[95] The Fourteenth Amendment extends this prohibition to states and municipalities.[96] Many state constitutions also contain a takings clause closely mirroring the federal clause[97] or providing broader protections by prohibiting the government from merely damaging private property without compensation.[98]

In the most clear-cut cases, a private property owner may sue the government for physically occupying his property or for completely depriving him of the economically beneficial use of his property.[99] Such cases are known as *per se* takings and, if proven, require the government to pay the owner "just compensation" for the deprivation.

94 See, e.g., *Jordan v. St. Johns Cnty.*, 63 So. 3d 835 (2011).

95 U.S. Const. amend. V.

96 *Chicago, Burlington & Quincy Railroad Co. v. City of Chicago*, 166 U.S. 226 (1897).

97 See, e.g., Conn. Const. art. I, § 11 ("The property of no person shall be taken for public use, without just compensation therefor"); Wis. Const. art. I, § 13 (same).

98 See, e.g., Alaska Const. art. I, § 18 ("Private property shall not be taken or damaged for public use without just compensation."); Ariz. Const. art. II, § 17 ("No private property shall be taken or damaged for public or private use without just compensation having first been made . . . "); Ark. Const. art. II, § 22 (" . . . private property shall not be taken, appropriated or damaged for public use, without just compensation therefor").

99 *Loretto v. Teleprompter Manhattan CATV Corp.*, 458 U.S. 419 (1982); *Lucas* v. *South Carolina Coastal Council*, 505 U.S. 1003 (1992).

In contrast, governments often institute regulations that decrease the range of uses available to a property owner, and possibly decrease property values, but do not completely deprive the owner of the use of his property. For example, in response to encroaching sea levels, a local government might impose a setback requirement, such that property owners cannot build on the portions of their plot closest to the shore. If an owner sued the local government for imposing the setback requirement,[100] his "regulatory takings" challenge would be governed by the multi-factor test announced in *Penn Central Transportation Co.* v. *New York City*.[101] The *Penn Central* test requires courts to inquire as to (1) the economic impact of the regulation on the property owner, (2) the character of the governmental action, and (3) the extent to which the regulation has interfered with the economic expectations of the property owner.[102] In other words, courts determining the merit of a regulatory takings claim must balance the interests of the parties, with a focus on the magnitude and character of the burden a particular regulation imposes upon private property rights.

### 6.3.1.1 Claims against Governments in Response to Adaptation Plans and Policies

Several cases demonstrate the challenges that property owners face when bringing takings claims in response to climate change adaptation policies adopted by the federal or local government. In one such case, the Supreme Court of South Carolina rejected a takings claim brought by a property owner who had been prohibited from developing since its land was located within a floodplain.[103] This prohibition is an example of an adaptation measure designed to reduce exposure to hazards by decreasing the amount of development within an area likely to experience flooding. The developer, Columbia Venture, LLC, purchased more than 4,000 acres of land along a riverbank. At the time of purchase, the Federal Emergency Management Agency (FEMA) was in the process of designating most of the area a regulatory floodway. The South Carolina county in which Columbia Venture's property was located adopted the FEMA maps, and, through an existing local ordinance,

[100] See, e.g., A *Piece of Paradise v. Borough of Fenwick Zoning Bd. of Appeals*. No. LNDCV136047679S, 2015 WL 10285888 (Conn. Super. Ct. Dec. 23, 2015).

[101] *Penn Central Transportation Company v. City of New York*, 438 U.S. 104 (1978).

[102] *Connolly v. Pension Ben. Guar. Corp.*, 475 U.S. 211, 212 (1986) (citing Penn Central, 438 U.S. at 124).

[103] *Columbia Venture, LLC v. Richland Cty.*, 413 S.C. 423, 431, 776 S.E.2d 900, 904 (2015), reh'g denied (Oct. 9, 2015), cert. denied, 136 S. Ct. 1458 (2016).

restricted construction in the floodway. The County's land-use standards were more restrictive than those required by the federal government in recognition of FEMA's failure to project the likelihood of increased flooding due to climate change.[104]

A Special Referee was appointed to consider Columbia Venture's regulatory taking claim. Applying the *Penn Central* factors, the Special Referee concluded that FEMA's designation of Columbia Venture's property as a regulatory floodway caused a significant decrease in the property's value.[105] This factor was outweighed, however, by the fact that Columbia Venture's expectation of being able to develop the property was unreasonable in light of the foreseeable potential regulatory bar on floodplain development. Moreover, the county's floodplain regulations served an important purpose – flood protection.

The South Carolina Supreme Court affirmed the Special Referee's decision that the county's floodway development restrictions did not constitute a taking. With respect to Columbia Venture's investment-backed expectations, the court held that the developer faced an "uncertain path forward" in light of the pending FEMA flood maps and associated county regulations. The county was a long-time participant in federal flood planning programs, and the developer was a sophisticated party with notice of the county's floodplain development restrictions, as well as the pendency of FEMA's revised floodway designation. The planned development was, therefore "purely speculative in nature," and the developer's expectation of being able to successfully pursue the development was unreasonable. The Supreme Court also acknowledged "the important public purposes of mitigating the social and economic costs of flooding [] served by the County's ordinances."[106] Thus, the County was not responsible for paying the developer "just compensation," despite the impact of its floodplain regulations on the value of the property.[107]

As the court recognized in *Columbia Venture*, there is no "magic formula" for applying the *Penn Central* factors in determining whether a government regulation causing interference with property is a taking.[108] In light of the

104 Ibid. at 431, n.4. 105 Ibid. at 441. 106 Ibid. at 452.

107 A Connecticut court reached a similar result in *A Piece of Paradise* v. *Borough of Fenwick Zoning Board of Appeals*. No. LNDCV136047679S, 2015 WL 10285888 (Conn. Super. Ct. Dec. 23, 2015). In that case, the plaintiff owned a parcel of land on which it planned to build a single family home. The lot was located within a coastal boundary, which required the municipality to undertake a coastal site plan review. The lot owner applied for a variance from a fifty-foot setback requirement imposed by the local government, and the board denied the plaintiff's variance request. Ibid. at *4. The court held that the board's denial did not constitute a taking without just compensation. Ibid. at *8.

108 Columbia Venture, 413 S.C. 423, at 448.

"nearly infinite variety of ways in which government actions or regulations can affect property interests," climate change adaptation policies' vulnerability to a takings claim will depend on the circumstances of the particular landowner's case. One conclusion that can be gleaned from *Columbia Venture*, however, is that governments choosing to pursue climate change adaptation policies should do so as soon as possible. As a report prepared by the University of Maine's Marine Law Institute observed in the context of a coastal retreat policy proposal, "[t]he earlier that the public is on notice of the ... policy choice ..., the more likely the regulations are to withstand legal challenge. Property that is purchased after the regulations are adopted will be bought subject to the expectations that the development restrictions will be applied ... "[109]

#### 6.3.1.2 Claims against Governments for Failure to Adapt the Built Environment

Property owners may also initiate litigation against the government for taking private property where the government fails to prevent the impacts of climate change, thereby indirectly causing damage to the owner's property.[110]

For example: Starting in the 1950s, the United States Army Corps of Engineers (the Corps) built the Mississippi River Gulf Outlet (MRGO), a 76-mile channel between New Orleans and the Gulf of Mexico intended to provide a shorter shipping route.[111] Although MRGO was designed to be 500 feet wide, decades of use eroded the channel to more than triple its design width.[112] MRGO's increased width allowed the channel to carry a much greater volume of water, exposed a greater surface area to wind causing more severe waves, and carried saltwater inland, destroying buffer wetlands. When Hurricane Katrina hit New Orleans in 2005, several levees and storm walls surrounding MRGO were destroyed, and the city was devastated by the resultant flooding.[113]

In the aftermath of the hurricane, hundreds of plaintiffs sued the United States government to recover damages for flooded property. Although a series

[109] U.S. EPA, Office of Policy, Planning, & Evaluation, *Anticipatory Planning For Seal-Level Rise Along the Coast of Maine* (1995).

[110] For an extremely informative discussion of government liability under the Takings Clause for failure to act to protect private property, see Christopher Serkin, *Passive Takings: The State's Affirmative Duty to Protect Property*, 113 Mich. L. Rev. 345 (2014).

[111] Michael Gerrard, *Hurricane Katrina Decision Highlights Liability for Decaying Infrastructure*, N.Y. L. J. (May 10, 2012).

[112] Ibid. [113] Ibid.

of cases brought under the Federal Tort Claims Act (FTCA) were unsuccessful because the government was found to be immune from suit,[114] a more recent decision in a case brought under the Takings Clause found in favor of the property owners.

In *Saint Bernard Parish Government* v. *United States*,[115] Judge Susan Braden of the United States Court of Federal Claims found that the Corps' negligent design and failure to maintain MRGO exacerbated flood damage in parts of New Orleans. The increased flooding, although temporary, wrongfully deprived landowners of the use of their property, requiring compensation. Judge Braden's decision relied heavily on a 2012 Supreme Court case, *Arkansas Game and Fish Commission* v. *United States*, which held that temporary flooding caused by government action is not categorically exempt from Takings Clause liability.[116] The plaintiffs in *Saint Bernard Parish* avoided the sovereign immunity issues that prevented the FTCA litigants from recovering damages, because the United States has waived sovereign immunity for claims brought under the Takings Clause through the Tucker Act.[117]

That the government was held liable for inadequately preparing federally constructed and maintained infrastructure for severe weather events in *Saint Bernard Parish* is significant in light of the increasing risk of such events due to climate change. Notably, *Saint Bernard Parish*, if it survives appeal, expands government liability from situations in which the government deliberately causes flooding, for example by releasing water from a dam, to include situations in which inaction by the government exacerbates flooding from severe weather through its failure to properly design or maintain federally owned infrastructure.[118] Professor Christopher Serkin has argued that the

[114] In those cases, the Fifth Circuit Court of Appeals at first affirmed the trial court's finding of liability, but then issued a subsequent ruling finding that the government was immune from the plaintiffs' claims, because its actions in connection with the design and maintenance of MRGO were largely discretionary. In *re Katrina Canal Breaches Litigation*, 673 F.3d 381 (5th Cir. 2012) (finding the federal government liable for Hurricane Katrina flood damage caused by the Corps' failure to armor the banks of MRGO); In *re Katrina Canal Breaches Litigation*, 696 F.3d 436 (5th Cir. 2012) (reversing the Circuit Court's prior decision and finding the federal government immune from suit under the discretionary function exception to the FTCA).

[115] *St. Bernard Parish Gov't v. U.S.*, 121 Fed. Cl. 687 (2015). The case has been appealed to the Federal Circuit Court for Federal Claims. Notice of Docketing, *St. Bernard Parish v. U.S.*, Case 16-2301 (Fed. Cir. July 6, 2016). As of August 2017, that proceeding is ongoing.

[116] *Arkansas Game & Fish Comm'n* v. *United States*, 133 S. Ct. 511, 515 (2012).

[117] 28 U.S.C.A. § 1491 (2012) (granting jurisdiction to the Court of Federal Claims for claims for "damages in cases not sounding in tort").

[118] Notably, *Saint Bernard Parish* is not the first case to support the idea that a government can commit a taking through the failure to act. An earlier case in Florida state court held that "government inaction – in the face of an affirmative duty to act – can support a claim for inverse condemnation." *Jordan* v. *St. Johns Cty.*, 63 So. 3d 835, 839 (Fla. Dist. Ct. App. 2011);

Takings Clause can serve as a basis for affirmative governmental obligations to protect private property more generally and cites sea level rise as "an ideal illustration" of an environmental shift that could require governments to act.[119] While governments obviously do not have an obligation to protect all property from all intrusions, a duty arises where the state exercises regulatory control over the injury-causing condition or where the state is complicit in creating the conditions responsible for harm to the property.[120] This developing area of law will have broad implications for state governments seeking to prepare for – or deliberately deciding not to prepare for – climate change impacts.

Some legal scholars have expressed concern that *Saint Bernard Parish* and *Arkansas Game and Fish Commission* allow the takings doctrine to improperly invade the traditional domain of tort law.[121] The apparent expansion of takings liability ushered in by these cases, however, will likely be tempered by the fact-specific analysis required in cases asserting claims for temporary takings due to flooding or other natural disasters. As Judge Braden explained, a plaintiff asserting a claim for a temporary taking must establish: (1) a protectable property interest under state law; (2) the character of the property and the owners' "reasonable-investment backed expectations"; (3) foreseeability; (4) causation; and (5) substantiality.

The first prong merely requires the plaintiff to show that he has an ownership interest in the property allegedly taken through government inaction.[122] With respect to the owners' reasonable investment-backed expectations, as discussed in the previous section, courts inquire whether the plaintiff was aware of the risks facing his property. In *Saint Bernard Parish*, the court concluded that "although Plaintiffs' properties were in a floodplain and 'had experienced flooding in the

but see *Harris Cty. Flood Control Dist.* v. *Kerr*, 499 S.W.3d 793, 799 (Tex. 2016), *reh'g denied* (Oct. 21, 2016) (rejecting landowners' takings claim where the defendant county allegedly caused their homes to flood by failing to fully implement a stormwater management plan and by approving private development upstream).

119 Serkin, *Passive Takings*, *supra* note 110, at 388.

120 Ibid. at 377–378 (noting that "if the government . . . were responsible for global warming then the duty to act would be stronger still"); see also *City of St. Petersburg v. Collom*, 419 So. 2d 1082, 1086 (Fla. 1982) (once a governmental entity creates a known dangerous condition that may not be readily apparent to one who could be injured by the condition . . . the governmental entity must take steps to avert the danger or warn [of] that danger); *Teall v. City of Cudahy*, 386 P.2d 493 (Cal. 1963) (defendant city may be held liable if it created a dangerous or defective condition); *Delarosa v. State*, 21 Ariz. App. 263, 265, 518 P.2d 582, 584 (1974) (state not entitled to notice of a dangerous condition that it has created or caused to be created).

121 John Echeverria, *Ruling in MR-GO Takings Lawsuit*, Takings Litigation: A Blog About Takings Law, May 1, 2015, https://perma.cc/U9ZM-NKP9.

122 See *St. Bernard Parish*, 121 Fed. Cl. 687, at 719.

past,' that flooding was not 'comparable' to the flooding during Hurricane Katrina and subsequent hurricanes and severe storms giving rise to the temporary takings claim at issue."[123] Professor Serkin has posited that "an 'ecological change' can interfere with owners' expectations just as much as an explicit legal transition."[124] As climate change causes increasingly severe natural disasters, more courts could find that a property owner's past experience with severe weather does not adequately put him on notice of future risk.

The foreseeability prong of a temporary takings claim invites an inquiry into "the degree to which the [government's] invasion is . . . the foreseeable result of government action."[125] In *Saint Bernard Parish*, the court found that it was foreseeable that MRGO would intensify flooding in New Orleans based on a variety of environmental factors, including increased erosion on MRGO's banks and increased storm surge.[126] The plaintiffs were also able to establish a causal connection between the Corps' failure to maintain MRGO and flooding during Hurricane Katrina, since the Corps' inaction was the cause of the erosion, increased storm surge, and other exacerbating factors.[127] As discussed in the previous section, developing adaptation measures requires that governments assess vulnerabilities. As governments compile data and develop models projecting future vulnerabilities, potential impacts due to government action or inaction may become increasingly foreseeable.

Finally, a plaintiff alleging a temporary taking must show a sufficiently severe economic impact on his property to constitute a legally cognizable interference.[128] Moreover, the plaintiff must establish that the government's inaction caused the diminished property value; in other words, the government must have had the ability to protect the property at issue.[129] In *Saint Bernard Parish*, the Plaintiffs established this element by showing that their properties were flooded because of the Corps' negligent maintenance of MRGO and that they lost their ability to access or use their properties for a "significant" time period – ranging from a few weeks to a few months – following Hurricanes Katrina and Rita.[130]

[123] Ibid. at 720 (citing *Arkansas Game & Fish, supra* 133 S. Ct. 511, at 522).
[124] Serkin, *Passive Takings, supra* note 110, at 352.
[125] *St. Bernard Parish*, 121 Fed. Cl. 687, at 720 (citing *Arkansas Game & Fish*,133 S. Ct. 511, at 522).
[126] Ibid. at 721–722. [127] Ibid. at 724–736.
[128] Ibid. at 745 (citing *Penn Cent. Transp. Co.* v. *City of New York*, 438 U.S. 104, 124 (1978)).
[129] Serkin, *Passive Takings, supra* note 45, at 382–383, 395 (noting that a taking does not occur where "the action that the government could have taken but did not would have been within the range of appropriate governmental actions" or where the action would not have been "financially plausibly considering the property").
[130] *St. Bernard Parish*, 121 Fed. Cl. 687, at 746.

In December 2016, the Corps appealed Judge Braden's ruling, arguing that the plaintiffs' case "would at most establish a potential tort claim, not a Fifth Amendment taking."[131] The Corps contends that flooding from Hurricane Katrina did not constitute a taking, because the government did not *intend* to flood plaintiffs' properties. Instead, according to the Corps, the flooding was "the sort of 'incidental or consequential injury' that has never been regarded as a taking."[132]

Until the Corps' appeal is finally decided – and there is a good likelihood the case will go to the Supreme Court – there is reason to believe that the argument for using the Takings Clause to impose an affirmative duty to protect private property, at least in cases where the government's past actions create vulnerabilities to natural disaster risk, is emerging. Such cases could promote climate change adaptation by encouraging governments to weigh the costs and benefits of both action and inaction in the face of the increasing risk of natural disasters.

### 6.3.2 *Negligence Claims*

On April 18, 2013, Illinois Governor Pat Quinn declared a state of emergency after heavy rain caused extensive flooding in Chicago and the surrounding area.[133] Residents of some Chicago suburbs were evacuated, and people were urged to reduce household water use to prevent further flooding at the Chicago River.[134] Tens of thousands of people lost power, a sinkhole swallowed cars, and hundreds of flights were canceled at O'Hare airport.[135]

Several months later, a group of insurance companies sued Chicago and over 100 nearby local governments in a series of class action lawsuits, claiming that the municipalities did not do enough to prevent the flooding.[136] The insurance companies argued that the local governments were negligent in failing to prepare for the impacts of climate change and sought to be

131 Juan Carlos Rodriguez, *Army Corps Asks Fed. Circ. to Undo Katrina Liability Ruling*, Law 360, Dec. 12, 2016, https://perma.cc/99X4-XPAT.

132 Ibid.

133 *Chicago Flooding: Heavy Rain Storm Prompts Emergency, Road Closures in Cook Cnty.*, Huffington Post Chicago (Apr. 18, 2013), https://perma.cc/N2MQ-L8A2.

134 Ibid.

135 Ibid.; Staff Report, *Storms Cause Damage, Power Outages*, NBC Chicago, June 13, 2013, www.nbcchicago.com/weather/stories/Storms-Chicago-Severe-Weather-211009451.html.

136 *Illinois Farmers Insurance Company v. Metropolitan Water Reclamation District of Greater Chicago*, No. 14-CH-06608 (Ill. Cir. Ct. Cook Co.); Robert McCoppin, *Insurance Company Drops Suits Over Chicago-Area Flooding*, Chicago Tribune, June 3, 2014, http://trib.in/2sPL9LY.

reimbursed for claims paid to property owners.[137] The cases were quickly withdrawn,[138] but similar suits have been brought in Illinois[139] and Australia.[140] If litigated, such cases would likely require inquiry into governments' efforts to implement various adaptation measures, such as requiring elevation of structures to reduce flooding or investing in green infrastructure to decrease bypass events by draining severe rainfall.

#### 6.3.2.1 Government Immunity

State governments are generally protected from litigation by the doctrine of "sovereign immunity."[141] Overcoming sovereign immunity presents a significant hurdle in negligence cases in both federal and state courts by

137 Ibid.; *Illinois Farmers Insurance Company* v. *County of DuPage*, No. 14-L-385 (Ill. Cir. Ct. DuPage Co.); *Illinois Farmers Insurance Company* v. *County of Lake*, No. 14-L-281 (Ill. Cir. Ct. Lake Co.); *Illinois Farmers Insurance Company* v. *County of Will*, No. 14-L-314 (Ill. Cir. Ct. Will Co.). The plaintiffs' negligence claims were brought under an Illinois statute codifying the government's potential civil liability. Specifically, the plaintiffs claimed that the local government defendants violated (1) 745 Ill. Comp. Stat. Ann. 10/3-102, which provides that "a local public entity has the duty to exercise ordinary care to maintain its property in a reasonably safe condition . . . " and (2) 745 Ill. Comp. Stat. Ann. 10/3-103, which provides that a local government may be liable where, after executing a plan to construct or improve public property, "it appears from its use that it has created a condition that is not reasonably safe." The plaintiffs also brought claims under the state and federal takings clause.

138 McCoppin, *Insurance Company Drops Suit*, *supra* note 136.

139 E.g., *Tzakis v. Berger Excavating Contractors, Inc.*, No. 09 CH 06159 (Ill. Cir. Ct. Cook Co.) (class action alleging, in part, that local governments caused flood damage to plaintiffs' property by failing to enact new standards reducing rainwater runoff despite the increasing risk of severe precipitation. Tzakis was ultimately dismissed on the basis of Illinois' Public Duty Rule, which provides that a public entity is not liable for its failure to provide adequate "governmental services," because the duty to provide such services is owed to the general public at large and not to any particular plaintiff or plaintiffs).

140 Several cases regarding the government's duty of care to protect citizens from risk, albeit not specifically from climate change risk, have been pursued in Australia, See, e.g., *Brodie* v. *Singleton Shire Council* 206 C.L.R. 512 (2001) (holding that in certain circumstances the government yields a special measure of control over the safety of citizens so as to impose a duty of care and an obligation to exercise its powers to avert a danger or to bring it to the knowledge of citizens); *Alec Finlayson Pty Ltd v. Armidale City Council* 123 A.L.R. 155 (1994) (holding that if a public authority creates or increases a risk to another person, he is obligated to take reasonable action to prevent injury unless statute precludes the duty to act).

141 Although the Eleventh Amendment bar to suit against states in federal courts does not extend to counties and municipalities, local governments also enjoy some measure of immunity based on state statutory and common law. *Falk* v. *Perez*, 973 F. Supp. 2d 850, 861 (N.D. Ill. 2013) (citing *Mt. Healthy City Sch. Dist. Bd. of Educ.* v. *Doyle*, 429 U.S. 274, 280 (1977)); *Crouch* v. *City of Kansas City*, 444 S.W.3d 517, 521 (Mo. Ct. App. 2014) (municipalities are entitled to sovereign immunity only when engaged in governmental functions); *Am. Home Assur. Co.* v. *Nat'l R.R. Passenger Corp.*, 908 So. 2d 459, 478 (Fla. 2005) (state statute granted cities immunity from judgments above certain limits); *City of Chesapeake* v. *Cunningham*,

virtue of their sovereignty.[142] Sovereign immunity extends both to the government entity itself and to government officials sued in their official capacities.[143] States may consent to be sued, and many have waived sovereign immunity under certain circumstances through legislative enactments. Waivers are, however, interpreted narrowly by courts.[144]

Nonetheless, it is sometimes possible to formulate a claim such that it falls outside the scope of state and local governments' immunity.[145] While the nuances of sovereign immunity vary from state to state, certain exceptions are common. First, while sovereign immunity generally bars suits seeking monetary damages against government agencies, actions for declaratory judgment or injunctive relief are permissible in some states.[146] Second, within

268 Va. 624, 634, 604 S.E.2d 420, 426 (2004) (Sovereign immunity protects municipalities from tort liability arising from the exercise of governmental functions).

142 *U.S. v. Lee*, 106 U.S. 196 (1882) (doctrine is derived from the laws and practices of English ancestors); *Coll. Sav. Bank v. Florida Prepaid Postsecondary Educ. Expense Bd.*, 527 U.S. 666, 669 (1999); *Alden v. Maine*, 527 U.S. 706 (1999) (part of the very nature of sovereignty to be immune from unconsented suits).

143 *HCMF Corp. v. Gilmore*, 26 F. Supp. 2d 873, 878 (W.D. Va. 1998) (Virginia's Eleventh Amendment immunity "extends to state officials when they are merely the nominal defendants and 'the state is the real, substantial party in interest.'"); *Illinois Health Care Ass'n v. Walters*, 303 Ill. App. 3d 435, 438, 710 N.E.2d 403, 405 (Ill. App. Ct. 1999) ("If a suit is brought against a state official, yet the judgment could operate to control the actions of the state or subject it to liability, then the suit is, in actuality, against the state."); *Olavarria v. Wake Cnty. Human Servs.*, 763 S.E.2d 18 (N.C. Ct. App.), appeal dismissed, review denied, 763 S.E.2d 394 (N.C. 2014) ("governmental immunity shields municipalities and the officers or employees thereof sued in their official capacities from suits based on torts committed while performing a governmental function").

144 *New Orleans Tanker Corp.* v. *Dep't of Transp.*, 728 A.2d 673, 675 (Me. 1999) (we start from the premise that immunity is the rule and exceptions to immunity are to be strictly construed); *Lockwood* v. *City of Pittsburgh*, 751 A.2d 1136, 1139 (Pa. 2000) (exceptions to immunity are to be strictly construed); *Guillen* v. *City of San Antonio*, 13 S.W.3d 428, 433 (Tex. App. 2000) (the Texas Tort Claims Act is a limited waiver of absolute common law immunity ... construed strictly on the side of preserving immunity).

145 Roxanne Conlin and Gregory Cusimano, *Litigating Tort Cases vol.* 6 § 66:2 ("whether immunity applies is often a matter of how the claim is characterized rather than the reality of the claim itself").

146 *Atl. Specialty Ins. Co.* v. *Webster Cnty., Miss.*, No. 140-CV-23, 2014 WL 3437019, at *6 (N.D. Miss. July 11, 2014) (while municipalities are immune from certain claims for monetary damages, governmental immunity does not prevent plaintiffs from seeking declaratory relief); *Roland* v. *Epps*, 10 So. 3d 972, 974 (Miss. Ct. App. 2009) ("a state official may be sued for injunctive relief in his or her official capacity"); *Legal Capital, LLC.* v. *Med. Prof'l Liab. Catastrophe Loss Fund*, 750 A.2d 299, 302 (Pa. 2000) (sovereign immunity does not apply because it is not applicable to declaratory judgment actions); *Texas Dep't of Banking* v. *Mount Olivet Cemetery Ass'n*, 27 S.W.3d 276, 281 (Tex. App. 2000) (holding that sovereign immunity did not bar suit for declaratory relief); *Penland* v. *Redwood Sanitary Sewer Serv. Dist.*, 956 P.2d 964, 965 (Or. 1998) (discretionary immunity does not bar a suit for injunctive relief).

certain jurisdictions, governmental immunity for tort claims is waived when the government purchases liability insurance covering such claims.[147]

Many states immunize discretionary functions, but allow suit against governments and government officials for "ministerial" actions.[148] In distinguishing between discretionary acts and ministerial functions, "the key factor is the presence of basic policy formulation, planning, or policy decisions which are characterized by an exercise of a high degree of official judgment or discretion."[149] For example, a city exercises discretion when it selects and adopts a public improvement plan, but carrying out the plan involves ministerial actions that must be carried out in a reasonably safe manner.[150]

Notably, many states include a "dangerous conditions exception" to sovereign immunity in their tort claims acts, allowing law suits arising from a government entity's maintenance of property it owns.[151] For example, Pennsylvania waives immunity for tort claims arising out of its control of, among other things, "utility

147 *Napier* v. *Town of Windham*, 187 F.3d 177, 190 (1st Cir. 1999) (Under Maine law, if a governmental entity procures insurance that provides coverage in areas where the governmental entity is immune under the state's Tort Claims Act, the entity waives its immunity, but only to the limits of the insurance coverage); *City of Caddo Valley* v. *George*, 9 S.W.3d 481, 484 (Ariz. 2000) ("a municipal corporation's immunity for negligent acts only begins where its insurance coverage leaves off"); *Gilbert* v. *Richardson*, 452 S.E.2d 476, 481 (Ga. 1994) (Georgia Tort Claims Act waives only sovereign or governmental immunity of local governmental agency to extent of liability insurance coverage).

148 *Trotter* v. *Sch. Dist.* 218, 733 N.E.2d 363, 375 (Ill. 2000); *Willow Creek Ranch, L.L.C.* v. *Town of Shelby*, 611 N.W.2d 693, 700 (Wis. 2000) (Under the Wisconsin Tort Claims Act, a municipality is immune from any suit for liability arising from discretionary acts); *Rivera* v. *City of Worcester*, No. 12-CV-40066, 2015 WL 685800, at *6 (D. Mass. Feb. 18, 2015) (in Massachusetts, the discretionary function exception bars government liability for claims based upon the exercise or performance or the failure to exercise or perform a discretionary function or duty); *Williams* v. *Mayor & City Council of Baltimore*, 753 A.2d 41, 60–61 (Md. 2000) (in Maryland, government actors generally immune from liability where tortious conduct occurred while performing discretionary as opposed to ministerial acts); *Chirieleison* v. *Lucas*, 72 A.3d 1218, 1224 (Conn. App. Ct. 2013) ("a municipal employee is liable for the misperformance of ministerial acts, but has a qualified immunity in the performance of governmental acts").

149 Roxanne Conlin and Gregory Cusimano, *Litigating Tort Cases vol. 6*, *supra* note 145, at § 66:47.

150 *Trotter*, 733 N.E.2d 363, *supra* note 23.

151 Dan B. Dobbs, et al., *The Law of Torts*, § 336 (2nd ed.) ("States also tend to eliminate immunity for injuries resulting from badly maintained government property."). Statutory exceptions to sovereign immunity for public property include: 42 Pa. Cons. Stat. Ann. § 8542 (a local entity may be liable for "the care, custody or control of real property in the possession of the local agency"); Colo. Rev. Stat. Ann. § 24-10-106 (sovereign immunity is waived by a public entity in an action for injuries resulting from a dangerous condition of any public property, including buildings, highways, and power facilities); Mo. Ann. Stat. § 537.600 (immunity of the public entity is expressly waived for injuries caused by the condition of a public entity's property); and Mich. Comp. Laws Ann. § 691.1406 (governmental agencies liable for injury resulting from dangerous condition of a public building if agency had

service facilities," which include stormwater management systems.[152] Statutes carving out such an exception generally impose "a broad duty . . . to maintain safe public places."[153] Under this exception, however, the plaintiff must show that the government either created the dangerous condition causing the plaintiff's injury or should have known of the condition.[154]

### 6.3.2.2 Negligence Elements

Once a plaintiff overcomes a sovereign immunity defense, a governmental entity is generally subject to the same rules of liability that apply to nongovernmental entities.[155] Litigants seeking to establish a negligence claim against governments that refuse to prepare for climate change would need to show that (1) the official had a duty to prepare for extreme weather events; (2) the official breached that duty by failing to prepare or causing others to fail to prepare; (3) the litigant suffered harm; and (4) this harm was caused or worsened by the government official's breach of duty. While climate change adaptation litigation is a new phenomenon,[156] analyzing the elements of a negligence claim in this context does not require novel legal theories.[157]

knowledge of the defect and failed to remedy the condition or take action reasonably necessary to protect the public against the condition).

152 42 Pa. Cons. Stat. Ann. § 8542(5); *Rooney* v. *City of Philadelphia*, 623 F. Supp. 2d 644, 653 (E.D. Pa. 2009).

153 Roxanne Conlin and Gregory Cusimano, *Litigating Tort Cases vol 6, supra* note 145, § 66:66.

154 *Bonilla v. Starrett City at Spring Creek*, 704 N.Y.S.2d 619, 620 (N.Y. 2000) ("To impose liability upon the defendants, there must be evidence tending to show the existence of a dangerous or defective condition and that the defendants either created the condition or had actual or constructive notice of it and failed to remedy it within a reasonable time."); *Willis v. City of New Bern*, 529 S.E.2d 691, 693 (N.C. 2000) (granting summary judgment for the defendant city where the plaintiff did not offer proof that the city had notice of the defect causing her injury); *Isbell* v. *Maricopa Cnty.*, 9 P.3d 311, 314 (Ariz. 2000) (A plaintiff need not establish "notice" if a government agency itself creates or causes the dangerous condition.); *Hawks v. City of Westmoreland*, 960 S.W.2d 10, 15 (Tenn. 1997) (The government has "constructive notice" of a dangerous condition where it could have been discovered by proper diligence and it had a duty to inquire into it.).

155 Indeed, courts have imposed liability on the basis of private property owners' failure to act to prevent damage from natural disasters. For example, in California, the owners of an unreinforced building were found to be negligent when they failed to retrofit the building, and two people inside were killed during an earthquake. *Myrick* v. *Mastagni*, 185 Cal. App. 4th 1082, 111 Cal. Rptr. 3d 165 (Cal. Ct. App. 2010).

156 Maxine Burkett, *Litigating Climate Change Adaptation: Theory, Practice, and Corrective (Climate) Justice*, 42 Envtl. L. Rep. News & Analysis 11144, 11156 (2012) (observing that no climate change adaptation cases had been filed through the end of 2012).

157 Ibid. at 11146 ("[T]ort law is well-equipped in both purpose and function to address the challenges of adapting.").

First, the extent of a government's obligation to protect people and ecosystems from the impacts of climate change may be determined by state statute. For example, in Illinois and many other states, local governments have an explicit duty to exercise ordinary care to maintain public property in a reasonably safe condition.[158] Plaintiffs should consider basing allegations of government negligence for failing to prepare for climate change on this prevalent statutorily prescribed government obligation. Many states have enacted statutes enumerating the specific obligations of local and state governments.[159]

Courts may also consider "compelling policy concerns"[160] to expand the scope of the government's duty beyond those expressed in a state statute.[161] Such policy concerns include the foreseeability of harm to the plaintiff, the capacity of the parties to bear the loss, and the consequences to the community of imposing a duty to exercise care.[162] As scientists work to refine predictions of the risk of extreme weather and associated damage to life and property, as manifested in vulnerability assessments, these impacts are increasingly foreseeable. Moreover, governments will almost always have a greater capacity to bear losses than individuals. And imposing a duty on governments to prepare for climate change will generally yield positive outcomes for communities that will otherwise be vulnerable to devastation from natural disasters and other climate risks. In short, policy concerns may support the imposition

158 745 Ill. Comp. Stat. Ann. 10/3-102. For additional examples of statutes defining a state's duty to maintain public property, see note 26.

159 See National Conference of State Legislatures, *State Sovereign Immunity and Tort Liability*, www.ncsl.org/research/transportation/state-sovereign-immunity-and-tort-liability.aspx (listing state tort claims acts); see, e.g., Ill. Ann. Stat. ch. 745, §§ 5/1 et seq.; Pa. Cons. Stat. Ann. tit. 42, §§ 8521 et seq.; Cal. Gov. Code §§ 815, et seq.

160 Maxine Burkett, *Duty and Breach in an Era of Uncertainty: Local Government Liability for Failure to Adapt to Climate Change*, 20 Geo. Mason L. Rev. 775, 786 (2013).

161 See *Norris v. Borough of Leonia*, 734 A.2d 762, 768 (N.J. 1999) (Even in cases where a common law immunity has been incorporated into or codified by statute, it remains subject to judicial modification); *Donaca v. Curry Cnty.*, 734 P.2d 1339 (Or. 1987) (finding that a county's liability for failure to maintain the grass at an intersection where an automobile accident occurred depended on "the existence and magnitude of the risk at the intersection ... [and the] feasibility and cost of avoiding the risk ... "); *Fazzolari By & Through Fazzolari v. Portland Sch. Dist. No. 1J*, 717 P.2d 1210 (Or. Ct. App. 1986), aff'd, 734 P.2d 1326 (Or. 1987) (finding that a jury could reasonably conclude that a public school had a duty to protect a student who was attacked on school grounds, after another person had been raped on campus two weeks earlier); *Williams v. Mayor & City Council of Baltimore*, *supra* note 23 (in case against city, listing variables to be considered in determining if a duty exists, including the foreseeability of the harm and the burden on the city of imposing a duty to exercise care).

162 See, e.g., *Ameriwood Indus. Int'l Corp. v. Arthur Andersen & Co.*, 961 F. Supp. 1078, 1090 (W.D. Mich. 1997); *Torres v. Graves*, No. 92-CV-4449, 1993 WL 19753, at *5-6 (E.D. Pa. Jan. 26, 1993); *Vu v. Singer Co.*, 538 F. Supp. 26, 29 (N.D. Cal. 1981).

of a governmental duty to act to protect citizens from potential harm by implementing appropriate climate change adaptation measures.[163]

Some jurisdictions adhere to the "public duty doctrine," which provides that governmental entities and their agents owe duties only to the general public, not to individuals, absent a "special relationship" or "special duty" between the entity and the injured party.[164] The Circuit Court of Cook County, Illinois, did so in a case brought against several municipalities by plaintiffs alleging a failure to prepare for climate change: The court held that the city defendants did not owe a duty of care to the plaintiffs in connection with their performance of public duties.[165] A litigant may overcome this public duty defense by showing that the government voluntarily assumed a duty to the plaintiff in particular.[166] Doing so creates a special relationship between the plaintiff and a governmental agency and with it a duty of care toward the individual even for discretionary functions.[167] A special relationship may arise when the government performs an affirmative act, or makes a specific promise or representation that under the circumstances creates a justifiable reliance on the part of the person injured.[168] Notably, many states have abandoned the public duty doctrine altogether.[169] Given its potential application to

163 Duty and Breach, *supra* note 448+, at 786.

164 *City Of Toccoa* v. *Pittman*, 648 S.E.2d 733, 736 (Ga. Ct. App. 2007); *Stone* v. N. *Carolina Dep't of Labor*, 495 S.E.2d 711, 714 (N.C. 1998).

165 *Tzakis*, No. 09 CH 06159, April 3 Order re PDR Decision as to LPES and other issues, https://perma.cc/LD7X-94QX (finding the public entity defendants immune under Illinois' Public Duty Rule [citing Harinek v. 161 N. Clark St./Ltd Partnership, 181 Ill. 2d 335, 345-47 (Ill. 1998)]).

166 Roxanne Conlin and Gregory Cusimano, *Litigating Tort Cases vol 6, supra* note 145, at § 66:48; see also *Souder v. Cannon*, 235 S.W.3d 841, 852 (Tex. App. 2007).

167 *Japan Airlines Co. v. Port Auth. of N.Y & N.J.*, 178 F.3d 103, 111 (2d Cir. 1999) (there is no governmental immunity where a special relationship exists between the governmental entity and the injured party); *Hartley v. Floyd*, 512 So. 2d 1022, 1024 (Fla. Dist. Ct. App. 1987) (Once defendant agreed to perform certain tasks, "his actions ceased to be discretionary actions and became merely operational level activities which must be performed with reasonable care and for which there is no sovereign immunity").

168 6 Roxanne Conlin and Gregory Cusimano, *Litigating Tort Cases vol 6, supra* note 145, at § 66:48; *1515-1519 Lakeview Boulevard Condo. Ass'n v. Apartment Sales Corp.*, 43 P.3d 1233, 1240 (Wash. 2002) (plaintiff demonstrated a special relationship with respect to the government's maintenance of a storm drain system by showing direct contact between a government official and herself, express assurances given by a public official, and justifiable reliance on those assurances).

169 *Jean W. v. Com.*, 610 N.E.2d 305 (Mass. 1993) (abolishing public duty doctrine in Massachusetts); *Adams v. State*, 555 P.2d 235, 243 (Alaska 1976) (abolishing public duty doctrine and applying traditional negligence analysis to government); *Ryan v. State*, 656 P.2d 597 (Ariz. 1982) (same); *Leake v. Cain*, 720 P.2d 152 (Colo. 1986) (public duty doctrine held no longer applicable in Colorado); *Commercial Carrier Corp. v. Indian River Cnty.*, 371 So. 2d 1010 (Fla. 1979) (holding governmental negligence to be determined by non-public entity standards); *Wilson v. Nepstad*, 282 N.W.2d 664 (Iowa 1979) (abolishing public duty

circumstances in which state or local government actors seek to protect individuals from heatwaves or floods but fall short, climate change might push the remaining states to do so as well.

Establishing that a government has an affirmative duty to act is more difficult than establishing that the government must exercise due care when it chooses to act. For example, showing that the government was negligent for failing to build a levee may present greater challenges than showing that the government was negligent in building a levee that was poorly designed or inadequately maintained, since the former might be said to require a greater degree of discretion. Even where a litigant ostensibly seeks damages for a government's "failure to act," however, it may be possible to characterize the government's obligation in other terms.[170] Since state governments make decisions in the context of an existing web of infrastructure, such as sewer systems and levees, the distinction between the duty to maintain government-owned property and to build new structures can be blurry.[171]

Importantly, even where a government action is shielded by discretionary immunity, if that action ultimately creates a dangerous condition known to the government but not readily apparent to people who could be injured by the condition, the governmental entity must take steps to avert the danger or properly warn people of the danger.[172] For example, in *City of St. Petersburg* v. *Collom*, three individuals drowned when they fell into open storm drainage ditches owned by the city.[173] The court expressed doubt that the city defendant could be held liable for defects in its "overall plan for the drainage system," since such planning constitutes a discretionary function.[174] The *St. Petersburg* court held, nonetheless, that the plaintiffs had stated a cause of action against

doctrine); *Schear* v. *Bd. of County Comm'rs*, 687 P.2d 728 (N.M. 1984) (public duty doctrine abolished by statute); *Brennen* v. *City of Eugene*, 591 P.2d 719 (Or. 1979) (abolishing public duty doctrine); *Coffey* v. *Milwaukee*, 247 N.W.2d 132 (Wis. 1976) (same); *DeWald v. State*, 719 P.2d 643 (Wyo. 1986) (same).

170 *See Jean W.*, 610 N.E.2d 305, *supra* note 169, at 312 (providing examples of cases that could be characterized either as cases of misfeasance or nonfeasance).

171 See Serkin, *Passive Takings*, *supra* note 45, at 345, 348 (arguing, in the Takings Clause context, that "[p]reexisting regulatory intervention means that the government should not be able to wash its hands of responsibility now").

172 *City of St. Petersburg*, 419 So. 2d 1082, *supra* note 120, at 1086; *Jezek v. City of Midland*, 605 S. W.2d 544, 548; see also *Larson v. Twp. of New Haven*, 165 N.W.2d 543, 546 (Minn. 1969); *Teall* v. *City of Cudahy*, 386 P.2d 493 (Cal. 1963); *Lowman* v. *City of Mesa*, 611 P.2d 943 (Ariz. App. 1980).

173 *City of St. Petersburg*, 419 So. 2d 1082, *supra* note 120.

174 Ibid. at 1086 ("defects inherent in the overall plan for an improvement, as approved by a governmental entity, are not matters that in and of themselves subject the entity to liability") (citing *Dep't of Transp.* v. *Neilson*, 419 So. 2d 1071 (Fla. 1982)).

the city defendant for its failure to either warn people of the open drain hazard or to correct the dangerous condition by adding fences or other barriers around the ditches.[175] According to the *St. Petersburg* court, "a governmental entity may not create a known hazard or trap and then claim immunity from suit for injuries resulting from that hazard on the grounds that it arose from a judgmental, planning-level decision."[176]

Assuming that a government entity has a duty to take action to protect citizens from the impacts of climate change, what would constitute a breach of that duty? The failure to provide complete protection from harm clearly does not itself constitute a breach of duty.[177] Instead, establishing a breach of the duty to prepare for the impacts of climate change would likely require a showing that the government's inaction exposed people to unnecessary risks. In this context, significant deference to the government in weighing the costs and benefits of its actions is certainly appropriate. Deference should not, however, "amount to abdication of oversight in the context of either action or inaction."[178]

The well-known "Hand formula" is a useful starting point in determining whether a defendant has breached his duty of care,[179] but would likely be difficult for courts to apply amid climatic changes. Generally, the formula dictates that a person breaches his duty where the likelihood of harm multiplied by the magnitude of harm is greater than the cost of preventing that harm.[180] In the context of a case alleging failure to prepare for climate change,

175 Ibid. at 1085–1087.

176 Ibid. at 1086; see also *Rooney*, 623 F. Supp. 2d 644, *supra* note 152 (finding that a city may not be held liable for an inadequate stormwater management system, but it may be liable for damages resulting from negligence in the construction or maintenance of the sewer system).

177 Instead, courts may find that a government has fulfilled its duty where its actions were justified by the information and resources available at the time of the action or omission at issue. *Cootey v. Sun Inv., Inc.*, 718 P.2d 1086, 1090 (Haw. 1986) ("Government is not intended to be an insurer of all the dangers of modern life, despite its ever-increasing effort to protect its citizens from peril."); *Jean W*, 610 N.E.2d 305, *supra* note 169, at 314–315 ("Police departments are no more responsible for every harm that befalls victims of crime than fire departments are responsible for every sparking of a fire, and neither should be an insurer of every loss sustained in those contexts.").

178 See Serkin, *Passive Takings*, *supra* note 45, at 385.

179 Dobbs, *The Law of Torts*, *supra* note 151, at § 161 ("If the defendant's cost of preventing the harm is less than the expected value of the harm itself, he is definitely negligent and liable under the Hand formula"); *American Law Institute, Restatement (Second) of Torts*, § 291 (1965) ("Where an act is one which a reasonable man would recognize as involving a risk of harm to another, the risk is unreasonable and the act is negligent if the risk is of such magnitude as to outweigh what the law regards as the utility of the act or of the particular manner in which it is done.").

180 In *re City of New York*, 475 F. Supp. 2d 235, 242 (E.D.N.Y. 2007), aff'd, 522 F.3d 279 (2d Cir. 2008) (applying the Hand Formula to determine whether the City of New York was negligent in connection with an accident on the Staten Island Ferry); *Bhd. Shipping Co. v. St. Paul Fire*

the likelihood of harm is the chance of a particular event, such as a heatwave or a 100-year flood,[181] at the time of the event. Since climate change is causing global alterations in the frequency, severity, and geographic distribution of significant adverse events, historical data will become less predictive of future events' probabilities. Instead, best estimates of the likelihood of harm must also draw on expert weather and climate projections.[182] Moreover, the probability of a natural disaster should be based on its cumulative chance over a relevant period of time – perhaps from the time the government defendant should have been aware of the relevant climate projections to the date of the event – rather than in one particular year;[183] since resiliency measures are implemented for the long term, it is irrelevant to the question of breach whether a natural disaster occurs one year versus the next.

For the purposes of applying the Hand formula, the magnitude of the harm should be measured by the predicted loss of property and life likely to result from a particular event. Hurricane Sandy caused over $50 billion in damage,[184] and Hurricane Katrina left in its wake over $100 billion in

& *Marine Ins. Co.*, 985 F.2d 323, 327 (7th Cir. 1993) (Under the Hand formula, a defendant is negligent if the burden (cost) of the precautions that he could have taken to avoid the accident ... is less than the loss that the accident could reasonably be anticipated to cause . . ., discounted (i.e. multiplied) by the probability that the accident would occur unless the precautions were taken."); *Levi* v. *Sw. Louisiana Elec. Membership Co-op. (SLEMCO)*, 542 So. 2d 1081, 1087 (La. 1989) ("When the product of the possibility of [injury] multiplied times the gravity of the harm, if it happens, exceeds the burden of precautions, the failure to take those precautions is negligence"); *U.S. v. Carroll Towing Co.*, 159 F.2d 169, 173 (2d Cir. 1947) (L. Hand, J.) ("if the probability be called P; the injury, L; and the burden, B; liability depends upon whether B is less than L multiplied by P: i.e. whether B less than PL").

181 FEMA defines a 100-year flood as a flood with a 1 percent likelihood of occurring or being exceeded in any given year, based on historical data. FEMA, *Flood Zones*, www.fema.gov/flood-zones. Climate change has increased the risk of a 100-year flood in many areas such that the actual chance of such a flood is great than 1 percent.

182 Long-term projections of future weather and climate conditions often provide a wide range of possible outcomes. See, e.g., C. Rosenzweig, et al., *Responding to Climate Change in New York State: The ClimAID Integrated Assessment for Effective Climate Change Adaptation*, *Ann.* N.Y. Acad. Sci., 1244 (2011), 2–649, https://perma.cc/KJ6X-3Y8L (predicting sea level rise in New York City between 15 and 75 inches by 2100). Some courts may be unpersuaded that governments should be held liable for failing to act on uncertain projections. Nonetheless, such projections may play a larger role in courts' liability determinations as scientists continue to refine climate models, increasing both their accuracy and precision.

183 The cumulative risk of a particular event increases as the time span increases. For example, while the risk of a 100-year flood may be approximately 1 percent within the next year, there is at least a 26 percent chance of a 100-year flood over the next 30 years. United States Geological Survey, *100-Year Flood–It's All About Chance* (April 2010), https://perma.cc/9H5N-JFNE.

184 US Department of Commerce, National Oceanic and Atmospheric Administration (NOAA), *Service Assessment Hurricane/Post-Tropical Cyclone Sandy, October 22–29, 2012* (May 2013), https://perma.cc/2F3B-H38Z.

damage.[185] Even if such events are infrequent, the extent of the devastation they cause justifies taking precautionary measures to minimize potential damage.[186]

Other factors may be relevant to whether a governments' failure to act was reasonable, including the precision and accuracy of available climate projections, access to technical and monetary resources, the extent to which the precautions would have reduced or eliminated the damage, any negative consequences of the precautionary measures beyond their expense, and alternative measures taken by the state or city to adapt to climate change. Ultimately, the plaintiff "need[s] to prove the unreasonableness of [the] defendant's actions in light of the well-established science of climate change."[187] While certainly a "formidable task," increasing knowledge of climate change risks should increase the potential for liability.[188]

Establishing the damage element of a negligence claim in this context would not differ materially from a typical negligence case. A litigant would have to show that he suffered an injury to his person or property. These types of injuries are "especially present in the climate adaptation context."[189] For example, where flooding causes widespread property damage or a heatwave increases mortality rates, many people will have suffered a clear and legally cognizable injury. Nonetheless, some states impose statutory dollar limitations to limit the amount that can be recovered against a government entity.[190]

To establish the causation prong of a claim for negligent failure to prepare for climate change, the plaintiff would need to show, as in any negligence case, that the defendant's breach of duty was the proximate cause of the plaintiff's injury. The question in this context is whether the government's

185 NOAA, *The Deadliest, Costliest, And Most Intense United States Tropical Cyclones from 1851 to 2010 (And Other Frequently Requested Hurricane Facts)* (Aug. 2011), https://perma.cc/QC T6-KL6K.

186 *In re City of New York*, 475 F. Supp. 2d 235, *supra* note 180, at 242 (comparing, without quantifying, "[t]he probability . . . of a scenario where the pilot [of the Staten Island Ferry] would become incapacitated . . . [with] [t]he gravity of . . . resulting injury to its passengers" against the burden of enforcing a safety measure that would have prevented the accident to finding New York City negligent); *Burkett, Duty and Breach, supra* note 160, at 781–782 ("Sandy also underscores the need for local governments to appreciate fully the costs of, to date, low probability yet unprecedented and devastating events.").

187 Burkett, *Litigating Climate Change Adaptation, supra* note 156, at 11145.

188 Ibid.; see also Attribution of Extreme Weather Events in the Context of Climate Change, National Academies of Sciences, Engineering, and Medicine (2016).

189 Ibid. at 11148.

190 See, e.g. Colo. Rev. Stat. Ann. § 24-10-114; Ga. Code Ann. § 50-21-29; Minn. Stat. Ann. § 466.04.

failure to take reasonable measures to protect people from the natural disaster at issue caused the damage.

The plaintiff must identify measures the government could have taken to prevent the injury.[191] Plaintiffs should challenge the city's failure to upgrade or build specific infrastructure, rather than the city's failure to consider climate change impacts in its planning documents, since the latter theory would also require the plaintiff to establish that the infrastructure would have been upgraded if the planning had been carried out. The litigant would *not* need to show that the natural disaster at issue was caused by climate change.[192] Instead, the effect of climate change on the likelihood of the weather event would be a factor in the determination of breach and foreseeability, as discussed earlier.[193]

As natural disasters become more likely – and, therefore, more foreseeable – due to climate change, governments face the risk of being found liable for refusing to take reasonable actions to prepare for climate-driven impacts. This type of litigation can serve the dual purposes of compensating plaintiffs for injuries and encouraging governments to better adapt to climate change.

## 6.4 CONCLUSION

The built environment transmits several of the most notable effects of climate change on public health and often shapes those effects. This makes the design and maintenance of its elements – buildings, infrastructure, streetscapes, and more – critical to climate change adaptation efforts. This point is not new, but its implications continue to unfold slowly in the context of land use policies

[191] See *Rooney*, 623 F. Supp. 2d 644, *supra* note 152 (plaintiffs' claim that city's negligent maintenance of sewer system caused flood damage survived summary judgment motion where plaintiffs presented evidence that clogged sewers caused the flooding and city had notice of the condition); *Gaylord ex rel. Gaylord v. Morris Twp. Fire Dep't*, 853 A.2d 1112, 1116 (Pa. Commw. Ct. 2004) (Municipal liability arising from real property ownership only available where the defect of the land itself causes the injury).

[192] Issues of causation raise more difficult questions in the context of a climate nuisance suit, since "current atmospheric levels of GHGs result from the cumulative emissions of millions or billions of emitters since the onset of the industrial revolution[, and] no specific injury can be attributed to any specific polluter." Michael Gerrard, *What Litigation of a Climate Nuisance Suit Might Look Like*, Yale L. J. Online (Sept. 2011); Burkett, *Litigating Climate Change Adaptation*, *supra* note 156, at 11145 at 11145 (observing that "establishing the causal link between a defendant's emissions and the alleged harms" would be the most challenging task for a plaintiff seeking tort remedies from greenhouse gas emitters).

[193] Burkett, *Litigating Climate Change Adaptation*, *supra* note 156, at 11150 ("Attributing extreme weather events to climate change . . . will occur at the state of establishing defendant's breach of duty . . . ").

and other state and local legal obligations on public and private actors. This pace should accelerate as climate-driven public health impacts related to the built environment become harder to ignore, and as the roles and responsibilities of stakeholders come into greater focus. This could theoretically result from judicial determinations in cases brought by parties seeking relief from the harmful effects of action or inaction on the part of governments or private actors. However, as the law provides governments with ample tools to avoid liability for many forms of inaction, it is perhaps more likely that transforming or merely modifying the built environment in service to public health goals will follow from efforts by proactive policymakers and citizens who push them to act. In nearly all jurisdictions, such efforts would need to amend the siloed approaches that currently prevent agencies responsible for public health and different elements of the built environment from consulting with each other, much less collaborating thoroughly toward integrated adaptation planning and implementation efforts.

# PART III

# IMPACTS AND INTERVENTIONS

# 7

# Heat

Sara Hoverter

*Increased temperatures and heatwaves are already serious public health problems in the United States and will worsen as climate change increases their frequency and intensity. Heat's direct public health impacts include heat stress and stroke, death, and worsening of cardiovascular and respiratory diseases. Heat also degrades air and water quality, causing secondary impacts as well. Heat does not affect everyone equally – some populations are physically more susceptible to its effects, and others can be more vulnerable due to their environments. Governments owe particular legal duties to some of these populations under federal and state law.*

*Interventions to reduce the negative health effects of heatwaves and higher temperatures fall into two categories: emergency response strategies that include mostly traditional local public health functions, and built environment strategies that use a wider array of government powers, such as changes to zoning and building codes.*

## 7.1 INTRODUCTION

Increased temperatures and extreme heat are typically the first impacts that one thinks of when one thinks about climate change. Temperatures in the United States are already increasing – average temperatures in the northern hemisphere in the second half of the twentieth century were "likely the highest in at least the past 1,300 years."[1] The first decade of this century has been recorded as the hottest ever on record.[2] Most

[1] Intergovernmental Panel on Climate Change (IPCC), *Climate Change 2007: Synthesis Report*, 1.1 (2007).

[2] J. Balbus, A. et al., Ch. 1: Introduction: climate change and human health, In: Alison Crimmins et al., *The Impacts of Climate Change on Human Health in the United States: A Scientific Assessment*, 25–42 (2016).

regions of the United States are also expected to be hotter in the future than they are now.[3]

Heatwaves, defined as multiple days in a row over a certain temperature threshold, are projected to be longer, more frequent, and more intense; extreme temperatures may increase faster than average temperatures do.[4] Although projecting extremes is difficult, models are improving, and projections should become more and more accurate.[5] In addition to more heatwaves, day-to-day temperatures are also likely to rise. This day-to-day increase poses a public health response challenge, given that most traditional public health responses focus on acute emergency situations.

Most data looking at heat projections focus on maximum temperatures during the day, but high minimum or nighttime temperatures are important values to track as well. High daytime temperatures are dangerous because the body only has so much capacity to adapt to the heat; high nighttime temperatures are dangerous because the body does not have time to recover from the stress of surviving the daytime high temperatures.[6] With a heatwave, where several high maximums and minimums follow one another closely, heat becomes a critical public health risk.[7]

Absolute temperature is not the only measure of the impact of heat, however: humidity combined with heat, usually referred to as the "heat index," better reflects how heat impacts on the human body.[8] Because high humidity inhibits the body's ability to cool itself through the evaporation of sweat, high humidity mixed with high temperatures presents

[3] M. C. Sarofim et al., Ch. 2: Temperature-related death and illness, In: Alison Crimmins et al., *The Impacts of Climate Change on Human Health in the United States: A Scientific Assessment*, 43–68 (2016); Summary for policymakers, In: C.B. Field et al., *Climate Change 2014: Impacts, Adaptation, and Vulnerability, Part A: Global and Sectoral Aspects. Contribution of Working Group II to the Fifth Assessment Report of the Intergovernmental Panel on Climate Change*, 1–32 (2014).

[4] B. Orlowsky and S. I. Seneviratne, *Global Changes in Extreme Events: Regional and Seasonal Dimension*, 10 Climatic Change 669 (2012) at 694. Definitions of heatwaves also vary.

[5] S. Herring et al., *Explaining Extreme Events of 2015 From a Climate Perspective*, 97(12) Bull. Am. Meteor. Soc. (2016), www.ametsoc.org/ams/index.cfm/publications/bulletin-of-the-american-meteorological-society-bams/explaining-extreme-events-from-a-climate-perspective/.

[6] U.S. Environmental Protection Agency. *Excessive Heat Events Guidebook* (2006). www.epa.gov/heatisland/about/heatguidebook.html.

[7] Joacim Rocklöv et al., *Mortality Related to Temperature and Persistent Extreme Temperatures: A Study of Cause-Specific and Age-Stratified Mortality*, 68(7) Occ. and Env. Med. 531–536 (2011).

[8] Sarofim et al., *Temperature-Related Death and Illness*, *supra* note 3, at 46.

a more serious health risk than high temperatures alone.[9] Changes in precipitation patterns that increase ambient humidity could in some cases compound the increases in temperatures to produce a heightened effect on the heat index.

People in urban and rural areas experience heat increases differently. Cities face particular challenges for several reasons. First, the high concentration of buildings and impervious surfaces such as concrete tend to trap heat and increase temperatures relative to surrounding areas with more vegetation. Traditional pavements and rooftops can be up to 90 degrees hotter than the air, while surfaces common to rural areas are likely to be close to or below air temperature.[10] Cities are also home to higher numbers of people in the same amount of space, and can have high proportions of low-income communities and communities of color. As discussed in this chapter, these communities are especially vulnerable to heat effects, even relative to others within the same city.[11]

Rural areas face other challenges. Unlike cities, where people live in relative density, rural populations are far more dispersed and may therefore be harder to track, check up on, and assist with heat-related illnesses.[12] Local authorities in rural areas may also not have as many resources to respond and adapt to heat-related health challenges, or climate challenges in general, as urban areas do.[13]

Heat and its public health impacts are challenging for public health practitioners for several reasons. First, the public health effects of heat vary widely for different populations within the same geographic area, impeding the ability to have an overarching, one-size-fits-all set of policies to address them. Second, the legal duties that local governments and other actors owe to each of these populations vary, again making a consistent policy challenging to implement jurisdiction-wide. Third, the solutions directly available to public health departments constitute only a relatively small subset of the appropriate responses. Strategies that focus on changing the built environment are likely managed by other

9 G. Luber and M. McGeehin, *Climate Change and Extreme Heat Events*, 35(5) Am. J. Prev. Med. 429–435 (Nov. 2008).

10 See U.S. Environmental Protection Agency (EPA), *Heat Island Impacts*, www.epa.gov/heat-islands/heat-island-impacts.

11 J. Klein Rosenthal et al., *Intra-Urban Vulnerability to Heat-Related Mortality in New York City*, 30 Health Place 45–60 (2014).

12 Janet L. Gamble et al., Chapter 9: Populations of concern, In: Alison Crimmins et al., *The Impacts of Climate Change on Human Health in the United States: A Scientific Assessment* (2016).

13 David Hales and William Hohenstein, Chapter 14: Rural communities, In: J.M. Melillo et al., *Climate Change Impacts in the United States: The Third National Climate Assessment*, 340 (2014).

agencies, leaving public health departments without authority over things such as building codes, zoning rules and decisions, and the structure and operation of the electricity grid. Thus, for public health departments to influence decisions taken by zoning commissions and local public utility commissions, they must coordinate with disparate agencies that may have not worked together previously (or at least well).

The next section discusses in more detail (1) the public health impacts of hotter temperatures and heat waves, (2) the disproportionate impacts on particular populations; and (3) the particular legal duties that local governments and others owe to some of these populations.

## 7.2 PUBLIC HEALTH EFFECTS AND RELEVANT LEGAL DUTIES AND PROTECTIONS

Public health effects from heatwaves and increased temperatures come in two varieties: direct and indirect. Direct effects include increases in heat stroke and stress, heat-related deaths, and hospitalizations for respiratory diseases like asthma that are exacerbated by higher temperatures. Indirect effects include environmental health effects such as those from decreased air quality or water quality due to increased temperatures. As the climate changes, all of these public health impacts are expected to increase in prevalence, as well as extending into seasons of the year that public health practitioners are not used to managing for them. Many of these impacts also do not affect all people equally – some populations such as the elderly, very young, homeless, and others are at greater risk.[14] Some debate exists over whether governments owe affirmative duties to the public to act to protect them from public health hazards. (See Chapter 6.) But there is no disputing that governments and providers do owe special duties to certain groups of citizens pursuant to obligations created by the Constitution or statute. The legal duties owed to specific groups are discussed in Section 7.2.3.

### 7.2.1 *Direct Health Effects from Heat*

Heatwaves kill. Heat caused more deaths between 1979 and 2003 in the United States than all other natural disasters combined.[15] Climate change is projected to exacerbate this problem. One study anticipates that by 2057 the current

[14] Gamble, *Chapter 9: Populations of Concern, supra* note 13.

[15] Ctrs. For Disease Control and Prevention (CDC), *Extreme Heat: A Prevention Guide to Promote Your Personal Health and Safety*, www.cdc.gov/disasters/extremeheat/heat_guide.html.

average annual rate of heat-related deaths just on the east coast of the United States will increase from 700 to as many as 3,500.[16] Heatwaves, where hot days and nights follow one another multiple days in a row, have a cumulative effect that can be especially dangerous.[17] Not only is the physical effect cumulative, but some research suggests that people may not be as aware of the risks of heatwaves as they are of cold snaps, and so take fewer precautions, such as staying inside and staying hydrated.[18]

In addition to heat-related deaths, higher temperatures can also cause heat stroke and heat stress, and can increase hospitalizations for those conditions as well as respiratory diseases like asthma that are exacerbated by higher temperatures.[19]

### 7.2.2 *Indirect Health Effects from Heat*

Communities will also experience indirect environmental health effects. For example, higher temperatures correlate with higher production of ground-level ozone, worsening air quality and intensifying allergens.[20] Increased temperatures can also degrade water quality, allowing the growth of cryptosporidiosis and cholera bacteria, among others.[21]

A Milwaukee outbreak of legionellosis in 2013 illustrates the challenge for local public health departments: After recording fifty-eight cases in a three-month summer period, compared with fifteen the previous year, the Milwaukee Health Department (MHD) initiated an epidemiological and environmental investigation to determine the cause.[22] Several contaminated HVAC cooling towers were the culprits; cooler-than-average springs had delayed both start-up of the cooling

[16] J. Wu et al., *Estimation and Uncertainty Analysis of Impacts of Future Heat Waves on Mortality in the Eastern United States*, 122 Environ. Health Perspect. 10–16 (2014). This estimate used a high emissions scenario (RCP8.5); estimates were lower using lower emissions scenarios.

[17] W. Dong et al., *New Climate and Socio-Economic Scenarios for Assessing Global Human Health Challenges Due to Heat Risk*, Clim. Change (2015) http://dx.doi.org/ 10.1007/s10584-015-1372-8.

[18] Barnett article.

[19] N. M. Lugo-Amador et al., *Heat-Related Illness*, 22 Emerg. Med. Clin. North Am. 315–327 (2004);

[20] K. Knowlton et al., *Impacts of Heat and Ozone on Mortality Risk in the New York City Metropolitan Region Under a Changing Climate*, 30 Advances in Global Change Research 143–160 (2008); J. A. Patz, *Climate Change and Health: New Research Challenges*, 6 Ecosystem Health 52–58 (2000); N. Fann et al. Ch. 3: Air quality impacts, In: Alison Crimmins et al., *The Impacts of Climate Change on Human Health in the United States: A Scientific Assessment*, 69–98 (2016).

[21] Patz, *Climate Change and Health, supra* note 20, at 52–58.

[22] Jill Krueger et al., *Human Health Impacts of Climate Change: Implications for the Practice and Law of Public Health*, 43(79) J. Law, Med., & Ethics 79–80 (2015).

towers and their usual biocide treatment, giving the Legionella an opportunity to colonize the cooling towers.[23] Because of the projected variability in temperatures expected with climate impacts, more incidents like this one are possible.

### 7.2.3 *Disproportionate Impacts on Special Populations and Legal Duties*

Some members of local communities are inherently more susceptible to heat due to either physical characteristics, such as age or chronic disease, or social and environmental factors, such as being on a low-income or working outdoors. This section explores each group's vulnerabilities to different aspects of increased heat, and any legal duty owed by governments and private actors to that group's members to protect them from heat-related illnesses and death.

Governments in the United States, absent an explicit constitutional provision or statute creating a specific duty, do not have a generalized affirmative duty to protect the public health.[24] (See Chapter 2.) The US Constitution primarily guarantees US citizens negative rights, or the right to be free from government interference in their lives in specific ways such as freedom of religion or the prohibition on cruel and unusual punishment.[25] The seminal case for this issue is a public health case: *DeShaney* v. *Winnebago County Department of Social Services*. In this tragic case, young Joshua DeShaney was beaten into a coma by his father after a long pattern of abuse, during which the county was aware of the problem but did not remove him from the home.[26] In holding that the county owed no duty to protect Joshua, the Court underscored that the due process protections of the 14th Amendment are limits on the state's power to act, not affirmative duties for them to protect citizens from one another.[27] While this interpretation has been widely contested,[28] it is the current interpretation under which US courts operate and so must be understood.

The parallels for the public health system in adapting to climate change are interesting. Do governments assume an affirmative duty to protect the public

23 Ibid.

24 Lawrence O. Gostin, *Public Health Law & Ethics: A Reader 2nd edn.*, 106 (Berkeley, Los Angeles & London: University of California Press, 2010).

25 U.S. Const. Amends. 1 and 8.

26 *DeShaney v. Winnebago County Department of Social Services*, 109 S.Ct. 998, 1001–1002 (1989).

27 Ibid. at 195–196. Justice Brennan in his dissent frames the issue not as one of allowable inaction, but instead focuses on the affirmative actions that the state has already undertaken in establishing a system for protecting children from abuse.

28 See, e.g., Aviam Soifer, *Moral Ambition, Formalism, and the "Free World" of DeShaney*, 57 Geo. Wash. L. Rev. 1513 (1989).

from the health impacts of climate change, simply by investigating what those impacts will be and adapting public health law or practice accordingly? Current case law, including *DeShaney*, suggests no. It is possible, however, that local governments will still face lawsuits for failure to take adequate adaptation action under a negligence approach. (See Chapter 6.)

### 7.2.3.1 Children

Children are particularly vulnerable to heat-related illnesses for several reasons. Their physiological immaturity,[29] including relatively poor thermoregulation and lower rates of sweating, increases their vulnerability.[30] Unlike most adults, they also are dependent on adults to keep them cool and safe during a heatwave.[31] Children often spend more time outside than adults, especially in the afternoons when temperatures are high, exposing them to heat and to air pollution at higher rates than adults.[32] Children of color are often even more highly exposed to environmental hazards due to their neighborhoods' exposure to sources of pollution.[33]

Children's legal relationship with government is qualitatively different than adults'. The government withholds some rights from children until they reach certain ages (i.e. driving, voting, and drinking alcohol) and for common law matters such as contracts, children are assumed not to be competent to act on their own behalf.[34] Nevertheless, children are still regarded as "persons" under the Constitution and are subject to its protections.[35] Federal and state statutes protect children from specific harms, including child labor[36] and sexual exploitation.[37]

General state duties of care for children, however, only arise from situations in which the state assumes an affirmative role in caring for the child *in loco*

29 J. R., Bytomski and D. L. Squire. *Heat Illness in Children,* 2 Curr Sports Med Rep 320–324 (2003).

30 Sarofim et al., *Temperature-Related Death and Illness, supra* note 3, at 54.

31 Ibid.

32 V. Kun et al., *Out of Breath: Children's Health and Air Pollution in Southern California* (Los Angeles: Natural Resources Defense Council).

33 See generally, Lawrie Mott, *The Disproportionate Impact of Environmental Health Threats on Children of Color,* 103 Envtl. Health Persp., Suppl. 6, 33 (1995).

34 Thomas R. Young, *The Rights of a Child to Make Contracts and Disaffirm Them,* 1 Leg. Rts. Child. Rev. 2D § 10: 1(3rd edn.) (2016).

35 *Tinker* v. *Des Moines Independent Community School Dist.*, 393 U.S. 503 (1969).

36 29 U.S.C.A. §§ 201 et seq. For state child labor laws see, e.g., 820 Ill. Comp. Stat. Ann. 205/1 et seq.; Md. Code Ann., Lab. & Empl. §§ 3–201 to 3–216; N.J. Stat. Ann. 34:2–21.1 et seq.

37 18 U.S.C.A. § 2256(1) et seq.

*parentis*.[38] These situations can arise when the state or local government has temporary custody or control over the children; for example, when children are at school or in an institutional setting such as a state-run medical or mental health facility, in foster care, or in a juvenile detention center.[39] For purposes of protecting these children's health from extreme heat, children in correctional situations are entitled to a safe living environment that includes adequate ventilation and temperature control. Although there has not been litigation on this issue, analogies from cases determining the definition of "adequate shelter" suggest that juvenile correctional facilities will likely need to keep temperatures cool enough not to endanger children's health during a heatwave.[40]

Minors are entitled to "reasonably safe" school environments under state constitutions. As with juvenile corrections environments, no case law yet directly addresses this issue. Related cases addressing adequacy of funding of schools under state constitutional requirements, asserting a minimum baseline of adequacy for safe school facilities, suggest that schools may need to be maintained at a temperature that does not endanger students' health.[41] For children with disabilities, the federal Individuals with Disabilities in Education Act (IDEA) requires school districts to make reasonable accommodations for those children to be able to access a "free appropriate public education."[42] For children whose disabilities, or medication to manage those disabilities, make them more susceptible to heat-related illnesses than their peers, school districts may need to ensure temperature-controlled environments under the IDEA as well.

IDEA also pertains to transportation to and from school for children with disabilities. The federal rules implementing the IDEA defined transportation to and from school and between school buildings as a "related service" to education.[43] School districts must provide related services such as transportation if an individual child's education plan requires it.[44]

In California, the state education agency has implemented these requirements by creating guidelines for school districts to follow in creating

38 67A C.J.S. Parent and Child § 367. 39 Ibid.

40 See, e.g., *Del A. v. Roemer*, 777 F. Supp. 1297 (E.D. La. 1991); *Nelson v. Heyne*, 355 F. Supp. 451, 457–458 (N.D. Ind. 1972), judgment aff'd, 491 F.2d 352 (7th Cir. 1974); *Santiago v. City of Philadelphia*, No. 74-2589 (E.D. Pa. Dec. 22, 1978) (stipulation in partial settlement).

41 See, e.g., *Campaign for Fiscal Equality v. State of* N.Y., 86 N.Y.2d 307 at 316 (1995).

42 Individuals with Disabilities Education Act, § 601 et seq., 20 U.S.C.A. § 1400 et seq.; *Bd. of Educ.* v. *Rowley*, 458 U.S. 176, 200 (1982).

43 34 CFR § 300.34(c)(16). 44 34 CFR § 300.105(a)(2).

a compliant and effective special education transportation program.[45] The California guidelines include as a primary consideration the pupil's needs, including not only physical accessibility and timing of transportation but also issues related to the pupil's medical diagnosis – specifically, temperature control and pupil's needs during very hot weather.[46] This thoughtful approach will be critical in keeping students with disabilities safe and healthy as climate change makes high temperatures and heatwaves more frequent and intense, and more common during the school months.

#### 7.2.3.2 Older Adults

Elderly members of the community, particularly those who are socially isolated, are also more vulnerable to extreme heat than younger adults. Heat impacts are projected to be located in parts of the country where older adults tend to live, further increasing their exposure.[47] Due to their poorer thermoregulation, older adults have an increased risk of death as a result of acute cardiovascular or respiratory distress during heatwaves.[48] They are also more likely to have chronic conditions and take medications that make them more sensitive to increased temperatures.[49] Living in poverty, with impaired mobility, or living alone all make older adults much more likely to develop a heat-related illness due to lack of access to air conditioning, difficulty in leaving their homes for a cooler location, and sometimes lack of information or understanding of heat emergency warnings.[50]

These risk factors all interact and can be synergistic. Recent research has found that social isolation, access to green space, and age of housing stock all increase older adults' risk of heat-related mortality.[51] Similarly, an analysis of the 1995 Chicago heatwave found that the highest risk of death was for people who had medical conditions, were socially isolated, were confined to bed, and did not have air conditioning.[52]

45 California Dep't. of Ed., *Special Education Transportation Guidelines*, www.cde.ca.gov/sp/se/lr/trnsprtgdlns.asp.

46 Ibid.

47 Janet L. Gamble, et al., *Climate Change and Older Americans: State of the Science*, 121(15) Envtl. Health Persp. 16 (2003).

48 Ibid. at 17. 49 Sarofim et al., *Temperature-Related Death and Illness*, *supra* note 3, at 54.

50 Gamble, *Climate Change and Older Americans*, *supra* note 47, at 19.

51 See, generally, Carina J. Gronlund, et al., *Vulnerability to Extreme Heat by Socio-Demographic Characteristics and Area Green Space Among the Elderly in Michigan,1990–2007*, 136 Environ Res. 449 (2015).

52 Jan C. Semenza, et al., *Heat-Related Deaths During the July 1995 Heat Wave in Chicago*, 335 N. Engl. J. Med. 84 (1996).

Residents of nursing homes or other institutional care facilities may be owed duties that older adults living at home are not. Nursing homes must be licensed by state governments and therefore must follow state regulations. Many nursing homes also choose to seek federal certification so that they can receive reimbursement for services from Medicaid and Medicare. Under the federal rules, a nursing home must "care for its residents in such a manner . . . as will promote maintenance or enhancement of the quality of life of each resident."[53] The Department of Health and Human Services has the "duty and responsibility . . . to assure that requirements which govern the provision of care . . . and the enforcement of such requirements, are adequate to protect the health, safety, welfare, and rights of residents."[54] Federal law requires creation of an individualized plan of care for each resident that must be periodically updated, and provides the specific right for residents to reside and receive services with "reasonable accommodation of individual needs and preferences."[55]

Federal rules also specifically require "comfortable and safe temperatures," and nursing homes must keep temperatures between 71°F and 81°F.[56] Many states also require maintenance of similar temperature ranges for older adults in nursing homes.[57] Rhode Island regulations require either air conditioning in resident bedrooms or in a common area when the external temperature exceeds 80°F.[58]

These requirements combine to create a set of duties for nursing homes to protect the health and safety of each resident as an individual, taking into account individual needs, medical conditions, and potentially disabilities. This approach suggests that nursing homes should not only follow the letter of state requirements to maintain temperature ranges, but also take individual residents' heat sensitivity into account and ensure that indoor temperatures are appropriate for individuals, including those with an impaired capacity to thermoregulate.

53 Omnibus Budget Reconciliation Act of 1987 (OBRA) 87 §§ 1819(b), 1919(b); 42 U.S.C.A. § 1395i-3(b)(1)(A); 42 U.S.C.A. § 1396r(b)(1)(A).

54 42 U.S.C.A. § 1395i-3(f)(1); 42 U.S.C.A. § 1396r(f).

55 42 U.S.C.A. § 1395i-3(b), (c); 42 U.S.C.A. § 1396r(a); 42 C.F.R. § 483.10.

56 42 CFR 483.10. Facilities first certified before 1990 are exempt from this requirement.

57 See, e.g., AR ADC 016.06.9–713.7 (specifying temperature ranges for days under 95 degrees and over 100 degrees external temperature); MN ADC 4658.4510 (requiring heating and cooling equipment to be installed in new construction allowing an indoor temperature range of 71 to 81 degrees, and requiring humidity control).

58 See, e.g., R.I. Admin. Code 31–4-8:19.23, implementing R.I. Gen.Laws 1956, § 23–17.5–27 (2017).

### 7.2.3.3 Low-income Communities and Communities of Color

Members of low-income communities and communities of color face heightened vulnerability to heat. Non-Hispanic Black persons, particularly, are more vulnerable to negative health effects from exposure to heat,[59] possibly because of chronic medical conditions that are more common among the Black population and the higher prevalence of Black persons living in neighborhoods with high concentrations of impervious surfaces and environmental justice burdens.[60] As with older adults, the correlation between these factors – as well as low levels of education and income and limited access to healthcare – and heightened risk is clear but those factors' causal mechanisms are not yet well understood.[61]

Federal civil rights laws in the United States are grounded in both the Constitution and in statutes such as the Civil Rights Act of 1964.[62] Federal law generally guarantees equal treatment by state actors and, in some cases, by "places of public accommodation" such as restaurants or hotels.[63] While almost no climate change-related cases have been filed under civil rights laws to date, local governments addressing climate change could potentially be open to suit if policies are not implemented in a way that treats all people or communities equally.[64] While a full analysis of the potential for enforcement of climate equity using civil rights law is beyond the scope of this chapter, it is worth noting that many states also have civil rights provisions in their constitutions and statutes, suggesting that local governments should pay close attention to whether communities of color are benefitting equally from climate change policies.

### 7.2.3.4 Homeless People and People with Psychiatric Illnesses

In many urban environments, homeless individuals are constantly exposed to the city's hottest temperatures during heatwaves, with little respite unless the

59 Carina Gronlund, *Racial and Socioeconomic Disparities in Heat-Related Health Effects and Their Mechanisms: A Review*, 1 Current Epi. Reports 165 (2014).

60 C. K. Uejioet al., *Intra-Urban Societal Vulnerability to Extreme Heat: The Role of Heat Exposure and the Built Environment, Socioeconomics, and Neighborhood Stability*, 17 Health Place 498–507 (2011).

61 Gronlund, *Racial and Socioeconomic Disparities*, *supra* note 59.

62 U.S. Const. amend. XIII, XIV, & XV; Civil Rights Act of 1964, 42 US.C. § 2000a.

63 Civil Rights Act of 1964, 42 U.S.C. § 2000a. *Heart of Atlanta Motel, Inc.* v. *U.S.*, 379 U.S. 241, 85 S. Ct. 348, 13 L. Ed. 2d 258, 1 Empl. Prac. Dec. (CCH) P 9712 (1964).

64 *Coal. for a Safe Env't v. Ca. Air Resources Bd.* was an administrative complaint filed with EPA under Title VI of the Civil Rights Act of 1964 and was rejected for lack of ripeness in 2012. *Coal. for a Safe Env't v. Ca. Air Resources Bd.*, Docket 09R-12-R9 (2012). Description and case documents available at http://climatecasechart.com/principle-law/civil-rights-act-of-1964/.

city's homeless shelters stay open.[65] This exposure combines with underlying chronic conditions among the homeless population such as respiratory and mental illnesses, along with poor access to healthcare in general, to make heatwaves dangerous.[66] Put simply, homeless people's mortality amid heatwaves can be high: "[d]uring the 2005 Phoenix extreme heat event, two-thirds of coroner-classified heat mortality cases were homeless people."[67]

Those with chronic mental and behavioral illnesses, including people without permanent residences, are also susceptible to heat-related illnesses at a greater rate. Individuals with mood disorders, substance abuse problems, and those taking psychotropic drugs (which interfere with thermoregulation) can have their symptoms worsen during high temperature days.[68]

Duties to homeless individuals in the United States stem primarily from state law: notably, seventeen states' constitutions include a duty that has been interpreted as requiring the state to provide support for "the needy."[69] In some states, duties have developed out of litigation in the 1970s and 1980s that established a "right to shelter."[70] While fewer cases have addressed the issue of the adequacy of the shelter provided, some standards have been created in limited circumstances addressing health and hygiene concerns.[71] Arguably, keeping a shelter at a safe temperature during a heatwave would similarly be considered to be a part of the standard for adequacy of shelter and could perhaps be enforced at least in states with those existing legal duties.

65 Brodie Ramin and Tomislav Svoboda, *Health of the Homeless and Climate Change*, 186 J. Urban Health 654 (2009).

66 Ibid.

67 F. Y. Yip et al., *The Impact of Excess Heat Events in Maricopa County, Arizona: 2000–2005*, 52 Int'l J. of Biometeorology 765–772 (2008).

68 Karin Martin-Latry et al., *Psychotropic Drugs Use and Risk of Heat-Related Hospitalization*, 22 Eur. Psych. 335 (2007).

69 See, e.g., N.Y. Const., art. XVII, §1 ("The aid, care and support of the needy are public concerns and shall be provided by the state and by such of its subdivisions, and in such manner and by such means, as the legislature may from time to time determine."). For a discussion of those state constitutional provisions, see James K. Langdon II and Mark A. Kass, *Homelessness in America: Looking for the Right to Shelter*, 19 Colum. J.L. & Soc. Probs. 305 (1985).

70 See, e.g., *Callahan v. Carey*, No. 79-42582 (N.Y. Sup. Ct. Dec. 5, 1979), reprinted in N.Y. L.J., Dec. 11, 1979, relying on N.Y. Const., art. XVII, §1; *Hodge v. Ginsberg*, 303 S.E.2d 245, 247–250 (W. Va. 1983).

71 See *Lamboy v. Gross*, 493 N.Y.S.2d 709, 710–711 (1985), citing the placement of a homeless family in an Emergency Assistance Unit, sleeping on tables with "inadequate food and bathing facilities" as a situation causing harm to the family.

### 7.2.3.5 Outdoor Workers

Outdoor workers, whether working in agriculture or other areas such as transportation or construction, are exposed to heat and poor air quality during heatwaves.[72] Respiratory diseases and heat stroke/stress are the most obvious health consequences of this type of exposure,[73] but outdoor workers whose break time and water intake are restricted can also experience acute kidney disease.[74] Furthermore, outdoor workers are exposed more to other health-related climate impacts such as vector-borne disease, water-borne disease, and extreme weather events.[75]

Worker safety in the United States is the subject of the Occupational Safety and Health Act (OSHA) of 1970, which requires employers to (1) keep workplaces free from hazards[76] and (2) follow federal regulations on worker safety.[77] OSHA interprets "hazard" to include extreme heat.[78] While formal OSH standards for heat exposure for workers have not been promulgated, the National Institute of Occupational Safety and Health (NIOSH) has created a set of criteria for employers to use to determine when workers are at risk of heat illness and death, and guidelines for managing heat for outdoor workers.[79] The criteria incorporate the most up-to-date research on heat stress and other illness, and recommend that more research be done on the potential impacts of climate change on occupational health and heat.[80]

Although OSHA rules apply to heat as an occupational hazard, the statute applies only to "employees," and so excludes contractors. In situations where employers might try to claim that agricultural, construction, or other outdoor

72 Tord Kjellstrom et al., *Heat, Human Performance, and Occupational Health: A Key Issue for the Assessment of Global Climate Change Impacts*, 37 Annual Review of Public Health 97–112 (2016).

73 Karin Lundgren et al., *Effects of Heat Stress on Working Populations When Facing Climate Change*, 51 Industrial Health 3 (2013).

74 Catharina Wesseling et al., *Mesoamerican Nephropathy. Report from the First International Research Workshop on MeN*. (2013) (cited in Kjellstrom et al., *Heat, Human Performance, and Occupational Health*, *supra* note 72).

75 Lundgren et al., *Effects of Heat Stress*, *supra* note 73, at 3.

76 29 U.S.C.A. § 654(a)(1). Unlike the Fair Labor Standards Act, agricultural workers are covered under OSHA. 29 C.F.R. § 1975.4(b)(2).

77 29 U.S.C.A. § 654(a)(2).

78 U.S. Occupational Safety and Health Administration, *Standards*, www.osha.gov/SLTC/heatstress/standards.html.

79 National Institute for Occupational Safety and Health, *NIOSH Criteria for a Recommended Standard: Occupational Exposure to Heat and Hot Environments*, DHHS (NIOSH) Publication 2016–106; US Dep't. of Labor, Occup. Safety and Health Admin., *Heat Index*, www.osha.gov/SLTC/heatillness/heat_index/.

80 Ibid.

workers are independent contractors, litigation or advocacy may be necessary to enforce OSHA protections.[81] The Occupational Safety Health Review Commission has created clear standards for determining whether a worker is an employee or independent contractor that involve both who pays the workers and who controls their work.[82]

The California Occupational Safety and Health Administration has adopted standards to protect California workers from heat illness.[83] They apply to agricultural, construction, landscaping, oil and gas extraction, and transportation workers and require employers to, among other things, provide shade and water, have a heat illness prevention plan, and create procedures for high-heat days and emergencies.[84] Particularly in the absence of a federal standard to protect the health of outdoor workers, California could serve as a model of public health regulation for other states to follow.

For state employees, the state of Delaware has created a program called Climate-Ready Workforce to protect the state's outdoor workers from heat and other climate hazards.[85] A collaboration between the state's environmental, transportation, emergency management, and public health agencies, Climate-Ready Workforce is a pilot project intended to demonstrate ways to protect the state's outdoor workers from heat and other hazards and to encourage cross-agency collaboration.[86]

### 7.2.3.6 Prisoners

Prisoners occupy a unique category with respect to climate and health impacts. Dependent on corrections staff and unable to relocate to cooler environments, prisoners in aging and overcrowded prison buildings with poor or no air conditioning face serious exposure to high temperatures.[87] Chronic conditions such as respiratory illnesses, an aging prisoner population, and a high rate of cognitive impairment would make the prisoners vulnerable to heat even if their imprisonment did not also expose them to heightened

[81] Occ. Safety & Health L. § 2:3 (2017 ed.).

[82] *Sec. of Labor* v. *Griffin & Brand of McAllen*, 6 O.S.H. Cas. (BNA) 1702 (O.S.H.R.C.), 1978 O.S.H.D. (CCH) P 22829, 1978 WL 7060 (1978).

[83] 8 CA ADC § 3395 [84] Ibid.

[85] Delaware Department of Natural Resources and Environmental Control, *Climate Action in Delaware: 2016 Progress Report*, 8 (2016) biwww.dnrec.delaware.gov/energy/Documents/2016%20Climate%20Action%20Progress%20Report/Climate%20Action%20in%20Delaware%202016%20Progress%20Report.pdf.

[86] Ibid.

[87] See Daniel W. E. Holt, Sabin Center for Climate Change Law, *Heat in U.S. Prisons and Jails: Corrections and the Challenge of Climate Change*, (2015).

environmental risks.[88] Indeed, inmates have died from heat exposure in prisons with inadequate cooling and ventilation systems, and many more are at risk.[89] Prisoners who work outdoors also experience the risks discussed in relation to workers.[90]

Both the Eighth Amendment to the U.S. Constitution and several federal statutes protect the health and safety of prisoners. Federal and state governments owe a particular legal duty to keep prisoners, over whom they have legal custody, healthy and safe.[91] The Constitution's Eighth Amendment guarantee against cruel and unusual punishments has been interpreted to require conditions that provide "the minimal civilized measure of life's necessities."[92] The Tenth Circuit has further interpreted this requirement to mean a "healthy habilitative environment" for prisoners, meaning that the prisoner's environment "does not cause his degeneration or threaten his mental and physical well being."[93] Recognized features of this environment include adequate space, light, ventilation, heating, and cooling, among others.[94]

In the context of heatwaves and high temperatures, courts have recognized that protection from high temperatures falls under the government's requirement to provide shelter.[95] In *Russell* v. *Johnson*, the mere risk of heat-related illness from high temperatures in the prison was enough for the court to identify a duty of care for the prison.[96] Remedies required by courts have included access to ice, fans, open windows with window screens, and cool showers.[97] As climate change increases temperatures throughout the country and as prison populations continue to age, prisons will need to be proactive in controlling temperatures for prisoners or risk Eighth Amendment litigation.

Federal statutes provide another source of legal duties and protection, in particular the Americans with Disabilities Act (ADA) and section 504 of the

88 Ibid. at 19–23.

89 E.g., *Brock v. Warren Cnty, Tn.*, 713 F. Supp. 238 (E.D. Tenn. 1989); *Hinojosa v. Livingston*, 807 F.3d 657 (5th Cir. 2015); *Gibson* v. *Moskowitz*, 523 F.3d 657 (6th Cir. 2008).

90 *Mays* v. *Rhodes*, 255 F.3d 644 (8th Cir. 2001).

91 See *Estelle* v. *Gamble*, 429 U.S. 97 (1976).

92 *Rhodes* v. *Chapman*, 452 U.S. 337, 347–48 (1981), at 347.

93 *Battle* v. *Anderson*, 564 F.2d 388 (10th Cir. 1977), at 395; *Ramos* v. *Lamm*, 639 F.2d 559 (10th Cir. 1980).

94 *Helling* v. *McKinney*, 509 U.S. 25, 31 (1993).

95 E.g., *Walker* v. *Schult*, 717 F.3d 119, 128 (2d Cir. 2013); *Hinojosa v. Livingston*, 807 F.3d 657 (5th Cir. 2015); *Gates v. Cook*, 376 F.3d 323 (5th Cir. 2004); *Hathaway* v. *Holder*, 491 F. App'x 207, 208 (2d Cir. 2012);

96 See *Russell v. Johnson*, No. 1:02CV261-JAD, 2003 WL 22208029, at *2 (N.D. Miss. May 21, 2003), aff'd in part, vacated in part on different grounds sub nom. *Gates v. Cook*, 376 F.3d 323 (5th Cir. 2004).

97 *Cole* v. *Collier*, No. 4:14-CV-1698, slip op. at 44–46 (S.D. Tex. July 19, 2017).

Rehabilitation Act of 1973.[98] Remedies are available to prisoners under Title II of the ADA and section 504 if they can show that they have a disability that interferes with a major life activity, that the prison was aware of that activity, and that the prison failed to make reasonable accommodations for that disability.[99] Both laws would apply whenever a prisoner with a disability (mental illness, chronic disease, etc.) would need relief from high temperatures to stay healthy – a special accommodation.[100] The Fifth Circuit Court of Appeals has held that thermoregulation is a "major life activity" under the meaning of the ADA and section 504; this qualifies people living with a disability, including prisoners, as being entitled to reasonable and effective accommodations.[101]

## 7.3 PUBLIC HEALTH INTERVENTIONS

One of the challenges of addressing heatwaves and unseasonable heat in the age of climate change is the diversity of interventions necessary to protect public health. Acute or emergency health impacts of heat require local public health departments to exercise core public health functions such as activating early warning systems, opening cooling centers, and tracking heat-related morbidity and mortality. Longer-term, gradual increases in everyday temperatures, in contrast, require changes to the built environment that counter the urban heat island effect and improve the resilience of indoor environments to the hotter and more humid days of the future. (See Chapter 6.) Effecting these changes with public health outcomes in mind is especially challenging because public health departments lack authority over building codes, zoning rules and decisions, and the structure and operation of the electricity grid. Thus, for public health departments to influence decisions taken by zoning commissions and local public utility commissions they must coordinate with disparate agencies that have not worked together much or at all in the past.[102]

The interventions described below relate to both acute responses for emergencies and longer-term changes to the built environment. This description notes sources of legal authority available to justify those interventions and includes examples from around the country of local governments, researchers, and community-based organizations that have tried each approach. Consistent with public health's essential commitment to protect society's

98 *Holt, Heat in U.S. Prisons and Jails*, *supra* note 87, at 50. 99 42 U.S.C. § 12102(1)(A).

100 42 U.S.C. §§ 12131–12134 (2014); 29 U.S.C.A. § 705(20)(B).

101 *Ball* v. *LeBlanc*, 792 F.3d 584 (5th Cir. 2015).

102 P. L. Kinney et al., *Approaches for Estimating Effects of Climate Change on Heat-Related Deaths: Challenges and Opportunities*, 11 Environ Sci Policy 87–96 (2008).

most vulnerable populations, all of these interventions should prioritize populations that are either most vulnerable to impacts or have the least adaptive capacity to respond to those impacts.[103] Local and state governments are increasingly incorporating this emphasis into their interventions, but more work needs to be done to ensure that sociodemographic vulnerabilities are considered in heat management strategies from beginning to end.

Many of the interventions below rely on state governments' police powers.[104] Those sovereign powers of state government include the authority to regulate and to act for the benefit of the public in several contexts. In the context of public health, that authority is construed broadly.[105] State police powers have been interpreted to include public health actions from isolation and quarantine to vaccination and nutrition regulation.[106]

Police powers must generally be applied uniformly and in a way related to a legitimate public purpose.[107] Local governments can exercise these police powers if the state government delegates them authority to do so through the state constitution or statute.[108] Local governments' interventions must generally be consistent with related state laws.[109] Authority that is explicitly conferred on local governments by constitution or statute may be discretionary (i.e. local governments can choose whether to exercise it) or mandatory (i.e. state law directs local governments to exercise it).[110] To illustrate: in *Petrushansky* v. *State of Maryland*, the county health commissioner was described as having a duty to exercise charter powers, including the power to enforce health and sanitation ordinances.[111]

Other interventions listed here rely on areas of state or local power such as the power to tax.

### 7.3.1 *Emergency Response Interventions*

Public health departments have a duty to design and implement emergency response plans to protect residents from heat-related illness and death resulting from heatwaves and heat emergencies. Interventions can include setting up and managing cooling centers for local residents, establishing early warning systems and outreach networks to communicate risk to residents, training

103 B. Fernandez Milan and F. Creutzig, *Reducing Urban Heat Wave Risk in the 21st Century*, 14 Curr. Opin. Envtl. Sustain. 221–231 (2015).

104 McQuillin Mun. Corp. vol. 6A, § 24:10 (3d ed.)

105 Gostin, *Public Health Law*, *supra* note 24, at 93–95.

106 Ibid.

107 McQuillin Mun. Corp. vol. 6A, § 24:9 (3d ed.)

108 Ibid. at 24:37.

109 Ibid. at 24:9.

110 McQuillin Mun. Corp. vol. AA, § 10:37 (3d ed.)

111 *Petrushansky* v. *State*, 182 Md. 164, 172 (1943).

and assisting the medical community in keeping vulnerable populations healthy and safe, and performing public health's epidemiological functions to determine whether interventions are having the desired effects.

### 7.3.1.1 Cooling Centers

Cooling centers provide a cool space for residents who do not have ready access to one during heat emergencies. Many also give away cold water. Cooling centers can be run by local governments, community-based organizations, or by coalitions of both together. Establishing cooling centers is discretionary, but, as described in this section, many local governments have chosen to make them available to residents – particularly low-income residents and other vulnerable community members.

Best practices for cooling center siting and management are beginning to emerge. Cooling centers work best if they are sited in central locations where community members are comfortable, such as libraries or recreation centers.[112] Low-income, socially isolated, and otherwise vulnerable residents are more likely to make use of cooling centers if they can easily travel there and find the environment welcoming and comfortable. Thus, ideally, cooling centers should be accessible by transit and/or the community should offer transportation for residents, as well as activities, snacks, and first aid to encourage attendees to remain for the hottest parts of the day.[113]

Some localities use Geographic Information Systems (GIS) mapping technologies to ensure that they site their cooling centers where the most vulnerable people are located. The city of Toronto, Ontario, for instance, has used both exposure (built environment) and sensitivity (sociodemographic vulnerabilities) data to spotlight the most vulnerable areas in the city and to determine cooling center locations.[114] The city's climate staff, public health staff, and planning staff collaborated on this effort, coordinating to establish criteria, find relevant data, and map the city.[115]

112 S. G. Nayak et al., *Surveying Local Health Departments and County Emergency Management Offices on Cooling Centers as a Heat Adaptation Resource in New York State*, 42 J. Community Health 43–50, 43 (2017).

113 Nayak, *Surveying Local Health Departments*, at 46.

114 C. Rinner et al., *Implementation of a Map-Based Heat Vulnerability Assessment and Decision Support System* 2011, www.climateontario.ca/doc/ORAC_Products/TPH/Mapping%20Tool%20-%20User%20Manual%20for%20Heat%20Vulnerability%20Mapping%20Tool.pdf

115 Ibid.

### 7.3.1.2 Early Warning Systems and Outreach

Early warning systems rely on weather monitoring and health surveillance to establish a threshold for communicating with the public that a heat emergency is occurring or is likely to occur.[116] Generally run by public health or emergency management agencies, early warning systems usually also trigger a variety of health communication strategies to inform the public about the risks of extreme heat and the actions they can take to reduce their risks. Without that health communication and outreach, early warning systems may fail to save lives and reduce disease as intended.[117] A meta-study of the effectiveness of early warning systems concludes that they are effective in reducing heat-related mortality, and possibly also cost-effective when comparing money spent implementing the system against lives saved.[118] More research is needed, however, on the effectiveness of early warning systems for vulnerable populations particularly.[119]

The Philadelphia Hot Weather–Health Watch/Warning System (PWWS), for example, is a comprehensive program developed in 1995 after a series of heatwaves over previous summers.[120] PWWS takes into account more sophisticated factors than just heat and humidity, including air mass types and how many consecutive days of hot weather are occurring.[121] In the event of a PWWS alert, several things occur:

* Media outlets (including radio, television, and newspapers) are asked to announce a heat warning along with recommendations about how to reduce heat risk and encouragement to check on elderly and vulnerable relatives and neighbors;
* The city opens a "Heatline" phone-based information service in cooperation with the Philadelphia Corporation on Aging to provide further information to callers on staying cool and healthy during the emergency;

[116] K. Price, et al., *Implementation of the Montreal Heat Response Plan During the 2010 Heat Wave*, 104 Can. J. Public Health Rev. Can Santé Publique e96–e100 (2013).

[117] World Meteorological Org. and World Health Org., *Heat Waves and Health: Guidance on Warning-System Development* 38–47 (G. R. McGregor, lead ed., 2015)

[118] Ghasem Toloo et al., *Are Heat Warning Systems Effective?* 12 Env'l Health 27 (2013). The authors stress, however, the need for more research on the cost-effectiveness question particularly.

[119] Ibid.

[120] K. L. Ebi et al., *Heat Watch/ Warning Systems Save Lives: Estimated Costs and Benefits for Philadelphia* 1995– 98, 85 Bulletin of the Am. Meteorological Soc'y 1067–1073 (2004).

[121] Ibid.

* The public health department reaches out to nursing homes and other facilities where populations vulnerable to heat are located to make sure that they're aware of the heat wave and know what steps to take;
* The local water and electric utilities cease suspending service for delinquent accounts;
* Emergency medical teams increase staffing to be able to respond to emergencies;
* The homeless services agency increases outreach to the city's homeless residents; and
* Senior centers run extended hours and provide cooling services for the elderly.[122]

A review of the program found that it likely saved 117 lives between 1995 and 1998, at a minimal cost to the city.[123]

Philadelphia's warning and outreach system incorporates several best practices seen in other cities as well: engagement with local media and agencies on aging, and collaboration with community-based organizations to help engage difficult-to-reach communities including communities of color and low-income communities. The City of Milwaukee's Heat Task Force similarly engages with various city agencies and community groups to implement a comprehensive outreach plan that builds on mapping of the most vulnerable communities throughout the county.[124]

### 7.3.1.3 Public Health Surveillance

Surveillance of disease and injury is a core part of the public health responsibilities that states and localities are empowered to carry out and can include tracking hospitalization rates, diagnoses, and causes of death in a way that allows for cross-tabulation with demographic data such as age, gender, and zip code.[125] (See Chapter 5.) Tracking these data enables local public health departments to better understand the impact of extreme heat on morbidity and mortality, decide whether to trigger early warning systems, and prioritize areas or populations for direct

122 Ibid.

123 Ebi et al., *Heat Watch/Warning Systems Save Lives*, *supra* note 120, at 1067–1073.

124 City of Milwaukee, WI, Health Department, *Excessive Heat Event Coordination Plan* (June 2014) http://city.milwaukee.gov/ImageLibrary/Groups/healthAuthors/DCP/PDFs/ExtremeHeat/2014MilwaukeeExcessiveHeatEven.pdf.

125 Gostin, *Public Health Law*, *supra* note 24, at 286. L. Josseran et al., *Assessment of a Syndromic Surveillance System Based on Morbidity Data: Results from the Oscour Network During a Heat Wave*, 5 PLos ONE e11984.1–8 (2010), www.plosone.org.

outreach and intervention.[126] A robust surveillance system will also help to quantify the increase in morbidity and mortality risks that will accompany future heatwaves and average temperature increases, and to ensure that resources are adequate and deployed in areas of highest need.

Surveillance of heat-related morbidity and mortality is challenging, however. Medical records do not always track the weather conditions coincident with a particular diagnosis.[127] Similarly, deaths due in whole or in part to heat can be attributed to other causes, such as cardiovascular disease, if an underlying health condition or chronic disease was also present.[128] Therefore, surveillance of heat-related disease, while important, may underestimate the real disease and mortality burden of heat. Suggestions have also been made to ensure that public health tracking systems incorporate not only sociodemographic data but also geolocation data in order to be able to map problems and interventions, and to track whether the interventions are effective.[129]

While many local governments already implement emergency response strategies and more are beginning to follow the best practices mentioned in this section, climate change projections suggest that these emergency response strategies could be unsustainable for the number of heat emergency days that localities will face in the future. Washington, DC, for example, has historically experienced approximately thirty days per year where the heat index is over 95 degrees.[130] By the 2080s, the District is projected to experience forty to seventy-five such days: potentially more than half of the ninety-day summer.[131] Emergency response strategies will save lives and protect health amid the increasing numbers of excessive heat events, but health departments making them a day-to-day practice over large portions of the summer seems impractical due to both financial and staffing constraints.

126 M. Pascal et al., *How Can a Climate Change Perspective be Integrated Into Public Health Surveillance?* 126 Public Health 660–667 (2012).

127 Sarofim et al. *Temperature-Related Death and Illness*, *supra* note 3, at 46.

128 Jeffrey Berko et al., *Deaths Attributed to Heat, Cold, and Other Weather Events in the United States, 2006–2010*, National Health Statistics reports No. 76 (2014).

129 A. Houghton et al., *Climate Change-Related Vulnerabilities and Local Environmental Public Health Tracking Through GEMSS: A Web-Based Visualization Tool*, 33 Applied Geography 36–44, 44 (2012).

130 District of Columbia, *Climate-Ready DC: The District of Columbia's Plan to Adapt to a Changing Climate*, 2 (2016).

131 Ibid.

### 7.3.2 *Long-term Adaptation/Prevention*

Heat problems are particularly acute in urban environments, where density of buildings and prevalence of dark-colored and impervious surfaces absorbs heat during the day and radiates it at night, keeping both daytime and nighttime temperatures higher than surrounding, more vegetated environments.[132] Protecting public health effectively amid future heatwaves and high temperatures will require changes to the built environment that reduce ambient urban temperatures and reduce exposure to extreme heat both during heat emergencies and day to day.[133] (See Chapter 6.) Increasing open space, vegetation, and tree canopy and decreasing pavement and dark-colored surfaces (such as roofs) wherever possible will cool urban areas and protect cities' most vulnerable residents.[134] These strategies, in contrast to emergency response strategies, also help to address broader consequences of higher temperatures, including higher energy use (and higher energy bills), air pollution, and water quality degradation – and thereby support both climate change mitigation and adaptation strategies.[135]

This section explores the built environment and collaboration strategies that enable local governments to cool their jurisdictions and protect public health in advance of a heat event. These strategies are a necessary complement to emergency response strategies and will require public health agencies to partner with and to influence other agencies that have jurisdiction over the built environment.

The following sections are organized by government power: Planning and Zoning; Building Codes; Control over Streets; Tax Powers; and Utility Regulation. Legal authority and challenges for each are discussed, along with examples of implementation from around the country. As with emergency response, built environment responses will also need to focus on vulnerable populations to the extent possible. Research shows a marked disparity in access to green space in urban areas for both low-income

[132] U.S. Environmental Protection Agency. *Excessive Heat Events Guidebook* (2006). www.epa.gov/heatisland/about/heatguidebook.html.

[133] B. Fernandez Milan and F. Creutzig, *Reducing Urban Heat Wave Risk in the 21st Century*, 14 Curr. Opin. in Envtl. Sustain. 221–231 (2015).

[134] Y. Xu, et al., *Differences on the Effect of Heat Waves on Mortality by Sociodemographic and Urban Landscape Characteristics*, 67 J. Epidemiol. Commun. Health 519–525 (2013).

[135] S. L. Harlan and D.M. Ruddell, *Climate Change and Health in Cities: Impacts of Heat and Air Pollution and Potential Co-Benefits from Mitigation and Adaptation*, 3 Curr. Opin. Envtl. Sustain. 126–134 (2011).

communities and communities of color;[136] interventions taking an equity approach should attempt to remedy this disparity through capital investment and other strategies.[137]

#### 7.3.2.1 Built Environment Strategies

The basic methods of altering buildings and landscapes to reduce heat include (1) green and cool roofs; (2) cool pavements (and less pavement); and (3) urban forestry, including all types of vegetation.[138] Each has benefits and drawbacks and different legal tools to implement them:

* Green and cool roofs, often encouraged or mandated by building or zoning codes, lower both interior temperatures and exterior temperatures over a limited geographic area.[139] While cool roofs reflect the sun's energy away from the building (instead of absorbing it), the vegetation on green roofs cools the building through shading and cools the air through evapotranspiration.[140]
* Cool pavements come in both light-colored and permeable varieties. As with green and cool roofs, cool pavements work by either reflecting the sun's energy (light-colored) or by evapotranspiration of the water that infiltrates the permeable surface.[141] Local governments can also simply use less pavement by increasing the size of tree boxes and installing sidewalks that are less wide where pedestrian traffic allows. In neighborhoods where foot traffic is less dense, the paved portion of the right-of-way can be less wide without sacrificing walkability or safety, and can be paired with bioretention or bigger tree boxes to both reduce heat and manage stormwater. Pavements on private property can be regulated through tools such as zoning and building codes. Patches of pavement on public property can be changed by local governments at

[136] D. Dai, *Racial/Ethnic and Socioeconomic Disparities in Urban Green Space Accessibility: Where to Intervene?* 102 Landsc Urban Plan 234–244 (2011).

[137] T. Kjellstrom et al., *Urban Environmental Health Hazards and Health Equity*, 84 J. Urban Health 86–97 (2007).

[138] See generally U.S. EPA, *Reducing Urban Heat Islands: Compendium of Strategies* (2008).

[139] U.S. EPA, *Reducing Urban Heat Islands: Compendium of Strategies*, Green Roofs 1 (2008) and U.S. EPA, *Reducing Urban Heat Islands: Compendium of Strategies*, Green Roofs 2–5 (2008).

[140] Ibid.

[141] U.S. EPA, *Reducing Urban Heat Islands: Compendium of Strategies*, Cool Pavements 2 (2008); see also Hui Li, Pavement Materials for Heat Island Mitigation 79, 79, In: Hui Li, *Design and Management Strategies* (2016).

will and public pavements can also be addressed systemically by altering the state or local government's street design standards.

* Urban forestry, including trees and other vegetation, can reduce heat both through shading and evapotranspiration.[142] Legal and policy levers to increase tree canopy can take the form of landscape ordinances, building codes, tax and other incentives for property owners, and public education and outreach programs.
* Somewhat differently, resilience of the energy grid in the face of a heatwave can help to protect public health by providing power during brownouts and blackouts that would otherwise leave buildings and residents overheated.

#### 7.3.2.2 Strategies for Facilitating Collaboration

These types of interventions can be challenging from a public health perspective given local public health departments' historic lack of jurisdiction over local government roles such as planning, zoning, building code specification, and streetscape design. Public health departments that want to influence these processes will need to work with other government agencies and to prevail upon them to incorporate public health and climate change perspectives into their decisions. Two tools in particular can facilitate such collaboration:

HEALTH IMPACT ASSESSMENTS Some governments are now requiring health impact assessments (HIAs) to be performed before certain projects can be permitted. Health impact assessments are rarely specifically required by law (Massachusetts is a limited exception),[143] but considerations of public health are often part of environmental reviews under state and federal laws such as the National Environmental Policy Act (NEPA).[144] Where HIAs are required or where environmental review processes emphasize public health, they can help inject public health priorities into built environment decision-making. (See Chapter 14.)

[142] D. Armson et al., *The Effect of Tree Shade and Grass on Surface and Globe Temperatures in an Urban Area*, 11 Urban Forest Urban Green 245–255 (2012).

[143] Massachusetts' Healthy Transportation Compact requires health impact assessments to be performed for transportation projects in order to ensure that the projects support healthy outcomes and take vulnerable populations into account. Mass. Gen. Laws ch. 6C, § 33.

[144] J. G. Hodge et al., *Legal Review Concerning the Use of Health Impact Assessments in Non-Health Sectors* (2010) www.humanimpact.org/wp-content/uploads/2010/05/Legal_Review_of_HIA_report.pdf.

CLIMATE ADAPTATION PLANS/CLIMATE ACTION PLANS If a state or local government has a legal obligation to take climate action, either through a climate adaptation plan adopted by the legislature or through a state or local executive order,[145] government agencies will need to work together to carry out that legal obligation. Public health agencies may find friendlier responses from allied agencies if leadership is encouraging or requiring action from both of them.

Whether or not government performs HIAs or operates pursuant to a climate action plan, funding scarcity will be an issue. Governments benefit from being able to identify multiple benefits when justifying expenditures, whether program funds or capital investments.[146] For example, tree planting can improve air quality, protect public health, lower temperatures, and manage stormwater. Many of these interventions improve air quality, water quality, or energy efficiency in addition to protecting public health from extreme heat. This focus on co-benefits, along with funding scarcity, may be persuasive to other agencies who manage, for example, the building or zoning codes.

More and better weather, demographic, and GIS data make localities better able to understand and respond to the public health risks of extreme heat. The metro government of Louisville, Kentucky, is tackling its urban heat island using data on both physical and human vulnerability throughout the city, with a focus on addressing the problem in the city's most vulnerable areas first.[147] While not every municipality will have the data and projections that Louisville has, any municipality can use basic GIS data to identify high concentrations of their vulnerable populations, discussed earlier, and target interventions first to areas of highest need.

### 7.3.2.3 Planning and Zoning

Local governments plan and control land use in their jurisdictions through exercise of their police powers. While this chapter deals primarily with planning and zoning powers, this cluster of police powers can also include local ordinances such as subdivision regulations and landscape ordinances, among others. Planning and zoning at their cores are an exercise in a community's setting a vision and goals for itself, and then deciding where to locate particular

145 See, e.g., Delaware Executive Order 41: "Preparing Delaware for Emerging Climate Impacts and Seizing Economic Opportunities from Reducing Emissions" (2013).

146 Graham Floater et al., *Co-Benefits of Urban Climate Action: A Framework for Cities* (Sept. 2016).

147 See generally Urban Climate Lab, *Louisville Urban Heat Management Study* (2016) https://louisvilleky.gov/sites/default/files/advanced_planning/louisville_heat_mgt_revision_final_prelim.pdf.

land uses in specific areas to achieve those goals. This proposition is extremely difficult to extricate from public health goals. For example, one of the original uses of zoning ordinances was to protect residential areas from more intensive and polluted industrial areas, in order to keep those residential areas healthier, safer, and quieter.[148] Local planning and zoning powers are limited by constitutional principles as well as by the scope of authority delegated by the state.

Planning and zoning can play a role in adapting to the public health impacts of heat in several ways. Comprehensive plans and zoning together can increase density and promote transit-oriented development to reduce how much residents are driving individual cars;[149] because waste heat from vehicles contributes to increased temperatures, less driving has a positive impact on heat.[150] Similarly, zoning codes can help address the urban heat island effect by promoting investments in energy efficiency and the installation of green infrastructure. In recent years, localities have increasingly cited public health and environmental priorities to justify zoning regulations,[151] supporting, for example, smart growth regulations on urban sprawl due to the air quality impacts and destruction of open green space.[152]

Local governments' comprehensive plans are frameworks for land use decision making that specify goals and policies.[153] Planning has long reflected public health and environmental concerns, and local governments are now beginning to include elements in their comprehensive plans that address health, climate, and other features together.[154] Other types of plans such as climate action plans, sustainability plans, and climate adaptation plans can guide local governments' efforts to protect people and infrastructure from

148 Julian Conrad Juergensmeyer and Thomas E. Roberts, *Land Use Planning and Development Regulation Law*, 23 (Thomson West 2003).

149 Arefeh Nasri and Lei Zhang, *The Analysis of Transit-Oriented Development (TOD) in Washington, DC and Baltimore Metropolitan Areas*, 32 Transport Policy 172–179 (2014).

150 Nick Earl et al. *Weekly Cycles in Peak Time Temperatures and Urban Heat Island Intensity*, 11 (7) Envtl Research Letters 074003 (2016).

151 Juergensmeyer and Roberts, *Land Use Planning, supra* note 148, at 23.

152 Juergensmeyer and Roberts, *Land Use Planning, supra* note 148, at 321; Brian Stone Jr., *Urban Heat and Air Pollution: An Emerging Role for Planners in the Climate Change Debate*, 71 J Am Plan Assoc. 13–25 (2005).

153 See generally, W. Goodman and E. Freund, *Principles and Practice of Urban Planning*, 349 (1968). Authority for planning comes from state government, and is sometimes mandatory and sometimes optional but generally encouraged; Juergensmeyer and Roberts, *Land Use Planning, supra* note 148, at 31.

154 The consolidated metro government of Louisville/Jefferson County, Kentucky is currently amending its comprehensive plan to include not only transportation and housing goals, but also health, sustainability, and equity goals woven together. Louisville Metro Government, *Comprehensive Plan 2040: Draft Goals and Objectives*, https://louisvilleky.gov/sites/default/files/planning_design/draftgoalsobjectivesmay172017.pdf.

projected climate impacts. These plans can bring together multiple priorities as a complement to the formal planning process. The city of Cleveland, Ohio, has produced a Tree Plan, formally adopted by the Planning Commission, to increase tree canopy in the city by 2040.[155] The Tree Plan incorporates not only impervious surface, air quality, and watershed protection data in identifying priority areas for tree planting, but also community sociodemographic data in order to reduce asthma rates and keep the city's most vulnerable residents cool.[156] The city of Detroit's Climate Action Plan includes an entire section on public health impacts and responses, including use of Health Impact Assessments.[157]

Because zoning is an expression of a state's police power, it must be exercised for the purpose of promoting the health, safety, and general welfare of the public.[158] While these purposes are broad, zoning actions must be careful to advance an appropriate public purpose, such as public health, managing growth, or traffic safety and to be within the delegated statutory authority, or else they could be found to be *ultra vires*.[159] Zoning can implement more specifically the broad goals established by the various planning processes: zoning to reduce heat is most often focused on increasing green roofs, cool roofs, and green infrastructure such as rain gardens and trees. Zoning regulations can be more prescriptive or can simply provide frameworks for property owners to follow, and can either use mandates and incentives to spur change. The city of Seattle has a zoning ordinance known as the Green Factor that requires new construction in commercial and multifamily residential zones to include a higher proportion of planted area, potentially including green roofs, permeable pavements, and green walls.[160] The green elements are required, but the exact methods the property owners choose to use are discretionary.

Zoning powers are also limited by constitutional requirements, including property rights. Takings challenges, specifically regulatory takings challenges, are the most significant constitutional stumbling block to exercise of zoning

155 City of Cleveland, *The Cleveland Tree Plan* (2015) www.city.cleveland.oh.us/sites/default/files/forms_publications/ClevelandTreePlan.pdf.

156 Ibid.

157 Detroiters Working for Environmental Justice, *Detroit Climate Action Plan*, 28–37 (2017) http://detroitenvironmentaljustice.org/wp-content/uploads/2017/10/CAP_WEB.pdf.

158 Zoning is an exercise of the state police power and is usually delegated through states' Zoning Enabling Acts or home rule acts, which vary in their requirements and generally require conformity with the comprehensive plan. Juergensmeyer and Roberts, *Land Use Planning*, *supra* note 148, at 46–47.

159 Ibid. at 53–61. See also *Griepenburg v. Twp. Ocean*, 105 A.3d 1082, 1090 (2015).

160 See, e.g., Seattle Municipal Code 23.49.031 (2017).

authority. Zoning regulations (and comprehensive plan provisions) that deprive landowners of all economically beneficial use of their property can be considered a "regulatory taking" requiring compensation for that diminishment.[161] While regulatory takings remain a significant concern in the coastal context, zoning regulations designed to lower temperatures generally do not meet this high threshold; property owners can still enjoy many economically beneficial uses of their properties, and so takings is unlikely to be a serious problem for these types of zoning regulations.

#### 7.3.2.4 Building Codes

Building codes are subject to limitations based on the public purposes they are designed to achieve and by potential regulatory takings challenges. Building codes are a major source of authority for local governments for encouraging or requiring "green building," a term encompassing a whole suite of sustainability, energy efficiency, waste reduction, water conservation, and climate-resilient practices.

While many state and local governments adopt model codes from, for example, the International Code Council, some choose also to adopt "green" codes that encourage or require higher performance standards for buildings.[162] These green codes often encourage energy efficiency, shading, or other heat-reducing strategies.

States vary widely in the extent to which they delegate authority to local governments for adopting individual building codes. Models of delegation range from:

* Washington, which creates a statewide building code that serves as a floor but allows local governments to adopt standards that are more restrictive.[163]
* California, which creates a statewide building code that serves as a floor for local governments, which can create standards that are more restrictive (but not less) than the state standards.[164] The code explicitly allows local governments to create more-restrictive "green building standards, reasonably necessary because of local climatic, geological, or topological

[161] *Lingle* v. *Chevron U.S.A. Inc.*, 544 U.S. 528, 125 S. Ct. 2074, 161 L. Ed. 2d 876 (2005).

[162] See, e.g., D.C. Green Construction Code Supplement, D.C. Mun. Regs. Tit. 12, § 101K et seq. (2014).

[163] See Wash. Rev. Code § 19.27.040 (1990). One exception includes amendments that affect residential properties, which are subject to review by the state building code council. Wash. Rev. Code § 19.27.060(1)(a) (2002).

[164] Cal. Health & Safety Code § 18941.5 (2010).

conditions."[165] A statewide commission has authority to disapprove these changes.[166]

* Massachusetts has a statewide building code but allows local governments to apply to a State Board to adopt changes based on climate conditions.[167] The Board then has discretion to approve or disapprove those changes.[168]
* Minnesota has a statewide building code from which local governments may not deviate.[169]

Minnesota's approach, which is not unusual,[170] obviously limits local governments' ability to adopt provisions that address the urban heat island effect and other sources of public health concern. Because adaptation is an inevitably local activity, this chapter recommends providing local governments with floors but not ceilings regarding climate change adaptation in building codes, particularly with respect to heat management and other issues that are directly relevant to public health outcomes.

The main building code strategies to deal with high temperatures focus on promoting green and cool roofs. As with zoning regulations, building code models vary in how prescriptive they are and the types of buildings to which they apply. The city of Los Angeles, for example, explicitly references the urban heat island effect as the reason for requiring cool roofs on all new or substantially renovated residential properties.[171] The local water and energy utility offers a rebate for cool roofs to offset some of the costs of installation.[172] The District of Columbia has adopted, as part of a green building supplement to its building code, a section requiring property owners to explicitly mitigate urban heat by, for example, providing shade and plantings.[173] Similar requirements that do not necessarily explicitly mention reducing heat can still achieve public health benefits. For example, New York City's requirement, adopted in 2012, that new buildings of a certain size reduce stormwater runoff is focused on reducing that runoff and improving water quality, but will reduce temperatures and improve

[165] Cal. Health & Safety Code § 18941.5(b) (2010).
[166] Cal. Health & Safety Code § 17958.7 (1997).
[167] See Mass. Gen. Laws ch. 143, § 98 (1989). [168] Ibid.
[169] See Minn. Stat. Ann. § 326B.121, Subd. 1 (2013).
[170] See, e.g., VA Code Ann. § 36–98 (2007).
[171] Los Angeles Municipal Code 99.04.106.5 et seq. (2014).
[172] Los Angeles Department of Water and Power, *Rebates & Programs*, www.ladwp.com/ladwp/faces/wcnav_externalId/r-sm-rebt-prog?_adf.ctrl-state=vfnrj3w8g_4&_afrLoop=215554573997560.
[173] See, e.g., D.C. CODE MUN. REGS tit. 12 § 408k (2014).

public health nonetheless.[174] Other stormwater ordinances and building code amendments can have similar multiple benefits, even if intended only for water quality purposes.

#### 7.3.2.5 Streetscapes

Local governments most often have control over streetscapes and sidewalks through delegation of their home rule powers.[175] Streets, like buildings, are one of the largest contributors to impermeable surfaces in an urbanized area. Reducing the proportion of dark-colored and impervious surfaces and planting trees in the right-of-way for shade can reduce temperatures and help to manage stormwater. Although streets are traditionally conceived as existing to move vehicles from one place to another, local governments are empowered to regulate streets to also protect pedestrians[176] and regulate the width of streets.[177]

Local governments can require or promote cool pavements and depaving strategies on private property through regulation or financial incentives; however, given the proportion of street and sidewalk paving performed by local and state governments themselves, changes in subregulatory policy (such as changing street design standards) may have the most direct impact on cool pavement usage and depaving. Street design standards allow local governments to provide clear direction for employees and contractors who may be installing green infrastructure and other heat-reducing strategies in rights-of-way along roadways. Washtenaw County, Michigan, has created a guidebook for its local governments outlining the best techniques for installing green infrastructure such as permeable pavements, including how to maintain those installations over time.[178] These guidelines will help their local governments to exercise their authority over the streets in a new way, reducing temperatures and helping to manage stormwater.

174 Department of Environmental Protection Promulgation of Chapter 31 of Title 15 Of the Rules of the City of New York Governing House/Site Connections to the Sewer System, 139 The City Record 9, 15 (Jan. 4, 2012).

175 10A McQuillin Mun. Corp. § 30:40 (3d ed.)

176 *Shuttlesworth* v. *City of Birmingham*, 43 Ala. App. 68, 180 So. 2d 114 (1965).

177 *City of Fulton* v. *Penny*, 273 Ky. 465, 116 S.W.2d 963 (1938).

178 Washtenaw County Water Resources Commissioner, *Rule and Guidelines: Procedures and Design Criteria for Stormwater Management Systems (rev. 2016)*, www.ewashtenaw.org/government/drain_commissioner/dc_webPermits_DesignStandards/dc_Rules/wcwrc-rules-2014-08-06_book.pdf.

TAX POWERS As with police powers, states delegate tax powers to local governments through either state constitutions or statutes; unlike police powers, however, tax powers generally come with strict limits on that authority.[179] The property tax provides roughly 25 percent of local revenue, and so is fiercely protected by local governments.[180] Property taxes are subject to state constitutional limits, including that the tax must be general, meaning for the local government's general funds, instead of for a particular purpose,[181] and state constitutions often place limits on local property tax rates.[182] Under these limits, however, local governments retain discretion over how to design "reasonable systems of taxation."[183] If specifically allowed by statute, local governments can create incentives through the tax code for private property owners to implement heat-reducing strategies.

New York City has instituted a tax abatement for green roofs, in which property owners can get a property tax reduction of $4.50 per square foot for installing a green roof that meets the specified requirements.[184] A state property tax law makes this tax abatement possible for New York municipalities.[185] This tax abatement, while well-intentioned, has been less successful than proponents had hoped, with only five green roofs using the tax abatement to date.[186] Tax credits and abatements are also not ideal for encouraging adoption in low-income neighborhoods that may be more vulnerable to climate impacts; low-income households may lack the up-front capital to make changes, which a tax abatement does not address, While local governments with similar authority to New York's should be able to use their tax power to create similar incentives, tax policy may not be the first tool to reach for in the toolbox.

ENERGY GRID RESILIENCE Most of our emergency response strategies for heatwaves are dependent on electricity. If a heatwave overtaxes the power grid and results in brown outs or black outs with temperatures at dangerous levels, the whole community is at risk. Critical infrastructure such as hospitals and wastewater treatment plants may have backup generators that can keep them running

179 Richard Briffault and Laurie Reynolds, *Cases and Materials on State and Local Government Law*, 649 (2009).

180 Ibid. at 649–650. 181 Ibid. 182 See, e.g., AR Const. Art. 12, § 4.

183 *Lehnhausen* v. *Lake Shore Auto Parts Co.*, 410 U.S. 356, 359 (1973).

184 N.Y.C. Dep't of Bldg., *NYC Green Roof Property Tax Abatement Program* (Jan. 2010), www.nyc.gov/html/dob/html/sustainability/green_roofs.shtml.

185 New York Real Property Tax Law, Title § 4-B et seq.

186 Justin Gundlach, *Putting Green Infrastructure on Private Property in New York City*, 10 (2017), http://columbiaclimatelaw.com/files/2017/06/Gundlach-2017-05-GI-in-NYC.pdf.

for some amount of time, but many community resources such as community centers that might otherwise serve as cooling centers will be unavailable.

The energy grid in the United States is outdated and in desperate need of maintenance and retrofitting, even without the looming impacts of climate change.[187] The American Society of Civil Engineers gave our energy infrastructure a grade of D+ in 2017 based on aging infrastructure and continued underinvestment in systems of all kinds.[188] Even if our infrastructure were in excellent shape today, it would likely not be prepared to handle the increased extreme weather events, heatwaves, and drought that climate change promises to bring.

Hurricane Sandy and the Washington, DC, derecho each illustrate some of the coming challenges for our energy grid. In 2012, Hurricane Sandy hit the coast of New Jersey, New York, and Connecticut, flooding electrical power stations and resulting in 8 million people without power.[189] Many jurisdictions have electrical infrastructure in the floodplain or close to sea level; this exposes the infrastructure to the impacts of extreme weather now, and to both extreme weather and sea-level rise in the future.

As devastating as Hurricane Sandy was, it could have been much worse if it had happened in July or August, when temperatures are hottest, instead of in the fall. A less-significant storm in hotter weather occurred with the 2012 Washington, DC, derecho.[190] The derecho traveled "over 700 miles in 10 hours, resulting in 4.2 million customers being without power for up to a week."[191] This power outage unfortunately occurred at the same time as a heatwave, creating health risks for both residents and for crews working to restore power; 34 people died of heat-related causes in the region during this period, and power restoration was unusually slow.[192] Although fewer people lost power than during Hurricane Sandy, the concurrent heatwave made the

187 Union of Concerned Scientists, *Power Failure How Climate Change Puts Our Electricity at Risk – and What We Can Do*, 2 (USC, 2014), www.ucsusa.org/global_warming/science_and_impacts/impacts/effects-of-climate-change-risks-on-our-electricity-system.html.

188 See American Society of Civil Engineers, *2017 Infrastructure Report Card: Energy* (2017) www.infrastructurereportcard.org/wp-content/uploads/2017/01/Energy-Final.pdf.

189 Gov. Accountability Office, *CLIMATE CHANGE: Energy Infrastructure Risks and Adaptation Efforts*, 2 (2014) GAO-14–74.

190 A derecho is a windstorm that has winds like a tornado, but tends to travel in a straight line, survive for much longer, and therefore cause serious damage. National Oceanic and Atmospheric Administration, *Derecho Facts Page*, (2017), www.spc.noaa.gov/misc/AbtDerechos/derechofacts.htm.

191 Kristina LaCommare et al., *Evaluating Proposed Investments in Power System Reliability and Resilience: Preliminary Results from Interviews with Public Utility Commission Staff*, 16 (Lawrence Berkeley Laboratories, 2017).

192 Ibid. at 16–17.

energy outages that much more dangerous. The derecho demonstrates that as climate impacts become more frequent and unpredictable, they may not always happen in isolation, but may co-occur with cascading effects.

Further discussion of steps involved in adapting the electricity grid is beyond the scope of this chapter. This underscores the challenge that states and localities face when trying to deal effectively with public health threats like heatwaves and hotter weather. If they are to be effective, the tasks of adapting buildings, infrastructure, and the grid cannot be undertaken in isolation from one another, nor should they be treated as beyond the bounds of a public health-driven agenda. Yet the organization of government agencies and the nature of legal obligations and institutions make greater integration difficult.

## 7.4 CONCLUSION

Extreme heat and hotter weather remain among the most commonly recognized effects of climate change, with serious impacts on the public's health. State and local governments have myriad legal tools at their disposal to prepare for and react to hotter temperatures. The challenges for public health agencies in deploying these legal tools come from both (1) the variation in the public health impacts on different populations and (2) the diversity of agencies with responsibility over those legal tools. The wide variation in vulnerability and sensitivity of different populations impedes the ability of public health agencies to develop overarching policies that will be effective for all individuals. For public health to be able to influence the actions of other agencies such as zoning commissions, for example, they will need to better coordinate with and persuade those agencies in order to influence them to take into account the public health implications of their decisions.

# 8

# Oceans and Coasts

Robin Kundis Craig

*Coastal areas are some of the first areas to require climate change adaptation, a result of sea-level rise, exacerbated coastal storms, changing coastal chemistry and biodiversity, and increasing human populations and infrastructure. Coastal adaptation should include a public health perspective, because climate change is simultaneously increasing the risks of coastal disasters, toxic contamination, loss of water supply, and disease. Given the multiple threats facing coastal areas, a public health response must arise from multiple sectors simultaneously, from building codes and land use planning to more traditional disaster and disease preparedness to more finely tuned application of environmental and natural resources laws.*

*This chapter examines climate change's impacts on the oceans and coasts, then focuses on two types of coastal adaptation. Increasing numbers of increasingly severe coastal storms pose one set of disaster-type public health risks to coastal communities, and this chapter examines prudent disaster preparedness actions to reduce them. However, coasts are also subject to longer-term and often more subtle public health threats as a result of rising seas and warming temperatures, requiring overlapping but distinct public health responses to adapt to salinization and new disease threats.*

## 8.1 INTRODUCTION: COASTAL CLIMATE CHANGE DISASTERS AS HEALTH DISASTERS

Coasts are some of the most densely populated regions of the world. As a result, when something goes wrong in the coastal zone, the risk is great that many people will be harmed. While reducing such risks has long been the goal of disaster law and emergency planning, changing ocean conditions, rising sea levels, and increasing vulnerability to coastal disasters all pose considerable public health risks that public health law and public health practitioners should also anticipate and address, with help from property and environmental regulation.

Climate change is increasing the risks to coastal communities. Indeed, these communities face a synergistic constellation of climate change impacts, as warming seas and sea-level rise combine with growing numbers of increasingly severe coastal storms to threaten public health in a number of ways. For example, warming and rising seas create conditions that promote many kinds of diseases, from cholera to mosquito-borne illnesses such as malaria and dengue fever. Coastal flooding and inundation threaten coastal food production and water supplies by contaminating soil and fresh water with salt. Increasing numbers of increasingly severe coastal storms can overwhelm sewage treatment systems and contaminate drinking water supplies with *E. coli* and other pathogens. Coastal flooding also fills city streets with a toxic soup of hazardous chemicals collected from businesses, land, and homes. All of these public health threats should be addressed in climate change adaptation strategies and disaster planning, including significant adaptation and planning in the public health sector.

Ocean acidification, often referred to as climate change's "evil twin," and climate change-induced changes to ocean chemistry simultaneously create a different set of threats to public health, particularly in terms of food supply. Ocean acidification results from the ocean absorbing anthropogenic carbon dioxide, which lowers the ocean's pH. Changing the ocean's pH, in turn, makes it difficult for organisms that live in calcium-based shells, like clams and oysters, to form their shells. Because such creatures are at the base of ocean food chains, their impairment disrupts ocean food webs and coastal aquaculture alike. Ocean acidification also exacerbates threats to coral reefs, which support the ecosystems that supply many island and coastal communities with their primary sources of protein.

This chapter explores climate change and storm disasters along the coasts as public health problems and considers how the law might help to mitigate those public health impacts. Section 8.2 begins by reviewing trends in coastal populations, both in the United States and around the world, and climate change projections relevant to coastal communities. Section 8.3 then examines in more detail the immediate public health consequences of severe coastal storms and their associated storm surge. Notably, the public health risks from these storm events do not differ significantly between developed and developing nations – although the capacity to meaningfully respond can. While coastal storms tend to attract media attention, however, the longer-term consequences of rising and warming seas pose more insidious public health threats to coastal communities. These threats are the subject of Section 8.4 and provide the subjects that public health law might most effectively address for the future.

## 8.2 COASTAL COMMUNITIES AND CLIMATE CHANGE PROJECTIONS

The public health consequences of warming oceans, coastal storms, sea-level rise, and changing ocean chemistry depend primarily on two factors: (1) how many people live in the coastal zone; and (2) how frequent, intense, and pervasive these disturbances become. This section examines projections for the future of the world's coastal communities with respect to both factors.

### 8.2.1 *Humans in the Coastal Zone: Projections for the Future*

In many parts of the world, coasts are crowded places. In the United States, for example, coastal populations have been increasing and are projected to continue to increase, making coastal communities some of the most densely populated areas of the country. Indeed, although shoreline counties constitute less than 10 percent of the total land area (not including Alaska), they already account for 39 percent of the total population – and that percentage is expected to continue increasing, as it has been since at least 1970.[1]

These trends also hold true for the rest of the world. According to Neumann et al.'s 2015 study, the most comprehensive and authoritative projection available as this volume goes to press, population growth and coastal urbanization trends indicate that low elevation coastal zones (LECZs) are growing more crowded and at a faster rate than other areas; by 2060 they will be home to between 1 and 1.4 billion people, or 12 percent of the global population, despite comprising only 2.3 percent of coastal countries' land mass.[2] To date, these trends have been most visible in China, India, and the world's megacities, and the majority (83 percent) of the global LECZ population lives in less developed countries.[3] However, Neumann et al. also project that the densest coastal populations will shift, both within Asia[4] and to Africa, which "will be the continent to experience the highest rates of growth and urbanisation in the LECZ across all scenarios."[5] European and North American LECZ

[1] National Ocean Service, National Oceanic & Atmospheric Administration, US Department of Commerce, *What Percentage of the American Population Lives Near the Coast?* http://oceanservice.noaa.gov/facts/population.html (as revised Feb. 29, 2016).

[2] Barbara Neumann et al., *Future Coastal Population Growth and Exposure to Sea-Level Rise and Coastal Flooding – A Global Assessment,* 10(3) PLoS ONE e0118571 (Mar. 11, 2015) (citations omitted). doi:10.1371/journal.pone.0118571. http://journals.plos.org/plosone/article/file?id=10.1371/journal.pone.0118571&type=printable. The LECZ is commonly defined as the contiguous and hydrologically connected zone of land along the coast and below 10 meters of elevation. Ibid. at 2 (citation omitted).

[3] Ibid. at 10. [4] Ibid. [5] Ibid. at 17.

populations will also continue to increase, but at the slowest rates,[6] while Central and South American coastal populations will grow more quickly than European and North American, but still much slower than Asia and Africa.[7]

According to these projections, much of the future coastal risk to human communities will occur in less developed nations.[8] There is thus reason for concern about future coastal disasters in public health terms, given these nations' more limited (on average) capacities for both climate change adaptation and public health delivery. Nevertheless, the United States presents a special case, with a large number of people living in the LECV as of 2000 (23 million), of whom 61 percent were located in dense urban areas, and a recent history of major coastal disasters like Hurricane Katrina in 2005, Superstorm Sandy in 2012,[9] and Hurricanes Harvey and Irma in 2017. These conditions may indicate increasing vulnerability in coastal areas.[10] Given their resources and the demand for risk reduction, the United States and other developed coastal nations have a critical role to play in developing and disseminating coastal climate change adaption strategies, including best practices, policies, and legal approaches to protect coastal public health in a changing future.

### 8.2.2 *Climate Change Projections for the Coast*

In 2014, the Intergovernmental Panel on Climate Change (IPCC) published its Fifth Assessment Report on climate change.[11] Although the IPCC's reports are often conservative, particularly where the ocean is concerned, the Fifth Assessment Report provides a good starting point for considering the probable future of coastal communities and the potential health consequences of climate change for them.

6 Ibid. at 18. 7 Ibid. at 19. 8 Ibid. at 20, 28. 9 Ibid. at 24.

10 See AECOM, *The Impact of Climate Change and Population Growth on the National Flood Insurance Program through 2100* (June 2013), available at http://bit.ly/2qRR9GH (anticipating that the special flood hazard area in coastal areas will grow 55 percent on average by 2100, and further that roughly 30 percent of this change will owe to population growth and 70 percent to climate change).

11 The IPCC's Fifth Assessment Report consists of four documents: *Intergovernmental Panel on Climate Change, Climate Change 2013: The Physical Science Basis (2013)* www.ipcc.ch/report/ar5/wg1/ [hereinafter 2013 IPCC Physical Science Report]; *Intergovernmental Panel on Climate Change, Climate Change 2014: Impacts, Adaptation, and Vulnerability* (2014) www.ipcc.ch/report/ar5/wg2/ [hereinafter 2014 IPCC Adaptation Report]; *Intergovernmental Panel on Climate Change, Climate Change 2014: Mitigation of Climate Change* (2014) www.ipcc.ch/report/ar5/wg3/ [hereinafter 2014 IPCC Mitigation Report]; and *Intergovernmental Panel on Climate Change, Climate Change 2014: Synthesis Report* (2014) www.ipcc.ch/report/ar5/syr/ [hereinafter 2014 IPCC Synthesis Report].

#### 8.2.2.1 Warming Seas

An increase in water temperature is one of the most direct impacts of climate change on the ocean. The world's ocean has been absorbing most of the extra heat produced as a result of increasing atmospheric concentrations of greenhouse gases. Indeed, according to the IPCC, ocean warming accounts for more than 90 percent of the energy accumulated between 1971 and 2010, as compared with only about 1 percent stored in the atmosphere.[12]

Globally, ocean warming has been largest near the surface: the upper 75 meters warmed by 0.11°C per decade from 1971 to 2010.[13] Nevertheless, ocean warming has continued to push deeper in recent years, reaching 3000 meters in depth at the beginning of the twenty-first century.[14] Moreover, "[t]he global ocean will continue to warm during the 21st century, with the strongest warming projected for the surface in tropical and Northern Hemisphere subtropical regions ... "[15] However, "[a]t greater depth the warming will be most pronounced in the Southern Ocean."[16]

Ocean warming has a number of potential consequences. First, it is likely to weaken major ocean currents,[17] which can result in changes to weather patterns and crop viability. Second, warming oceans and changing currents are changing marine ecosystems,[18] with implications for both food production and disease, both of which will be discussed in more detail in the next section.

#### 8.2.2.2 Sea-Level Rise

According to the IPCC, global mean sea level rose by 0.19 meters over the period 1901 to 2010, and "the rate of sea level rise since the mid-19th century has been larger than the mean rate during the previous two millennia."[19] The IPCC also concluded that the mean rate of global average sea-level rise for the period from 1993 to 2010 was twice what occurred from 1901 to 2010.[20] Moreover, "extreme sea levels (for example, as experienced in storm surges) have increased since 1970, being mainly a result of rising mean sea level."[21]

Sea-level rise has two main components: melting land-based ice (glaciers and ice shelves) and expanding volume as the ocean warms.[22] Although the two contributors have been roughly equal until recently, melting ice and disintegrating ice shelves have become significantly more

[12] 2014 *IPCC Synthesis Report*, *supra* note 11, at 4. For the sake of simplicity and consistency with the rest of this volume, this chapter has omitted the IPCC's confidence notations.

[13] Ibid. [14] Ibid. at 40. [15] Ibid. at 11. [16] Ibid. at 60. [17] Ibid. at 60, 62.

[18] Ibid. at 67. [19] Ibid. at 4. [20] Ibid. at 42. [21] Ibid. at 8.

[22] 2014 *IPCC Synthesis Report*, *supra* note 11, at 42.

important.[23] Sea-level rise will continue to accelerate through the twenty-first century and beyond, affecting a projected 95 percent of the ocean area and approximately 70 percent of coastlines worldwide.[24] However, sea-level rise will not be uniform across regions. For example, "[s]ince 1993, the regional rates for the Western Pacific are up to three times larger than the global mean, while those for much of the Eastern Pacific are near zero or negative."[25]

The future of the planet's ice presents a worrisome uncertainty. The increasing pace of polar ice melt has added significant volatility to the art of sea-level rise prediction.[26] Studies repeatedly indicate that the Greenland ice sheet and Antarctic ice are melting faster than expected,[27] and an August 2007 study published in *Science* suggested "that future sea-level rise may be larger than anticipated and that the component due to GIC [glaciers and ice caps] will continue to be substantial."[28] The IPCC noted in 2014 that the Greenland and Antarctic ice sheets were losing mass, likely at an increasing rate, and that glaciers around the world have continued to shrink.[29] It projects that these ice sheets and glaciers will continue to decrease throughout the twenty-first century, shrinking 15 percent to 85 percent by 2100.[30] "Greenland's land ice is already thawing fast enough to raise worldwide seas 0.74 millimeter per year," and its complete thawing (probably over several centuries) could raise sea level by seven meters.[31] The IPCC concluded that knowledge concerning "[a]brupt and irreversible ice loss from the Antarctic ice sheet … is insufficient to make a quantitative assessment" of its

23 Mark F. Meier et al., *Glaciers Dominate Eustatic Sea-Level Rise in the 21st Century*, 317 Science 1064, 1065 (2007) (arguing that glaciers and ice caps "contribute about 60% of the eustatic, new-water component of sea-level rise"); Antarctic Climate & Ecosystems Cooperative Research Centre, *Australia, Report Card: Sea Level Rise 2012*, 4, 10 (2012), http://acecrc.org.au/wp-content/uploads/2015/03/2012-ACE-RC-Sea-Level-Rise.pdf (noting that while "[t]his present sea-level rise is due to a combination of thermal expansion of a warming ocean and the melting of glaciers and ice sheets," "[o]ver the last decade, the contribution to sea-level rise from melting ice has exceeded that due to thermal expansion of the ocean").

24 *2014 IPCC Synthesis Report*, *supra* note 11, at 13. 25 Ibid. at 42.

26 Anny Cazenave, *How Fast are the Ice Sheets Melting?* 314 Sci. 1250, 1250–1251 (2006).

27 Ibid. at 1251; J. L. Chen et al., *Satellite Gravity Measurements Confirm Accelerated Melting of Greenland Ice Sheet*, 313 Sci. 1958, 1958 (2006).

28 Meier et al., *Glaciers Dominate*, *supra* note 23, at 1066.

29 2014 IPCC Synthesis Report, *supra* note 11, at 4. 30 Ibid. at 12.

31 Annie Sneed, *How Is Worldwide Sea Level Rise Driven by Melting Arctic Ice?* Scientific American, www.scientificamerican.com/article/how-is-worldwide-sea-level-rise-driven-by-melting-arctic-ice/ (June 5, 2017). See also Vivian Gornitz, *Sea Level Rise, After the Ice Melted and Today* (Goddard Institute for Space Studies, NASA, 2007), biwww.giss.nasa.gov/research/briefs/gornitz_09/.

likelihood.[32] However, the West Antarctic Ice Sheet contains enough ice to raise sea level by five to seven meters (17–23 feet).[33] If all of Antarctica melts, sea level will rise approximately 60 meters, or almost 200 feet.[34] If both Greenland and Antarctica melt completely, sea level would rise about 65 meters,[35] or approximately 215 feet.

#### 8.2.2.3 Coastal Storms

Coastal storms are costly climate disasters, both economically and in terms of human lives. In 2016, Hurricane Matthew caused about $6 billion in damage in the United States,[36] while a month later Tropical Storm Otto killed six people in Nicaragua and nine in Costa Rica, which had not experienced such a storm since at least 1851.[37] The Philippines is generally considered the world's most storm-exposed country: December 2012's Typhoon Bopha "killed more than 1,900 people and cost the country some $1 billion."[38] Typhoon Haiyan in 2013 killed an estimated 6,300 people and caused almost $2 billion in damages.[39]

Nevertheless, while the IPCC concludes that "it is virtually certain that intense tropical cyclone activity has increased in the North Atlantic since 1970," scientists continue to have difficulty attributing coastal storm events to anthropogenic climate change.[40] Even so, in its 2014 Fifth Assessment Report, the IPCC highlighted the world's increasing risks of climate disasters, including coastal storms. For example, it concluded with very high confidence that "[i]mpacts from recent climate-related extremes, such as heat waves, droughts, floods, [and] cyclones ... reveal significant vulnerability and exposure of ...

32 2014 *IPCC Synthesis Report*, *supra* note 11, at16. 33 Id.

34 Antarctic Treaty Consultative Meeting XXIX, *The Antarctic and Climate Change*, 3 (2006) www.asoc.org/portals/0/pdfs/Climate%20Change%20IP.pdf.

35 Cazenave, *How Fast*, *supra* note 26, at 1250.

36 Doyle Rice, *Hurricane Matthew Economic Damage Nears $6 Billion*, USA Today, www.usatoday.com/story/weather/2016/10/08/hurricane-matthew-economic-damage-cost-6-billion/91783304/ (Oct. 8, 2016).

37 *Tropical Storm Otto Kills Nine in Costa Rica*, BBC News, www.bbc.com/news/world-latin-america-38103330 (29 Nov. 2016).

38 Sophie Brown, *The Philippines Is the Most Storm-Exposed Country on Earth*, Time, http://world.time.com/2013/11/11/the-philippines-is-the-most-storm-exposed-country-on-earth/ (Nov. 11, 2013).

39 National Disaster Risk Reduction & Management Council, *Republic of the Philippines, Final Report re Effects of Typhoon "Yolanda" (Haiyan)*, 3, 5 (2015), www.ndrrmc.gov.ph/attachments/article/1329/FINAL_REPORT_re_Effects_of_Typhoon_YOLANDA_(HAIYAN)_06-09NOV2013.pdf.

40 2014 *IPCC Synthesis Report*, *supra* note 11, at 53.

many human systems to current climate variability . . . "[41] In addition, "[d]irect and insured losses from weather-related disasters have increased substantially in recent decades, both globally and regionally."[42] Coastal storms are part of these climate disasters and are becoming increasingly costly and increasingly common risks, demanding an effective legal response.

Rising seas make these storm events worse. Indeed, the exacerbation of storm surge is the most immediate and significant consequence of sea-level rise. According to the IPCC, "extreme sea levels (for example, as experienced in storm surges) have increased since 1970, being mainly the result of mean sea level rise."[43]

#### 8.2.2.4 Changes to Ocean Chemistry

The ocean is the world's largest carbon sink, and it has been absorbing about half to two-thirds of all anthropogenic emissions of carbon dioxide since the Industrial Revolution. While this absorption helps to keep the impacts of climate change at bay, it comes at a price: Carbon dioxide chemically reacts with seawater to reduce the ocean's pH, a phenomenon known as ocean acidification. According to the IPCC, absorption of $CO_2$ has decreased the pH of ocean surface water by 0.1 on average, which corresponds to a 26 percent increase in acidity.[44] The IPCC projects that ocean acidification will continue throughout the twenty-first century under all emissions scenarios, with a further average drop in sea surface pH of 0.06 to 0.32, representing increases in the ocean's acidity of 15 to 109 percent.[45] Moreover, "ocean acidification will increase for centuries if $CO_2$ emissions continue, and will strongly affect marine ecosystems."[46]

Acidifying sea water interferes with the ability of shell-forming marine species to form their shells. Moreover, ocean acidification is tied to reductions in the ocean's dissolved oxygen content, which is critical to marine animal life.[47] The IPCC projects that the ocean's dissolved oxygen content will decrease by "a few percent during the 21st century in response to surface warming, predominantly in the subsurface mid-latitude oceans."[48] These seemingly small changes in ocean chemistry could have profound impacts on the ocean's capacity to act as a human food supply: "Due to projected climate change by the mid-21st century and beyond, global marine species redistribution and marine biodiversity reduction in sensitive regions will challenge the sustained provision of fisheries productivity and other ecosystem services."[49]

[41] Ibid. [42] Ibid. [43] Ibid. [44] *2014 IPCC Synthesis Report*, *supra* note 11, at 4.
[45] Ibid. at 12. [46] Ibid. at 16. [47] Ibid. at 41. [48] Ibid. at 62. [49] Ibid. at 13.

#### 8.2.2.5 Cumulative Risks

In its Fifth Assessment Report, the IPCC identified four cumulative and systemic risks for coastal and other communities. First is the "[r]isk of severe ill-health and disrupted livelihoods resulting from storm surges, sea level rise and coastal flooding ... "[50] Thus, the IPCC acknowledged that climate change impacts on coastal communities pose direct threats to public health.

The other risks that the IPCC identified pose *indirect* threats to public health. For example, its second category is "[s]ystemic risks due to extreme weather events leading to breakdown of infrastructure networks and critical services."[51] These critical services include emergency responders, hospitals, and more basic medical care, as well as supplies of treated fresh water and food. The IPCC's third category is the "[r]isk of food and water insecurity and loss of rural livelihoods and income, particularly for poorer populations."[52] As will be discussed in the next section, sea-level rise and storm surge threaten coastal water supplies and cropland with salt contamination, creating these kinds of insecurities for coastal communities – most of which, as discussed earlier, will be more severe in poorer developing nations. Finally, the IPCC identified the "[r]isk of loss of ecosystems, biodiversity and ecosystem goods, functions and services" as a cumulative risk.[53] As will be discussed in the next section, changed coastal ecosystems often benefit pathogens such as cholera and mosquito-borne diseases and again undermine local food security, especially where subsistence fishing and aquaculture constitute a large percentage of coastal communities' diet.

## 8.3 THE IMMEDIATE HEALTH IMPACTS OF EXTREME COASTAL STORMS

### 8.3.1 *Introduction*

Coastal storms and flooding can become costly disasters, both economically and in terms of human lives and well-being. In terms of economics, for example, the IPCC concluded that "[f]or the 21st century, the benefits of protecting against increased coastal flooding and land loss due to submergence and erosion at the global scale are larger than the social and economic costs of inaction ... "[54] Moreover, "protecting against flooding and erosion is considered economically rational for most developed coastlines in many countries under all socio-economic and sea level rise scenarios analyzed."[55]

[50] Ibid. at 65. [51] Ibid. [52] Ibid. [53] Ibid.
[54] *2014 IPCC Coastal Systems*, *supra* note 11, at 3. [55] Ibid.

Of more relevance to this chapter, coastal flooding and storm surge pose significant and immediate public health, as well as physical, threats to coastal inhabitants. Many of the public health risks are the same in developed and developing nations. However, the extent and severity of the damage, in terms both of economics and human health, can depend in many circumstances on the extent and type of coastal infrastructure present as well as the preparedness of public health and emergency response personnel and facilities.

Preparing for these public health disasters should thus be part of the emergency planning laws of all coastal communities. Importantly, however, public health preparedness for coastal disasters in a climate change era requires a multivalent legal and policy response that incorporates a number of areas of law and budgeting: public health training, staffing, and facilities; building codes and land use planning, including facility siting and construction requirements; environmental law, including investment in coastal contamination cleanup; and more traditional emergency and disaster planning.

### 8.3.2 *Examples from the United States and Japan*

Many of the public health threats from coastal storms affect even developed nations. For example, as Neumann and her colleagues noted, the United States is unusually vulnerable among developed nations to coastal disasters, and those vulnerabilities include threats to public health. During Hurricane Katrina in 2005, Superstorm Sandy in 2012, and Hurricanes Harvey and Irma in 2017, sewage treatment systems failed, contaminating waters – including drinking water supplies – with raw sewage[56] and

[56] Sewage contamination as a result of Katrina and Sandy is well documented in the scientific, public health, and regulatory literature. E.g., Alyson Kenward et al., *Sewage Overflows from Hurricane Sandy*, Climate Central 1, 2 (April 2013), www.climatecentral.org/pdfs/Sewage.pdf; John S. Furey, Herbert Fredrickson, Chris Foote, and Margaret Richmond, *Post-Katrina Fecal Contamination in Violet March Near New Orleans*, 4(2) Int J Envtl. Res. Public Health 84, 86 (Feb. 2007), www.ncbi.nlm.nih.gov/pmc/articles/PMC3728572/; U.S. Department of Commerce, NIST Technical Note 1476, Reconnaissance Report – Hurricanes Katrina and Rita 134 (June 2006); John Manuel, *In Katrina's Wake*, 114 Envtl. Health Persp. A32, A34 (2006), available at www.ncbi.nlm.nih.gov/pmc/articles/PMC1332683/; Office of the Inspector General, U.S. Environmental Protection Agency, *Evaluation Report: EPA Provided Quality and Timely Information Regarding Wastewater after Hurricane Katrina*, 1, 2 (March 28, 2006), www.epa.gov/sites/production/files/2015-11/documents/20060328-2006-p-00018.pdf. While Hurricanes Harvey and Irma are too recent to have yet generated peer-reviewed scientific analyses, newspapers reported similar releases of raw sewage. Sammy Roth, *Hurricane Harvey Floodwaters Brimming With Raw Sewage, Toxic Chemicals*, USA Today, Sept. 5, 2017,

increasing the risks of outbreaks of waterborne diseases such as cholera and typhoid fever.[57]

However, coastal storms pose myriad public health threats, as both Hurricane Katrina and Superstorm Sandy well illustrated. First and foremost are the people killed (117 during Sandy) or injured by the storm itself.[58] "The most severe acute effect of hurricane landfall is death from drowning, electrocutions, or physical trauma. Older age increases the risk of death – nearly 85% of people killed during and in the immediate aftermath of Hurricane Katrina were aged 51 and older, and almost half were older than 75 years of age."[59] Second, Sandy's unusual timing – late fall rather than summer – and the resulting widespread loss of electricity

www.usatoday.com/story/weather/2017/09/05/hurricane-harvey-floodwaters-brimming-raw-sewage-toxic-chemicals/632937001/; John Patrick Pullen, *Raw Sewage is Now Bubbling Up In These Hurricane-Irma-Battered Cities*, Fortune, Sept. 15, 2017, http://fortune.com/2017/09/15/hurricane-irma-damage-update-sewage/; Dianna Wray, Millions of gallons of raw sewage released during Hurricane Harvey, *Houston Press*, Sept. 20, 2017, www.houstonpress.com/news/hurricane-harvey-damaged-multiple-wastewater-treatment-plants-in-houston-southeast-texas-9808645.

57 Alyson Keyword et al., *Overflow: Climate Change, Heavy Rain, and Sewage* (Sept. 2016), http://assets.climatecentral.org/pdfs/Overflow_sewagereport_update.pdf (surveying recent sewage bypass events in the United States and noting their relationships to climate change impacts). While no sewage-related illnesses occurred after Hurricane Katrina in Louisiana, Mississippi, or Alabama, the US Environmental Protection Agency was nevertheless concerned about a wide range of public health threats that can result from human exposure to raw sewage:

> Human exposure to raw sewage may cause illness ranging in severity from mild gastroenteritis to life-threatening ailments such as cholera. Potential pathways of exposure to raw sewage include:
>
> * Direct contact, such as contact with floodwater or ingestion during recreation;
> * Drinking water contaminated with raw sewage; and
> * Consumption of shellfish harvested from areas contaminated by raw sewage.

Office of the Inspector General, US Environmental Protection Agency, *Evaluation Report: EPA Provided Quality and Timely Information Regarding Wastewater after Hurricane Katrina* 2 (March 28, 2006), www.epa.gov/sites/production/files/2015-11/documents/20060328-2006-p-00018.pdf. Similarly, the editorial staff of Contagion warned that Hurricane Harvey's floodwaters could cause several infectious diseases, especially typhoid fever, cholera, and three other Vibrio infections. Editorial Staff, *Hurricane Harvey Puts Health Officials On Alert for Water-Borne Infections*, ContagionLive, Aug. 28, 2017, www.contagionlive.com/news/hurricane-harvey-health-officials-waterborne-infections.

58 Umair Irfan, *Superstorm Sandy May Have Long-Term Public Health Impacts*, Scientific American, www.scientificamerican.com/article/superstorm-sandy-may-have-long-term-public-health-impacts/ (Nov. 6, 2012).

59 Kathryn Lane et al., *Health Effects of Coastal Storms and Flooding in Urban Areas: A Review and vulnerability assessment*, 2013 J. Envtl. & Public Health, Article ID 913064, 2 (21 April 2013) http://dx.doi.org/10.1155/2013/913064.

left people vulnerable both to cold exposure[60] and to carbon monoxide poisoning from generators as they tried to power alternative sources of heat.[61] The power losses had numerous other public health impacts, as well, from failure of at-home medical equipment like ventilators and oxygen machines to increased food-borne disease incidences as a result of lack of refrigeration.[62] Evacuations of hospitals posed their own additional risks to patients both during hurricanes and afterward as a result of the loss of power,[63] while in New Orleans during Hurricane Katrina "[p]risoners locked in cells were left alone

60 "Power outages persisted through a cold snap that put temperatures in the single digits." John Manuel, *The Long Road to Recovery: Environmental Health Impacts of Hurricane Sandy*, 121(5) Envtl. Health Persp. a152, a152 (2013), www.ncbi.nlm.nih.gov/pmc/articles/PMC3673205/. See also Michael T. Schmeltz, *Lessons From Hurricane Sandy: A Community Response in Brooklyn*, 90(5) J. of Urban Health 799, 800 (Oct. 2013), https://link.springer.com/article/10.1007/s11524-013-9832-9 ("There was no heat for Red Hook Houses residents for 17 days, between October 30 and November 15, during which time the city faced average temperatures in the mid-40s (degrees Fahrenheit) and a winter Nor'easter a week after Sandy. Exposure to cold temperatures for long periods of time can lead to hypothermia and can increase the risk of a heart attack and pneumonia.").

61 Lane et al., *Health Effects of Coastal Storms*, *supra* note 59, at 4. According to the Centers for Disease Control (CDC),

> "In the wake of Sandy, state and federal public health agencies have observed an increase in the number of exposures to carbon monoxide (CO) reported to poison centers. CO is imperceptible and can cause adverse health effects ranging from fatigue and headache to cardiorespiratory failure, coma, and death. CO poisoning is a leading cause of mortality and morbidity in post-disaster situations, when widespread power outages occur and risky behaviors, such as improper placement of generators and indoor use of charcoal grills, increase . . . As of November 6, a total of 263 CO exposures related to Hurricane Sandy had been reported to poison centers in eight states: 80 in New York, 61 in New Jersey, 44 in Connecticut, 39 in Pennsylvania, 27 in West Virginia, eight in Virginia, three in Maryland, and one in Delaware. Four of the reported exposures, all in Pennsylvania, resulted from the use of a generator in a garage and were fatal. This likely is an underestimation of the total number of fatal cases; larger numbers of CO-related deaths have been reported in the media."

Centers for Disease Control and Prevention, *Notes from the Field: Carbon Monoxide Exposures Reported to Poison Centers and Related to Hurricane Sandy – Northeastern United States, 2012*, Morbity & Mortality Weekly Report, www.cdc.gov/mmwr/preview/mmwrhtml/mm6144a5.htm?s_cid=mm6144a5_w (Nov. 9, 2012). See also Manuel, The Long Road to Recovery, *supra* note 60, at a155 ("Every storm in which power is lost for an extended period of time seems to result in tragedies related to carbon monoxide (CO) poisoning. People resort to using gas stoves or ranges to heat their dwellings and portable gas generators to provide homes with electricity. The situation was exaggerated after Sandy, given the sheer number of gas generators purchased and the threat to safety that presented.").

62 Lane et al., *Health Effects of Coastal Storms supra* note 59, at 4.

63 Ibid. at 3; Tia Powell et al., *Emergency Preparedness and Public Health: The Lessons of Hurricane Sandy*, 308(24) JAMA 2569, 2569–2571 (Dec. 26, 2012).

without power, food, water, or even sufficient ventilation during the storm. Prisoners described lack of access to medical care and interruptions in care for serious chronic illness during and after the storm."[64]

In addition, hurricanes in the United States have illuminated the latent toxicity of coastal zone infrastructure and revealed the risks that invading seawater can create both a toxic soup of sewage, oil, and hazardous chemicals from coastal businesses (such as dry cleaners and auto repair facilities), industrial sites, Superfund sites, and toxic waste facilities[65] and unusually high emissions of hazardous air pollutants. Luckily, the "toxic soup" risk has not yet been fully realized as a result of a major US coastal storm, but "[f]ollowing Hurricane Katrina, hazardous substances such as volatile organic compounds (VOCs), lead, and arsenic were detected in the air, soil, and sediment samples," and "the potential for a toxic release of hazardous substances after a storm exists."[66] Indeed, "[a]fter Hurricane Sandy devastated New York and New Jersey in 2012, officials had to monitor 247 Superfund sites – one of which, the Gowanus Canal, overflowed into people's homes."[67]

Hurricane Harvey's 2017 flooding of the Houston area may produce one of the most toxic legacies of US hurricanes. Harvey inundated thirteen of the Houston area's forty-one hazardous waste sites (eighty such sites stood in Hurricane Irma's path).[68] The *New York Times* described the floodwaters as "a stew of toxic chemicals, sewage, debris and waste ... Runoff from the city's sprawling petroleum and chemicals complex contains any number of hazardous compounds.

64 Lane et al., *Health Effects of Coastal Storms, supra* note 59, at 4.

65 Danny D. Reible et al., *Toxic and Contaminant Concerns Generated by Hurricane Katrina*, 36 (1) The Bridge, Spring 2006, www.nae.edu/NAE/bridgecom.nsf/weblinks/MKEZ-6MYQQP?OpenDocument; Lane et al., *Health Effects of Coastal Storms, supra* note 59, at 5.

66 Lane et al., *Health Effects of Coastal Storms, supra* note 60, at 5. See also Steven M. Presley et al., *Assessment of Pathogens and Toxicants in New Orleans LA Following Hurricane Katrina*, 40(2) Envtl. Science & Technology 468, 468 (Dec. 14, 2005) ("Concentrations of aldrin, arsenic, lead, and seven semivolatile organic compounds in sediments/soils exceeded one or more United States Environmental Protection Agency (USEPA) thresholds for human health soil screening levels and high priority bright line screening levels.").

67 Emily Atkin, *America Has a Toxic Waste Hurricane Problem*, The New Republic, https://newrepublic.com/article/144737/america-toxic-waste-hurricane-problem (Sept. 12, 2017). See also Elizabeth A. Harris, *In Brooklyn, Worrying About Not Only Flooding But What's in the Water*, The New York Times, www.nytimes.com/2012/11/06/nyregion/gowanus-canal-flooding-brings-contamination-concerns.html?mcubz=1 (Nov. 5, 2012) (reporting on the Gowanus Canal Superfund site overflow).

68 Atkin, *Toxic Waste Hurricane Problem, supra* note 67.

Lead, arsenic and other toxic and carcinogenic elements may be leaching from some two dozen Superfund sites in the Houston area."[69] The EPA is still assessing three reported spills at US Oil Recovery, described by news outlets as "a former petroleum industry waste processing plant contaminated with a dangerous brew of cancer-causing chemicals."[70] Flooding of petroleum processing and chemical facilities caused other kinds of exposures to toxic substances, as well. For example, a flooded chemical plant in Houston caught fire, prompting an evacuation of nearby residents.[71] In late August, ExxonMobil acknowledged "that Hurricane Harvey damaged two of its refineries, causing the release of hazardous pollutants" – specifically, high emissions of volatile organic compounds and over 1 million pounds of sulfur dioxide,[72] both of which are regulated air pollutants under the Clean Air Act. Initial reports from Texas regulators indicate that because of Hurricane Harvey, "the region's massive petrochemical industry released more than 2 million pounds of harmful pollutants into the air as of Aug. 29" – "roughly 40 percent of what the entire Houston released in 2016 . . . "[73]

Most directly relevant to the immediate spread of human disease, testing of post-storm floodwaters often reveals bacterial concentrations of great concern. For example, "[h]igh numbers of *Aeromonas* spp., pathogenic *Vibrio* spp., and other coliform bacteria were found in floodwater samples" after Hurricane Katrina.[74] Concentrations of total coliform and *E. coli* were particularly high, evidencing extensive sewage contamination, while "the number of *Vibrio* infections reported

69 Hiroko Tabuchi and Sheila Kaplan, *A Sea of Health and Environmental Hazards in Houston's Floodwaters*, The New York Times, www.nytimes.com/2017/08/31/us/houston-contaminated-floodwaters.html?mcubz=1 (Aug. 31, 2017).

70 Michael Biesecker and Frank Bajak, *Hurricane Harvey: Floodwaters Caused Chemical Spill' at Houston's Dirtiest Toxic Waste Plant*, The Independent, www.independent.co.uk/news/world/americas/hurricane-harvey-latest-houston-chemical-plant-toxic-spill-floodwater-us-oil-recovery-superfund-epa-a7954586.html (19 Sept. 2017).

71 Nicole Chavez, *Harvey Aftermath: Region Begins Long Road to Recovery*, CNN News (Sept. 4, 2017) http://edition.cnn.com/2017/09/03/us/houston-texas-flood-aftermath/index.html (Sept. 4, 2017).

72 Steven Mufson, *ExxonMobil Refineries are Damaged in Hurricane Harvey, Releasing Hazardous Pollutants*, Washington Post, www.washingtonpost.com/news/energy-environment/wp/2017/08/29/exxonmobil-refineries-damaged-in-hurricane-harvey-releasing-hazardous-pollutants/?utm_term=.d8ccfb69eaf7 (August 29, 2017).

73 Adam Allington, *Flooded Houston Facing Air Threat, Too, With Toxic Gas Releases*, Bloomberg BNA, *Natural Disasters Cast Unexpected Light on Environmental Hazards, Regulatory Gaps* 1, 1 (Oct. 2017) (quoting Elena Craft).

74 Presley et al., Pathogens and Toxicants in New Orleans, *supra* note 66, at 468.

in the month following Hurricane Katrina was higher than normal."[75] Initial problems with diarrhea can become the means of transmission of waterborne diseases[76] rather than direct contact with floodwaters, and diarrhea was not uncommon after Hurricane Katrina.[77]

Notably, the risk and variety of waterborne diseases both generally increase in less developed countries; for example, leptospirosis and hepatitis A become risks in addition to typhoid fever and cholera.[78] A 2005 epidemiological review noted that "published studies ... have reported postflood increases in cholera, cryptosporidiosis, nonspecific diarrhea, poliomyelitis, rotavirus, and typhoid and paratyphoid," with documented substantial increases in the relative risks of contracting these diseases during flooding in Indonesia.[79] Flooding can also increase mosquito-borne diseases like West Nile virus and malaria and rodent-borne diseases like Hantavirus Pulmonary Syndrome and leptospirosis.[80]

75 C. D. Sinigalliano et al., *Impacts of Hurricanes Katrina and Rita on the Microbial Landscape of the New Orleans Area*, 104(21) PNAS 9029, 9029 (May 22, 2007).

76 Ling-Ya Huang et al., *Water Outage Increases the Risk of Gastroenteritis and Eyes and Skin Diseases*, 11(726) BMC Public Health 1 (2011), https://link.springer.com/article/10.1186/1471-2458-11-726 ("Huge diarrhea outbreaks cause cholera and *Escherichia coli* infections after a flood.").

77 "CDC received reports of clusters of diarrheal disease among persons in evacuation centers in Louisiana, Mississippi, Tennessee, and Texas. In Louisiana, approximately 20 clusters of diarrheal illness in evacuation centers were reported and investigated. In Memphis, Tennessee, gastrointestinal illness was the most common acute disease complaint among evacuees. Approximately 1,000 cases of diarrhea and vomiting were reported among adult and child evacuees in Mississippi and Texas; tests detected norovirus in stool specimens from patients in Texas. Sporadic nontyphoidal Salmonella, nontoxigenic V. cholerae O1, and other infections were identified. No confirmed cases of *Shigella* dysentery, typhoid fever, or infection by toxigenic V. cholerae O1 were reported in evacuees from Hurricane Katrina." Centers for Disease Control, *Infectious Disease and Dermatologic Conditions in Evacuees and Rescue Workers After Hurricane Katrina – Multiple States, August–September, 2005*, Morbidity & Mortality Weekly Report, www.cdc.gov/mmwr/preview/mmwrhtml/mm5438a6.htm (Sept. 30, 2005). See also Karen Levy et al., Untangling the impacts of climate change on waterborne diseases: A systematic review of relationships between diarrheal diseases and temperature, rainfall, flooding, and drought, 50(10) Envtl. Science & Technology 4905, 4908 (2016) ("In reviews of large national databases, positive associations were noted between flood days and diarrhea incidence in China, and in Finland the most common causes of waterborne disease outbreaks were floods and surface runoff. Several articles report changes in the percentage of diarrhea cases attributable to particular pathogens during or after floods, including increased detection of pathogenic Escherichia coli, and Vibrio cholera; and both increased and decreased detection of rotavirus.").

78 World Health Organization, *Flooding and Communicable Diseases Fact Sheet*, www.who.int/hac/techguidance/ems/flood_cds/en/ (as viewed Oct. 27, 2017).

79 Mike Ahern et al., *Global Health Impacts of Floods: Epidemiological Evidence*, 27 Epidemiological Rev. 36, 38 (2005).

80 Ibid.

Other kinds of coastal infrastructure pose different public health threats as a result of coastal disasters. Many coastal states in the United States (Florida, California, Texas, Louisiana, and many states along the Atlantic coast[81]), and many other developed nations, site nuclear power plants in their coastal zones,[82] another potential source of toxic contamination and continuing public health risk, as the Fukushima plant in Japan so tragically illustrated. On March 11, 2011, a magnitude 9 earthquake occurred off the east coast of Japan, causing a tsunami that struck the coastline, in turn causing severe damage to the nuclear core at the Fukushima Daiichi Nuclear Power Station and the release of radioactive water. While this particular nuclear disaster resulted from an earthquake, violent coastal storms pose many of the same risks.

Remarkably, "[t]here were no acute radiation injuries or deaths among the workers or the public due to exposure to radiation resulting from the FDNPS accident,"[83] but there were several other reasons to worry about public health in the wake of the disaster. In February 2013, the World Health Organization (WHO) released its report assessing the public health risks to Japanese citizens and the rest of the world from the Fukushima disaster.[84] WHO concluded that the cancer risks for most of these populations – including to most people in Japan – were quite low, and it expected to see no observable increase in baseline cancer rates as a result of the disaster.[85] However, people living in the most contaminated areas nearest the plant site are still subject to significant increased risks of cancer and leukemia.[86] Food contamination remains a concern, both in Japan and elsewhere. The Ministry of Health, Labour, and Welfare of Japan still monitors the radioactivity of food supplies within Japan,[87] while the rest of the world continues to worry about contaminated fish, such as contaminated Pacific

81 U.S. Nuclear Regulatory Commission, *Operating Nuclear Power Reactors*, www.nrc.gov/info-finder/reactors/.

82 Mostly because of these plants' need for large amounts of cooling water, nuclear power plants are often sited on coastlines. "A review of nuclear reactors around the world found that most are located adjacent to the sea, especially outside the United States and Russia." David D. Caron and Harry N. Scheiber, *The Oceans in the Nuclear Age: Legacies and Risks* 541 (Leiden & Boston: Brill Nijhoff, 2014).

83 World Health Organization, *FAQs: Fukushima Five Years On*, www.who.int/ionizing_radiation/a_e/fukushima/faqs-fukushima/en/.

84 World Health Organization, *Health Risk Assessment From the Nuclear Accident After the 2011 Great East Japan Earthquake and Tsunami, Based on a Preliminary Dose Estimation* (Feb. 2013) www.who.int/ionizing_radiation/pub_meet/fukushima_risk_assessment_2013/en/.

85 World Health Organization, *Global report on Fukushima nuclear accident details health risks*, www.who.int/mediacentre/news/releases/2013/fukushima_report_20130228/en/ (Feb. 28, 2013).

86 Ibid.

87 World Health Organization, *FAQs: Fukushima Five Years On*, www.who.int/ionizing_radiation/a_e/fukushima/faqs-fukushima/en/.

Bluefin tuna, even though all studies indicate that the risks from eating these fish are miniscule despite their incorporation of radioactive isotopes traceable to Fukushima.[88]

Many of the public health issues following Fukushima are similar to public health effects that follow many coastal disasters: trauma from evacuation; impacts from loss of healthcare for chronic conditions; and mental health problems.[89] As WHO explains:

> There were public health consequences related to the response actions to the disaster, such as evacuation and relocation of people. These measures were taken based on radiation safety considerations and the massive damage to the infrastructure and facilities following the earthquake and tsunami. These measures resulted in a wide range of social, economic, and public health consequences. A sharp increase in mortality among elderly people who were put in temporary housings has been reported, along with increased risk of non-communicable diseases, such as diabetes and mental health problems. The lack of access to health care further contributed to deterioration of health.
>
> Similar to what was observed and reported for the Chernobyl population, the displaced Fukushima population is suffering from psycho-social and mental health impact following relocation, ruptured social links of people who lost homes and employment, disconnected family ties and stigmatization. A higher occurrence of post-traumatic stress disorder (PTSD) among the evacuees was assessed as compared to the general population of Japan. Psychological problems, such as hyperactivity, emotional symptoms, and conduct disorders have been also reported among evacuated Fukushima children. While no significant adverse outcomes were observed in the pregnancy and birth survey after the disaster, a higher prevalence of postpartum depression was noted among mothers in the affected region.[90]

### 8.3.3 *Recommendations for a Multivalent Legal Approach to Public Heath Preparedness*

These examples of coastal disasters suggest both that multivalent legal preparedness and response are necessary to adequately address the public health risks from coastal storms and flooding and that public health preparedness should be tailored to the particular risks inherent in specific locations (e.g. the prevalence

88 Steven Novella, *Health Risk From Fukushima Radiation*, 38(3) Skeptical Inquirer, www.csicop.org/si/show/health_risk_from_fukushima_radiation (May/June 2014).

89 See Chapter 11.

90 World Health Organization, *FAQs: Fukushima Five Years On*, www.who.int/ionizing_radiation/a_e/fukushima/faqs-fukushima/en/ (as viewed Feb. 27, 2017).

of toxic industries in Houston, different potentials for disease exposure). New York City, for example, learned not to house hospital back-up generators in basements, a new requirement to be pursued in local and state building codes and specific requirements for hospitals. Faulty generators (or generators used improperly) present a consumer protection law problem. Construction standards for sewage treatment plants, and requirements for how to handle overflows, should be the subject of coordinated building codes and environmental requirements. Provision of emergency water supplies, needed medications, and emergency medical treatment, as well as evacuation plans, all generally fall within the province of disaster preparedness. This preparation, moreover, should include the ability to detect escaped radiation and to treat radiation sickness in any community with a nuclear power plant nearby, although the design and construction requirements for nuclear power plants are the subject of separate regulatory laws implemented (and preferably updated and enforced) often long before any disaster occurs. Use and storage of toxic materials and wastes by local businesses is often the province of environmental laws, many of which could be updated and tailored to the specific vulnerabilities of coastal communities with more stringent requirements to limit the exposure of toxic materials to flood waters. Finally, the mental health issues that arise in the aftermath of such disasters are a relatively newly recognized component of coastal disasters, requiring longer-term post-disaster preparedness and specialized forms of public health assistance.

Katrina, Sandy, Harvey, and Fukushima thus suggest a series of recommendations for public health preparedness in the coastal zone. While the ability to pursue these kinds of adaptive capacities will vary considerably among coastal nations, the list here can provide starting guidelines for WHO, national public health agencies, and international aid and capacity-building organizations.

* Invest in coastal cleanup efforts, especially to eliminate toxic soil and aquifer contamination.
* Update (or enact) building codes to anticipate the worst projected storm surge with an additional margin of safety, addressing such topics as the location of generators in hospitals, storm overflow prevention for sewage treatment plants, building sealing requirements for other hazardous waste facilities, and basic coastal storm construction design.
* Revise siting requirements for long-lived but potentially hazardous infrastructure such as sewage treatment and nuclear power plants to reflect local sea-level rise projections and the potential for increasing storm surge.
* Update public health training to include protocols for addressing mass exposure to waterborne disease vectors and mixed toxic contamination.

* Invest in emergency water supply, emergency heating capacity, and facilities, supplies, and capabilities for routine, emergency, and psychological care during coastal storm and flooding disasters, including by disaster-proofing medical records and other patient information.

## 8.4 THE LONGER-TERM PUBLIC HEALTH IMPLICATIONS OF RISING, WARMING SEAS AND COASTAL INUNDATION

### 8.4.1 *Physical Inundation from the Sea and Coastal Public Health*

Sea-level rise poses two challenges for legal leaders trying to address climate change-related public health risks. First, sea-level rise is slow, measured in millimeters per year, and the full extent of climate change-driven sea-level rise is expected to take centuries to manifest. This is a planning horizon outside the ken of most governmental bodies. Second, scientists are still uncertain as to the extent of the problem, particularly because sea-level rise varies depending on location and because ice melt has proven difficult to predict. This uncertainty means that legal public health responses will have to be tailored to the particular experiences and vulnerabilities of specific coastal areas. There is, unfortunately, no globally appropriate single public health adaptation strategy for the world's coastal communities with respect to sea-level rise.

Some vulnerabilities, moreover, are more amenable to legal prevention than others. This section provides an overview of three public health vulnerabilities from sea-level rise and its associated worsening storm surge: coastal aquifer contamination with salt water, salt water inundation of coastal farmland, and human displacement (discussed further in Chapter 11).

#### 8.4.1.1 Saltwater Intrusion and Inundation into Coastal Water Supplies

Sea-level rise and storm surge pose direct threats to coastal fresh water supplies, raising public health concerns and potentially precipitating mass migrations from coastal regions. Loss of a clean, fresh water supply is clearly a public health issue, and coastal aquifers are more critically the focus of legal attention than coastal rivers and streams.

"Groundwater is generally the most important freshwater resource in many coastal regions which are threatened by seawater intrusion ... "[91] Coastal

[91] Hameb Ketabchi et al., *Sea-Level Rise Impacts On Seawater Intrusion in Coastal Aquifers: Review and Integration*, 535 J. Hydrology 235, 235 (Feb. 2016).

freshwater aquifers are put at risk by sea-level rise in two ways. First, sea-level rise expedites saltwater intrusion – the drawing in of saltwater as a result of coastal groundwater pumping. Second, it leads to seawater inundation.[92]

Overpumping of coastal aquifers puts them at risk of saltwater intrusion, and a 2012 study in *Nature Climate Change* concluded that water extraction remains the major salinity risk for most coastal aquifers.[93] Nevertheless, in some places, sea-level rise is the more direct threat to coastal aquifers. For example, "aquifers with very low hydraulic gradients are more vulnerable to sea-level rise and these regions will be impacted by saltwater inundation before saltwater intrusion."[94] Moreover, sea-level rise and increasingly severe coastal storms can increase the vulnerability of coastal aquifers to salinity contamination,[95] even when aquifer pumping is the more important factor. Such vulnerabilities are arising in coastal nations around the world, from South Korea[96] to Egypt[97] to the southeastern United States.[98] Research also indicates that aquifers' responses to sea-level rise are nonlinear and that they can cross thresholds or tipping points where the aquifer rapidly transforms from mild levels of salinity to complete saltwater intrusion.[99]

Threats of saltwater intrusion caused primarily by human pumping are readily amenable to legal solutions. Coastal communities can and should modify their laws governing groundwater pumping – generally included as part of the relevant government's water law – to account for saltwater intrusion, limiting pumping to levels that avoids such intrusion.[100] For example, governing entities could directly limit total coastal groundwater pumping to levels that do not cause saltwater intrusion, taking local sea-level rise and storm surge risks into account.

92 Ibid. at 253. See also Katerina Mazi et al., *Tipping Points for Seawater Intrusion for Coastal Aquifers Under Rising Sea Level*, 8 Envtl. Research Letters 1, 2 (9 Jan. 2013).

93 Grant Ferguson and Tom Gleeson, *Vulnerability of Coastal Aquifers to Groundwater Use and Climate Change*, 2 Nature Climate Change 342, 342 (May 2012).

94 Ibid. at 342.

95 Ketabchi et al., *Seawater Intrusion in Coastal Aquifers, supra* note 91, at 253.

96 Sung-Ho Song and Gil Zemansky, *Vulnerability of Groundwater Systems with Sea-Level Rise in Coastal Aquifers, South Korea*, 65 Envtl. Earth Sciences 1865–1876 (Mar. 2012).

97 Adrian D. Werner and Craig T. Simmons, *Impact of Sea-Level Rise on Sea Water Intrusion in Coastal Aquifers*, 47(2) Ground Water 197, 198 (March-April 2009).

98 U.S. Environmental Protection Agency, *Climate Impacts in the Southeast*, www.epa.gov/climate-impacts/climate-impacts-southeast (as updated Dec. 22, 2016, and viewed Jan. 27, 2017).

99 Mazi et al., *Tipping Points for Seawater Intrusion, supra* note 92, at 6.

100 Ferguson and Gleeson, *Vulnerability of Coastal Aquifers, supra* note 93, at 344 ("Human water use is a key driver in the hydrology of coastal aquifers, and efforts to adapt to sea-level rise at the expense of better water management are misguided.").

Such an approach, however, might simultaneously require investment in alternative sources of coastal fresh water supply, such as brackish water or sea water desalination, as is common throughout the Middle East.

However, the more unpredictable risks to coastal aquifers from sea-level rise and storm surge, coupled with the possibility of rapid threshold crossing, suggests that even more precautionary measures in coastal groundwater law and management are warranted. For example, the Orange County Water District in southern California uses treated wastewater to recharge coastal aquifers, in part to stop saltwater intrusion.[101] In addition, water managers in coastal states should begin to evaluate seriously the potential impact on fresh water supplies of a range of sea-level rise scenarios, taking into account the region's population dynamics as well. In areas where saltwater intrusion is already a threat to water supplies, and even in areas where sea-level rise of a meter or less will lead to saltwater intrusion, managers should at least begin to plan for, and perhaps begin to implement, water conservation measures, including changes to water pricing. In addition, water managers should identify and, where fiscally and politically possible, secure alternative sources of water supply. Finally, managers and the relevant government agencies should begin to identify and plan for potential water supply infrastructure needs (e.g. transportation, desalination, water treatment) for a range of sea-level rise scenarios. Such planning should build on consultation with public health officials, which might in turn require minor changes to state or local laws and budgeting decisions.

#### 8.4.1.2 Inundation and Salination of Coastal Farmland

Saltwater is no better for coastal agriculture than it is for coastal drinking water. Inundation of coastal agricultural land thus presents a food security risk for both developing and developed coastal nations. In central Java in Indonesia, for example, sea-level rise and coastal inundation threaten rice paddies, dryland agriculture, and aquaculture operations.[102] More generally, Indonesia is predicted to outright lose approximately 5 percent of its coastal agricultural land (rice fields) to sea-level rise and to have

[101] Water Replacement District of Southern California, *Groundwater Recharge with Recycled Water*, www.wrd.org/water_quality/recycled-water-groundwater-recharge.php (as viewed Jan. 27, 2017).

[102] Muh Marfai, *Impact of Coastal Inundation on Ecology and Agricultural Land Use Case Study in Central Java, Indonesia*, 30(3) *Quaestiones Geographicae*, DOI: https://doi.org/10.2478/v10117-011-0024-y (Sept. 2011).

more than 23 percent of its agricultural land impacted.[103] On the other side of the globe, effects of storm surge on English farmland have already been significant enough to prompt new research into reclaiming salt-water-contaminated soil,[104] and the United Kingdom's Department of the Environment, Food and Rural Affairs "has estimated that 35,000 hectares of high-quality horticultural and arable land will be flooded at least once every three years by the 2020s. This could rise to around 130,000 hectares by the 2080s if there is no change to current flood defense provision."[105]

Coastal inundation of farmland and sea-level rise thus add to climate change's threats to food security – and, hence, to long-term public health. They pose particular food security risks to island nations – especially, perhaps, to the Pacific Island nations that face a panoply of existential threats from sea-level rise.[106] Food security planning for coastal regions, as part of a public health strategy, does not differ excessively from food security planning elsewhere, although the loss of both coastal farmland and coastal fisheries (see next section) may limit a particular coastal community's future food options severely and require either migration or increased dependence upon imported purchased foods. In areas where inundation is slow and salinization occurs primarily as a result of infrequent but severe ("freak") storms, soil reclamation techniques and their regulation will likely become important to maintaining both food security and public health, and the United Kingdom, as discussed, and China[107] may become leaders in developing these techniques.

More broadly, however, loss of coastal farmland raises the same questions regarding appropriate coastal adaptation legal strategies that coastal inundation does. The most basic of these questions is whether coastal communities should

103 Jimy Kalther et al., *Predicting the Economic Values of Sea Level Rise's Impact on Agricultural Sector at Subang Coast, Indonesia: A Basic Reference for Adaptation and Mitigation Policies, Conference Proceedings: International Conference on Agricultural, Environmental and Biological Sciences (AEBS-2014) April 24–25, 2014, Phuket (Thailand)*, at 60 (2014), http://iicbe.org/upload/8546C414054.pdf.

104 Elizabeth Allen, *Research Aims to Support Farmers to Bring Back Saltwater-Damaged Land After Coastal Flooding*, https://phys.org/news/2016-12-aims-farmers-saltwater-damaged-coastal.html (Dec. 13, 2016).

105 Jamie Doward et al., *Severe Floods Threaten Food Security', Say Farmers and Environmental Groups*, The Guardian, www.theguardian.com/environment/2014/feb/08/severe-floods-threaten-food-security-climate-change (Feb. 8, 2014).

106 United Nations Food & Agriculture Organization, *Pacific: Climate Change Threatens Food Security*, www.fao.org/news/story/en/item/8658/icode/.

107 See, e.g., Y. N. Yang et al., *Desalination of Saline Farmland Drainage Water Through Wetland Plants*, 156 Agricultural Water Management 19–29 (1 July 2015).

pursue an armoring or a retreat strategy. Armoring is expensive,[108] has a number of coast-damaging impacts,[109] and is rarely an effective long-term solution, but retreat is often politically unviable – until, as is true for many Pacific island nations,[110] it becomes absolutely necessary.[111] Food security issues and the public health risks they raise should, however, become important factors in any coastal community's evaluation of these two basic coastal adaptation strategies.

#### 8.4.1.3 Periodic and Permanent Displacement

Loss of coastal fresh water supplies and croplands are two reasons why sea-level rise, storm surge, and coastal storms can lead to periodic and permanent displacement of coastal communities. Indeed, James Titus has predicted that "[t]he impact of sea level rise on groundwater salinity could make some areas uninhabitable even before they were actually inundated, particularly those that rely on unconfined aquifers just above sea level."[112]

However, the triumvirate of coastal climate change effects – sea-level rise, storm surge, and increasing numbers of increasingly severe coastal storms – can also physically destroy coastal communities. Titus has identified eight physical effects from sea-level rise: "A rise in sea level would (1) inundate wetlands and lowlands, (2) erode shorelines, (3) exacerbate coastal flooding, (4) increase the salinity of estuaries and aquifers and otherwise impair water quality, (5) alter tidal ranges in rivers and bays, (6) change the locations where rivers deposit sediment, (7) increase the heights of waves, and (8) decrease the amount of light reaching the bottoms."[113] All of these physical results have

[108] As one example, "[t]he Dutch government has set aside one billion euros a year through 2100 to strengthen dunes and dams throughout the country." Danny Vinik, *Climate Change Will Force Us to Abandon Coastal Cities. We'd Better Start Preparing Right Now,* The New Republic, https://newrepublic.com/article/117799/climate-change-will-force-us-abandon-cities-if-we-dont-prepare-now (May 18, 2014).

[109] E.g. Gary B. Griggs, The effects of armoring shorelines – The California experience, In: *U.S. Geological Survey, Puget Sound Shorelines and the Impacts of Armoring: Proceedings of a State of the Science Workshop* 77–84 (2010), https://pubs.usgs.gov/sir/2010/5254/pdf/sir20105254_chap8.pdf.

[110] *Five Pacific Islands Lost to Rising Seas as Climate Change,* The Guardian, www.theguardian.com/environment/2016/may/10/five-pacific-islands-lost-rising-seas-climate-change (May 10, 2016).

[111] See, e.g., Vinik, *We'd Better Start Preparing, supra* note 108; Justin Gillis and Kenneth Chang, *Scientists Warn of Rising Oceans From Polar Melt,* The New York Times, www.nytimes.com/2014/05/13/science/earth/collapse-of-parts-of-west-antarctica-ice-sheet-has-begun-scientists-say.html (May 12, 2014).

[112] James G. Titus, *Greenhouse Effect, Sea Level Rise and Land Use,* 7 Land Use Pol'y 138, 144 (1990).

[113] Ibid. at 140.

potential public health impacts, including periodic and permanent displacement of coastal communities.

In Chapter 11, Maxine Burkett discusses the more general public health consequences that arise or potentially arise when communities and even larger populations are displaced from their homelands. The point here is simply that rising seas and eroding shorelines displace coastal communities.[114] Indeed, erosion actually makes the effects of sea-level rise worse. As Professor Titus has explained, "[i]n many areas the total shoreline retreat from a 1 m[eter] rise would be much greater than suggested by the amount of land below the 1 m[eter] contour on a map because shores would also erode."[115] Across the globe, the combined processes of sea-level rise and erosion could be devastating. "A 1 m[eter] rise in sea level would inundate 17% of Bangladesh ... Shanghai and Lagos – the largest cities of China and Nigeria – are less than 2 m[eters] above sea level, as is 20% of the population and farmland of Egypt."[116]

Indeed, human displacement as a result of coastal storms and inundation is already occurring, including in the United States. Hurricane Katrina displaced almost the entire population of New Orleans (454,000 people), and while about 53 percent of residents returned to the area within the first year, 47 percent remained displaced one year later, concentrated in Texas, other parts of Louisiana, and other parts of the southern United States.[117] A decade later, New Orleans' population remained reduced by about 71,000 people compared to pre-Katrina levels (with greater migration by the black population),[118] and many hurricane-displaced migrants continued to struggle in their new homes.[119]

Coastal inundation and storm surge also threaten a number of Native Alaskan and Native American groups, and much of this displacement has been directly attributed to the impacts of climate change. For example, climate change is affecting the Arctic more severely than much of the rest of the world, and the Government Accountability Office has identified at least twelve Alaska Native Villages that will need to relocate from the

114 Ibid. at 148; Orrin H. Pilkey and Andrew G. Cooper, *Society and Sea Level Rise*, 303 Science 1781 (2004).

115 Titus, *Greenhouse Effect*, *supra* note 112, at 141.

116 Ibid. at 140 (citations omitted).

117 Narayan Sastry and Jesse Gregory, *The Location of Displaced New Orleans Residents in the Year After Hurricane Katrina*, 51(3) Demography 753–775 (June 2014), doi: 10.1007/s13524-014-0284-y, available at www.ncbi.nlm.nih.gov/pmc/articles/PMC4048822/.

118 Malcom Gladwell, *Starting over*, The New Yorker, www.newyorker.com/magazine/2015/08/24/starting-over-dept-of-social-studies-malcolm-gladwell (Aug. 24, 2015).

119 E.g., Id.; Eli Saslow, *Ten Years After Katrina*, The Washington Post, www.washingtonpost.com/sf/national/2015/08/29/youre-one-of-us-now/?utm_term=.be04a1d1f70e (Aug. 29, 2015).

coast.[120] In late February 2008, the Inupiat Eskimo coastal village of Kivalina filed a public nuisance lawsuit against twenty-four oil and energy companies, seeking as damages up to $400 million necessary to relocate the village, because melting sea ice and thawing permafrost are eroding away the village's foundation.[121] The Village eventually lost its climate change-based public nuisance action.[122] Nevertheless, the plight of Alaska Native Villages needing to relocate remains, leading the Village of Newtok in January 2017 to seek federal disaster relief and funding.[123]

Other areas of coastal inundation in the United States are also creating displacements. Louisiana's Gulf coast presents an extreme case, with inundation occurring as a result of a combination of land subsidence, loss of sediment, rising seas, and coastal hurricanes.[124] As a result of these interacting forces, "[s]ince 1955, the Isle de Jean Charles band of the Biloxi-Chitimacha-Choctaw tribe has lost 98 percent of its land to the encroaching Gulf waters. Of the 22,400-acre island that stood at that time, only a 320-acre strip remains. The tribe's identity, food, and culture have slowly eroded with the land."[125] In January 2016, the US Department of Housing and Urban Development awarded the Tribe $48 million to relocate.[126] A 2017 report from the Center for Progressive Reform identified seventeen native communities in Louisiana, Washington State, and Alaska that are in the process of relocating and at least eighteen more in Alaska that will need to relocate.[127]

120 Robin Bronen, Brookings Institute, *Climate-Induced Displacement of Alaska Native Communities iii* (Jan. 30, 2013) www.brookings.edu/wp-content/uploads/2016/06/30-climate-alaska-bronen-paper.pdf.

121 Yereth Rosen, *Village in Alaska Sues Energy Companies Over Erosion Linked to Warming Climate*, St. Env't Daily (Feb. 29, 2008), http://pubs.bna.com/ip/bna/sed.nsf/eh/A0B6D2P9D7.

122 *Native Village of Kivalina v. ExxonMobil Corp.*, 696 F.3d 849, 857–58 (9th Cir. 2012), cert. denied, 133 S. Ct. 2390 (2013).

123 Rachel Waldholz, *Alaska Village, Citing Climate Change, Seeks Disaster Relief in Order to Relocate*, National Public Radio, www.npr.org/2017/01/10/509176361/alaskan-village-citing-climate-change-seeks-disaster-relief-in-order-to-relocate (Jan. 10, 2017).

124 "The major causes of this land loss include the effects of climate change, sea level rise, subsidence, hurricanes, storm surges, disconnection of the Mississippi River from coastal marshes, and human impacts." State of Louisiana, Louisiana's Comprehensive Master Plan for a Sustainable Coast ES-2 (June 2, 2017), http://coastal.la.gov/wp-content/uploads/2017/04/2017-Coastal-Master-Plan_Web-Single-Page_CFinal-with-Effective-Date-06092017.pdf.

125 Carolyn Van Houten, *The First Official Climate Refugees in the U.S. Race Against Time, National Geographic*, http://news.nationalgeographic.com/2016/05/160525-isle-de-jean-charles-louisiana-sinking-climate-change-refugees/ (May 25, 2016).

126 Ibid.

127 Maxine Burkett et al., *Reaching Higher Ground: Avenues to Secure and Manage New Land for Communities Displaced by Climate Change*, 6–7 (May 2017), http://progressivereform.org/articles/ReachingHigherGround_1703.pdf.

Displacement in the United States is only likely to increase throughout the twenty-first century, because "upwards of 13 million Americans will be at risk of displacement from a projected 0.9 m[eter] rise in sea levels by 2100."[128] Worldwide, that number only increases – but displacement from sea-level rise critically depends on the climate change future we live through. One study, for example, projects that an increase of global average temperatures of 4° Celsius would bring about sea level increases of 35 feet, displacing 760 million people globally – about 10 percent of the world's population.[129]

### 8.4.2 *Increasing and Changed Disease Vulnerabilities*

Lindsay Wiley addresses appropriate legal responses to new and increasing disease threats in Chapter 9, and much of her discussion is relevant for coastal communities. This chapter complements that work by emphasizing the disease threats arising from warming and rising seas and coastal storms and recommending legal responses.

#### 8.4.2.1 Increased Cholera Outbreaks

The ocean is a reservoir of disease bacteria and viruses, and warming and rising sea levels could expose new and more extensive populations to these diseases. For example, cholera outbreaks are "associated with drinking or bathing in unpurified river or brackish water" but also appear to be linked to climate and temperature.[130] Moreover, the cholera bacterium (*Vibrio cholerae*) has a sea stage, during which copepods (a type of tiny animal, or zooplankton) act as host organisms.[131] According to researchers investigating the link between climate change and cholera, "[c]limate, seasonal weather changes and

[128] Ibid. at 7.

[129] Danny Clemens, *Climate Change by the Numbers: 760 Million Displaced by Rising Sea Levels*, Discovery, www.discovery.com/dscovrd/nature/climate-change-by-the-numbers-760-million-displaced-by-rising-sea-levels/ (Nov. 10, 2015).

[130] Brad Lobitz et al., *Climate and Infectious Disease: Use of Remote Sensing for Detection of Vibrio Cholerae by Indirect Measurement*, 97 Proc. Nat'l Acad. Sci. 1438, 1438 (2000).

[131] Kate Madin, Woods Hole Oceanographic Institute, *Bacteria Hitchhike on Tiny Marine Life*, Oceanus Magazine, www.whoi.edu/oceanus/feature/cholera (Feb. 1, 2013) (noting that studies have proven that these copepods can be the disease-carrying agents of cholera); see also Rita R. Colwell et al., *Reduction of Cholera in Bangladeshi Villages by Simple Filtration*, 100 (3) PNAS 1051, 1051–55 (Feb. 4, 2003) (proving that filtering out copepods from water reduced cholera infection).

seasonal changes in ocean currents affect the growth of copepods."[132] The links among climate change, alteration of marine parameters such as temperature and pH, and cholera dynamics are now accepted, if not yet completely understood.[133] Researchers hope that by measuring ocean parameters such as temperature and plankton blooms, they will be able to provide "an early warning system for cholera, enabling an effective deployment of resources to minimize or prevent cholera epidemics . . . "[134]

Cholera-carrying copepods "live[] in salt or brackish waters, including rivers and ponds, and travel[] with currents and tides. Copepods harbour both dormant, nutrient-deprived and culturable *Vibrio*. The bacteria can survive as an inactive sporelike form – dormant but still infectious – in the gut and on the surfaces of copepods in between epidemics."[135] Moreover, ships transport a very large number of copepods – and other disease organisms – in ballast water.[136]

Evidence indicates that "cholera outbreaks occur shortly after sea-surface temperature and sea-surface height are at their zenith."[137] Thus, sea-level rise, in connection with changes in currents and sea temperatures, could promote the spread of cholera. Moreover, cholera spreads through drinking water and, as has already been discussed, one consequence of sea-level rise is contamination of drinking water supplies. Perhaps not coincidentally, therefore, in this century – that is, within the same timeframe that climate change has begun to affect ocean temperatures and ocean currents – cholera has reemerged in epidemic form in the coastal areas of Southeast Asia, Central America, and South America.[138]

In the United States, an interdisciplinary effort to understand the relationships among climate change, ocean acidification, and cholera is focused on Chesapeake Bay.[139] This effort is using cholera as the initial bacterium to create models capable of predicting "ocean-related public health risks from pathogens," and this effort "represents a step toward developing a robust ecosystem modeling capability . . . "[140] The researchers use daily sea-surface temperature and salinity measurements to generate predictions of *Vibrio cholerae* probabilities in the Bay.[141] While cholera outbreaks in the United States have become extremely rare, such prediction models could become important public health tools in other

132 Rita R. Colwell, A *Voyage of Discovery: Cholera, Climate, and Complexity*, 4 Envtl. Microbiology 67, 67 (2002).

133 Guillaume Constantin de Magny and Rita R. Colwell, *Cholera and Climate: A Demonstrated Relationship*, 120 Transactions of the American Clinical & Climatological Ass'n 119, 119–120 (2009).

134 Lobitz et al., *Climate and Infectious Disease, supra* note 130, at 1438.

135 Colwell, A *Voyage of Discovery, supra* note 132, at 68. 136 Ibid. 137 Ibid. at 67.

138 Ibid. 139 Magny and Colwell, *Cholera and Climate, supra* note 133, at 122–125.

140 Ibid. at 125. 141 Ibid.

countries and for other diseases, and they could help people in the United States to continue to avoid cholera infection despite changing ocean conditions.

#### 8.4.2.2 Other Ocean-Mediated Diseases and Toxic Blooms

Cholera is not the only disease agent connected to the ocean. For example, a related species, *Vibrio vulnificus,* is a sea-dwelling bacterium that can cause disease in humans. These bacteria "colonize filter feeding animals such as oysters, crabs and mussels, but can also be found free-living in seawater."[142] While "[m]ost people become infected with *V. vulnificus* through eating raw shellfish," the bacterium "can also cause wound infections where an open wound is exposed to seawater."[143] In addition to unpleasant but less serious effects, septicemia leading to amputation or death is one potential outcome from either route of infection.[144]

The disease potential of *Vibrio vulnificus* appears to be linked to sea temperature, and during the twentieth century most identified infections occurred along the very warm Gulf of Mexico, especially in Florida. However, the emergence of *Vibrio vulnificus* disease in other parts of the world, notably Israel, has been linked to climate change and increasing temperatures.[145] Similarly, in the United States in the early twenty-first century, there has been an increase in the number of *Vibrio vulnificus* infections along the Atlantic coast, stretching as far north as Delaware, New Jersey, and Rhode Island, linked to increasing sea temperatures.[146] Sea-level rise, in concert with this temperature increase, will expose new populations of humans to the *Vibrio vulnificus* bacterium, potentially increasing the disease's incidence.

Contaminated sea water is also already the source of increasingly frequent toxic algae blooms. A variety of factors spur marine algae blooms, including temperature, nutrients from agricultural run-off, and other oceanic properties.[147] Some of these algae produce toxic chemicals, and when the algae are present in high concentrations, these toxins can affect humans and other animals. For example, sea lions in California have died when blooms of certain marine algae produce domoic acid.[148] As with *Vibrio vulnificus* disease, sea-level rise

142 Carina Blackmore, Florida Division of Disease Control, *Vibrio Vulnificus* (1999), www.doh.state.fl.us/Disease_ctrl/epi/htopics/reports/vib_vul.pdf.

143 Ibid. 144 Ibid.

145 S. Paz et al., *Climate Change and the Emergence of* Vibrio Vulnificus *Disease in Israel,* 103 Envtl. Res. 390, 390–391 (2007).

146 Jessica Forres, *Vibrio Bacteria a Bigger Threat to Swimmers Than Sharks as Northern Waters Warm, Stories That Matter,* May 23, 2007, www.storiesthatmatter.org/index.php?option=com_content&task= view&id=123.

147 The *Rising Tide of Ocean Plagues: How Humans Are Changing the Dynamics of Disease,* EurekAlert, Feb. 17, 2006, www.eurekalert.org/pub_releases/2006-02/s-trt021206.php.

148 Ibid.

will take these public health threats inland to new populations, while temperature changes can spread new types of algae blooms to new locations.

Finally, melting ice can expose people to disease. In 2006, Dr. Scott Rogers, a Bowling Green State University biologist, reported "the potential for long-dormant strains of influenza, packed in ice in remote global outposts, to be unleashed by melting and migratory birds."[149] As a result, melting ice could expose human populations to strains of flu, such as the virus that caused the 1918 flu pandemic, against which human immunity has died out.[150] Dr. Scott contends this "information could be used to help develop inoculation strategies for the future."[151] As "tabloid press" as this concern may sound, ice-based disease transmission appears to have already occurred. In August 2016, a 12-year-old boy died from and several other people living on the Yamal Peninsula, Siberia, were hospitalized with anthrax, apparently infected when melting permafrost exposed an infected reindeer that had been frozen for about seventy-five years.[152] Moreover, scientists found the 1918 H1N1 pandemic influenza virus in frozen corpses in Alaska, which they could then study both *in vitro* and *in vivo*, including sequencing its genome.[153]

#### 8.4.2.3 Increasing Vector-Borne Disease

In combination with higher temperatures in many coastal areas, sea-level rise will contribute to the expected resurgence of certain mosquito-borne diseases such as malaria and the introduction of new mosquito-borne diseases, such as dengue fever. As James Titus has noted, "[b]y deepening shallow bodies of water, a sea level rise could cause them to stagnate."[154] Warm, stagnant bodies of brackish water are perfect breeding grounds for disease-bearing mosquitoes. Worldwide, malaria and dengue fever are spreading, both by emerging into

149 *Flu Can Bide Time in Icy Limbo Before Re-Emerging, Biologist Believes*, Science Daily, Nov. 28, 2006, www.sciencedaily.com/releases/2006/11/061127210430.htm.

150 Ibid. 151 Ibid.

152 Jasmin Fox-Skelly, *Long-Dormant Bacteria and Viruses, Trapped in Ice and Permafrost for Centuries, Are Reviving as Earth's Climate Warms*, BBC Earth, www.bbc.com/earth/story/20170504-there-are-diseases-hidden-in-ice-and-they-are-waking-up (4 May 2017). See also Boris A. Revich and Marina A. Podolnaya, *Thawing of Permafrost May Disturb Historic Cattle Burial Grounds in East Siberia*, 4 Global Health Action, DOI: 10.3402/gha.v4i0.8482, at 1–7 (2011) (detailing the increasing risk of anthrax infections in cattle and reindeer as a result of melting permafrost).

153 Jeffery K Taubenberger et al., *Discovery and characterization of the 1918 pandemic influenza virus in historical context*, 12 Antivirus Theory 581, 581–91 (2007).

154 Titus, *Greenhouse Effect*, *supra* note 112, at 145.

new areas and by returning to areas where the diseases had been under control. For example, malaria has returned to countries like Peru, largely as a result of climate change and deforestation. In Peru, malaria was almost eradicated forty years ago, but in 2008 64,000 cases were registered in the country, half in the Amazon region. It is thought there are many more unregistered cases deep within the massive and humid rainforest, where health authorities find it almost impossible to gain access.[155] Malaria is also endemic in the United States, if currently essentially under control;[156] thus, the resurgence of the disease in Peru provides a cautionary note for Americans.

Dengue fever is also spreading worldwide. According to WHO in 2017,

> The incidence of dengue has grown dramatically around the world in recent decades. The actual numbers of dengue cases are underreported and many cases are misclassified. One recent estimate indicates 390 million dengue infections per year (95% credible interval 284–528 million), of which 96 million (67–136 million) manifest clinically (with any severity of disease).1 Another study, of the prevalence of dengue, estimates that 3.9 billion people, in 128 countries, are at risk of infection with dengue viruses.[157]

Dengue fever epidemics spread through South America,[158] and the disease has spread up both coasts of Mexico to the United States border, with a small number of local cases occurring in South Texas[159] and, more recently, Florida.[160]

#### 8.4.2.4 Dealing with New and Increasing Coastal Disease Vulnerabilities

Around the world, disease threats arising from warming and rising seas and coastal storms warrant significant changes to public health laws governing appropriate training for medical professionals and response planning and

[155] Andrés Schipani and John Vidal, *Malaria Moves in Behind the Loggers*, The Guardian, Oct. 30, 2007, www.guardian.co.uk/world/2007/oct/30/environment.climatechange.

[156] Centers for Disease Control and Prevention, *Eradication of Malaria in the United States (1947–1951)*, www.cdc.gov/malaria/history/eradication_us.htm (last visited April 9, 2008).

[157] World Health Organization, *Dengue and Severe Dengue*, www.who.int/mediacentre/fact sheets/fs117/en/.

[158] Marcela Valente, *South America: Climate Change Fuels Spread of Dengue Fever*, Inter Press Service, Mar. 19, 2007, http://ipsnews.net/news.asp?idnews=36994.

[159] Centers for Disease Control and Prevention, *Dengue Fever*, http://www.cdc.gov/ncidod/dvb id /dengue.

[160] From 1934 to 2009, dengue fever occurring in Florida was the result of international travel. However, in 2009, 2010, and 2013, the state experienced locally acquired outbreaks of dengue fever, which affected sixty-six individuals in 2010. Florida Department of Health, *Dengue Fever: Dengue Occurrence in Florida*, www.floridahealth.gov/%5C/diseases-and-conditions/ dengue/index.html (as viewed June 30, 2017).

preparation. Coastal medical professionals may increasingly be coping with diseases with which they have little previous familiarity. For example, the medical profession in the United States is generally unaccustomed to dealing with malaria, dengue fever, and cholera, especially in epidemic proportions, and *Vibrio vulnificus* and algae blooms are already presenting new medical challenges to many coastal communities. Therefore, training medical personnel in these communities to recognize and treat these diseases and other sea-related health issues should be a legal priority.[161] Such training will help to ensure that coastal medical professionals can recognize and treat these diseases if and when they emerge. However, public health laws and regulations must also help to ensure that the right medications and treatment facilities are available for new and emerging diseases in coastal communities, allowing public health officials to implement control measures and to engage in larger-scale public health preparation in case of epidemics. These laws must also coordinate public education, public notifications, and vector control, ensuring that public health concerns receive their proper due. During the 2013 outbreak of dengue fever in Florida's Key West, for example, "the Florida Keys Mosquito Control District (FKMCD) acted aggressively, using all means at their disposal to reduce the populations of" the mosquito vectors.[162] However, members of the public were less than cooperative:

> The FKMCD also embarked upon a direct public information campaign focused on residents, visitors, and business owners, explaining the importance of eliminating exposed water holding containers to reduce the abundance of dengue vectors, and of personal protection to minimize the probability of bites by remaining ones. Local reception to the latter efforts, however were lukewarm at best as once again fears of economic repercussions, in this case damage to the all-important tourist industry, became paramount. In many cases, citizens and business owners made no effort to mitigate mosquito production from their properties, some denied that disease transmission was taking place, and some went as far as accusing the FKMCD of exaggerating the situation in order to boost the agency's funding.[163]

161 For example, in response to the emergence of the Zika virus, the Centers for Disease Control stated that its goals included ensuring that "state and territorial health departments have the support needed to adequately test specimens from suspect Zika cases," informing "healthcare providers on the risks, recognition, diagnosis, reporting, clinical management, outcomes, and prevention of Zika virus infections," and informing "the public about Zika virus and related birth defects or illnesses, especially microcephaly and other harmful effects to pregnant women and their fetuses," as well as preventing its further spread. Centers for Disease Control, *Top 10 Zika Response Planning Tips: Brief Information for State, Tribal, Local, and Territorial Health Officials*, www.cdc.gov/zika/public-health-partners/tips.html.

162 Jorge R. Ray, *Dengue in Florida (USA)*, 5(4) Insects 991–1000 (Dec. 2014), www.ncbi.nlm.nih.gov/pmc/articles/PMC4592614/.

163 Ibid.

Florida's dengue fever experience, therefore, highlights that even the most prepared and proactive of public health responses can falter in the face of recalcitrant or uncooperative citizens.

### 8.4.3 *Cumulative Impacts on Coastal and Marine Food Supplies*

As outlined in the previous section, climate change is creating significant threats to farming and terrestrial food supplies globally. (See also Chapter 10.) Often neglected in public health preparedness, however, are the impacts of climate change on marine food supplies, which can similarly contribute to public health problems resulting from food insecurity.

Of the seven billion (and counting) people who live on the planet, one billion rely on fish (including freshwater fish) as their *primary* source of protein.[164] As of 2010, 2.9 billion people get about 20 percent of their protein from fish, while 4.3 billion people – well over half the world's human population – get at least 15 percent of their protein from fish.[165] In addition, fish is a critical source of protein (30 percent or more of protein consumed) in many specific countries around the world, including several countries in Africa, Japan, Greenland, Taiwan, Indonesia, and several South Pacific island nations.[166]

According to the IPCC, "all ocean basins show global-scale changes including large-scale distribution shifts of species … and altered ecosystem composition … "[167] Many species of fish and invertebrates have shifted their ranges toward the North and South Poles and/or deeper, seeking the cooler waters to which they are accustomed.[168] In tropical waters, however, coral reefs cannot migrate fast enough to keep up with climate change. As a result, coral reefs and their attendant – and historically highly productive – ecosystems have experienced significant stress, and "[s]ome warm-water corals and their reefs have responded with species replacement, bleaching and a decreased coral cover causing habitat loss."[169] The most profound demonstration of coral reefs' fragility in the face of climate change came with the devastation of the Great Barrier Reef through water temperature increases, coral bleaching, and widespread coral death in 2015 through 2017, with experts concluding

164 Marine Stewardship Council, *Fish as food*, www.msc.org/healthy-oceans/the-oceans-today/fish-as-food (as viewed June 29, 2017).

165 Ibid. 166 Ibid. 167 *2014 IPCC Summary Report, supra* note 11, at 3. 168 Ibid.

169 Ibid.

in May 2017 that the Reef can no longer be saved in its former form, although they hope to preserve at least some of its ecological function.[170]

In addition, fish and other important commercial, artisanal, and subsistence species are increasingly no longer where they are "supposed" to be, and food security issues – with their attendant public health consequences – are arising both locally and globally as a result. According to the IPCC, "By the mid 21st century, the spatial shifts of marine species will cause species richness to increase at mid and high latitudes ... and to decrease at tropical latitudes ... resulting in global redistribution of catch potential for fishes and invertebrates, with implications for food security ... "[171] Moreover, under the right circumstances, "the overall fisheries catch potential may also decrease."[172] As a result, given the dependence of many communities on fish and seafood as food sources, climate change impacts to the oceans could exacerbate global inequalities in food distribution, increase rates of starvation in certain parts of the world, and, possibly, reduce the total availability of high protein food sources.[173]

As noted earlier, changing issues in food security for coastal communities – both loss of coastal farmland and loss of critical fisheries – should be addressed as part of an overall coastal health and climate change adaptation strategy. Unfortunately, climate change impacts on marine fisheries, like global ice melting, present many uncertainties for coastal nations, but Arctic nations are likely to benefit through the mid-twenty-first century while tropical nations, the continental United States, China, and Chile see their marine food supplies dwindle over that same

170 Ian Johnston, *Great Barrier Reef Can No Longer Be Saved, Australia Experts Concede*, The Independent, www.independent.co.uk/news/world/australasia/great-barrier-reef-dying-coral-bleaching-global-warming-australia-climate-change-a7761351.html (May 29, 2017). The expert panel was a bit more reserved, noting that "[t]here is great concern about the future of the Reef, and the communities and businesses that depend on it, but hope still remains for maintaining ecological function over the coming decades. Members agreed that in our lifetime and on our watch, substantial areas of the Great Barrier Reef and the surrounding ecosystems are experiencing major long-term damage which may be irreversible unless action is taken now." Independent Expert Panel on the Great Barrier Reef, Reef 2050 Plan: Communiquè (May 5, 2017) www.environment.gov.au/system/files/pages/abffod5e-b9 4d-4495-b79b-90dc52274f69/files/expert-panel-communique-5may2017.pdf.

171 *2014 IPCC Summary Report*, *supra* note 11, at 3–4. 172 Ibid. at 4.

173 See Ibid. at 5 (summarizing how climate change impacts to marine species will increase socioeconomic vulnerability in tropical countries and emphasize disparities in food security); Matthew Huelsenbeck, Oceana, *Ocean-Based Food Security Threatened in a High CO2 World: A Ranking of Nations' Vulnerability to Climate Change and Ocean Acidification* (Sept. 2012) http://oceana.org/sites/default/files/reports/Ocean-Based_Food_Security_Threat ened_in_a_High_CO2_World.pdf.

period.[174] In specific coastal nations, therefore, public health considerations may dictate radically different short-term (avoid food shortages) and long-term (avoid or deal with fisheries collapses) legal approaches to marine fisheries regulation. Nevertheless, a more intensely precautionary approach to fishery catch limits and diversification of food supply will be appropriate public health strategies for many countries to pursue, and public health officials should become educated participants in fisheries management, law, and policy decision-making, which are currently often driven by short-term economics rather than food security or public health planning.

## 8.5 CONCLUSIONS AND RECOMMENDATIONS

The legal consequences of taking a public health perspective on the climate change future of the world's coastal communities is multivalent and demands significant coordination among a variety of different types of laws – disease control and prevention laws, emergency preparedness and response laws, training requirements for medical personnel, land use planning and building codes, environmental and pollution control requirements, water law, food safety laws, and fisheries management laws and requirements. Most basically, a public health assessment reveals significant vulnerabilities in coastal communities that go far beyond the simple fact of physical inundation by seawater and the policy debate of "retreat versus seawalls" – although, of course, that fact and that debate remain part of sea-level rise and coastal storm response. Instead, even with respect to physical inundation, a public health perspective suggests that coastal governments have to pay far more legal attention to public water supplies, to the existing and potential toxic and nuclear contamination of their communities, to the siting and construction requirements for coastal infrastructure, to the potential for new and reintroduced diseases, and, of course, to the emergency response and disaster preparedness requirements that they impose. Looking a bit more broadly, sea-level rise, coastal storms, and the more general changes occurring in ocean ecosystems mean that coastal governments should also be concerned about food security, particularly in coastal communities that depend heavily on coastal agriculture and/or subsistence or artisanal fishing for the bulk of their calories and nutrition.

[174] William W. L. Cheung et al., *Large-Scale Redistribution of Maximum Fisheries Catch Potential in the Global Ocean Under Climate Change*, 16(1) *Global Change Biology* 24–35 (Jan. 2010).

Vulnerability assessments that include public health vulnerabilities are an important first step in adapting the various kinds of relevant laws to sea-level rise and coastal storms. Again, there is no universal coastal preparedness formula that can be developed or offered, because the rate and severity of the coastal impacts from climate change will differ among coastal communities and because the specific public health risks that a community faces can vary not only because of those varying climate change impacts but also because of: (1) the constellations of pathogens and vectors in that region; (2) the types of and reliability of the various kinds of infrastructure in place; (3) the source(s) of local water supply and changes occurring in it/them; (4) the sources of local food supply; (5) existing and evolving land use patterns; and (6) the existing and evolving human populations present – including the extent of crowding and the options for temporary and permanent relocations.

Vulnerability assessments are not enough, however. Instead, the vulnerability assessment should become the basis for effective law and policy interventions and innovations. In Chapter 6, Justin Gundlach and Jennifer Klein discussed opportunities for and limits of such interventions in relation to the built environment. In Chapter 9, Lindsay Wiley addresses more generally many of the innovations needed in the medical profession with respect to training, disease diagnosis, and treatment, and these legal improvements apply with equal force in the coastal zone. In Chapter 10, Margot Pollans addresses legal innovations that may be necessary to deal with food security and nutrition issues; again, many of her recommendations apply with equal force in the coastal zone. In Chapter 11, Maxine Burkett addresses the public health impacts of climate change displacements, whether permanent or temporary, and, again, the recommendations from that chapter apply with equal force in the coastal zone.

While the coastal zone is thus not unique regarding the potential issues that climate change poses to public health, coastal storms make particularly sensible the public health vulnerabilities that outdated infrastructure designed for milder environments can create. Sometimes, as noted, the legal solutions are as simple as building codes that require hospital back-up generators to be housed on higher floors (not the basement), that require sewage treatment plants to have sufficient storage capacity to avoid discharges of raw sewage even during a 1000-year (or 10,000-year?) storm event, and that require coastal houses and businesses to be built (or retrofitted) to the highest hurricane/typhoon standards. Sometimes they are as complex and potentially expensive as creating and enforcing requirements for decommissioning nuclear power plants and moving them out of the coastal zone. In either case, however, coastal governments can no longer afford to ignore the many interactions among coastal infrastructure (particularly long-lived infrastructure), sea-level rise, and coastal storms and increased storm surge.

As noted in this chapter, environmental and natural resources laws also have important roles to play in reducing the public health vulnerabilities of the coastal zone in a climate change era. As discussed, water law has an important role to play in protecting coastal aquifers and finding and managing new sources of potable drinking water if necessary. Similarly, coastal communities are best advised to address their potential toxic exposures *before* coastal flooding and coastal inundation occur. Coastal communities should enact legal requirements for the handling and storage of toxic materials that take exposure to saltwater flood waters into account and consider toxic facility siting requirements in their land use planning efforts. State governments and the US EPA may want to prioritize cleanup of existing contamination under the federal Comprehensive Environmental Response, Compensation, and Liability Act (CERCLA or Superfund)[175] and equivalent state statutes; cleaning up already-contaminated coastal land is a worthy public health strategy under any circumstance, but it is particularly critical in reducing residents' risk of toxic exposure during and after a coastal flooding event. The EPA and Congress might also consider increasing the stringency of requirements for handling hazardous waste in the coastal zone under the Resource Conservation and Recovery Act (RCRA)[176] and for pesticide use under the Federal Insecticide, Fungicide, and Rodenticide Act (FIFRA).[177] Regulation of how most other toxic and hazardous materials are handled is generally up to the states, which could do more to address toxic exposure in the coastal zone through state coastal zone management plans[178] and state tort liability – but both aspects of state law would benefit from the expertise of public health professionals and toxicologists regarding these often-ignored public health risks facing coastal communities.

Coastal governments should also consider the extent to which they are encouraging populations to migrate to the coastal zone, in an awareness of the projections that many of these populations are likely to be displaced by the end of this century. As Justin Gundlach and Jennifer Klein discuss in Chapter 6, retreat strategies may become increasingly necessary, and they will be much less disruptive and expensive if vulnerable coastal zones are less populated to begin with.

175 42 U.S.C. §§ 9601–9675 (2012). 176 42 U.S.C. §§ 6901-6992k (2012).

177 7 U.S.C. §§ 136-136y (2012).

178 See 16 U.S.C. §§ 1454, 1455, 1456a, 1456b, 1456c (2012) (allowing for federal approval of, funding for, and technical assistance regarding federally approved state coastal zone management plans).

Overall, a public health perspective on the impacts of sea-level rise and coastal storms can provide more politically and popularly salient, not just additional, reasons for engaging in active climate change adaptation efforts. A culture of continuously updating, reevaluating, and preparing for emerging disease threats promotes awareness among both medical personnel and the general public regarding relevant risks, potential exposures, techniques for avoiding transmission, and symptoms, aiding the entire public health enterprise in avoiding, minimizing, and responding early to potential disease crises. Reduction of toxic and hazardous materials and contamination in the coastal zone has numerous public health, safety, and environmental benefits even if a severe storm never occurs, generally rendering such efforts "no regrets" climate adaptation strategies – particularly when the financing can come from responsible corporations rather than governments. Rational flood insurance policies, climate-resilient building codes, and land use planning can avoid perpetuating coastal risk-taking into the future, while a public health perspective on those risks can often justify the expense of retrofitting better than property value approaches. Finally, in countries where "takings" of private property by governments is a real legal concern, a public health perspective on coastal retreat and increased regulation of private property can underscore the public benefits being pursued, helping to insulate such measures from governmental liability.[179]

The impacts of climate change on the coasts strongly suggest that the expertise of public health personnel should be sought on many more subjects than disease prevention, control, and treatment. Public health experts and toxicologists already often play roles in the design and execution of environmental law, although that expert influence could still be fine-tuned to the particular hazards of coastal storms and inundation. Moreover, prudent climate change adaptation efforts in and for coastal communities should seek to include public health experts in discussions and decision-making about coastal building codes, facilities siting, land use planning, fisheries management, water supply, and coastal agriculture. At bottom, effective public health and disaster preparedness for coastal communities in a climate change era requires people with the knowledge base to effectively implement a "big picture" approach to risk management, and public health officials may become particularly well-positioned to assume those roles.

[179] See generally, Robin Kundis Craig, *Of Sea Level Rise and Superstorms: The Public Health Police Power as a Means of Defending Against "Takings" Challenges to Coastal Regulation*, 22 N.Y.U. Envtl. L. Rev. 84 (2014).

# 9

# Infectious Disease

Lindsay F. Wiley

*Climatic conditions are affecting interactions between the agents, vectors, and hosts of disease within our changing environment, leading to changing patterns of vector-borne, zoonotic, food-borne, and waterborne illness. The public health law toolkit for infectious disease control includes laws that facilitate surveillance, access to medical countermeasures, vector control, health education and communication, social distancing, and public health emergency preparedness and response. Additionally, two public health principles should guide adaptation to the infectious disease impacts of climate change: One Health and Health in All Policies.*

## 9.1 INTRODUCTION

Our vulnerability to infectious diseases reminds us of how interconnected we are – to each other and to the organisms that share our environment. Because living in a bubble is impractical, we develop protective infrastructure (sanitation, surveillance, and vector control systems) and medical countermeasures (vaccines and antimicrobials). Because pathogens often outpace our ability to combat them, we use social controls to separate the sick and exposed (isolation and quarantine) and urge individuals to adopt protective behaviors (e.g. hand hygiene, appropriate use of antimicrobials). Each of these strategies relies – to varying degrees – on public health law and policy. Each will be crucial as we adapt to the infectious disease impacts of climate change.

In the United States and many other countries, legal frameworks and infrastructure for controlling the spread of infectious disease date back to the turn of the twentieth century. Public health laws developed to combat cholera, smallpox, tuberculosis, and polio using nineteenth-century technology continue to provide the basic framework for determining the obligations and authority of health officials and constraints on that authority derived from individual rights. These legal frameworks and the infrastructure on which they rely remain outdated and underfunded in spite of increased attention to the emergence of new threats, such as HIV/AIDS and bioterrorism, and renewed concerns about old threats, such as pandemic influenza. The infusion of funding and legal reforms to enhance public preparedness for bioterrorism and pandemic influenza in the early twenty-first century has limited relevance to climate change adaptation.

The fragmentary nature of the US public health system hinders its capacity to respond effectively and efficiently to infectious disease threats. Public health powers are still predominantly exercised at the local and state level, notwithstanding the federal government's growing role in public health emergency preparedness and response. Health departments at every jurisdictional level play a principal role in infectious disease control, but they are increasingly occupied by the gargantuan task of regulating and supporting the healthcare system. By comparison, public health functions (e.g. surveillance, contact tracing, epidemiological investigation, and social distancing) receive little attention. Through a process of specialization within the administrative state, health departments have become more focused on the medical model, while other agencies have emerged to focus on environmental protection, sanitation systems, agriculture, food and drug safety, the workforce, and other social concerns relevant to preventing and responding to disease outbreaks.

The public health response to environmentally sensitive infectious disease threats such as Dengue fever, hantavirus, West Nile virus, and Zika virus provide a window into how well the system is adapting to climate change. In the words of United Nations Environment Programme Executive Director Achim Steiner, "The spread of Zika, just as with Ebola, has sent a strong signal to the international community that there is a need for increased attention to the linkages between environment and health."[1] The warning signs are clear. Infectious disease threats that are unresponsive to available medical countermeasures, that emerge slowly, those that disproportionately affect people living in low-income households, neighborhoods, and countries, or that relatively well-off people can use their personal resources to protect themselves against tend to get short shrift.

This chapter discusses public health law as a tool for adapting to the infectious disease impacts of climate change. In Section 9.2, I discuss how climatic conditions are affecting interactions between the agents, vectors, and hosts of disease within our changing environment, leading to changing patterns of

[1] UNEP Newscentre *From Asthma to Zika: UNEP Tackles Links between Health and Environment*, Feb. 18, 2016, www.unep.org/newscentre/asthma-zika-unep-tackles-links-between-health-and-environment.

vector-borne, zoonotic, food-borne, and waterborne illness. Section 9.3 focuses on the public health law toolkit for infectious disease control, including laws that facilitate surveillance, access to medical countermeasures, vector control, health education and communication, social distancing, and public health emergency preparedness and response. In Section 9.4, I introduce two principles that should guide adaptation to the infectious disease impacts of climate change: One Health and Health in All Policies. Finally, I offer some concluding reflections on the potential influence of climate change adaptation on the development of public health law and policy.

## 9.2 INFECTIOUS DISEASE RISKS ASSOCIATED WITH CLIMATE CHANGE

Epidemiologists describe patterns of infectious disease as being produced by the interaction between the agent of disease or pathogen (viruses, bacteria, fungi, protozoa, helminths, or prions) and the host (the somewhat eerie scientific term used to describe the infected individual) within an environment (broadly defined so as to include social, economic, and cultural factors in addition to features of the built and natural environment).[2] Vectors may mediate the relationships among host, agent, and environment (see Figure 9.1). Experts typically categorize infectious diseases in terms of type of agent (e.g. viral, bacterial, fungal, and parasitic diseases), the infectivity and transmissibility of the pathogen within a specified host population under specified environmental conditions, and the routes by which it is transmitted. Table 9.1 provides definitions for these and other key terms.

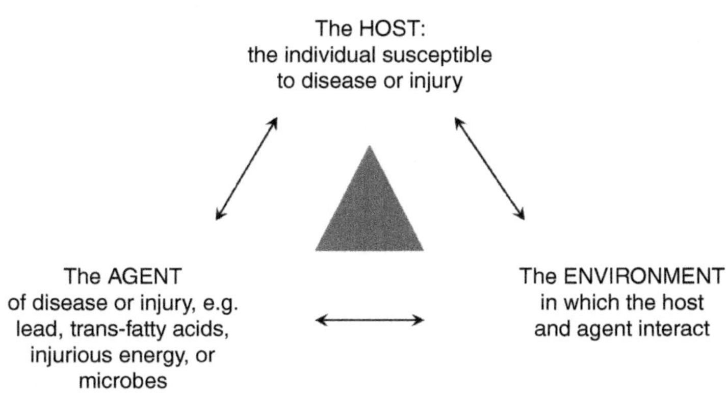


FIGURE 9.1 The epidemiological triad
*from Gostin and Wiley,* Public Health Law: Power, Duty, Restraint, *3rd edn. (2016).*

[2] See, e.g., Leon Gordis, *Epidemiology*, 19 (5th edn. 2013).

TABLE 9.1 *Key terms in infectious disease prevention and control*

| | |
|---|---|
| Agent/Pathogen | Infectious diseases are caused by *agents* (also known as *pathogens*) and/or toxins they produce. Most infectious disease pathogens are viral, bacterial, or fungal. Parasitic diseases are caused by protozoa or helminths ("worms"). With the exception of prions (misfolded proteins that cause some types of encephalopathies), all known infectious disease pathogens are microorganisms. |
| Virulence/Pathogenicity | The *virulence* (also known as *pathogenicity*) of an agent refers to its capacity to produce overt disease within a host. Virulence is often a function of the pathogen's ability to replicate within the body of the host. It is sometimes expressed in terms of case-fatality rate: the proportion of infected individuals who die of the disease. Virulence and case-fatality rate are not inherent to a pathogen; they are influenced by characteristics of the host population and the environment in which agent and host interact. |
| Transmissibility/Infectivity | The *transmissibility* (also known as *infectivity*) of an agent refers to its capacity to spread from host to host. |
| Routes of Transmission | Depending on the pathogen, the *route of transmission* may be respiratory (via aerosolized droplets that are directly inhaled or ingested or transmitted indirectly via "fomites," such as door handles or telephones), fecal-oral contact (usually indirectly via contaminated water or food), sexual contact, oral contact (direct or indirect via sharing drinks or utensils), or skin contact (direct or indirect via shared towels or clothing).<br>Some food- and waterborne pathogens and zoonotic diseases are transmitted directly to human hosts but are not transmitted further from person to person. |

TABLE 9.1 (*continued*)

| | |
|---|---|
| Vector-borne disease | Vector-borne diseases are transmitted via a *vector* organism (e.g. mosquito, tick, or rodent) that carries the pathogen from person to person. |
| Zoonotic disease | Zoonotic diseases are transmitted from an animal to a person. An infected animal becomes an *intermediary host*. An animal population susceptible to a pathogen that is transmissible to humans is sometimes referred to as an *animal reservoir* of disease. It is difficult to control zoonotic disease within a human population if the disease remains present in an animal reservoir. In some cases, a disease first emerges within an animal population and then develops the capacity to be transmitted from animal to person. |
| Waterborne disease | Waterborne diseases are caused by pathogens that live in contaminated water, which the human host ingests. They may be deposited in water by infected humans or animals. |
| Water-related disease | Water-related diseases encompass waterborne diseases as well as water-washed diseases (those that are transmitted easily from person to person in the absence of good hygiene practices) and some vector-borne diseases (those transmitted by vector organisms that are affected by how wet or dry the climate is). |
| Food-borne disease | Food-borne diseases are caused by pathogens that live in contaminated food, which the human host ingests. They may or may not also be transmissible from person to person. |
| Endemic | The *endemic* level of a disease refers to its usual prevalence in a given geographic area. A disease is endemic to an area if it is steadily present there. |
| Outbreak | A disease *outbreak* occurs when a disease is present in greater numbers than expected in a given place during a given time. |

TABLE 9.1 *(continued)*

| | |
|---|---|
| Epidemic | An infectious disease *epidemic* occurs when a disease spreads rapidly to many people. |
| Pandemic | A *pandemic* occurs when an epidemic has spread across a wide region of the globe. |

adapted from Gostin & Wiley, *Public Health Law: Power, Duty, Restraint*, 3rd. edn. (2016).

Climate change is expected to change the geographic and seasonal distribution of a wide range of infectious diseases.[3] Pathogens, vector organisms, and intermediary animal hosts are all living organisms. They typically depend on particular environmental, climatic and weather conditions to support the various stages in their lifecycle.[4] Climate-related factors can affect the number and type of pathogens, vector organisms, and animal hosts in any given geographic area. These factors can also affect the virulence and transmissibility of many pathogens. Globally, epidemiologists estimate that about half of all vector-borne disease risk and 94 percent of diarrheal illness risk is attributable to modifiable environmental factors.[5] For example, higher average temperatures or rainfall can create more favorable breeding conditions for a vector population like mosquitos, increasing the risk that they will transmit diseases like Zika virus and West Nile virus to humans living in the area. Patterns of alternating drought and flooding can cause waterborne pathogens like cholera to become concentrated in small pools of water during times of drought before being washed into larger bodies of water (where people get their drinking water, wash their clothes, or play) during periods of flooding. An animal reservoir population may be weakened by malnutrition or loss of habitat, creating ripe conditions for the emergence of a new zoonotic disease, such as Ebola virus. Changes in the natural environment – whether caused by climate change, deforestation, desertification, or any combination of human-caused or natural processes – also affect the social, economic, cultural, and built environments. The impacts of climate change on food and water security, natural disasters, and national security could also make human populations

[3] K. R. Smith et. al., Human health: Impacts, adaptation, and co-benefits. In: *Climate Change 2014: Impacts, Adaptation, and Vulnerability. Part A: Global and Sectoral Aspects. Contribution of Working Group II to the Fifth Assessment Report of the Intergovernmental Panel on Climate Change*, 722–726.

[4] Xiaoxu Wu et al., *Impact of Climate Change on Human Infectious Diseases: Empirical Evidence and Human Adaptation*, 86 Envtl. Int'l 14, 15 (2016).

[5] A. Prüss-Üstün and C. Corvalán, *Preventing Disease through Healthy Environments* 9 (World Health Organization 2006).

more vulnerable to infection as a result of social, economic, and cultural instability, degradation of the built environment, and malnutrition.

The effects of climate change on the natural environment vary from region to region. Some places become more arid, others become wetter. Many experience both extremes – drought and flooding – more frequently. Many places become warmer while some experience greater extremes of hot and cold temperatures. In many places, weather patterns are less stable, with unseasonable conditions occurring more frequently. In many cases, climate change will redistribute weather patterns, causing vector and reservoir populations to move to higher or lower altitudes or latitudes. Weather conditions and landscape features can generate highly localized patterns of infectious disease transmission.[6] The result may be a reduction in endemic infectious diseases for some regions and increases in others, shifts in the seasonal distribution of outbreaks, and increased risk of emerging infectious diseases with pandemic potential.[7]

### 9.2.1 *Vector-borne*

Vector-borne diseases, especially mosquito-borne diseases, are highly sensitive to environmental changes.[8] Globally, climate change is expected to expand the geographic and seasonal range of mosquito-borne illness while also shifting it in some places. For example, mosquitoes may be able to thrive at higher altitudes – where temperature and moisture conditions were previously inhospitable to mosquitoes. This is concerning because infectious disease outbreaks tend to be more pathogenic when they occur in a population that has not previously experienced much exposure to the disease.[9] If the southeastern United States were to experience a rapid increase in Dengue fever, for example, we would expect to see higher rates of morbidity and mortality because only a small percentage of the population (those who have resided or traveled in areas where Dengue is endemic) will have previously been exposed. Other mosquito-borne diseases that may be affected by climate change include malaria, Zika virus, West Nile virus, and chikungunya.[10] Hantavirus (transmitted by rodents)

6 George Luber et al., Ch. 9: Human health. In: J.M. Melillo et al., *Climate Change Impacts in the United States: The Third National Climate Assessment*, 220–256, 225 (U.S. Global Change Research Program, 2014) doi. 10.7930/J0PN93H5.

7 Ibid.; Paul Epstein, *The Ecology of Climate Change and Infectious Disease: Comment*, 91(3) Ecology 925–928 (2010); Pim Martens, *Health and Climate Change* 153 (2013).

8 Wu *Infectious Diseases, supra* note 4, at 15.

9 Martens, *Health and Climate Change, supra* note 7, at 149.

10 Charles Beard et al., Ch. 5 Vector-borne diseases, In: Alison Crimmins et al., *The Impacts of Climate Change on Human Health in the United States: A Scientific Assessment* 129–156 (U.S. Global Change Research Program, 2016).

and Lyme disease (transmitted by ticks) are also sensitive to changing weather patterns.[11] There is also a risk that changing climatic conditions could create opportunities for the emergence of new vector-borne diseases.[12]

### 9.2.2 *Zoonotic*

Diseases that can be spread from animals to people, known as zoonotic diseases or zoonoses, are also highly sensitive to environmental changes that alter the habitats of animal populations and bring them into closer contact with human populations. The most pathogenic infectious diseases that emerged in the late twentieth century – HIV, Ebola virus, and avian influenza – began with animal to human transmission before the pathogens became transmissible from person to person. Newly emerging zoonotic diseases warrant particular concern because of the risk that they will become efficiently transmissible from person to person before effective medical countermeasures can be developed and disseminated. Any infectious disease that can thrive in an animal reservoir population is extremely difficult to eradicate. Smallpox, the only infectious disease to be successfully eradicated, was a relatively simple target in part because it lacked an animal reservoir. Poliovirus and measles are similarly attractive targets for eradication because they, too, lack an animal reservoir (though they are less susceptible to eradication because unlike individuals infected with smallpox, who exhibit visible signs of disease during the entire period in which they are contagious, those infected with polio and measles are contagious prior to the onset of visible symptoms).[13] In contrast, Ebola virus continues to circulate in animal populations between outbreaks in human populations, making it virtually impossible to eradicate even after an effective vaccine becomes widely available.[14]

### 9.2.3 *Food-borne*

Epidemiological studies have documented that warmer temperatures are associated with an increased likelihood of food-borne disease outbreaks.[15]

[11] Wu *Infectious Diseases, supra* note 4, at 15.

[12] Beard et al., *Ch. 5: Vector-Borne Diseases, supra* note 10, at 132.

[13] World Health Organization, *The Global Eradication of Smallpox: Final Report of the Global Commission for the Certification of Smallpox Eradication*, 4 *History of Int'l Pub. Health* 122 (1980).

[14] Xavier Pourrut et al., *The Natural History of Ebola Virus in Africa*, 7 Microbes and Infection 1005 (2005).

[15] Luber, *Ch. 9: Human Health, supra* note 6, at 225; Lewis Ziska et al., Food Safety, nutrition, and distribution, In: Alison Crimmins et al., *The Impacts of Climate Change on Human*

Warmer temperatures promote proliferation of pathogens in hospitable food items. Additionally, higher average ocean surface temperatures combined with increased runoff from sewage systems create favorable conditions for bacteria, increasing the risk of contaminated seafood.[16] Warmer conditions at facilities for agricultural production, livestock processing, and food manufacturing and packaging could also increase the risk of food-borne illness.[17]

### 9.2.4 *Waterborne*

Changes in moisture and patterns of drought and flooding affect a number of diseases carried by water.[18] Some water-related illnesses sensitive to climatic conditions, such as those caused by toxin-producing algae, are not technically infectious.[19] Others, such as cholera, are highly virulent and pathogenic bacterial infections. The risk of giardiasis, which is caused by a parasite typically deposited in water by infected small mammals and then ingested by humans, is also linked to climate change. Rising surface water temperatures may also increase the risk of acquiring infectious illnesses through swimming and other water sports.[20]

Many waterborne diseases sensitive to climate are exacerbated by increased agricultural run-off. Nitrogen and phosphorous, common fertilizer ingredients, create ripe conditions for harmful algae blooms in bodies of water near or downstream from concentrated agricultural operations, posing a risk to bathers who inadvertently ingest water when they swim or contract skin infections, in addition to posing a risk of food-borne illness for those who consume contaminated shellfish. Concentrated animal confinement facilities can also contribute to contamination of ground and surface water, increasing the prevalence of infectious agents that have developed resistance to antibiotics commonly used at subtherapeutic doses in animal feed for growth promotion purposes.[21]

*Health in the United States: A Scientific Assessment*, 189–216 (U.S. Global Change Research Program, 2016).

16 Luigi Vezzulli et al., *Climate Influence on Vibrio and Associated Human Diseases During the Past Half-Century in the Coastal Northern Atlantic*, 113 Proceedings of the Nat'l Academy of Sci. E5062 (2016).

17 Ziska et al., *Ch. 7: Food Safety, Nutrition, and Distribution*, *supra* note 15, at 192.

18 Luber, *Ch. 9: Human Health*, *supra* note 6, at 245; Juli Tranj and Lesley Jantrasami, Ch. 6: Climate impacts on water-related illness, In: Alison Crimmins et al., *The Impacts of Climate Change on Human Health in the United States: A Scientific Assessment*, 157–188 (U.S. Global Change Research Program, 2016).

19 Tranj and Jantrasami, *Water-Related Illness*, *supra* note 18, at 159.

20 Luber, *Ch. 9: Human Health*, *supra* note 6, at 245.

21 Tranj and Jantrasami, *Water-Related Illness*, *supra* note 18, at 161–162.

When infrastructure crumbles – because of a natural disaster, state collapse, or some combination of the two – water-related disease outbreaks soon follow. Cholera in particular is the characteristic disease indicator of a failing state.[22] Cholera and other diarrheal illnesses are especially sensitive to breakdowns in basic sanitation infrastructure for the removal of human waste, as seen in Haiti following an earthquake in 2010. Because climate change stresses governments, economic systems, and infrastructure, it heightens the probability of social failures that lead to disease outbreaks, even as it also makes environmental conditions more hospitable for the spread of disease.

## 9.3 PUBLIC HEALTH LAW TOOLS FOR ADAPTING TO THE INFECTIOUS DISEASE IMPACTS OF CLIMATE CHANGE

Climate change will make the work of "ensur[ing] the conditions required for people to be healthy"[23] considerably more challenging in coming decades. Although public health law has broadened its scope to encompass noncommunicable disease threats and injuries, the law of infectious disease control remains at its core. As policymakers seek to adapt to the health consequences of climate change they are turning to legal frameworks, human resources, and technical infrastructure devoted to public health surveillance, development of and access to medical countermeasures, vector control, health education and communication, social distancing, and public health emergency preparedness and response.

### 9.3.1 *Surveillance*

Responding effectively to infectious disease threats requires good information. Public health authorities conduct continuous and systematic surveillance to monitor outbreaks in human and animal populations.[24] Laws mandate that healthcare providers, laboratories, and other parties report individual cases of disease to health department registries. Case reports may trigger epidemiological investigations to identify the source of an outbreak. Analysis and interpretation of case reports and diagnostic test results also inform planning, policy formation, and evaluation of public health interventions.[25]

22 See, e.g., Andrew Price-Smith, *Cholera and the Collapse of Government in Zimbabwe*, Harvard Int'l Rev., Feb. 2, 2009, http://hir.harvard.edu/article/?a=1819.

23 Lawrence O. Gostin and Lindsay F. Wiley, *Public Health Law: Power, Duty, Restraint* 4 (2016).

24 Kavita Marfatia Berger, *The Role of Science in Preparedness and Response*, 6 U. St. Thomas L.J. 622, 623 (2009).

25 Gostin and Wiley, *Public Health Law: Power, Duty, Restraint*, *supra* note 23, at 36.

Surveillance will be particularly important to ensure detection and tracking of gradual changes in the geographic and seasonal distribution of infectious disease as a consequence of climate change.

Rapid detection and response to infectious disease outbreaks is a global priority. In response to concerns about pandemic influenza, the World Health Assembly adopted revised International Health Regulations (IHR) in 2005, which obligate member states to ensure adequate infrastructure to detect infectious disease threats with potential for international spread.[26] Lack of adequate resources has resulted in the majority of member states failing to satisfy IHR requirements. In 2014, only sixty-four member states reported that they were in full compliance, while eighty-one requested an extension and forty-eight failed to file a report.[27] Full implementation and strengthening of the IHR, with an eye toward the health impacts of climate change, should be a global health and climate adaptation priority.

Surveillance raises concerns about health information privacy. An infectious disease diagnosis may affect an individual's employment, personal relationships, and other aspects of life. Much of public health law with respect to surveillance systems is aimed at balancing between collective needs for accurate, traceable information to inform disease control efforts and individual interests in the privacy, confidentiality, and security of sensitive information.[28] Public health information infrastructure, and the regulatory regimes that govern it, which are discussed in Chapter 6 of this volume, are crucial to climate adaptation.

### 9.3.2 *Access to Effective Medical Countermeasures*

Medical countermeasures – including vaccines and antimicrobials – are powerful tools for preventing and treating infections and controlling the spread of communicable diseases associated with climate change. Because the market typically fails to provide adequate incentives to develop and produce vaccines and antimicrobials, law plays a role in promoting the development of new countermeasures, ensuring their safety, and securing appropriate access.[29]

In addition to mechanisms designed to encourage private investment and provide direct public investment in research and development of

26 Michael G. Baker and David P. Fidler, *Global Public Health Surveillance Under New International Health Regulations*, 12 Emerging Infectious Diseases 1058 (2006).

27 Lawrence O. Gostin and Rebecca Katz, *The International Health Regulations: The Governing Framework for Global Health Security*, 94 Milbank Quarterly 264–313 (2016).

28 Gostin and Wiley, *Public Health Law: Power, Duty, Restraint*, *supra* note 23, at 317–330.

29 Ibid. at 361–364, 404–409.

countermeasures for routine medical needs, the United States has adopted special measures to encourage development of new countermeasures for biological agents determined by the Department of Homeland Security to pose a material threat. Some of the targeted agents, such as Ebola, are climate-sensitive. For example, the Biomedical Advanced Research and Development Authority (BARDA), part of the U.S. Department of Health and Human Services, awards grants to private firms to assist with clinical trials and the development of manufacturing capacity to produce new vaccines, treatments, and diagnostic tools for public health emergencies.[30]

The countermeasures successfully developed with public investment are often stockpiled in government facilities for rapid deployment in response to a public health emergency. In an emergency, an official declaration following prescribed procedures may also trigger exceptions or enhanced flexibility with respect to legal requirements designed to ensure the safety of drugs, vaccines, and medical devices.[31] An emergency may also necessitate the development and implementation of protocols for rationing scarce medical resources.[32]

Most of the population is ready and willing to take advantage of medical countermeasures. In rare cases, however, individuals prefer to decline recommended vaccinations and treatment, due to distrust of government and scientific recommendations or unwillingness to accept the risks associated with vaccination, particularly when the benefits of herd immunity accrue to the community at large. In the United States, vaccination laws are focused primarily on ensuring that children are vaccinated for several preventable communicable diseases prior to entering school.[33] These provisions are regularly challenged in court on a variety of constitutional grounds, including First Amendment protection of religious freedom and Fourteenth Amendment substantive due process rights to bodily integrity and decisional privacy. Judges have consistently upheld school vaccination requirements, holding that the Due Process Clause does not protect a fundamental right of a parent "to refuse to have her child immunized before attending public or private school where immunization is a precondition to attending school."[34]

30 See Jonathan B. Tucker, *Developing Medical Countermeasures: From Bioshield to BARDA*, 70 Drug Development Research 224 (2009).

31 U.S. Food and Drug Administration, *Emergency Use Authorization of Medical Products and Related Authorities: Guidance for Industry and Other Stakeholders* (January 2017).

32 Gostin and Wiley, *Public Health Law: Power, Duty, Restraint*, *supra* note 23, at 414–415; Daniel J. Barnett, et al., *Resource Allocation on the Front Lines of Public Health Preparedness and Response: Report of a Summit on Legal and Ethical Issues*, 124 Pub. Health Reports 295 (2009).

33 Gostin and Wiley, *Public Health Law: Power, Duty, Restraint*, *supra* note 23, at 350–353.

34 See, e.g., *Boone* v. *Boozman*, 217 F.Supp.2d 938, 956 (E.D.Ark.2002); *Workman v. Mingo Cnty. Bd. of Educ.*, 419 F. App'x 348. (4th Cir. 2011) (accord).

Additionally, courts have consistently held that compulsory vaccination laws, which are laws of general applicability and do not target particular religious groups or views, need not allow for religious exemptions to be constitutional.[35] Adult vaccination mandates are rare, though many institutional healthcare providers (e.g. hospitals and long-term care facilities) have adopted policies requiring employees to be vaccinated.[36] Mandatory vaccination may be particularly fraught in an emergency context, when vaccines may be recently developed under loosened safety requirements. Dicta in the Supreme Court's landmark opinion in *Jacobson* v. *Massachusetts* (upholding a compulsory smallpox vaccination law applicable to adults as well as children) suggests that a compulsory vaccination law that made no exception for medical contraindications (such as an allergy or immune deficiency) would probably run afoul of the Fourteenth Amendment.[37] As a matter of balancing state interests against individual rights, however, a public health emergency (such as the frequent smallpox outbreaks during Jacobson's time or the outbreak of a novel infectious disease associated with changing environmental conditions) justifies more intrusive measures than maintenance of community immunity against more routine threats, such as measles.

Coercive government measures are rarely needed to compel compliance with recommended medical treatment because most individuals who are actually or potentially infected desire treatment. Where a patient is unwilling or unable to comply with treatment recommendations, however, health officials (primarily at the state and local level) may order compulsory treatment, including under conditions of confinement.[38] Although the US Supreme Court has recognized a constitutionally protected liberty interest in refusing unwanted medical treatment in cases involving abortion, terminal illness, and mental illness, it has not directly addressed the question of what the Fourteenth Amendment requires for confinement of individuals who are actually or

35 Ibid. Note that under *Employment Division, Dep't of Human Resources of Oregon* v. *Smith*, 494 U.S. 872 (1990), generally applicable laws that impose incidental burdens on religious practices are constitutional so long as they are rationally related to a legitimate government purpose. A federal statute, the Religious Freedom Restoration Act (RFRA), imposes a higher standard of judicial review for federal statutes, but not for state laws. Because vaccination requirements are imposed at the state and local level, they are subjected to heightened review only if the relevant state has adopted its own version of RFRA.

36 Ibid. at 348–349; Hilary M. Babcock et al., *Mandatory Influenza Vaccination of Health Care Workers: Translating Policy to Practice*, 50 Clinical Infectious Diseases 459 (2010).

37 *Jacobson* v. *Massachusetts*, 197 U.S. 11. (1905) (stating in dicta that in a case where, due to "a particular condition of his health or body," it would be "cruel and inhuman in the last degree" to subject a person to vaccination, the judiciary would be "competent to interfere and protect the health and life of the individual concerned").

38 Gostin and Wiley, *Public Health Law: Power, Duty, Restraint*, *supra* note 23, at 373–379.

potentially infected with a contagious disease. The lower courts have relied heavily on cases reviewing civil commitment of individuals with mental illness, which balance individual liberty against state interests by requiring a showing that the individual poses a danger to himself or others, that treatment is in the individual's medical interest, and that the medication is administered by a licensed physician acting in accordance with professional standards.[39] Additionally, treatment must be necessary to significantly further important government interests in light of less intrusive alternatives.[40] The Fourteenth Amendment also protects the patient's right to due process to ensure that confinement or conditional release is warranted on a case by case basis.[41] These standards may prove problematic in cases where a treatment is new and its effectiveness is unproven, such as with an emerging zoonotic disease like Ebola. On the other hand, lack of effective medical countermeasures heightens the state's interest in controlling the spread of disease, which would probably justify confinement, as discussed in the section on quarantine and isolation.

Compulsory application of effective medical countermeasures may be justified by the harm principle. Although we tend to conceive of vaccination and medical treatment in terms of their benefit to the infected individual, medical countermeasures also play a crucial role in slowing the spread of disease from person to person. When immunization levels are sufficiently high, outbreaks are quickly contained because there are too few vulnerable hosts for the pathogen to spread beyond a small number of individuals.[42] Similarly, antimicrobials typically reduce or eliminate a patient's ability to infect others.[43] In some cases, compulsory treatment may be necessary to ensure that the full course of treatment is completed, thus reducing the risk that the pathogen will develop resistance as a consequence of being exposed to subtherapeutic doses of medication.[44]

39 *Riggins* v. *Nevada* 504 U.S. 127 (1992). 40 *Sell v. U.S.*, 539 U.S. 166 (2003).

41 See, e.g., *Greene v. Edwards*, 263 S.E.2d 661 (W.V. 1980) (holding that individuals confined for treatment of communicable tuberculosis are entitled to the same due process protections as individuals involuntarily committed for the treatment of mental illness, including "(1) an adequate written notice detailing the grounds and underlying facts on which commitment is sought; (2) the right to counsel; (3) the right to be present, cross-examine, confront and present witnesses; (4) the standard of proof to warrant commitment to be by clear, cogent and convincing evidence; and (5) the right to a verbatim transcript of the proceeding for purposes of appeal").

42 The specific threshold required for community immunity is determined by the transmissibility of the pathogen, the effectiveness of the vaccine, and social characteristics of the community. Paul Fine et al., *Herd Immunity: A rough guide*, 52 Clinical Infectious Diseases 911 (2011).

43 Gostin and Wiley, *Public Health Law: Power, Duty, Restraint*, *supra* note 23, at 346.

44 Ibid. at 373–379.

Antimicrobial resistance is a mounting global health crisis. In addition to inappropriate use in human populations, the administration of subtherapeutic levels of antibiotics in animal feed as a growth-promotion strategy contributes to the evolution and transmission of resistant strains.[45] Regulation of veterinary use of antibiotics and industrial farming practices is thus crucial to maintaining the efficacy of available countermeasures. Unfortunately, and in spite of efforts by advocates and policymakers, such regulation has not been forthcoming in the United States.

Market forces privilege development of pharmaceuticals that can sustain routine use by a significant population of patients who can afford expensive treatment. Public health emergency preparedness programs like BARDA have prioritized countermeasures for agents with the potential to be weaponized by terrorists or enemy states. Investment in countermeasures for infectious diseases that disproportionately burden the world's poorest people has been paltry by comparison. Some of these historically neglected diseases, such as malaria, Zika virus, and Ebola virus, are particularly sensitive to changing environmental conditions. Climate change, by exacerbating health threats that disproportionately affect less privileged populations, is likely to expose longstanding injustices in health investment and governance.[46]

### 9.3.3 *Isolation, Quarantine, and Social Distancing*

For some infectious agents, such as Ebola virus in 2014–2015 and Zika virus in 2015–2016, effective medical countermeasures like vaccines and antivirals are limited or entirely unavailable. Under these circumstances, particularly if the agent is highly infectious and virulent, it may be necessary to isolate infected individuals, quarantine those who have been or may have been exposed, and implement measures to limit social contact among the population at large. Even in cases where medical countermeasures are being used, these tools may be useful to limit the spread of disease during the treatment period.

Like compulsory vaccination and treatment, these tools implicate individual rights to liberty, bodily integrity, and autonomy. They are understandably controversial and are hotly debated by lawmakers, health officials, and judges, who must balance individual rights against collective needs – and demands – for health security. Under emergency conditions, political considerations and

45 Wolfgang Witte et al., *Medical Consequences of Antibiotic Use in Agriculture*, 279 Sci. 996 (1998).

46 Wiley, *supra* note 55; Harris Ali, et al., The Social and Political Dimensions of the Ebola Response: Global Inequality, Climate Change, and Infectious Disease, in *Climate Change and Health* (2016).

public panic may come into play. In *Mayhew* v. *Hickox*, for example, a Maine court enjoined state health authorities from enforcing travel restrictions on a nurse who had recently returned from treating patients infected with Ebola virus in West Africa.[47] The court's order urged the nurse to respect public concerns, even while recognizing that those concerns were based on a misunderstanding of the actual risk posed by individuals who may have been exposed to Ebola virus but had not yet developed any symptoms of disease.

In the United States, most states have granted authority to health officials to initiate isolation and quarantine under routine circumstances on a case by case basis.[48] In some states, these statutes are limited to specified conditions, such as tuberculosis, requiring officials to reinvent the wheel when presented with a novel infectious disease outbreak. In other states, the statutes are drafted more broadly, allowing officials and courts greater flexibility. Most states also have special emergency procedures, which may grant health officials broader isolation and quarantine authority during a declared emergency. Federal isolation and quarantine authority is limited to prevention of international and interstate spread of disease, though new federal quarantine regulations released in early 2017 after a lengthy effort to modernize the rules governing federal authority for quarantine, isolation, and conditional release to prevent the interstate spread of communicable diseases take a fairly broad approach to determining when the risk of interstate transmission warrants federal involvement.[49] Virtually identical provisions apply to people arriving in the United States from foreign countries.[50]

Government authority to isolate the sick, quarantine the exposed, and enforce social distancing for those at risk of exposure is particularly relevant to the potential emergence of new pathogens with pandemic potential. If, as some experts warn, climate change will increase the rate at which new pathogens emerge, legal authorities that permit infringement of individual liberty in the interest of collective needs to contain the spread of disease will continue to be tested.

### 9.3.4 *Vector Control*

For diseases carried from host to host by insects or rodents, vector control activities are a crucial component of infectious disease prevention.

47 Order Pending Hearing, *Mayhew v. Hickox*, No. CV-2014–36, 3 (Me. Dist. Ct., Fort Kent, Oct. 31, 2014).

48 See National Conference of State Legislatures, *State Quarantine and Isolation Statutes*, (Oct. 29, 2014) www.ncsl.org/research/health/state-quarantine-and-isolation-statutes.aspx.

49 42 CFR § 70.1. 50 42 C.F.R. § 71.

Fumigation with insecticides sprayed by trucks and airplanes is the most visible form of vector control, but other methods – including the treatment of breeding areas with larvicides – play a more important role in controlling the mosquitoes most likely to spread disease. Removal of standing water where mosquitoes may lay eggs and land use plans that divert human settlements away from high-risk breeding areas are also important strategies.

In the United States, vector control is conducted primarily by local government authorities, often organized through special vector control districts.[51] Federal regulations under the Federal Insecticide, Fungicide and Rodenticide Act (FIFRA) govern appropriate use of specific insecticides and larvicides. Past outbreaks of West Nile virus and Zika virus in the United States have spurred disputes between local authorities and citizen groups concerned about the potential harmful effects of spraying. In *No Spray Coalition* v. *City of New York*, for example, an environmental and public health advocacy group sought to enjoin the city's pesticide use in response to an outbreak of West Nile virus.[52] After the courts dismissed the Coalition's claim under other city, state, and federal environmental protection statutes, a federal district court ruled that its claim under the federal Clean Water Act (CWA) survived summary judgment because pesticide spraying near bodies of water might constitute discharge of a pollutant under the CWA. The city then agreed to pay $80,000 to five environmental protection nonprofit organizations, cover the plaintiffs' legal fees, and hold two three-hour meetings between the city's Department of Health and Mental Hygiene and the plaintiffs to discuss "health and environmental concerns pertaining to . . . pesticides and proposals for alternative and non-toxic approaches."[53] No Spray Coalition continues to express concern about the potential health effects of pesticides, distrust for government-sponsored scientific assessments of risks and benefits, and preference for "natural" approaches to mosquito control using essential oils.[54]

[51] See, e.g., *Environmental Protection Agency (EPA) and the U.S. Centers for Disease Control and Prevention (CDC)* (Sept. 2012) www.epa.gov/mosquitocontrol/joint-statement-mosquito-control-united-states.

[52] *No Spray Coal., Inc. v. City of New York*, 2005 WL 1354041, 1 (S.D.N.Y. June 8, 2005).

[53] *No Spray Coalition*, Stipulation of Agreement and Order, https://newyorkeljp.files.wordpress.com/2016/10/4-12-07-final-settlement-stip-and-order-nospray.pdf.

[54] No Spray Coalition, http://nospray.org/.

### 9.3.5 *Food and Animal Safety*

Enforcement of sanitary agricultural, livestock processing, and food manufacturing practices is essential to the prevention of zoonotic and food-borne illness. Controlling the spread of zoonotic and food-borne illness sometimes requires destruction of infected animals (e.g. poultry, hogs) and contaminated products (e.g. grain). In other cases, products may need to be diverted to markets for uses other than human consumption, reducing their value (e.g. farmers may be limited to selling salmonella-contaminated eggs on the "breaker" market where they are used for limited purposes). Compensation of property owners could be prohibitively expensive. In the United States, courts have determined that compensation may be constitutionally required (under the Takings Clause) where uninfected animals are destroyed (e.g. for testing purposes), but not for diseased animals.[55] Food safety and agricultural regulations may also provide for compensation, but these statutory provisions may be waived in the case of a declared emergency.[56]

### 9.3.6 *Health Education and Communication*

Many infectious disease risks can be managed, at least in part, through protective individual behaviors. Health departments at every jurisdictional level – as well as departments with responsibility for other sectors, such as fishery management and recreation – play a role in developing and disseminating educational materials and public messages. Local authorities, for example, issue warnings urging individuals to avoid exposure to water and shellfish from designated areas due to recognized infectious disease risks. Health communication strategies also target specific groups, such as healthcare workers, to urge them to adopt more rigorous hand hygiene practices when working with patients in settings that are ripe for the spread of antibiotic-resistant strains of bacteria. Vector-control authorities also rely on communication strategies to inform residents about the need to survey their homes and property for small amounts of standing water (e.g. in potted plants, open containers outdoors) that might become breeding sites for mosquitoes. (See also Chapter 4).

55 See, e.g., *Yancey v. United States*, 915 F.2d 1534 (Fed. Cir. 1990).

56 Gostin and Wiley, *Public Health Law: Power, Duty, and Restraint*, *supra* note 23, at 205.

## 9.4 HEALTH GOVERNANCE PRINCIPLES FOR ADAPTING TO THE INFECTIOUS DISEASE IMPACTS OF CLIMATE CHANGE

Public health law frameworks are supplemented and informed by a wide range of policy principles. Two of these have particular relevance to the infectious disease impacts of climate change: One Health and Health in All Policies. These principles seek to overcome the challenges posed by the traditional division of social, economic, and environmental programs into distinct silos, each with its own mandates, expertise, and organizational structure. One Health promotes intersectoral work among agencies and experts focused on human health, animal health, agriculture, and the natural environment. Health in All Policies seeks to integrate public health goals into the work of all sectors – those focused on the natural environment, animal health, and agriculture, as well as transportation, built environment, economic development, national security, and more.

### 9.4.1 *"One Health"*

Global health, veterinary medicine, agriculture, environmental protection, and other experts are coming together to promote the principle that animal health, human health, and environmental health are interconnected in reality and policy efforts to protect them should also be joined up (see Figure 9.2).[57]

The One Health principle has increased engagement between government agencies responsible for environmental, animal, and human health. In the United States, for example, the Centers for Disease Control and Prevention hosts a National Center for Zoonotic, Vector-Borne, and Enteric Diseases. The center draws on the expertise of dozens of on-staff veterinarians, who support surveillance, epidemiological investigation, and response efforts. Other federal agencies also combine veterinary and human health expertise to inform pandemic and bioterror preparedness efforts.[58]

At the international level, the World Health Organization, Food and Agriculture Organization of the United Nations, and the World

[57] World Health Organization et al., *Contributing to One World, One Health: A Strategic Framework for Reducing Risks of Infectious Diseases at the Animal-Human-Ecosystems Interface* (2008); Am. Veterinary Med. Ass'n, *One Health: A New Professional Imperative* (2008).

[58] Berger *The Role of Science in Preparedness and Response, supra* note 24, at 624.

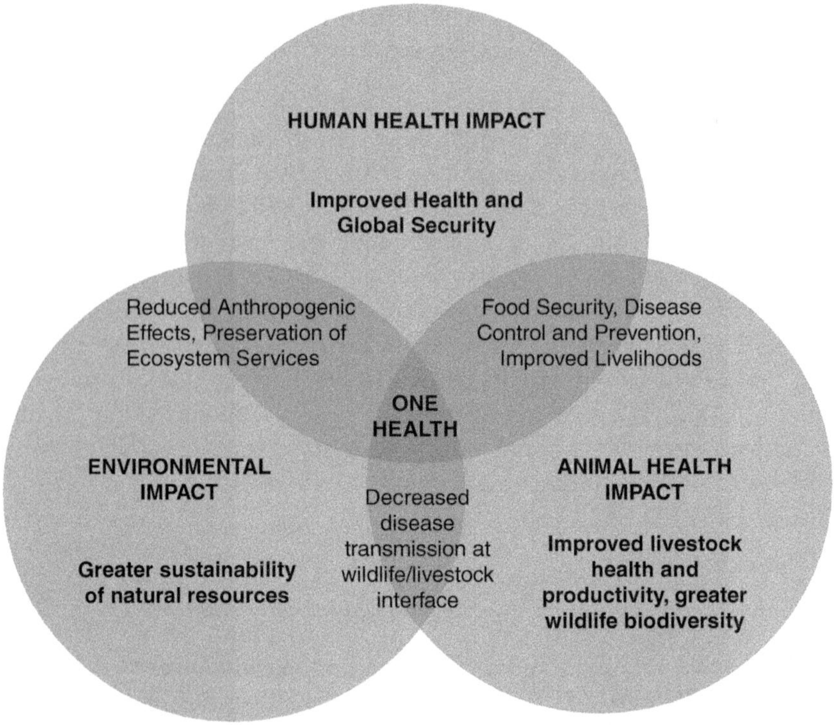


FIGURE 9.2 Thematic representation of One Health depicting potential added benefits of sectoral overlap.
*From Anna L. Okello, Kevin Bardosh, James Smith, and Susan C. Welburn, One Health: Past Successes and Future Challenges in Three African Contexts,* PLoS Neglected Tropical Diseases, *vol. 8 no. 5, e2884 (May 2014), p. 3.*

Organization for Animal Health coordinate their work on food safety and zoonotic diseases, as well as "other public health threats at the human-animal-ecosystem interface."[59] Commentators note that these efforts are characterized by "general enthusiasm" but that advocates are struggling to "translate theory into action."[60] Furthermore, these efforts have focused narrowly on diseases with pandemic potential, such as highly pathogenic avian influenza and severe acute respiratory syndrome (SARS), leaving other threats neglected.[61] To have a meaningful impact on climate change adaptation,

59 World Health Organization, *One Health*, April 2017, www.who.int/features/qa/one-health/en/.
60 Okello et al., *One Health: Past Successes and Further Challenges, supra* figure 9.2, at 1.
61 Ibid.

One Health will require greater investment in infrastructure and human resources to hire and train staff across multiple areas of expertise.

### 9.4.2 "Health in All Policies"

The Health in all Policies principle reflects growing recognition that many of the determinants of population health are governed by laws and policies developed and implemented in nonhealth sectors.[62] It recognizes that responding effectively to public health threats often requires integrated, whole-of-government approaches, with lawmakers and government officials from multiple sectors working to harmonize their goals and strategies for achieving them.[63]

Lawmakers in several countries across Asia and Europe have translated the Health in All Policies principle into a legal requirement that agencies conduct health impact assessments (HIAs) as part of the regulatory process. Some also require private commercial entities to assess the likely health impact of proposed infrastructure projects. (See Chapter 14.)

In other cases, policymakers and judges have incorporated consideration of health impacts into broader requirements for regulatory impact assessment. For example, in *Association of Irritated Residents* v. *San Joaquin Valley*, a California state court required a local air pollution district that encompasses eight counties to reassess a proposed regulation of confined animal agricultural facilities in light of anticipated impacts on public health, pursuant to a state statutory directive.[64]

Health impact assessments are appealing from a common sense standpoint, but they face many challenges in practice. Affected industries and organizations may deploy financial resources and political capital to oppose implementation of health impact assessments as a legal requirement. As a procedural exercise, HIAs run the risk of becoming one more in a series of bureaucratic hurdles that slow down, but fail to otherwise influence, development projects. Evaluating their effectiveness requires careful analysis of whether and how HIA recommendations are accepted and implemented and whether the aims and objectives of the HIA were ultimately served by the procedural requirement.[65]

62 Sandro Galeo, *Healthier* (Oxford: Oxford University Press, 2017).

63 Gostin and Wiley *Public Health Law: Power, Duty, and Restraint*, *supra* note 23, at 165–167; World Health Organization, *Adelaide Statement on Health in All Policies: Moving towards a Shared Governance for Health and Well-Being* (2012).

64 *Assoc. of Irradiated Residents v. San Joaquin Valley Unified Air Pollution Control Dist.*, 85 Cal. Rptr. 3d 590 (Cal. Ct. App. 2008).

65 NHS Health Development Agency, *Evaluating Health Impact Assessment*, 2.

## 9.5 CONCLUSION: CHALLENGES AND OPPORTUNITIES

Responding to the infectious disease impacts of climate change presents multiple challenges and opportunities.[66] The infectious disease and other impacts of climate change are occurring in the midst of an ongoing transformation in public health law. Two key shifts in the scope and nature of public health law are particularly significant for climate change adaptation. First, the shift toward increasing integration between the public health and healthcare sectors. Second, the shift away from the behavioral model of public health that held sway among scientists and practitioners during the second half of the twentieth century and toward the social-ecological model, which is now favored by scientists but has encountered cultural, political, and legal resistance. Both of these shifts have implications for longstanding tensions in public health law and governance between individual interests and collective needs and between personal responsibility and public responsibility for health.

### 9.5.1 *Integration of the Public Health and Healthcare Sectors*

To the extent that responding to the infectious disease impacts of climate change prompt public health authorities and healthcare providers to better define channels of communication, while also giving healthcare providers a greater stake in protecting environmental and animal health, adaptation may also reinforce an ongoing shift toward greater integration between the public health and healthcare sectors. In the United States, for example, budgetary crises have led to stagnating or declining investments in public health infrastructure,[67] prompting public health advocates to take a greater interest in partnering with better-financed healthcare providers.[68] At the same time, tax and health care reimbursement incentives are giving healthcare providers a greater financial stake in patient outcomes and population-level drivers of

[66] Lindsay F. Wiley, *Adaptation to the Health Consequences of Climate Change as a Potential Influence on Public Health Law and Policy: From Preparedness to Resilience*, 15 Widener L. Rev. 483 (2010); Lindsay F. Wiley, Climate change adaptation and public health law, In: Jonathan Verschuuren, *Research Handbook on Climate Change Adaptation Law* (2013).

[67] We need cites for statements of this sort. More details about budgetary changes etc. would be welcome as well. [MB: Agree. I don't think you want to get into ACA politics, but some reference to policies implicated there may be relevant?] Trust for America's Health, Investing in America's Health: A State-by-State Look at Public Health Funding and Key Health Facts, 2016, http://healthyamericans.org/assets/files/TFAH-2016-InvestInAmericaRpt-FINAL.pdf

[68] Lindsay F. Wiley and Gene Matthews, *Health Care System Transformation and Integration: A Call to Action for Public Health*, J. of L., Medicine & Ethics, 45 (supplement 1), 94–97 (2017).

disease and injury. In the context of a changing climate, these incentives could also prompt providers to take a more active interest in protecting the environment.

This integration presents opportunities to better serve healthcare and public health goals alike, but it also presents challenges. Public health advocates worry that the better financed and politically favored healthcare sector, with its emphasis on downstream treatment strategies (e.g. new drugs for malaria) will overwhelm the public health sector, with its emphasis on upstream prevention strategies (e.g. vector control and community siting strategies to minimize exposure to disease-carrying pests).[69]

### 9.5.2 *Controversy over the Social-Ecological Model of Health*

To the extent that climate change adaptation activities provide a forum for advancing One Health, Health in All Policies, and integration of the public health and healthcare systems as general principles for health governance, they could reinforce an ongoing shift toward the social-ecological model of health. For most of the twentieth century, infectious disease control strategies relied principally on the agent model of disease. The agent model focused attention on combatting disease with agent-specific countermeasures (i.e. vaccines and antimicrobials). In the late twentieth century, policymakers and public health practitioners began to apply the behavioral model (which arose in response to noncommunicable disease threats) to infectious disease threats as well. The behavioral model shifted the focus from the agents of disease to the behaviors of human hosts. It emphasized health communication and education as key interventions, putting the onus on individuals to reduce risky behaviors (e.g. tobacco use, raw shellfish consumption) and adopt protective behaviors (e.g. condom use, mosquito repellant use). In contrast, the social-ecological model that emerged around the end of the twentieth century emphasizes the environments – physical, social, economic, etc. – in which the host and agent interact. It points policymakers toward structural, community-level interventions to create healthy living conditions, rather than relying predominantly on clinical interventions that target discrete agents of disease or behavioral interventions that focus on individual choices.

The social-ecological model is favored by most public health scientists for its descriptive power and normative thrust. It is also particularly well suited for adaptation to the health impacts of climate change, the scale and scope of

69 Wiley and Matthews, *Health Care System Transformation and Integration*, *supra* note 68.

which defy the technological solutionism that characterizes the agent model of health (e.g. medical countermeasures tailored to combat specific climate-sensitive pathogens) and the emphasis on individual responsibility to avoid health hazards (e.g. advising individuals to avoid shellfish) that characterizes the behavioral model of health. The social-ecological model is broad enough to encompass deployment of medical countermeasures and behavioral interventions while also digging deeper to examine the structural, social, and environmental determinants that shape access to and use of medical countermeasures, acceptance (or nonacceptance) of scientific and government advice, and many other factors such as urban planning and community resilience. Yet the social-ecological model faces many challenges. The behavioral model, with its individualistic orientation toward behaviors and choices, is a better fit for most people's intuitive understanding of health as a primarily personal concern. The agent model, with its focus on medical countermeasures that allow private pharmaceutical companies to capture the benefits of public investment, enjoys political favor. In contrast, the social-ecological model more frequently points policymakers toward public investments that are not easily captured by politically powerful private interests (e.g. sanitation systems, infrastructure to support surveillance and epidemiological investigation) and toward regulation of private commercial activities (e.g. pollution controls).

#### 9.5.2.1 Tension between Individual and Collective Rights and Responsibilities

At the heart of both of these shifts – toward increased integration between public health and healthcare and toward the social-ecological model of health – is an intractable tension between individual interests and collective needs and between personal and public responsibility for health. Do individuals bear primary responsibility for avoiding exposure to mosquitoes and other disease-carrying pests or should pest control be publically financed? Should a minority of individuals who oppose the use of pesticides (e.g. because they do not trust the safety assurances of government scientists) be able to block public vector control efforts? Is it the responsibility of individuals who rely on local bodies of water for recreation and fishing to find alternatives, or is there a public responsibility to ensure that bodies of water and the animals that inhabit them are safe? Should individual property owners be barred from using their land in ways that contribute to water contamination? Why should governments stockpile influenza drugs to be distributed in case of emergency while allowing many who need HIV drugs to go without?

In theory, infectious disease threats are the paradigmatic example of why collective needs must sometimes outweigh individual rights and collective action is sometimes the only mechanism by which individuals may protect themselves. The US Supreme Court famously articulated these principles in *Jacobson v. Massachusetts*, in which it rejected the notion that an individual's right to refuse vaccination trumps the community's need to protect itself from a smallpox epidemic.[70] In practice, however, many forces still resist the notion that infectious disease threats justify infringement of individual rights (e.g. to bodily integrity, property, and autonomy) and trigger an obligation to provide mutual aid (e.g. through collective mechanisms to finance development and deployment of public health infrastructure, medical countermeasures, and healthcare services).

The vector-borne and diarrheal illnesses that are likely to be exacerbated by climate change have long been neglected by collective financing and health governance mechanisms at the international, domestic, and local levels. Governmental and nongovernmental investments in global health have tended to privilege threats such as influenza pandemics that spread rapidly across borders and affect the wealthy alongside the poor. These investments are typically justified as a form of "enlightened self-interest" in light of the notion that "disease knows no borders" and so the resources of wealthy states are well-spent containing disease threats at their point of origin, even if that means building infrastructure in far-flung places. Climate-sensitive diseases tend to serve as counter-examples to this well-worn trope. In fact, one could map cases of Dengue fever, cholera, and other illnesses that impose extraordinary burdens on human health and wellbeing and easily identify borders between governmental jurisdictions. Public spending on health infrastructure matters. Economic opportunities, which enable a higher standard of living, matter. Climate change threatens to exacerbate the unconscionable health disparities that have long plagued human society. If we rise to the challenges posed by the health impacts of climate change by investing in basic public health infrastructure and human resources, promoting intersectoral collaboration, and privileging upstream, social-ecological interventions over downstream interventions that put the onus on individuals to change their behaviors and rely on expensive and reactive clinical treatments, our world will be better for it. Failure, on the other hand, would be unsurprising in light of the history of health governance.

70 *Jacobson v. Massachusetts.*, 197 U.S. 11 (1905); Gostin and Wiley, *Public Health Law: Power, Duty, and Restraint, supra* note 23, at 120–131.

# 10

# Food Systems

Margot J. Pollans[1]

*Food is a fundamental component of public health. The consequences of climate change for the food system, and through that system for public health, are predictable yet complex. Complicating this story further is the fact that food production is itself a significant contributor to climate change, and mitigation strategies are, perhaps falsely, viewed as a threat to food security. This chapter uses a food security lens to map climate effects, then explores policy responses, some of which emerge from the food security paradigm, which focuses on the food needs of individuals. The chapter argues, however, that these responses, which tend to emphasize increased production and food aid, are inadequate because, while climate change will likely affect overall food production levels, the greater challenges lie with food access and affordability for low-income people. The alternate paradigm of food sovereignty considers control within the food system. As participation in control serves as a critical bulwark against climate related food system disruption this framework offers another set of useful tools that may help respond to the food system as both a cause and a victim of climate change.*

Food is a fundamental component of public health. As the American Public Health Association explains, threats to the food system "have enormous human, social, and economic costs that are growing, cumulative, and unequally distributed."[2] These threats include diet-related chronic disease; antibiotic resistance; chemical and pathogen contamination of food, air, soil, and water; and depletion of natural resources.[3] Climate change – which is already affecting rainfall levels, temperatures, spread of pests, and soil and water chemistry – will also affect the food system. Climate change may increase food-related public health concerns by damaging food production, undermining food quality, and impeding food access. In the United States, which currently produces more than enough food to feed its population, the

[1] Associate Professor, Elisabeth Haub School of Law at Pace University.

[2] American Public Health Assoc., *Toward a Healthy Sustainable Food System*, Policy No. 200712 (Nov. 6, 2007), www.apha.org/policies-and-advocacy/public-health-policy-statements/policy-database/2014/07/29/12/34/toward-a-healthy-sustainable-food-system.

[3] *Id.*

greatest risks relate not to food production itself but to deepening poverty. The populations most at risk are those already susceptible to food-related health problems including food borne illness, malnutrition, and other diet-related disease.

The consequences of climate change for the food system, and through that system for public health, are predictable yet complex. Complicating this story is the fact that food production is itself a significant contributor to climate change, and mitigation strategies are, perhaps falsely, viewed as a threat to food security. This chapter uses a food security lens to map these effects. It then explores ongoing and theoretical policy responses. Some of these responses emerge from the food security paradigm, which focuses on the food needs of individuals. The chapter argues, however, that these responses, which tend to emphasize increased production and food aid, are inadequate because, while climate change will likely affect overall food production levels, the greater challenges lie with food access and affordability for low-income people. The alternate paradigm of food sovereignty considers control within the food system, and participation in control serves as a critical bulwark against climate related food system disruption. This framework also offers an additional set of useful tools that may help respond to the food system as both a cause and a victim of climate change.

## 10.1 FOOD SECURITY AND PUBLIC HEALTH

Perhaps the most straightforward lens for understanding the relationship between climate, food, and public health is food security. The United States Department of Agriculture (USDA) defines food security as "the ability to obtain and use sufficient amounts of safe and nutritious food."[4] Food security provides a framework for understanding why a functioning food system is essential for public health and for evaluating how climate change might affect the functioning of that system. Historically, this framework was almost entirely productivist, focusing on levels of food production as the primary metric for evaluating the performance of the food system.[5] The productivist paradigm emerged during an era of extreme population growth, when policy makers were concerned that "under a business as usual scenario, entire regions would

[4] U.S. Dept. of Agric., *Climate Change, Global Food Security, and the U.S. Food System*, ix (Dec. 2015) www.usda.gov/oce/climate_change/FoodSecurity2015Assessment/FullAssessment.pdf. The UNFAO definition is similar: Food security is the state or condition "when all people at all times have physical, social and economic access to sufficient, safe, and nutritious food to meet their dietary needs and food preferences for an active and healthy life." United Nations Food & Agric. Org., *The State of Food Insecurity in the World*, 53 (2015), www.fao.org/3/a-i4646e.pdf.

[5] Roberta Sonnino et al., *Sustainable Food Security: An Emerging Research and Policy Agenda*, 21 Intl. J. of Agric. & Food 173, 174–175 (2014) ("Under this approach, food is reduced to the quantity produced and it is valued according to the efficiency of the production process.").

be facing starvation because agricultural output would be unable to catch up with demographic growth and shifting diets linked to urbanization."[6]

Over time, policy makers have added criteria to the framework to account for the fact that levels of food production are only one, and often not the most important, factor in food security. Today, both the USDA and the United Nations Food and Agriculture Organization (FAO) use four criteria to assess food security: availability, utilization, access, and stability.[7] Availability and utilization are both production-side metrics. *Availability* queries "the existence of food in a particular place at a particular time."[8] Primary factors for assessing availability are overall levels of food production and the functioning of food distribution networks. *Utilization* assesses "the ability to use and obtain nourishment from food. This includes a food's nutritional value and how the body assimilates its nutrients."[9] Utilization recognizes that nutrient content, and not just calorie count, is essential for an adequate diet.[10] Cultural appropriateness of available food is also a key component of utilization. *Access* is a demand-side characteristic, assessing "the ability of a person or group to obtain food."[11] Access may be constrained for physical, financial, or cultural reasons.[12] Finally, *stability* focuses on "[t]he absence of significant fluctuation in availability, access, and utilization."[13]

Food insecurity is a significant global problem. About 795 million people, or 1 in 9, were undernourished between 2014 and 2016.[14] At least 2 billion people live

6 Olivier De Schutter, *The Specter of Productivism and Food Democracy*, 2014 Wisc. L. Rev. 199, 199–200 (2014).

7 U.S. Dept. of Agric., *Climate Change, Global Food Security, and the U.S. Food System, supra* note 3, at 1.

8 Ibid.

9 Ibid.

10 U.N. Food & Agric. Org., *Trade Reforms and Food Security: Conceptualizing the Linkages* (2003) (describing the history of the definition of food security).

11 U.S. Dept. of Agric., *Climate Change, Global Food Security, and the U.S. Food System, supra* note 3, at 1.

12 Sonnino, *Sustainable Food Security, supra* note 4, at 175. Note that cultural appropriateness is relevant both to utilization, where the concern is about "the social, religious and cultural functions and benefits a food provides," and to food access, where the concern is about "social, religious or cultural norms and values that influence consumer demand for certain types of food." John Ingram, *A Food Systems Approach to Research Food Security and its Interactions with Global Environmental Change*, 3 J. Food Sec. 417 (2011).

13 U.S. Dept. of Agric., *Climate Change, Global Food Security, and the U.S. Food System, supra* note 3, at 1.

14 United Nations Food & Agric. Org., *The State of Food Insecurity in the World, supra* note 3, at 8 (noting also that, despite an increase in population of 1.9 billion people since 1992, the number of undernourished people has declined by 216 million in the same time period). Much depends, however, on how undernourishment is counted. Different counting methodologies can lead to dramatically different results. The UNFAO number, for instance, does not account for short-

with insufficient nutrients.[15] The vast majority of these people live in developing regions, but food security also remains a significant problem in the United States, where 12.7 percent of households are food insecure at least some of the time, and 5 percent of households have very low food security.[16]

Food insecurity is a core public health issue. Chronic food insecurity threatens physical and mental health. In a study commissioned by Bread for the World, researchers found that in the United States hunger costs $160 billion per year.[17]

> For an individual, chronic food insecurity may manifest as a reduced capacity to perform physically, diminished mental health and development, and an increased risk of chronic disease. Undernourishment, including inadequate caloric and/or nutrient intake, is a consequence of food insecurity and leads to outcomes such as stunting (short for one's age), wasting (thin for one's age), and micronutrient malnutrition.[18]

Specific health issues may include increased susceptibility to colds, migraine headaches, iron deficiency, higher incidence of upper gastrointestinal disorders, increased hospitalizations, and higher levels of anxiety, suicide, and depression.[19] In the United States, one particularly intractable public health cost of food insecurity is the spread of diabetes and heart disease. These illnesses correlate with excess calorie consumption of nutrient-poor food.[20] An indirect cost of food insecurity is lost productivity. According to the USDA, food insecurity

term undernourishment, because it assesses year-long energy-intake averages. De Schutter, *The Specter of Productivism and Food Democracy*, *supra* note 5, at 203 (also noting that the UNFAO does not account for inequalities related to intra-household distribution of food, "although it is known that women and girls within households are the first to sacrifice themselves in times of crisis and are generally discriminated against in their access to food"). In addition, the FAO assumes a sedentary lifestyle, but if it focused instead on the calorie requirements for the normal activity of the many poor people who "perform physically demanding activities in difficult conditions," it would need to revise its estimate upwards by 53 percent. Ibid. at 203–204.

15 U.S. Dept. of Agric., *Climate Change, Global Food Security, and the U.S. Food System*, *supra* note 3, at 17.

16 U.S. Dept. of Agric., *Household Food Security in the United States in* 2015, 5 (Sept. 2016), www.ers.usda.gov/publications/err-economic-research-report/err215.aspx ("the defining characteristic of 'very low food security' is that, at times during the year, the food intake of household members was reduced and their normal eating patterns were disrupted because the household lacked money and other resources for food").

17 Bread for the World Inst., *The Cost of Hunger in America*, http://hungerreport.org/costofhunger/.

18 U.S. Dept. of Agric., *Climate Change, Global Food Security, and the U.S. Food System*, *supra* note 3, at 19.

19 Bread for the World Inst., *The Cost of Hunger in America*, *supra* note 16; Donald S. Shepard et al., *Hunger in America: The Suffering We All Pay For* (*Center for American Progress*, Oct. 2011), www.americanprogress.org/wp-content/uploads/issues/2011/10/pdf/hunger_paper.pdf.

20 See De Schutter, The *Specter of Productivism and Food Democracy*, *supra* note 5, at 204 (explaining that "food intake itself cannot be assessed solely on the basis of its energy content").

"diminishes global economic productivity by 2–3% annually" and in some countries reduces gross domestic product by as much as 10 percent.[21]

A variety of nonclimate factors contribute to food insecurity around the world.[22] Some are supply-side, including food production levels, food handling practices, petroleum costs, and biotechnology. Others are demand-side, including population growth, food distribution systems, and economic and trade issues. Perhaps the most significant nonclimate determinant is poverty.[23] Indeed, while underproduction, associated with localized crop failure or with shift of arable land from food crops to biofuel crops, can lead to price shocks that exacerbate food insecurity, the global food system produces more than enough calories to feed the world. This is particularly true in the developed world.[24] Underproduction is rarely a problem. According to the UNFAO, the US produces 3,639 calories per capita per day.[25] This number far exceeds the recommended 2,000-calorie diet for adults.[26]

The food security framework is not a perfect assessment tool. It "avoid[s] discussing the social control of the food system."[27] Social control refers both to ownership of the means of food production and to access to democratic

[21] U.S. Dept. of Agric., *Climate Change, Global Food Security, and the U.S. Food System*, *supra* note 3, at 1.

[22] Howard Frumkin, et al., *Climate Change: The Public Health Response*, 98 Framing Health Matters 435, 436 (2008); U.S. Dept. of Agric., *Climate Change, Global Food Security, and the U.S. Food System*, *supra* note 3, at 21–23 (describing non-climate drivers of food systems and food security).

[23] Amartya Sen, *Poverty and Famines: An Essay on Entitlement and Deprivation* (1981); see also Rajeev C. Patel, *Food sovereignty: Power, gender, and the right to food*, 9 PLoS Medicine 1 (2012) (arguing that "Questions about hunger and its attendant pathologies, therefore, ought to begin with questions about social and political configurations around power over food, rather than about the mere presence or absence of food in the vicinity of a hungry individual.").

[24] Iain R. Lake et al., *Climate Change and Food Security: Health Impacts in Developed Countries*, 120 Envtl. Health Persp. 1520 (2012).

[25] United Nations Food & Agric. Org, Food Security Portal, *Calorie Supply Per Capita*, www.foodsecurityportal.org/api/countries/fao-calorie-supply-p.

[26] Alice H. Lichtenstein et al., *Diet and Lifestyle Recommendations Revision*, 114 Circulation 82, 87 (2006).

[27] Raj Patel, *What Does Food Sovereignty Look Like?* 36 J. of Peasant Studies 663, 665 (2009); Madeleine Fairbairn, *Framing Transformation: The Counter-Hegemonic Potential of Food Sovereignty in the U.S. Context*, 29 Agric. & Human Values 217, 221 (2012) (identifying as a core critique that food security "highlight[s] the microeconomic choices facing individual households in a free market, rather than the policy choices facing governments . . . [and], consistent with the neo-classical agenda, 'food security' came to be redefined, and institutionalized, in the [World Trade Organization] as international market relation . . . a system of 'free trade' in agricultural products was installed to privatize food security as a global, corporate relation" [internal citations and quotation marks omitted]).

participation in food governance.[28] The food security framework has historically facilitated policies designed to increase production, to redistribute food through surplus food donation, and to provide low-income individuals with food purchasing assistance. It rarely allows policy makers to address the root causes of poverty. It thus provides a tool to manage, but not solve, food insecurity. It has also been used to justify limited environmental regulation of food production despite significant harm. These critiques are not, however, fatal to the food security framework. They only caution against relying on it exclusively in understanding food system challenges. In the context of public health, the framework, which focuses on whether individuals are able to meet their nutritional needs, is particularly useful: It provides an effective metric both for assessing the success of the food system at protecting public health and for evaluating how climate change might aggravate food insecurity. When looking at solutions, however, it is essential to look beyond the traditional food security toolkit. Other food system paradigms, such as food sovereignty, provide useful supplements. Section 10.2 of this chapter, which evaluates the effects of climate change on the food system, focuses on food security. Section 10.3, which examines policy responses, introduces food sovereignty as an essential concept to drive a comprehensive policy response.

## 10.2 CLIMATE CHANGE AND FOOD SECURITY

Climate change is likely to aggravate all four food insecurity criteria, thus harming public health. People and communities already susceptible to food insecurity are likely to be affected most.[29] The scope of these consequences is highly uncertain, but their existence is not.[30] The UN Subsidiary Body for Scientific and Technological Advice identifies consequences for the food system as

[28] See Hans G. Bohle et al., *Climate Change and Social Vulnerability: Toward a Sociology and Geography of Food Insecurity* 4 Global Envtl. Change 37, 44–45 (1994) (connecting lack of control to vulnerability to climate change and noting that "groups with relatively low vulnerability include those with a high command over food through markets and distribution systems ... often these groups are in positions of control in the food system, which ensures their ability to extract surplus production and maintain economic relationships"); Patel, *What Does Food Sovereignty Look Like?*, *supra* note 26, at 665 (observing that "the power politics of the food system needed very explicitly to feature in the discussion" of food security and focusing on state control in the face of US hegemony).

[29] This is because climate change will tend to "superimpose on existing vulnerabilities." African Development Bank et al., *Poverty and Climate Change: Reducing the Vulnerability of the Poor Through Adaptation*, IX (2003) www.oecd.org/env/cc/2502872.pdf.

[30] U.S. Dept. of Agric., *Climate Change, Global Food Security, and the U.S. Food System*, *supra* note 3, at 25–39 (describing various scenarios depending on the severity of climate change and potential political responses).

an indirect effect of climate change.[31] First, climate change threatens food availability at a time when the global population is rapidly increasing. Because food production and distribution depend directly on natural systems, changes to those systems can be disruptive to overall production levels and to our ability to move food around the world.[32] Second, climate change threatens the safety, quality, and cultural appropriateness of the food supply. Finally, climate change threatens to disrupt access to the food supply by increasing food costs and impeding self-provisioning. Effects in each of these three categories may contribute to increases in overall levels of malnutrition and undernutrition.[33] All three of these effects will be distributed inequitably, with the most serious consequences for the 2.3 billion people residing in rural areas dominated by smallholder agriculture.[34]

Complicating these climate change effects is the fact that the food system is itself a significant contributor to climate change. This section begins with an overview of climate change affects in each of these three areas, offering brief summaries of the range of impacts.[35] Along the way, it will also identify some public health effects of climate change on the food system that are not directly related to food security. The section concludes with a brief description of the food system's greenhouse gas (GHG) footprint.

[31] Subsidiary Body for Scientific and Technological Advice (SBSTA), 46th session, Bonn, *Human Health and Adaptation: Understanding Climate Impacts on Health and Opportunities for Action*, FCCC/SBSTA/2017/2, para. 6 (May 2017) http://unfccc.int/resource/docs/2017/sbsta/eng/02.pdf.

[32] United Nations Food & Agric. Org., *The State of Food and Agriculture* 4 (2016), www.fao.org/3/a-i6030e.pdf (describing the mechanisms by which agriculture is inextricably linked to natural systems).

[33] SBSTA, *Human Health and Adaptation, supra* note 30, at para. 27 ("Malnutrition and undernutrition were highlighted as a concern for a number of developing countries in Africa, Asia and Latin America, which discussed the impacts of climate change on food security, particularly in relation to floods and drought.").

[34] Sonja J. Vermeulen et al., *Climate Change and Food Systems*, 37 Annual Rev. Environ. Resour. 195, 207 (2012).

[35] This chapter does not discuss separately the fourth criterion, stability, because it is not an independent issue. Rather, climate-related instability threatens each of the first three criteria. For more complete assessments of all four criteria, including in-depth discussion of stability-related issues, See U.S. Dept. of Agric., *Climate Change: Global Food Security, and the U.S. Food System, supra* note 3, Intergovernmental Panel on Climate Change, *Fifth Assessment Report: Climate Change 2014 Impacts, Adaptation, and Vulnerability*, www.ipcc.ch/pdf/assessment-report/ar5/wg2/WGIIAR5-Chap7_FINAL.pdf [hereinafter IPCC, 5th Assessment]; Vermeulen et al., *Climate Change and Food Systems, supra* note 33; United Nations Food & Agric. Org., *Food Security Portal, Calorie Supply Per Capita, supra* note 31.

### 10.2.1 *Availability: Food Production Levels and Location*

Availability "[a]ddresses the 'supply side' of food security, which is determined by food production, transportation, food stocks, storage, and trade."[36] Thanks to technological development, food production, measured both in absolute terms and as a function of yield per acre, has rapidly increased over the last century.[37] These increases have staved off Malthusian concerns that population growth would outpace food production leading to population collapse.[38] At the present moment, when food production growth is slowing and the population is rapidly growing, some policymakers have expressed concern that we are in our own Malthusian moment, and that we need desperately to figure out how to increase production levels again.[39] Former special rapporteur on the right to food, Olivier de Schutter, argues that the key question is not how much more food we need, but how we will produce it. He calls for a "reorientation of [] agricultural systems towards modes of production that are highly productive [and] highly sustainable," and he identifies resilience to climate change as a key component of sustainability.[40]

Climate change threatens all categories of food production, including crops, livestock, and fisheries.[41] Although climate change is likely to affect global food production, this factor is not likely to exacerbate food insecurity in the United States, where production far exceeds demand and adaptive capacity is high. Effects on domestic food production could, however, have consequences in other parts of the world that currently depend on US food exports and food aid. Domestic effects on trade and distribution networks are of more serious concern

36 U.S. Dept. of Agric., *Climate Change: Global Food Security, and the U.S. Food System*, *supra* note 3, at 18.

37 Pete Smith, *Malthus is Still Wrong: We Can Feed a World of 9–10 Billion, But Only by Reducing Food Demand*, 74 Proc. of the Nutrition Soc'y 187, 188 (2014).

38 Ibid. at 187.

39 See, e.g., United Nations Food & Agric. Org., 2050: A *Third More Mouths to Feed* (Sept. 23, 2009) www.fao.org/news/story/en/item/35571/icode/ (calling for an increase in food production by 70 percent by 2050); Jeffrey D. Sachs, *Shortages Coming True?* Sci. Am. (Sept. 1, 2008), www.scientificamerican.com/article/are-malthus-predicted-1798-food-shortages/ (arguing that while Malthus failed to predict that technological development would dramatically increase food production capacity we still do not know for sure if we have successfully avoided Malthusian collapse).

40 Olivier De Schutter, *Report Submitted by the Special Rapporteur on the Right to Food*, UN A/HRC/16/49, paras. 28–31 (Dec. 20, 2010) (discussing the role of agroecology in improving resilience to extreme weather events, drought, flood, and infestation of pests, weeds, and disease).

41 Specific attribution of production changes to climate change can be challenging because the number and strength of nonclimate drivers, including changes in production inputs such as fertilizer, makes establishing a baseline for comparison and isolating climate effects complicated. IPCC, *Fifth Assessment Report*, *supra* note 34, at 491.

for domestic food security. This section will briefly explore potential climate change effects on crops, livestock, fisheries, and distribution networks.

#### 10.2.1.1 Crops

There is a close relationship between crop yield and climate. Crop success is a function of temperature, precipitation, atmospheric carbon dioxide concentrations, soil moisture, and nutrient availability.[42] In a 2012 review of the "extensive literature on the tightly coupled relationship between climate change and food systems," climate scientists Sonja Vermeulen, Bruce Campbell, and John Ingram identified a variety of climate-related risks to land crop production including extreme weather events, "heat stress, water stress, waterlogging, and increases or reductions in pests and diseases."[43] Understanding the full scope of effects, i.e. the consequences for global crop yields, is extremely challenging due both to uncertainty about precise effects and to the complexity of the global food system, but some scientists have estimated that climate change has already affected global crop yields by diminishing yield increases by as much as 2.5 percent per decade.[44] Others predict negative impacts for many crops in the longer term.[45]

Distribution of these effects across the globe will be extremely variable. For instance, in warmer climates, higher average temperatures may decrease crop yield by reducing soil moisture and exacerbating drought conditions.[46] At some higher latitudes, higher temperatures may lengthen the growing season and thus increase yields by allowing for double cropping.[47] Warmer temperatures might also increase decomposition and thus increase soil nutrient availability. Even at higher latitudes, however, temperature increases can undermine productivity by reducing water availability.[48]

[42] U.S. Dept. of Agric., *Climate Change: Global Food Security, and the U.S. Food System*, *supra* note 3, at 55.

[43] Vermeulen et al., *Climate Change and Food Systems*, *supra* note 33, at 203.

[44] U.S. Dept. of Agric., *Climate Change: Global Food Security, and the U.S. Food System*, *supra* note 3, at 7.

[45] United Nations Food & Agric. Org., *The State of Food and Agriculture*, *supra* note 31, at 6, 26–27 (expecting net negative impacts after around 2030); Vermeulen et al., *Climate Change and Food Systems*, *supra* note 33, at 203 (listing predictions for various staple crops in Africa and South Asia); IPCC, 5th Assessment, *supra* note 34, at 505 (estimating net yield losses of 8 percent for major crops in Africa and South Asia by 2050 and estimating a mean reduction in yield of 1 percent for every decade of climate change).

[46] U.S. Dept. of Agric., *Climate Change: Global Food Security, and the U.S. Food System*, *supra* note 3, at 55.

[47] IPCC, *Fifth Assessment Report*, *supra* note 34, at 491.

[48] U.S. Dept. of Agric., *Climate Change: Global Food Security, and the U.S. Food System*, *supra* note 3, at 55.

Other factors affecting yield include changing precipitation levels, increased incidence of extreme weather, spread of pests and disease that thrive in warmer environments, and changes in atmospheric carbon dioxide. Increased incidence of extreme weather affects not only geographic yield patterns but also yield stability from season to season.[49] Changes to atmospheric carbon dioxide can increase yields in some places by increasing water efficiency.[50]

In sum, estimates suggest that between now and 2050, crop yields will decrease in water-constrained regions, but that because some regions will experience yield increases, there is the possibility that total global food availability will be only marginally affected.[51] Beyond 2050, however, climate change could result in major production effects.[52]

#### 10.2.1.2 Livestock

Although little research has focused specifically on the relationship between livestock production and global climate change, indirect consequences via "impacts on pasture and feed supplies, water, diseases, and genetic diversity" are expected.[53] In addition, heat stress can diminish livestock food intake and physical activity, affecting growth, survival, and reproductive rates.[54] Some studies also suggest that genetic selection for productivity, including increasing milk yield in dairy cattle and higher growth rates and leanness in pigs or poultry, "will tend to result in animals with reduced heat tolerance."[55] Heat stress can cause mortality and impair development and reproductive efficiency.[56] Related problems may arise with water stress, particularly in regions increasingly susceptible to drought.[57] As with crops, warmer temperatures may also lead to higher levels of loss due to pests and disease.[58] Livestock are also susceptible to extreme weather. For instance, in October 2016, flooding from Hurricane Matthew drowned tens of thousands of poultry and hogs in North Carolina.[59]

49 Ibid. at 56. 50 Ibid.

51 Vermeulen et al., *Climate Change and Food Systems*, *supra* note 33, at 201–202. 52 Ibid.

53 Ibid. at 204.

54 U.S. Dept. of Agric., *Climate Change: Global Food Security, and the U.S. Food System*, *supra* note 3, at 7.

55 IPCC, *Fifth Assessment Report*, *supra* note 34, at 502. 56 Ibid. 57 Ibid.

58 U.S. Dept. of Agric., *Climate Change: Global Food Security, and the U.S. Food System*, *supra* note 3, at 58; United Nations Food & Agric. Org., *The State of Food and Agriculture*, *supra* note 31, at 28.

59 Kirk Ross and Darryl Fears, *Flooded North Carolina Farms are Likely Littered with Drowned Livestock*, Wash. Post (Oct. 12, 2016), www.washingtonpost.com/news/animalia/wp/2016/10/11/flooded-north-carolina-farms-are-likely-littered-with-drowned-livestock/?utm_term=.e8aa37135b3c.

### 10.2.1.3 Fisheries

Fisheries, including wild capture and aquaculture, constitute a critical part of the global food system. About 3 billion people get 20 percent of their average per-capita animal protein from fish, and in some regions of the world, fish make up over half of all protein consumed.[60] Already under threat from overfishing, pollution, and habitat modification, fisheries are particularly vulnerable to the additional stressors of climate change.[61] Ocean temperature increase is a big problem. For instance, studies in the Northeast Atlantic show that temperature increases have led to increased abundance of fish species at the northern extent of their ranges and decreases at the southern extent.[62] Temperature increases also threaten coral reefs. Whereas many fish species are routinely subject to large daily and seasonal fluctuations of water temperature and are able to cope with them, temperatures of only one or two degrees Celsius over normal maximums for three to four weeks are sufficient to disrupt reefs and cause coral bleaching.[63]

Other climate change factors affecting marine fisheries are ocean acidification, sea-level rise, and changing precipitation patterns. Ocean acidification is expected to have negative effects on calcifying organisms such as mollusks, which constitute 23.6 percent of global aquaculture production.[64] Extreme weather events are also expected to disrupt fisheries and aquaculture operations, particularly in low-lying coastal areas.[65]

Climate change effects on aquaculture and freshwater fisheries are less well understood, but evidence suggests that inland fisheries are susceptible to water scarcity, increased incidence of flooding, and harmful algal blooms.[66] Recent research finds that rising temperatures, higher levels of

60 U.S. Dept. of Agric., *Climate Change: Global Food Security, and the U.S. Food System, supra* note 3, at 58.

61 Ibid.

62 IPCC, *Fifth Assessment Report, supra* note 34, at 493; United Nations Food & Agric. Org., *The State of Food and Agriculture, supra* note 31, at 28 (predicting losses of as much as 40 percent in the tropics and increases by 30 to 70 percent in high-latitude regions).

63 IPCC, *Fifth Assessment Report, supra* note 34, at 501; United Nations Food & Agric. Org., *The State of Food and Agriculture, supra* note 31, at 28.

64 IPCC, *Fifth Assessment Report, supra* note 34, at 500.

65 Ibid. "Ocean surface waters have become 30% more acidic over the last 250 years as they have absorbed large amounts of carbon dioxide from the atmosphere. This ocean acidification makes water more corrosive, reducing the capacity of marine organisms with shells or skeletons made of calcium carbonate (such as corals, krill, oysters, clams, and crabs) to survive, grow, and reproduce, which in turn will affect the marine food chain." US Global Change Research Program, *Overview: Climate Change Impacts in the United States* 9 (2014), www.globalchange.gov/browse/reports/overview-climate-change-impacts-united-states-third-national-climate-assessment.

66 Vermeulen et al., *Climate Change and Food Systems, supra* note 33, at 204.

carbon dioxide, and increased salinity due to drought all contribute to increased algal bloom occurrence and toxicity. These blooms in marine and freshwater cause damage to aquatic environments by blocking sunlight, depleting oxygen, and producing toxins with adverse effects for humans and wildlife.[67]

### 10.2.1.4 Distribution

Despite these threats to production, "[t]oday, and probably in the foreseeable future, the problem of food insecurity is principally one of distribution of food among nations, regions and households rather than insufficient overall production."[68] One-sixth of total global agricultural production is traded internationally.[69] "A globalized food system can ... buffer the local effects of weather events" and other climate-related productivity disruptions,[70] but climate change can also undermine availability by disrupting distribution networks themselves. The stability of distribution networks is a key issue in the United States, where, particularly in the short term, domestic food production levels are not likely to change significantly as a result of climate change.

Damage to roads, rail, and shipping infrastructure can disrupt the global supply networks that help to smooth shocks resulting from more localized production disruptions.[71] In the United States, for instance, where large volumes of grain are moved by water and there are few alternative pathways for transport, an extreme weather event affecting the waterway, such as flooding or drought, "presents an especially relevant risk to food [security] if [it] coincides with times of agricultural distribution, such as the fall harvest."[72] In 2005, after Hurricane Katrina, "where storm surges of 25 to 28 feet were recorded along parts of the Gulf Coast, grain transportation by rail or barge was severely slowed due to physical damage to infrastructure and the displacement of employees."[73]

The United States may also face increased pressure on its domestic food system as effects on productivity in other places increase demand for exports and food assistance and change the costs and types of food available for import.[74] For

[67] EPA, Office of Water, *Impacts of Climate Change on the Occurrence of Harmful Algal Blooms*, EPA 820-S-13-001 (May 2013) www.epa.gov/sites/production/files/documents/climatehabs.pdf.

[68] Brown et al., *supra* note 129, at 156.

[69] U.S. Dept. of Agric., *Climate Change: Global Food Security, and the U.S. Food System*, *supra* note 3, at 68.

[70] Ibid. at 18. [71] Ibid. at 9.

[72] Lewis Ziska et al., *The Impacts of Climate Change on Human Health in the United States: A Scientific Assessment, Report of the U.S. Global Change Research Program*, 199 (2016), https://s3.amazonaws.com/climatehealth2016/low/ClimateHealth2016_07_Food_small.pdf (internal citations omitted).

[73] Ibid. at 199.

[74] U.S. Dept. of Agric., *Climate Change: Global Food Security, and the U.S. Food System*, *supra* note 3, at 2.

instance, the United States imports about 80 percent of its seafood.[75] It is thus dependent on global production levels and distribution networks.[76]

### 10.2.2 *Food Utilization: Safety, Quality, and Culture*

Food utilization assessment provides a metric for evaluating whether available food can serve the needs of those consuming it. In other words, is available food adequately safe, adequately nutritious, and culturally appropriate? Climate change may impede food utilization by increasing the risk of food-borne illness, increasing levels of chemical contaminants in the food chain, lowering the nutritional value of food, and decreasing availability of culturally preferred foods.

Climate change will foster food-borne illness by increasing temperature and humidity and lengthening the warm season. These factors can improve growing conditions for and expand the seasonal and geographic ranges of microbial contaminants, fungi, and pests. To the extent these affect crops while still in the fields, they may contribute to crop loss. To the extent contamination occurs after food has entered the supply chain, it may make people sick. Experts predict more contamination by Salmonella, *E. coli,* and Campylobacter, because all thrive in warm humid conditions.[77] Further, "[f]ungal contamination resulting in the increase of mycotoxins in the food supply occurs due to high temperature moisture levels during pre- and post-harvest and during storage, transportation, and processing .... "[78]

Climate change can also increase incidence of chemical and metal contamination. These contaminants enter the food supply via floodwaters and precipitation.[79] Increased flooding may also increase the risk that agricultural chemicals spread into the environment, creating risk not for food security but for recreational users of

75 Nat'l Oceanic & Atmospheric Admin., *Global Wild Fisheries*, www.fishwatch.gov/sustainable-seafood/the-global-picture.

76 Currently seafood represents only about 5% of American animal protein consumption. Linda Kantor, *Americans' Seafood Consumption Below Recommendations*, Amber Waves (Oct. 3, 2016), www.ers.usda.gov/amber-waves/2016/october/americans-seafood-consumption-below-recommendations/ (but noting that this is just a quarter of what is recommended in the 2015–20 Dietary Guidelines for Americans).

77 Ziska et al., *The Impacts of Climate Change on Human Health in the United States, supra* note 71, at 195.

78 U.S. Dept. of Agric., *Climate Change: Global Food Security, and the U.S. Food System, supra* note 3, at 10; IPCC, *Fifth Assessment Report, supra* note 34, at 504; Ziska et al., The *Impacts of Climate Change on Human Health in the United States, supra* note 71, at 196.

79 Ziska et al., *The Impacts of Climate Change on Human Health in the United States, supra* note 71, at 196.

water bodies and for drinking water supplies.[80] Spread of microbial contaminants, pests, and disease also gives rise to the concern that farmers, food processors, and distributors will respond with increased use of pesticides and fungicides to compensate.[81] There is also heightened risk of pesticide resistance.[82] The risk of exposure here also has another health implication beyond the scope of food security – it puts farm and food system workers at direct risk.[83]

Increased temperatures can also contribute to food spoilage and put additional strain on the energy grid as a result of increased need for refrigeration.[84] In some regions, inadequate access to refrigeration may exacerbate food safety concerns and increase levels of food waste.

> Food waste that occurs as a result of climate-sensitive activities during food storage, processing packaging, and trade affects utilization rates. Estimates suggest that 30–50% of total global food production by mass is lost as waste. Food waste in retail, in food service, and at home accounts for most food waste in developed regions; in developing nations the absence of adequate food system infrastructure is a primary cause.[85]

Extreme weather events may also increase incidence of energy grid disruptions; power outages may result in food spoilage along the supply chain and in homes.[86] "For example, in August of 2003, a sudden power outage affected over 60 million people in the northeastern United States and Canada. New York City's Department of Health and Mental Hygiene detected a statistically significant citywide increase in diarrheal illness resulting from consumption of spoiled foods due to lost refrigeration capabilities."[87]

Climate change will also affect nutrition both by curtailing fruit and vegetable availability and by affecting nutrient levels in crops. One modeling study found that climate change would reduce fruit and vegetable consumption

80 Margot J. Pollans, *Drinking Water Protection and Agricultural Exceptionalism*, 77 Ohio State L.J. 1195 (2016).

81 Ziska et al., *The Impacts of Climate Change on Human Health in the United States*, *supra* note 71, at 197–198.

82 Ibid.

83 Thomas A. Arcury et al., *Pesticide Safety Among Farmworkers: Perceived Risk and Perceived Control as Factors Reflecting Environmental Justice*, 110 Envtl Health Persp. 233, 233 (2002).

84 S. J. James and C. James, *The Food Cold-Chain and Climate Change*, 43 Food Research Int'l 1944 (2010) (also observing that the cold chain is responsible for 1 percent of global carbon dioxide emissions).

85 U.S. Dept. of Agric., *Climate Change: Global Food Security, and the U.S. Food System*, *supra* note 3, at 10.

86 Ziska et al., *The Impacts of Climate Change on Human Health in the United States*, *supra* note 71, at 199–200.

87 Ibid. at 200.

worldwide by about 4 percent per person.[88] Further, elevated levels of carbon dioxide may reduce protein concentrations in staple crops such as wheat, barley, rice, and potatoes by 10–15 percent.[89] Increased carbon dioxide levels will also affect levels of micronutrients such as calcium, zinc, and iron.[90] In addition, rising temperatures may contribute to the spread of "illness-based diseases that affect the body's capacity for absorbing nutrients from food."[91]

Finally, utilization also depends on the cultural appropriateness of food and on regional food preferences. As climate change affects regional crop suitability, preferred foods may become less available.[92]

### 10.2.3 *Food Access: Financial and Physical*

Food access considers the demand side of food security: "the ability of a person or group to obtain food. Economic access to food (affordability) and allocation within society (including intranation and intrahousehold distribution) are integral to this component."[93] Concerns about physical access overlap with the aspects of food availability that focus on distribution, but while the focus there is on established food distribution networks, the focus here is on whether individual consumers have the means to get to distribution sites.[94]

Affordability is a measure of both food price and consumer income.[95] Because of the global consequences for food availability and quality, climate change may result in increased food prices.[96] "A globalized food system can ... buffer the

[88] Marco Springmann et al., *Global and Regional Health Effects of Future Production Under Climate Change: A Modelling Study*, 387 The Lancet 1937 (2016) (estimating that this and other food availability reductions will lead to 529,000 deaths worldwide by 2050).

[89] Vermeulen et al., *supra* note 33, at 204; US Dept. of Agric., *supra* note 3, at 10; IPCC, 5th Assessment, *supra* note 34, at 504; United Nations Food & Agric. Org., *supra* note 31, at 9.

[90] Ziska et al., *The Impacts of Climate Change on Human Health in the United States*, *supra* note 71, at 199; Samuel S. Myers et al., *Increasing $CO_2$ Threatens Human Nutrition*, 510 Nature 139 (2014).

[91] US Dept. of Agric., *Climate Change: Global Food Security, and the U.S. Food System*, *supra* note 3, at 10.

[92] Gina Ziervogel and Polly J. Ericksen, *Adapting to Climate Change to Sustain Food Security*, 1 WIREs Climate Change 525, at 529 (2010).

[93] U.S. Dept. of Agric., *Climate Change: Global Food Security, and the U.S. Food System*, *supra* note 3, at 18.

[94] Just as extreme weather may disrupt distribution systems, it may also disrupt consumer access to retail food establishments. Ziervogel and Ericksen, *Sustain Food Security*, *supra* note 91, at 528.

[95] U.S. Dept. of Agric., *Climate Change: Global Food Security, and the U.S. Food System*, *supra* note 3, at 9.

[96] See Ziervogel and Ericksen, *supra* note 91, at 525 (arguing that research has overfocused on this factor at the expense of other criteria); United Nations Food & Agric. Org., *The State of Food and Agriculture*, *supra* note 31, at 9 (noting that "most projections of the food price

local effects of weather events" and other climate-related productivity disruptions "but may also increase vulnerability by transmitting price shocks globally."[97] Because poor consumers spend a higher share of their income on food, this group faces "disproportionately negative effect[s] of price increases ... "[98] Indeed, according to the World Bank, increased food prices have pushed an additional 44 million into poverty in low- and middle-income countries since 2010.[99] One study found that climate change effects on production and distribution might increase food prices for staple crops by as much as 10 to 60 percent by 2030.[100] Increases in food prices exacerbate food insecurity for individuals who are net food purchasers (as opposed to those who rely primarily on self-provisioning or who are net food sellers). Net food buyers include urban consumers (a growing share of the population in most countries), rural nonfarming households (about 22 percent of the population in fifteen low-income countries), and food producers who are net buyers (the majority of poor rural households).[101] In the United States and other developed countries, increases in food prices may lead lower income individuals to shift to lower cost food items, such as energy-dense processed foods with high sugar and fat content.[102]

At the same time, and perhaps more importantly, climate change effects such as flooding and other extreme weather events may contribute to food insecurity by exacerbating poverty. For food producers, the same factors that affect food prices may also affect livelihoods. These consequences will be most severe in places where agriculture is a significant sector of the economy.[103] For instance, in addition to affecting food availability, climate effects on livestock production may aggravate food insecurity by undermining livelihoods for the billion people worldwide, and 600 million households in developing

impacts of climate change point to increases, although the magnitude and locations vary considerably across models and climate scenarios").

97 U.S. Dept. of Agric., *Climate Change: Global Food Security, and the U.S. Food System*, *supra* note 3, at 18.

98 IPCC, *Fifth Assessment Report*, *supra* note 34, at 502. 99 Ibid.

100 Thomas W. Hertel et al., *The Poverty Implications of Climate-Induced Crop Yield Changes*, 20 Global Envt'l Change 577, at 579 (2010).

101 IPCC, *Fifth Assessment Report*, *supra* note 34, at 503.

102 Lake et al., *Climate Change and Food Security: Health Impacts in Developed Countries*, *supra* note 23, at 1521–1522 (also noting that some shifts to lower cost foods could have neutral or even positive effects on nutrition).

103 African Development Bank et al., *Poverty and Climate Change*, *supra* note 28, at 6; United Nations Food & Agric. Org., *The State of Food and Agriculture*, *supra* note 31, at 29 (concluding that "the effect of climate change on the production and productivity of the agriculture sectors will translate into mostly negative economic and social impacts, with implications for all four dimensions of food security"). Globally, agriculture is only 2.4 percent of GDP, but in many poor countries it is greater than 40 percent. Hertel et al., *The Poverty Implications of Climate-Induced Crop Yield Changes*, *supra* note 99, at 579.

regions, that rely on livestock production.[104] "Poorer livestock keepers will be particularly susceptible to mortality of livestock in arid and semiarid regions where drought events are projected to become more frequent."[105] In the United States, however, less than 2 percent of the population farms, and robust federal crop insurance and disaster relief programs can help to insulate farmers from income effects of weather-related production shocks.[106]

For nonfood producers, climate change will also deepen poverty. Extreme weather events destroy homes, businesses, and infrastructure. The economic consequences of these events will be most severe in places with limited ability to cope with climate variability and extreme weather events. "Extreme weather events, which would cause limited damage and few casualties in a developed country, often cause extensive damage and substantial loss of life in a developing country. Poor people are particularly vulnerable to deviations from average climatic conditions such as prolonged drought and to natural disasters such as floods. In periods of stress they may be forced to sell off their physical assets such as land, bicycles, and farming implements, thereby undermining the sustainability of their livelihoods over the longer term."[107] Outside of developed countries, poor people "are often directly dependent on goods and services from ecosystems, either as primary or as supplementary sources of food, fodder, building materials, and fuel."[108] Although poor communities in the United States are less likely to be directly dependent on ecosystem goods and services, they are nevertheless also vulnerable to loss of life, loss of property, and loss of livelihood amid extreme weather events.[109] Thus the same phenomenon will play out within the United States. For instance, one study in the eastern United States found a correlation between poverty, access to air conditioning, and susceptibility to extreme heat-related mortality.[110]

104 US Dept. of Agric., *Climate Change: Global Food Security, and the U.S. Food System*, *supra* note 3, at 7.

105 Vermeulen et al., *Climate Change and Food Systems*, *supra* note 33, at 204.

106 Joseph W. Glauber, *Crop Insurance Reconsidered*, 86 Am. J. Agric. Econ. 1179 (2004) (describing growth of crop insurance programs); Megan Stubbs, *Agricultural Disaster Assistance*, Cong. Res. Serv. RS 21212 (Apr. 14, 2016) (summarizing details of USDA's agricultural disaster relief programs).

107 African Development Bank et al., *Poverty and Climate Change*, *supra* note 28, at 6.

108 Ibid. at 8.

109 George Luther and Michael McGeehin, *Climate Change and Extreme Heat Events*, 35 Am. J. Preventive Med. 429, 434 (2008).

110 Frank C. Curriero et al., *Temperature and Mortality in 11 Cities of the Eastern United States*, 155 Am. J. of Epidemiology 80, at 85–86 (2002) (finding that even adjusting for higher latitude, the indicators of socioeconomic status – percentage living in poverty and percentage with only high school education – were associated with higher heat-related mortality).

### 10.2.4 The Food System as a Cause of Climate Change

Although the primary focus of this chapter is on the consequences of climate change for the food system, it is worth briefly mentioning that the food system is itself a significant source of GHG emissions. The relationship is complicated because agricultural systems can emit GHGs and can serve as a carbon sink.[111] Altogether, the food system contributes between 19 and 29 percent of total anthropogenic GHG emissions.[112] Eighty to 86 percent of this total is from agricultural production itself, with the remainder coming from preproduction activities (fertilizer manufacture) and postproduction activities including processing, packaging, refrigeration, transportation, retail, catering, domestic food preparation, and food waste disposal.[113] Direct GHG emissions from agricultural production fall into five categories: (1) nitrous oxide emissions from soils, (2) methane emissions from enteric fermentation, (3) biomass burning, (4) rice production, and (5) manure management. Indirect emissions come from conversion of land from forest to agricultural use.[114]

The role of the food system as a contributor to GHG emissions is important not just because attention to these emissions is necessary for successful climate change mitigation but also because it complicates policy responses to the consequences described previously. Some mitigation strategies have adaptation co-benefits.[115] For instance, techniques to build up soil health can both improve crop resilience and reduce the need for application of nitrogen-based fertilizers, thereby reducing crop system nitrous oxide emissions.[116] Co-benefits may also occur where mitigation strategies have direct public health benefits. Some studies suggest that reduced meat consumption can both reduce GHG emissions and promote cardiovascular health.[117] Other mitigation policies instead present tradeoffs and may be viewed as threats to food security. For example, limits on land conversion may prevent indirect agriculture GHG emissions but

111 United Nations Food & Agric. Org., *The State of Food and Agriculture*, *supra* note 31, at 74 (discussing soil carbon sequestration potential); Ibid. at 82 (observing that this sequestration is often unprotected and is lost if the practices that cultivate it cease).

112 Vermeulen et al., *Climate Change and Food Systems*, *supra* note 33, at 198.

113 Ibid.

114 Ibid.

115 United Nations Food & Agric. Org., *The State of Food and Agriculture*, *supra* note 31, at 76–82 (describing various food security co-benefits of mitigation strategies).

116 R. Lal, *Soil Carbon Sequestration Impacts on Global Climate Change and Food Security*, 304 Science 1623 (2004).

117 See, e.g., Sharon Friel et al., *Public Health Benefits of Strategies to Reduce Greenhouse-Gas Emissions: Food and Agriculture*, 374 The Lancet 12 (2009); United Nations Food & Agric. Org., *The State of Food and Agriculture*, *supra* note 31, at 86 (noting that "there is increasing evidence that dietary patterns with low environmental impacts are also healthier").

could also limit the ability of a country to ramp up food production in the face of population growth and climate-induced lost productivity.[118]

## 10.3 TOWARD SECURITY AND SOVEREIGNTY IN THE FOOD SYSTEM

Traditional responses to food insecurity fall into two categories: those that focus on food consumers and those that focus on food producers, processors, and distributors. Food consumer-oriented policies respond primarily to access barriers and to the distribution element of availability. These policies include a variety of food security safety nets that provide either food itself or financial benefits for food purchasing. As climate change exacerbates food insecurity, these policies will become increasingly important. Pre-climate change, production-, processing-, and distribution-oriented policies focused on increasing food production, mainly through development of yield-increasing technology. In the climate change context, these policies focus on climate change adaptation in the agriculture sector. They address food availability and utilization issues. A third set of policies relies on a competing framework for evaluating the food system: food sovereignty, which introduces concerns about how and by whom food is produced. Food sovereignty policies also push for solidarity between producers and consumers. The following discussion of policy tools identifies how public health practitioners engage and could further engage with climate-related food policy to protect public health.[119]

### 10.3.1 *Developing Food Security Safety Nets*

Although "right to food" doctrines and policies are not related specifically to climate change, they serve as a powerful tool to assist food consumers in

118 See Pete Smith et al., *How Much Land-Based Greenhouse Gas Mitigation Can be Achieved Without Compromising Food Security and Environmental Goals?* 19 Global Change Biology 2285, 2287–2289 (2013) (describing tradeoffs in various land use decisions); see also Ingram, *Food Security and its Interactions with Global Environmental Change*, *supra* note 11 (explaining that the "synergies and tradeoffs need to be carefully considered, although the complexity of the food system makes analyses difficult"); United Nations Food & Agric. Org., *The State of Food and Agriculture*, *supra* note 31, at xiv.

119 Most public health focus on food in the United States focuses on diet-related disease such as diabetes and heart disease and on food-borne illness, but public health practitioners are increasingly recognizing the importance of sustainability in the food system as key to long-term public health. Academy of Nutrition & Dietetics, et al., *Principles of a Health, Sustainable Food System* (June 2010) www.apha.org/~/media/files/pdf/topics/healthy_sustainable_food_systems_principles_2012may.ashx (concluding that a "healthy, sustainable food system emphasizes, strengthens, and makes visible the interdependent and inseparable relationships between individual sectors (from production to waste disposal) and characteristics (health-promoting, sustainable, resilient, diverse, fair, economically balanced, and transparent) of the system").

circumstances in which climate change undermines food access.[120] Specifically, right to food policies can be used to combat declining food affordability associated with scarcity and to support geographic redistribution of food to areas most badly affected. Right to food policies may also focus on economic development, and other strategies for poverty alleviation. Overall, maintenance of robust safety nets is a critical tool to combat the potential effects of climate change on food-related public health.[121]

In the United States, there is no formal right to food. The United States has not ratified any of the international treaties guaranteeing the right to food, and no such right is codified in the US Constitution.[122] Nevertheless, a variety of federal programs provide financial support for food purchasing. These programs are not poverty alleviating; rather, they mitigate some of the worst effects of poverty. Most importantly, the Supplemental Nutrition Assistance Program (SNAP) serves as the backbone of US food security programs. As currently structured, SNAP can be a powerful tool to combat climate-related food security problems for several reasons. First, the program is designed to be countercyclical. As poverty increases, SNAP spending increases. Individuals affected by climate change can move into the program as necessary. Second, because SNAP provides a generic financial benefit, recipients can use it to purchase almost any type of food, thus insulating the program from the instability that might result from dependence on a specific set of foods. A variety of other programs provide

120 For the purposes of this chapter, I treat the "right to food" as a legal tool that can be used to implement food security by creating an individual right to food security. Food security itself "is not a legal concept per se." Hilal Elver, *The Challenges and Developments of the Right to Food in the 21st Century: Reflections of the Special Rapporteur on the Right to Food*, 20 UCLA J. Int'l & Foreign Aff. 1, 9 (2016). The relationship is in fact more complicated than this. Although the 1996 World Food Summit, which adopted both the multidimensional definition of food security used here and the concept of a right to food, critics argue that "fundamental ideological differences between the two concepts remain. Civil society organizations and developing countries continue to criticize the concept food security… as represent[ing] a supply-oriented economic view." Ibid. at 8–9 (noting that the UNFAO views "food security as being a precondition of the full enjoyment of the right to food"). Another way to think about this is that food security is a metric to describe the system, whereas the right to food is an individual legal right focusing on whether specific people can participate in that system.

121 Vermeulen et al., *Climate Change and Food Systems*, *supra* note 33, at 212.

122 Mariana Chilton, *A Rights-Based Approach to Food Insecurity in the United States*, 99 Am. J. Pub. Health 1203, 1203 (2009) (suggesting that the United States should adopt a right to food to facilitate "progress in resolving domestic food insecurity"). For a history and discussion of the contours and limits of the right to food in international law, see Smita Narula, *Reclaiming the right to food as a normative response to the global food crisis*, 13 Yale Human Rights & Development L.J. 403, 404–407 (2010) (explaining that individual states "occupy a central and critical role in ensuring the right to food and should discharge these obligations in a transparent, accountable, and non-discriminatory manner that ensures the participation of those most affected").

additional benefits for specific at-risk populations such as nursing mothers and young children, seniors, and the homeless.[123] Although these benefits are widely thought to be insufficient to truly meet the nutritional needs of participants (or to reach all those in need), they do make a significant difference in overall food insecurity.[124] SNAP and other food safety net programs are authorized in the farm bill, omnibus food and agriculture legislation passed every five to seven years.[125] Public health organizations already pay close attention to the bill because of its role in encouraging production and consumption of healthy foods.[126] Because the farm bill is regularly updated, it presents an opportunity to strengthen food safety nets in light of the threat of climate change.[127]

In addition to direct food support, many experts point to trade policy as a critical tool to combat climate-related food insecurity.[128] Because many production-related climate effects will be localized, "[p]olicies to manage local and international trade will be important in dampening the effects of localized climate shocks on food prices."[129] Removal of trade barriers can enhance food availability by allowing for redistribution from high production to low production areas.[130] It can also improve household access by driving down prices.[131] Accordingly, mainstream international development entities such as the World Bank, the International Monetary Fund, and the UNFAO

[123] For a complete guide to US food assistance programs, see Randy Alison Aussenberg and Kirsten J. Colello, *Domestic Food Assistance: Summary of Programs*, Cong. Res. Serv. R42353 (Feb. 17, 2016).

[124] See, e.g., Caroline Racliffe et al., *How Much Does the Supplemental Nutrition Assistance Program Reduce Food Insecurity?* 93 Am. J. Agric. Econ. 1082 (2014) (finding that "receipt of SNAP benefits reduces the likelihood of being food insecure by roughly 30% and reduces the likelihood of being very food insecure by 20%").

[125] For an overview of the farm bill see, Renée Johnson and Jim Monke, *What is the Farm Bill?*, Cong. Res. Serv. RS22131 (July 23, 2014).

[126] See, e.g., David Wallinga, *Agricultural Policy and Childhood Obesity: A Food Systems and Public Health Commentary*, 29 Health Affairs 405 (2010) (describing the links between farm bill agriculture policy and childhood obesity).

[127] Although increases to SNAP spending by expanding eligibility or increasing benefits may not be politically feasible at this time, lobbying is nevertheless extremely important to protect against budget cuts.

[128] By contrast, commodity derivatives trade may be a potential threat to food security. Some experts argue that a speculative bubble was partially responsible for the food price spike that caused a global food crisis in 2007–2008. See, e.g., Olivier de Schutter, *Food Commodities Speculation and Food Price Crises: Regulation to Reduce the Risks of Price Volatility*, UN Special Rapporteur in the Right to Food, Briefing Note 02 (Sept. 2010), www2.ohchr.org/english/issues/food/docs/Briefing_Note_02_September_2010_EN.pdf.

[129] Vermeulen et al., *Climate Change and Food Systems*, *supra* note 33, at 212.

[130] Molly E. Brown et al., *Do Markets and Trade Help or Hurt the Global Food System Adapt to Climate Change?* 68 Food Pol'y 154, 156 (2017).

[131] Ibid.

have espoused free trade as a basic tenet of global food security policy.[132] The benefits of trade are, however, widely contested, and Section 10.3.3 introduces the argument that trade liberalization hinders the economic development necessary to secure prosperity in rural communities.

Another set of consumer-oriented policies focus not on affordability but on consumer food choice. These policies are climate specific. They can target mitigation by directing consumers to select less carbon-intensive foods.[133] Or they can target adaptation by encouraging consumers to select crops less affected by climate change.[134] Education and "food environment" programs can target consumer food choice at the local, state, and national levels.[135] Numerous existing programs that focus on nutrition could serve as a platform for climate-related food choice education.[136]

A final key component to developing and improving food security safety nets is research. Dr. John Ingram, a soil scientist and food systems researcher, notes that "[t]here is … still a predominant research emphasis on increasing crop productivity … [R]esearch which considers multiple aspects of food security and food systems is needed."[137] Here, public health tools can play a key role to

[132] Marcia Ishii-Eiteman, *Food Sovereignty and the International Assessment of Agricultural Knowledge, Science and Technology for Development*, 36 J. of Peasant Studies 689, 695 (2009); Sonnino, *Sustainable Food Security, supra* note 4, at 175; Christina L. Davis, *Food Fights over Free Trade: How International Institutions Promote Agricultural Trade Liberalization* (2003).

[133] Smith, *How Much Land-Based Greenhouse Gas Mitigation Can be Achieved, supra* note 117, at 2290–2292 (assessing the mitigation potential of "demand-side" strategies such as switching to a vegetarian or vegan diet).

[134] Vermeulen et al., *Climate Change and Food Systems, supra* note 33, at 212.

[135] The Centers for Disease Control and Prevention defines the "food environment" as "the physical presence of food that affects a person's diet, a person's proximity to food store locations, the distribution of food stores, food service, and any physical entity by which food may be obtained, or a connected system that allows access to food." *General Food Environment Resources*, www.cdc.gov/healthyplaces/healthtopics/healthyfood/general.htm.

[136] At the national level, the most obvious place to address education is through the Dietary Guidelines, but an effort in 2015 to include recommendations based on environmental concerns was roundly rejected by the USDA and the Department of Health and Human Services. Tom Vilsack and Sylvia Burwell, *2015 Dietary Guidelines: Giving You the Tools You Need to Make Healthy Choices*, Oct. 6, 2015, www.usda.gov/media/blog/2015/10/6/2015-dietary-guidelines-giving-you-tools-you-need-make-healthy-choices.

[137] John Ingram, *A Food Systems Approach to Research Food Security and its Interactions with Global Environmental Change*, 3 J. Food Sec. 417, 418–419 (2011) (noting that production-oriented research is vitally important, but that it is insufficient to address food security problems"); see also Tim Wheeler and Joachim von Braun, *Climate Change Impacts on Global Food Security*, 341 Science 508, 510 (2013) (finding, after conducting a literature review, that studies assessing climate change and food security "heavily skewed toward food availability within 70% of the publications. Access, utilization, and stability dimensions of food security are represented by only 11.9, 13.9 and 4.2% of the total publications on food security and climate change, respectively.").

monitor ongoing health effects of food insecurity, evaluate effectiveness of food insecurity mitigation tools, and develop new and more effective tools.[138]

### 10.3.2 *Climate Change Adaptation in the Food Sector*

Climate-related producer-oriented policies recognize that food production itself and farmer livelihoods are vulnerable. The primary focus of these policies is climate change adaptation. "[T]he prevailing scientific view is that adaptation to the level that fully mitigates global climate-related losses in food availability is technically possible, although at a sizeable environmental and social cost for particular regions. Financial costs are not, however, high."[139] Estimates are around only $7 billion per year, with most investment in infrastructure, such as rural roads in Africa, and in agricultural research.[140] "Adaptive capacity is potentially high but also highly variable, and depends on decisions made at multiple levels throughout a diverse food system."[141]

Both in the United States and internationally, adaptation policies exist primarily within a productivist paradigm. Policies emphasize the ongoing importance of maintaining, and even increasing, food production levels.[142] In this context, the simultaneous need to address agriculture's GHG footprint becomes problematic. The 2015 Paris Agreement ignored calls for aggressive mitigation in the agricultural sector in favor of the concern that mitigation may undercut food security. The Preamble to the Agreement "[r]ecogniz[es] the fundamental priority of safeguarding food security and ending hunger, and the particular vulnerabilities of food production systems to the adverse impacts of climate change." Article 2, section 1 adds:

> This Agreement, in enhancing the implementation of the Convention, including its objective, aims to strengthen the global response to the threat of climate change, in the context of sustainable development and efforts to eradicate poverty, including by: * * * (b) Increasing the ability to adapt to the adverse

[138] Frumkin et al., *Climate Change: The Public Health Response*, *supra* note 21, at 438 (identifying public health tools to be deployed in response to climate change).

[139] Vermeulen et al., *Climate Change and Food System*, *supra* note 33, at 210.

[140] Ibid. (although this estimate appears to be at the low end); see also United Nations Food & Agric. Org., *The State of Food and Agriculture*, *supra* note 31, at 60 (summarizing results of studies in Uganda and Vietnam, which both found that the costs to agriculture of doing nothing vastly exceeded the costs of adaptation).

[141] U.S. Dept. of Agric., *Climate Change: Global Food Security, and the U.S. Food System*, *supra* note 3, at 10.

[142] United Nations Food & Agric. Org., *The State of Food and Agriculture*, *supra* note 31, at 13 (noting that "a key challenge in framing climate change responses is to ensure that they do not jeopardize food security or progress in poverty reduction").

> impacts of climate change and foster climate resilience and low greenhouse gas emissions development, in a manner that does not threaten food production.

This wording carefully avoids calls for mitigation from the agricultural sector. As New Zealand noted in its comments to the UNFCCC Subsidiary Body for Scientific and Technological Advice, "global mitigation efforts will need to include the agricultural sector," but because the UNFAO estimates that food production must increase 70 percent by 2050 to meet increased food demand, agricultural emissions should be expected to level out rather than to decrease to near zero as they are for other sectors.[143]

Participating countries under the agreement are free to include emissions in the agriculture sector as part of their "Intended Nationally determined contributions" (INDCs). As of July 2016, 148 out of 189 countries included agriculture in mitigation contributions, and 131 out of 134 included it as an adaptation priority area.[144] The United States was not included in either list even before the Trump Administration announced that it would withdraw the United States from the agreement.

Nevertheless, the United States is home to a variety of agriculture adaptation and mitigation programs developed under the auspices of President Obama's *Priority Agenda: Enhancing the Climate Resilience of America's Natural Resources*.[145] Pursuant to that agenda, the USDA launched two programs, one focused on

143 New Zealand, *Submission to the Subsidiary Body for Scientific and Technological Advice, Issues Related to Agriculture* (March 2016), www4.unfccc.int/submissions/Lists/OSPSubmissionUpload/55_84_131030897778822914-New%20Zealand%20SBSTA%20submission%20agriculture%20%2022%2003%202016%20FINAL.pdf; see also United Nations Food & Agric. Org., *The State of Food and Agriculture, supra* note 31, at 14 (prioritizing as objectives for "climate-smart agriculture" "sustainably increasing agricultural productivity" and "increasing adaptive capacity and resilience" over "reducing [GHG] emissions and increasing carbon sequestration" by qualifying the third objective with the phrase "where possible" but not so qualifying the first two).

144 United Nations Food & Agric. Org., *The State of Food and Agriculture, supra* note 31, at 12; Research Program on Climate Change, Agriculture and Food Security, *Agriculture's Prominence in the INDCs: Data and Maps* (Oct. 2016), https://ccafs.cgiar.org/agricultures-prominence-indcs-data-and-maps#.WUrqcHXyveQ; Rita Strohmaier et al., *The Agriculture Sectors in the Intended Nationally Determined Contributions: Analysis* xiv–xv (2016), www.fao.org/3/a-i5687e.pdf (finding that "the agriculture sectors are central to the response to climate change and in contributing to sustainable development").

145 The agenda was developed after the Climate and Natural Resources Working Group of the Interagency Council on Climate Preparedness and Resilience prepared an "inventory and assessment of land and water-related policies necessary to make the nation's natural resources and the communities that depend on them more resilient in the face of a changing climate." US Dept. of Agric., Climate Change and Land Sector: Improving Measurement, Mitigation, and Resilience of our Natural Resources, 21 (Dec. 2015), www.usda.gov/oce/climate_change/Climate_Change_and_Land_Sector_Report_2015.pdf. President Obama ordered the creation of the Interagency Council in Executive Order 13,653, Preparing the United States for the Impacts of Climate Change.

adaptation and the other on mitigation. The centerpiece of the adaptation plan is the USDA's system of Regional Climate Hubs, which "deliver science-based knowledge [and] practical information" to stakeholders.[146] These hubs provide technical resources, tools, and training modules for producers and land managers, and they conduct local research and demonstration projects.[147] They also offer farmers a variety of decision-making tools including access to historical crop and climate data. In 2015, the USDA published eight regional climate vulnerability assessments "that provide land managers and agency partners with an introduction to regional sensitivities and strategies for climate mitigation and adaptation."[148] USDA's agricultural extension programs, which offer county level research and farmer training, also support adaptation, and the USDA's National Institute of Food and Agriculture, the agency's research arm, provides grant support for research, case studies, and pilot projects on adaptation strategies.[149]

The mitigation plan, titled "Building Blocks for Climate Smart Agriculture and Forestry," calls for voluntary, incentive-based conservation, forestry, and energy programs to reduce GHG emissions from the agricultural sector, increased carbon sequestration, and expansion of agriculture-based renewable energy production.[150] The plan emphasizes the importance of "provid[ing] economic and environmental benefits through efficiency improvements, co-benefits, improved yields, or reduced risks."[151] The plan is "designed to meet the needs of producers ... USDA will encourage actions that enhance productivity and improve efficiency."[152] By contrast to other US climate initiatives, this plan includes no mandatory reduction measures. A variety of other non-climate change-specific USDA programs may support preservation of food production levels and farmer livelihoods in the face of climate change. They include federal crop insurance, research, and conservation programs. Federal crop insurance and other disaster relief programs help insulate farmers from the financial consequences of extreme weather. These programs are widely criticized for their failure to incentivize farmers to take steps to make

146 U.S. Dept. of Agric., *Climate Change and Land Sector, supra* note 144, at 14. As of October 2017, the Climate Hubs remain in operation.

147 Ibid. 148 Ibid.

149 Nat'l Inst. of Food & Agric., *Global Change and Climate Programs*, https://nifa.usda.gov/program/global-change-and-climate-programs (listing various grant programs that support climate research); U.S. Department of Agriculture, *The Role of Extension in Climate Change Adaptation in the United States* (June 2013), https://nifa.usda.gov/sites/default/files/resources/climate_ext_summit.pdf.

150 U.S. Dep't of Agric., *Building Blocks for Climate Smart Agriculture and Forestry: Implementation Plan and Progress Report*, at 2 (May 2016), www.usda.gov/sites/default/files/documents/building-blocks-implementation-plan-progress-report.pdf.

151 Ibid. 152 Ibid.

their operations more resilient to climate shock.[153] In addition, the majority of crop insurance benefits go to farmers with higher average household income, suggesting that the program does little to stem food insecurity.[154]

USDA also operates a variety of conservation programs that prioritize reducing soil erosion, increasing soil health, and conserving water resources, all goals that have the co-benefit of making agricultural operations more resilient to climate change.[155] The most important of these are the Environmental Quality Incentives Program and the Conservation Stewardship Program, through which the USDA provides financial support to help offset the costs of implementing conservation measures on working agricultural lands.[156] Although these programs are credited with success, they are vastly underfunded.[157]

Finally, USDA supports research and development on a variety of priorities including food safety and global food security. The agency supports research in crop science and input technology and supports development of strategies such as precision agriculture.[158] Precision agriculture, sometimes referred to as "as needed" farming, relies on modern technology to gather and respond to

153 Chad G. Marzen and J. Grant Ballard, *Climate Change and Crop Insurance*, 43 B.C. Envtl. Aff. L. Rev. 387 (2016) (arguing that the federal crop insurance program is well-positioned to "promote resilient agricultural practices").

154 Ibid. at 389.

155 Megan Stubbs, *Conservation Provisions in the 2014 Farm Bill (P.L. 113-79)*, Cong. Res. Serv. R43504 (Apr. 24, 2014), http://nationalaglawcenter.org/wp-content/uploads//assets/crs/R43504.pdf; Peter Lehner and Nathan A. Rosenberg, *Legal Pathways to Carbon-Neutral Agriculture*, 47 Envtl. L. Rep. 10845 (2017); see also Thomas L. Dobbs and Jules N. Pretty, *Agri-Environmental Stewardship Schemes and 'Multifunctionality'*, 26 Rev. of Agric. Econ. 220 (2004) (evaluating multifunctionality in agricultural green payment programs in the United Kingdom).

156 Stubbs, *supra* note 155, at 8–9 (summarizing the major components of both programs); Lehner and Rosenberg, *supra* note 155, at 10863–10865.

157 For instance, of the 389 million acres of cropland in the United States, only 25 million are enrolled in the Conservation Stewardship Program. Envtl. Working Group, *Conservation Stewardship Program in the United States*, https://conservation.ewg.org/csp.php?fips=00000&regionname=theUnitedStates. Sustainable agriculture advocacy organizations have also criticized the Environmental Quality Incentives Program for providing financial support for concentrated animal feeding operations. See, e.g., National Sustainable Agriculture Coalition, *CAFOs and Cover Crops: A Closer Look at EQIP Dollars* (Nov. 20, 2015) http://sustainableagriculture.net/blog/fy15-general-eqip-update/ (noting over $100 million in funding for animal feedlots in 2015 alone).

158 See generally, Committee on a Review of the USDA Agriculture and Food Research Initiative; Board on Agriculture and Natural Resources; Division on Earth and Life Studies; National Research Council, *Spurring Innovation in Food and Agriculture: A Review of the USDA Agriculture and Food Research Initiative Program* (2014), www.nap.edu/catalog/18652/spurring-innovation-in-food-and-agriculture-a-review-of-the (providing an in-depth analysis of the Agriculture and Food Research Initiative, the USDA's flagship competitive grants program).

localized information such as soil moisture and nutrient levels. Precision agriculture has a variety of goals including "reduction and more efficient use of production inputs, use of information technology to increase the size and scope of farming operations without increasing labor requirements, improved site selection and control of production processes ... and improved recordkeeping and production tracking for food safety, and environmental benefits."[159]

As with SNAP and other nutrition programs, most of the programs discussed here are authorized by the farm bill. Farm bill reauthorization (next due in 2018) thus presents a key intervention point for public health advocates. By contrast, the USDA's climate action plans were agency initiatives not mandated by statute. Insulating the climate action plans, via agency-level advocacy or codification in the farm bill, should thus be a public health priority. Producer-oriented programs currently receive less attention from public health practitioners than do the consumer-oriented programs described previously but will be essential to the long-term food system and public health. Although threats to production are not the main climate change challenge to food security in the United States, investing in adaptation is nonetheless essential to maintain existing levels of production, protect farm economies, and generate mitigation and other environmental co-benefits.

These producer-oriented policies tend to evaluate agriculture as a standalone issue, focusing primarily on how to maintain and increase production levels in the face of climate change. By contrast, the policies discussed in the next section call for a reorientation of food systems, recognizing the need to think about food production as one piece of the complicated food system puzzle.

### 10.3.3 *Beyond Food Security*

The food security lens provides a useful framework for understanding the effects of climate change on the food system and through it on public health, and for identifying and evaluating responsive policies. Particularly when the framework fully incorporates access and utilization metrics, it moves beyond its productivist origin to recognize the root causes of hunger, malnutrition, and other diet-related health problems. But the framework does not challenge

[159] Natural Resources Conservation Serv., *Precision Agriculture: NRCS Support for Emerging Technologies*, Agronomy Technical Note No. 1, 1 (June 2007), www.nrcs.usda.gov/Internet/FSE_DOCUMENTS/stelprdb1043474.pdf.

the status quo of the modern food system.[160] Modern agricultural systems generate significant environmental degradation. Although much of this degradation, except that associated with deforestation and other land degradation, is not related to climate change, the threat of climate change to the food system presents an opportunity not just to develop more climate-resilient agriculture but also to revamp food systems across the board to promote public and environmental health.[161] The same is true of the relationship between modern agricultural systems and corporate control of the food system.[162] The threat of climate change presents an opportunity to diversify scale in the system. Diversity of scale, like diversity of species, promotes resilience. It is also a key tool for combating poverty, the primary driver of food insecurity.

The food sovereignty framework offers an alternative lens that incorporates concerns about ownership and control of food production without losing sight of the primary health-related concerns at the center of the food security framework. La Via Campesina, an international peasant movement, first defined food sovereignty in 1996 as "the right of peoples to healthy and culturally appropriate food produced through ecologically sound and sustainable methods, and their right to define their own food and agriculture systems."[163] This framework offers a different set of tools to respond to the public health consequences of climate change.

Proponents of food sovereignty argue that the push, over the last several decades, "toward industrial agriculture and globalization – with an emphasis on export crops, lately transgenic crops, and with the rapid expansion of biofuel crops ... – is increasingly reshaping the world's agriculture and food supply, with potentially severe economic, social, and ecological impacts and risks."[164]

160 Fairbairn, *Framing Transformation, supra* note 26, at 222 (framing this criticism more pointedly by saying "the global elite who first frame food security did not question the political and economic structures within which they rose to power").

161 For discussion about the relationship between agricultural production, the environment, and public health, see J. B. Ruhl, Farms, *Their Environmental Harms, and Environmental Law*, 27 Ecol. L.Q. 263 (2000), Pollans, *Drinking Water Protection and Agricultural Exceptional, supra* note 79 (exploring the relationship between agricultural pollution and drinking water contamination), and De Schutter, *The Specter of Productivism and Food Democracy supra* note 5, at 206–211.

162 *See* Diana L. Moss and C. Robert Taylor, *Short Ends of the Stick: The Plight of Grower and Consumers in Concentrated Agricultural Supply Chain,* Wisc. L. Rev. 337 (2014) (describing concentration of ownership in the U.S. food system).

163 Andrew D. Jones et al., *Assessing the Potential and Limitations of Leveraging Food Sovereignty to Improve Human Health,* 3 Pub. Health 263, 1 (2015); see also Patel, *What Does Food Soverignty Look Like?, supra* note 27, at 663–665 (describing the origins of the concept of food sovereignty).

164 Miguel A. Altieri, *Agroecology, Small Farms, and Food Sovereignty*, 61 Monthly Review 34 (2009).

The food sovereignty movement critiques Green Revolution-type solutions to food insecurity on the ground that they overemphasize the role of production in food security and that they reduce self-sufficiency in developing countries by making farmers more dependent on foreign inputs.[165] Specifically, food sovereignty advocates seek to challenge the underlying assumption of the Green Revolution, that "achieving more economically efficient agricultural systems (i.e. systems that produce higher yields and greater profits) will translate into improved nutrition and health outcomes. Empirical evidence has consistently failed to support this assumption."[166] Although the food sovereignty movement has its roots in the global south, it has spread to the United States and other parts of the global north as an appealing alternative to "what some researchers see as the anemia of US agrifood politics."[167]

Food sovereignty has five core domains: "(1) use of agroecological management practices for food production, (2) the localizing of food production and consumption, (3) promotion of social justice and equity, (4) valuation of traditional knowledge, and (5) transforming economic and political institutions and structures to support self-determination."[168] The first of these, agroecology, is "the application of ecological concepts and principles to the design and management of sustainable agricultural ecosystems."[169] Agroecology requires enhancing habitat both above and below ground, and, importantly, relies on indigenous knowledge to improve understanding of the complexity of agroecosystems.[170] Agroecology promotes biodiversity in

165 Ibid. at 34. The Green Revolution, initiated in the 1940s–1960s, relied on technology and strong state support to increase agricultural productivity in Latin America and South Asia. De Schutter, *The Specter of Productivism and Food Democracy*, *supra* note 5, at 201.

166 Jones et al., *Limitations of Leveraging Food Sovereignty*, *supra* note 162, at 3.

167 Fairbairn, *Framing Transformation*, *supra* note 26, 217–218 (identifying as complaints about the existing agrifood movement that it "has been accused of reinforcing the hegemony of neoliberal economic ideology, failing to question the social injustices which sometimes characterize local food systems and agriculture in general, and generally confining their efforts to the 'politics of the possible'" (internal citations omitted)).

168 Jones et al., *Limitations of Leveraging Food Sovereignty*, *supra* note 162, at 2 (developing these domains from a set of six "pillars" of food sovereignty that were developed at the "International Forum for Food Sovereignty" in Mali in 2007); Forum for Food Sovereignty, Nyéléni Synthesis Report (Feb. 2007), https://nyeleni.org/IMG/pdf/31Mar2007NyeleniSynthesisReport-en.pdf.

169 Altieri, *Agroecology*, *supra* note 163, at 34; see also Jones et al., *Limitations of Leveraging Food Sovereignty*, *supra* note 162, at 3 (identifying agroecology as "one of the most consistent and conspicuous priorities of the food sovereignty movement"). Agroecology is also gaining traction in mainstream food policy circles. See, e.g., United Nations Food & Agric. Org., *The State of Food and Agriculture*, *supra* note 31, at 51 (identifying it as a strategy for climate adaptation for smallholder farmers).

170 Altieri, *Agroecology*, *supra* note 163, at 34–35.

agroecosystems, which, in turn, supports resilience to climate change.[171] In this way, agroecology is a response to concerns about food availability.[172] And because of its emphasis on soil health, it is a potential climate change mitigation strategy. By focusing on the link between environmental sustainability and livelihoods, agroecology thus merges adaptation and mitigation concerns. Some evidence also suggests that agroecology may have more direct health impacts related to diet, nutrition, and toxic exposures, but evidence on these effects is mixed, showing both costs and benefits associated with agroecology practices.[173]

Agroecology extends beyond food availability. By working hand in hand with the other pillars of food sovereignty, it simultaneously addresses access and utilization concerns by democratizing food system control. Food sovereignty advocates seek to dismantle corporate control of the food system with the goals of fostering local, regional, and national level self-sufficiency, reducing dependence on foreign aid and on foreign inputs.[174] Food sovereignty can provide a path out of poverty by targeting poor smallholder or landless farmers, providing improved market opportunities and greater solidarity with food consumers.[175] Even in the United States, where only a very small percentage of the population is engaged in smallholder agriculture, food sovereignty offers

[171] Miguel A. Altieri & Clara I. Nicholls, *The Adaptation and Mitigation Potential of Traditional Agriculture in a Changing Climate*, 140 Climatic Change 33 (2017); Altieri, *Agroecology*, *supra* note 163, at 35. For discussion of the connection between biodiversity in agricultural ecosystems and resilience of those ecosystems, see Lora Ann Thrupp, *Linking agricultural biodiversity and food security: The valuable role of sustainable agriculture*, 76 Int'l Affairs 265 (2000).

[172] See De Schutter, *The Right to Food*, *supra* note 39 (arguing that "scaling up agroecological practices can simultaneously increase farm productivity and food security, improve incomes and rural livelihoods, and reverse the trend toward species loss and genetic erosion"); Ibid. paras. 16–20 (arguing that agroecology can increase productivity at the field level).

[173] Jones et al., *Limitations of Leveraging Food Sovereignty*, *supra* note 162, at 3–4 (surveying available data and finding links between agroecology and more diverse diets and reduced synthetic toxins exposure, but also finding mixed evidence regarding nutritional content of individual foods and exposure to microbial pathogens and infectious microbes); De Schutter, *The Right to Food*, *supra* note 39, at paras. 26–27 (arguing that agroecology can improve nutrition).

[174] Although local control is a key component of food sovereignty, it is not necessarily an end in and of itself. Fairbairn, *Framing Transformation*, *supra* note 26, at 226–227 (criticizing the tendency of US food sovereignty advocates to "reify the local" at the expense of systemic change and observing that sovereignty may be exercised on a variety of levels).

[175] United Nations Food & Agric. Org., *The State of Food and Agriculture*, *supra* note 31, at 51 (identifying agroecology as a key strategy for adaptation in smallholder agriculture, and identifying adaptation as essential to rural prosperity); Altieri, *Agroecology*, *supra* note 163, at 39; De Schutter, *The Right to Food*, *supra* note 39, at paras. 21–25 (arguing that agroecology reduces rural poverty).

appealing prospects for addressing marginalization of Native American communities and gender inequality within the US food system.[176]

Various existing USDA programs already support or could be used to support elements of food sovereignty. First, a variety of agency programs support development of regional food systems by funding regional distribution infrastructure and marketing initiatives – including food hubs and farmers markets.[177] These programs promote farmer to consumer networks, a key tenet of food sovereignty.[178] They also chip away at concentrated corporate control of food system decision-making. Second, the USDA funds access to land for new farmers, female farmers, and minority farmers, allowing these groups to share in control of the food system.[179] Although these programs have expanded in scope in recent years, many question whether they are effective and call for even greater investment.[180] Third, USDA research, technical support, and

176 See Fairbairn, *Framing Transformation, supra* note 26, at 225–226 (identifying ongoing efforts to use the food sovereignty frame to address these issues and noting that it could also be used to address farmworker rights issues).

177 US Econ. Research Serv., *Trends in U.S. Local and Regional Food Systems, Report to Congress*, Admin. Pub. No. 068, 52-60 (Jan. 2015), www.ers.usda.gov/webdocs/publications/42805/51173_ap068.pdf?v=42083 (summarizing federal and state policies promoting development of regional food systems and connections between farmers and consumers).

178 Jones et al., *Limitations of Leveraging Food Sovereignty, supra* note 162, at 4 (noting the emphasis on "strengthening community connections through the localization of food sourcing"). The promotion of solidarity between farmers and consumers must be understood in the context of an industrialized and corporate food system, in which farmers and consumers have the greatest numbers but the least power, and are separated by a lengthy supply chain. Margot J. Pollans, *Farming and Eating*, 13 J. of Food Law & Pol'y 99 (2017); Moss and Taylor, *Short Ends of the Stick, supra* note 161. At the same time, however, there is a risk that overemphasis on consumers shifts food sovereignty from a radical movement to transform the food system to a neoliberal movement focused on consumer purchasing choices. Fairbairn, *Framing Transformation, supra* note 26, (concluding that such a neoliberal appropriation of the term food sovereignty in the United States has not yet occurred but could); Allison Condra, *Food Sovereignty in the United States: Supporting Local and Regional Food Systems*, 8 J. Food L. & Pol'y 281 (2012) (distinguishing local food protection laws in the United States, which focus on freedom from regulation, from food sovereignty).

179 For instance, the Section 2501 Program authorizes the USDA's Office of Advocacy and Outreach to engage in and fund outreach efforts to connect socially disadvantaged and veteran farmers to USDA resources. 7 U.S.C. § 2279 (originally authorized as section 2501 of the Food, Agriculture, Conservation, and Trade Act of 1990).

180 Non-white farm ownership increased by 15 percent between the 2007 and 2012 agricultural censuses, but non-whites still represent only about 7 percent of farmers despite constituting 36.3 percent of the population. See National Sustainable Agriculture Coalition, 2012 Census Drilldown: Minority & Women Farmers (June 3, 2014), http://sustainableagriculture.net/blog/census-drilldown-sda/; USDA Census of Agriculture, *2012 Census Highlights: Farm Demographics* (May 2014), www.agcensus.usda.gov/Publications/2012/Online_Resources/Highlights/Farm_Demographics/; Institute for Agriculture & Trade Policy, *Disadvantaged Farmers: Addressing Inequalities in Federal Farm Programs for Farmers of Color* (Mar. 28,

financial support programs could prioritize agroecology over adaptation approaches such as precision agriculture.[181] Finally, USDA administers the Packers and Stockyards Act, which, among other things, provides limited protections to poultry producers from "unfair, unjustly discriminatory, or deceptive practice[s]" in contracting with poultry processors.[182]

Food sovereignty advocates also call for an end to the liberalization of free trade. This is the most controversial element of food sovereignty policy because it is the sole component that is entirely incompatible with traditional food security and productivism. Food sovereignty advocates argue that neoliberal trade policies have prioritized market access to imports over domestic markets for domestic producers, undercutting local and regional economic development.[183] The growth of international agricultural trade facilitated "regional specialization in a relatively narrow range of products ... and because of associated technological and policy choices, the benefits were concentrated in the hands of large production units and landholders at the expense of smaller-scale producers and landless workers."[184] Additional

2012), www.iatp.org/documents/disadvantaged-farmers-addressing-inequalities-in-federal-programs-for-farmers-of-color.

181 Marcia S. DeLonge et al., *Investing in the Transition to Sustainable Agriculture*, 55 Envtl. Sci. & Pol'y 166 (2015) (analyzing allocation of approximately 10 percent of the USDA's 2014 Research, Extension, and Economics budget and finding that only about 10 percent of that was directed to agroecology research); Altieri, *Agroecology*, *supra* note 163, at 40 (calling for research into the principles of agroecology).

182 7 U.S.C. § 192(a). During the Obama administration, the USDA drafted rules that would strengthen protections for poultry producers, but those rules were rescinded by the Trump administration. Grain Inspection, Packers and Stockyards Administration (GIPSA), Final Rule, Scope of Sections 202(a) and (b) of the Packers and Stockyards Act, 82 Fed. Reg. 48594 (Oct. 18, 2017) (rescinding an interim final rule that would have allowed poultry producers to show statutory violations without establishing harm to competition); see also GIPSA, Unfair Practices and Undue Preferences in Violation of the Packers and Stockyards Act, 82 Fed. Reg. 48603 (Oct. 18, 2017) (deciding to take no further action on a December 2016 Proposed Rule that clarified the meaning of "unfair, unjustly discriminatory, or deceptive practice[s]"). In addition to the Packers and Stockyards Act, the Clayton and Sherman Antitrust Acts offer limited protection against concentration of power in the food system.

183 Peter Rosset, *Food Sovereignty and Alternative Paradigms to Confront Land Grabbing and the Food and Climate Crises*, 54 Development 21, 22 (2011); Carmen G. Gonzalez, *Institutionalizing Inequality: The WTO Agreement on Agriculture, Food Security, and Developing Countries*, 27 Colum. J. Envtl. L. 433, 438 (2002) (exploring the "food security implications of the [World Trade Organization] Agreement on Agriculture" and arguing that the agreement "systematically favors agricultural producers in industrialized countries at the expense of farmers in developing countries").

184 De Schutter, *The Specter of Productivism and Food Democracy*, *supra* note 5, at 212 (explaining that increased reliance on agritechnology and monocultures rewarded economies of scale and improved labor productivity, but "gave a premium to the largest landholders who were better positioned to achieve efficiency gains under this model").

research is needed to better understand the relationship between free trade, food sovereignty, climate change, and public health.[185]

The food sovereignty approach integrates concerns about climate change and the food system with other food system issues including equity and autonomy.[186] Its focus not on individuals as food consumers within a market economy, but on social solidarity resists the commodification of food.[187] The theoretical advantage of this approach is its more holistic understanding of the root causes of food-related health problems.[188] The framework seeks to eliminate the conditions that generate food insecurity in the first instance.

## 10.4 CONCLUSION

As climate scientists Tim Wheeler and Joachim von Braun pointed out in 2013, "the consequences for global undernutrition and malnutrition of doing nothing in response to climate change are potentially large and will increase over time ... [And] food inequalities will increase, from local to global levels."[189] The serious and inequitable public health consequences merit a robust response from the public health community. Food security is a traditional public health response tool, and, within the United States, where there is not yet a robust right to food, the framework has much remaining promise. Thus food security should not be

185 Brown et al., *Do Markets and Trade Help or Hurt the Global Food System Adapt to Climate Change*, *supra* note 129, at 156–158 (evaluating advantages and disadvantages of trade for food security in a changing climate and observing that the key differences arise depending on whether data is aggregated or whether analysis is more localized, but concluding that "continuous market integration and economic growth has helped dramatically increase global wealth and reduce food insecurity"); *Gonzalez, Institutionalizing Inequality*, *supra* note 182, at 438, 484–489 (offering proposals for revision of the WTO agreement on agriculture that would "achieve a more level playing field" for developing countries).

186 *But* see Eric Holt-Gimenez and Miguel Altieri, *Agroecology, Food Sovereignty, and the New Green Revolution*, 37 Agroecology & Sustainable Food Systems 90 (2013) (arguing that selective incorporation of principles of agroecology into "the corporate food regime" relegates agroecology to "the margins" and "erase[s] its history, stripping it of its political content").

187 Fairbairn, *Framing Transformation*, *supra* note 26, at 222. Versions of food sovereignty imported into the United States have, at times, been reframed to focus more on benefits to consumers from buying local, and, even, the right to buy local. Ibid. at 224 (noting that despite this reframing "food sovereignty as framed by US organizations has retained its critical stance towards neoliberal governance of food and agriculture").

188 One practical advantage of this approach is the ease with which it lends itself to local level experimentation, which can generate lessons to be applied at the national and international levels. See, e.g., Daniel R. Block et al., *Food Sovereignty, Urban Food Access, and Food Activism: Contemplating the Connections Through Examples From Chicago*, 29 J. Agric. & Human Values 203 (2012); Fairbairn, *Framing Transformation*, *supra* note 26, at 227 (emphasizing the importance of tying local food sovereignty programs to the national context and warning against "defensive localism").

189 Wheeler and von Braun, *Climate Change Impacts on Global Food Security*, *supra* note 136, at 512.

tossed aside.[190] It continues to provide an important set of metrics for evaluating the health of the food system. But, perhaps because the food security framework evolved within a neoliberal context, it remains overly focused on food production and on individual participation within food markets. It lacks systemic transformative potential. The food sovereignty framework merges concerns about productive sustainability with concerns about food system control, providing a set of tools that seek to achieve the same basic ends of food security – adequate, nutritious, and culturally appropriate food for all – while also addressing underlying inequality and oppression, including that related to the environmental externalities of food production. Because climate change will exacerbate existing inequalities, the food sovereignty lens provides a necessary antidote.

At first blush, food sovereignty appears further afield from public health. Not surprisingly, it is not a dominant lens adopted with the public health community in the United States. Food security, by contrast, is commonly used, and public health officials have paid increasing attention to food, particularly with regard to diet-related disease, food safety, and hunger.[191] Food sovereignty offers a new, and, for the most part, compatible, toolkit with much promise for public health.

190 See, e.g., Tina D. Beuchelt and Detlef Virchow, *Food Sovereignty or the Human Right to Adequate Food: Which Concept Serves Better as International Development Policy for Global Hunger and Poverty Reduction?* 29 Agric. & Human Values 259 (2012) (arguing that while food sovereignty is important for civil society movements, the right to food framework remains the better lens for achieving hunger and poverty reduction).

191 For instance, the New York City Department of Health and Mental Hygiene identifies healthy eating as a key component of a healthy life and has developed targets for reducing obesity, consumption of sugary drinks, and sodium intake. See *Take Care New York 2020: Every Neighborhood, Every New Yorker* 17 www1.nyc.gov/assets/doh/downloads/pdf/tcny/tcny-2020.pdf. Similarly, the Federal Department of Health and Human Services manages a variety of programs promoting healthy eating. Dept. Health & Human Servs., *Nutrition and Fitness*, www.hhs.gov/programs/prevention-and-wellness/nutrition-and-fitness/index.html. Of the top five public health graduate programs, four of the five have food-specific departments, but three of those four are nutrition departments that offer only a course or two on food systems issues. The website of the Michigan program characterizes the field as follows:

> "Nutrition has become a topic of growing national and global interest in the past decade, with intensifying research and scholarship related to topics like obesity, chronic disease risks, food safety and security, global hunger and undernutrition, and the environmental and health impacts of agriculture."

https://sph.umich.edu/ns/index.html. The most robust of the four is at the Johns Hopkins School of Public Health, which offers a certificate program called "Food Systems, the Environment, and Public Health." For listings of food-related courses, see *Course Catalog*, www.jhsph.edu/courses/list/?keyword=food&yearId=2017 (no classes relate specifically to food sovereignty).

# 11

# Migration

Maxine Burkett*

*The displacement, migration, and relocation of individuals and communities are likely consequences of a changing climate. There are contemporary instances of that movement as well as historical analogues that illustrate the impact climate change – both sudden and slow-onset events – may have. Further, based on projected increased severity of fast-acting and destructive coastal storms, as well as sea-level rise, droughts, and desertification, many more individuals and communities may be forced to move, with attendant impact on their livelihoods. The myriad pathways of population movement, the movement itself, as well as weaknesses or failures in related systems and services they provide, may negatively affect individual and community health. This chapter provides a survey of what climate change means for public health in the migration context, and describes how existing law supports (or fails to support) appropriately responsive interventions.*

## 11.1 INTRODUCTION

The displacement, migration, and relocation of people are visible and emerging consequences of a changing climate. There are contemporary instances of these kinds of population movements as well as historical analogues that illustrate the impact climate change may have. For example, Hurricane Katrina hit Louisiana in August of 2005; however, its legacy impacts, such as displacement from homes and longstanding communities, were felt for several years afterwards – with many who did not return home. Similarly, occurring more slowly but with no less disruptive effect, droughts like that experienced in the horn of Africa have, in part, produced famine, further triggering

* Professor of Law, William S. Richardson School of Law, and Global Fellow, Woodrow Wilson International Center for Scholars. I thank Arielle Kramer for her excellent research assistance. I thank the anonymous reviewer whose feedback greatly improved this chapter.

migration.[1] In addition, based on projected increased severity of fast-acting and destructive coastal storms, as well as sea-level rise, droughts, and desertification, many more individuals and communities will be forced to move, with attendant impact on their livelihoods. The myriad pathways of population movement, the movement itself, and weaknesses or failures in related systems and services those systems provide, may negatively impact individual and community health. This chapter provides a survey of what climate change means for public health in the migration context, and describes how existing law supports (or fails to support) appropriate interventions. While there are a number of potential climate-related drivers of displacement, this chapter refers most frequently to examples driven by sea-level rise and severe storms.

Climate change's impact on health is multiple. Researchers have described three types of effects, all of which arise in the context of population movement.[2] Primary or direct effects include, for example, injuries and deaths caused by extreme weather events such as cyclones. These sudden events displace individuals and communities, either for discrete periods of time, which is most common, or permanently. Secondary or indirect effects include the increased geographic range and population exposed to vector-borne disease. As new migrants, particularly from different temporal regions, cross borders or move within countries, so too might new diseases and means of transmission. Tertiary, diffuse, or delayed effects include disruptions to services, including health. In the case of migration, population movements might introduce new and increased illnesses – whether acute, chronic, or vector-borne – or overwhelm the sectors and services that support health and well-being. These probable negative impacts militate in favor of comprehensive, proactive, and compassionate planning to prepare both those who may need to move and host communities to ensure overall public health.

This chapter first describes the complex linkages between climate change's impacts on the movement of individuals and communities and then describes the adverse public health impacts that can result from this kind of movement. It then discuses in Section 11.2 the existing law and policy frameworks that are relevant to elements of the phenomenon of climate migration. These frameworks

[1] I use the term migration to describe, specifically, the movement of peoples that can occur proactively, in anticipation of a significant stressor, or reactively, in response to a climate-related impact. I also use it throughout as an umbrella term that includes the circumstances of forced displacement and planned relocation.

[2] L. McIver et al., *Health Impacts of Climate Change in Pacific Island Countries: A Regional Assessment of Vulnerabilities and Adaptation Priorities*, 124 Environ Health Perspect 1707–1714, 1707 (2016) http://dx.doi.org/10.1289/ehp.1509756.

are multi-scalar and include public health law, often the domain of local, state, and federal laws; federal immigration laws; and migration and refugee laws, which straddle federal and international legal regimes. They are also, however, inadequate to address the challenges posed by climate migration. Further, other existing laws that decision-makers might augment to meet the challenge are under threat. In the United States, these law and policy frameworks are compounded by an emboldened strain of xenophobia that might retard progress on meeting the health and relocation needs of those traveling across borders to the United States. Section 11.3 considers gaps and interventions to supplement this current body of law, proposing prudent policy changes that decision-makers can implement now and those that, while too nascent or controversial to introduce in today's political climate, are necessary in the foreseeable future.

### 11.1.1 *Causal Linkages between Climate Change, Migration, and Public Health*

#### 11.1.1.1 Understanding Climate Migration

The relationship between climate and human migration is longstanding. Indeed, climatic influences on "environmental conditions, air temperatures, food and water availability, social structures, and migratory routes" have induced and, in some instances, directed migratory flows.[3] Today, climate change serves as a significant driver of population movement resulting from the risks it introduces to "communities and livelihoods via impacts on ecosystem 'goods and services,' loss of arable land, and increased severity and frequency of climate-related disasters."[4] Whether individuals embrace migration as an adaptive strategy or resign themselves to it as a necessary, and sometimes forced, response to increased risks, climate-induced displacement, migration, and relocation will likely proliferate as the rate of change increases and climate change impacts worsen.

Climate change impacts act in concert with other drivers to induce people to move. Indeed, disaster displacement "occurs *in the context* of disasters."[5] In

[3] Celia McMichael et al., *An Ill Wind? Climate Change, Migration, and Health*, 120(5) Envtl. Health Persp. 646, 650 (May 2012). See also, A. Issar, *Climate Change as a Draw-Bridge Between Africa and the Middle East*, 72(4) Glob Planet Change 451–454 (2010).

[4] McMichael et al., *An ill wind?*, *supra* note 3, at 647.

[5] See generally, Burkett, *Justice and Climate Migration: The Importance of Nomenclature in the Discourse on 21st Century Mobility* (manuscript on file with author). See also, The Nansen Initiative, *Agenda for the Protection of Cross-Border Displaced Persons in the Context of Disasters and Climate Change* (2015) www.nanseninitiative.org/wp-content/uploads/2015/10/Agenda-Final-Draft.pdf (emphasis in original) ¶ 17.

other words, some of the most adverse impacts of a storm, including displacement, are as (or more) causally related to a community's lack of preparedness, for example, than the natural disaster that has struck. The pertinence of "context" has been extensively noted and discussed in environmental and climate justice literature on the phenomenon of the "second disaster."[6] The "second disaster" describes the aftermath of an environmental disaster in which failures of social infrastructure are glaring and equally, if not more, devastating than the storm, drought, or other initiating weather event.[7]

Migration related to environmental degradation, generally speaking, is not a new phenomenon, even if it is not fully understood. The International Organization for Migration (IOM) recognizes "environmental migrants" and defines them as "persons or groups of persons who, for compelling reasons of sudden or progressive changes in the environment that adversely affect their lives or living conditions, are obliged to leave their habitual homes, or choose to do so, either temporarily or permanently, and who move either within their country or abroad."[8] The link between environmental change and migration is not wholly understood, in many scenarios, because its causes and operation are extremely complex and thus only somewhat reducible to simplifying, general characterization.

It is important not to overstate climate change's primacy in contemporary population movements. Existing policy failures or poor and corrupt leadership, for example, exacerbate the experience of a changing climate. As noted in studies exploring the role of climate change in the ongoing war in Syria and concomitant refugee crisis, for example, drought made more likely because of climate change has devastating consequences "when coupled with preexisting acute vulnerability, caused by poor policies and unsustainable land use practices . . . perpetuated by the slow and ineffective response" by leadership.[9]

6 See IPCC, *Special Report of the Intergovernmental Panel on Climate Change, Managing the Risks of Extreme Events and Disasters to Advance Climate Change Adaptation*, 36, 80 (IPCC 2012) www.ipcc.ch/pdf/special-reports/srex/SREX_Full_Report.pdf.

7 See generally, Manuel Pastor et al., *In the Wake of the Storm: Environment, Disaster, and Race After Katrina* 17 (2006) (describing the phenomenon of the "second disaster" as slow recovery problems that often arise in the process of rebuilding and recuperation); see also, Maxine Burkett, *Just Solutions to Climate Change: A Climate Justice Proposal for a Domestic Clean Development Mechanism*, 56 Buffalo L. Rev. 169 (2008).

8 IOM, *Discussion Note: Migration and the Environment* 2–3 (2007) http://bit.ly/2mAz61h.

9 See Colin P. Kelley et al., *Climate Change in the Fertile Crescent and Implications of the Recent Syrian Drought*, 112(11) Proceedings of Nat'l Academy of Sci. 3241 (2015); See generally, Nives Dolšak and Aseem Prakash, *'Climate Change Did It!' Is a Convenient Excuse, Slate*, 21 October 2016,. www.slate.com/articles/health_and_science/science/2016/10/blaming_natural_disasters_on_climate_change_will_backfire.html accessed March 28, 2017. Dolšak and Prakash provide a critical reminder that blaming climate change solely makes it "easier to escape responsibility for not

Public and private actors behaving irresponsibly may be one of the most potent non-climate drivers of migration as the impacts of climate advance. A significant element of climate-induced migration, however, immediately sets it apart. Climate change is not a static event, it is a process of change, an increase in average global temperature, with concomitant impacts – and the *rate* of change is increasing over time. In other words, what may seem a deeply entangled multi-causal event today will not only expose poor policy and unsustainable management practices, but will also soon have quite a strong and undeniably disproportionate climate signal.[10]

Further complicating planning at the intersection of climate and migration are the varied kinds of movement that climate change induces. In their comprehensive and oft-cited policy paper, *Protecting People Crossing Borders in the Context of Climate Change*,[11] Walter Kälin and Nina Schrepfer identify five scenarios in which climate change may trigger population movements. They are: (i) sudden-onset disasters, such as flooding or storms; (ii) slow-onset degradation, such as rising sea levels and salinization of freshwater and arable land; (iii) the "special case" of slow-onset *disasters*, specifically the impact of rising seas on low-lying small-island States; (iv) governments prohibiting areas for human habituation as they become high risk, or else government-sanctioned relocation; and (v) violence, armed conflict, or unrest over dwindling resources that seriously disturbs public order and triggers migration. Climate change impacts engage with each quite differently.

The scenarios involving sudden and slow-onset disasters and government land use determinations are particularly salient at this time. News of those displaced by Superstorm Sandy, the planned relocation of the Biloxi-Chitimacha-Choctaw tribe of Isle de Jean Charles in Louisiana,[12] and the seemingly inevitable demise of atoll nations like Kiribati[13] have highlighted the surprisingly near-term impacts of climate change on island and coastal

enforcing zoning laws or allowing development on flood plains." Ibid. This critique is similar to earlier literature on the threats of the "second disaster." See IPCC (n. 3).

10 Burkett, *Justice and Climate Migration, supra* note 5.

11 Walter Kälin and Nina Schrepfer, *Protecting People Crossing Borders in the Context of Climate Change Normative Gaps and Possible Approaches*, (2012) Legal and Protection Research Series www.unhcr.org/4f33f1729.pdf.

12 Coral Davenport and Campbell Robertson, *Resettling the First American 'Climate Refugees'*, New York Times, May 3, 2016. www.nytimes.com/2016/05/03/us/resettling-the-first-american-climate-refugees.html.

13 Josh Haner and Mike Ives, *A Remote Pacific Nation, Threatened by Rising Seas*, New York Times, July 2, 2016. www.nytimes.com/2016/07/03/world/asia/climate-change-kiribati.html?_r=0.

communities. Sea-level rise and coastal flooding that compromise entire communities are present crises, with public health consequences. The different scenarios will introduce distinct, though sometimes overlapping, health impacts requiring tailored responses.

In this chapter, therefore, population movement or migration, generally, will describe three different iterations of the phenomenon: (forced) displacement, planned relocation, and migrating populations.[14] Displacement describes the forced, temporary or permanent, movement of people from a location due to one or several factors, most often resulting from natural disaster.[15] Relocation is the planned process of leaving a home community, locality, or country for another permanent settlement. Migration can also describe, specifically, the movement of people from one location to another for the purpose of permanent settlement to obtain comparatively more desirable living conditions.[16] Importantly, there is no single law and policy framework dedicated to the emergent scenarios that climate-induced migration presents. A law and policy framework to address the intersection of climate migration and health will have to draw from analogous circumstances in which migration, prompted by other phenomena, has presented challenges and opportunities for public health infrastructure and outcomes.

Public health infrastructure and positive health outcomes are squarely in the crosshairs of climate change. Researchers currently project that climate change will exacerbate morbidity and mortality, among other things, to differing degrees across varied regions and communities.[17] Generally speaking, two sets of circumstances are relevant to migration and public health: climate risks to health that spur movement, and the health outcomes resulting from forced displacement, planned relocation, and migration.[18] When climate

[14] For policy options, and relevant case studies that might aid communities in coping with displacement, migration, and relocation see, Burkett et al., *Reaching Higher Ground: Avenues to Secure and Manage New Land for Communities Displaced by Climate Change*, Center for Progressive Reform, May 2017.

[15] Like many of the migration scenarios, geographic and temporal scales will vary. So, too, will the degree of autonomy one can exercise in the decision and means for movement. As McMichael, et al. explain, "The spectrum of migration decisions ranges from forced to voluntary, yet most movements are neither entirely one nor the other. The spectrum in response to climate change is likely to be similar." McMichael et al., *An ill wind?*, *supra* note 3, at 647.

[16] Burkett et al., *Reaching Higher Ground*, *supra* note 14.

[17] McMichael et al., *An Ill Wind?*, *supra* note 3, at 646 (noting that "Climate-change–related migration is likely to result in adverse health outcomes, both for displaced and for host populations, particularly in situations of forced migration. However, where migration and other mobility are used as adaptive strategies, health risks are likely to be minimized, and in some cases there will be health gains.").

[18] McMichael et al. discuss both in *An Ill Wind?*, *supra* note 3, at 646.

risks negatively impact human health, spurring population movements, migrating voluntarily may serve as a form of adaptation that improves health outcomes by avoiding those adverse climate change impacts.[19] In some instances, however, it may prove to be a maladaptive or suboptimal response.[20] Migration often results in poor health outcomes for migrants because of a multitude of factors, including marginalization and discrimination, poverty, exposure to disease vectors, malnutrition, and crowding. Host communities might also experience increased stress on health services as they seek to accommodate migrants. This chapter focuses on the second set of circumstances – the health implications and outcomes of those who move as a result of forcible displacement, planned relocation, and voluntary migration as well as the communities that host them.

The relationship between migrating populations and the public health infrastructure that identifies risks and intervenes to prevent public health crises is especially relevant to the United States. There, low socioeconomic status and other forms of social exclusions can impede access to healthcare services.[21] Concerns about immigration status also currently lead people in the United States to avoid interaction with officials or healthcare providers.[22]

Also troubling is the absence of more thorough research on the consequences of climate-induced migration for the health of those impacted, given that the few researchers who have considered the consequences posit that they are "likely to become a major source of human suffering, disability, and loss of life."[23]

### 11.1.2 *Adverse Public Health Effects of Climate-Induced Migration*

#### 11.1.2.1 Climate Change and Public Health

Climate-induced migration will impact all elements of health and, on balance, create more health challenges than it resolves. Health, as defined by the WHO, is not merely the "absence of disease or infirmity," but is a "state

19 Ibid. at 648(citing migration rates associated with famine and drought across international borders in the horn of Africa).

20 See Ibid.

21 Clark Gray et al., *Population Displacement and Mobility in Sumatra after the Tsunami*, http://iussp2009.princeton.edu/papers/90318.

22 Jan Hoffman, *Sick and Afraid, Some Immigrants Forgo Medical Care*, New York Times, June 26, 2017 www.nytimes.com/2017/06/26/health/undocumented-immigrants-health-care.html?_r=0.

23 McMichael et al., *An Ill Wind?*, *supra* note 3, at 647. The authors also note lack of research on impacts to humanitarian concerns and equity.

of complete physical, mental and social wellbeing."[24] This definition includes mental health, and the WHO recognizes that failure to provide basic services and support challenges both mental and physical well-being. As mentioned in the previous section, the research on established links and impacts of climate-induced migration and health is inchoate. As a result, analogous scenarios are instructive: An examination of the health status of migrating and displaced populations during the early phases of movement, for example, might foretell the impacts of climate-induced migration on public health. Chronic diseases that can move with migrants and displaced populations are among the health concerns demonstrated.[25] If climate change spurs large-scale movements as projected, adverse health outcomes will be especially severe if affected migrants arrive in developing regions where public health resources are lacking or inadequate.[26] Negative health outcomes such as increased risk of maternal mortality and malnutrition compound the better known direct (primary) and indirect or delayed (secondary and tertiary) impacts of climate change.

Climate change's chief health maladies may be best understood as acute or chronic physical, mental health, and additional public health impacts.

PHYSICAL IMPACTS Acute physical impacts, as they relate to the movement of people, are diverse. They can entail contracting disease resulting from poor living and working conditions. Vector-borne diseases are of particular concern, as they can have grave impacts and have been the subject of public focus with the recent outbreak of Zika. Climate change can influence the geographic range, seasonality, and incidence rate of infectious diseases. Vector-borne diseases specifically are "augmented"[27] by climate change increasing the rate of vector reproduction, the frequency of vector interactions with humans, and the geospatial range of vector

[24] Sarah Whitmee et al., *Safeguarding Human Health in the Anthropocene Epoch: Report of The Rockefeller Foundation–Lancet Commission on Planetary Health,* 386 Lancet 1973–2028 (2015), http://dx.doi.org/10.1016/ S0140-6736(15)60901-1.

[25] See Seiji Yamada et al., *Sea-Level Rise and the Marshallese Diaspora,* 10(4) Envtl. Justice 94 (2017). See also, McMichael et al., *An Ill Wind?, supra* note 3, at 646 ("Given the nascent status of this research topic, we necessarily draw on previous studies of refugees, people in resettlement schemes, and migrants as analogs for the health issues associated with future climate-related movements.")

[26] Carballo M. Smith and K. Pettersson, *Climate Change and Displacement: Health Challenges,* Forced Migr Rev 31:32–33 (2008) http://ehp.niehs.nih.gov/1104375/.

[27] Kaleem Hawa, Health And Climate Security: interconnected Security Challenges of Climate Change and infectious disease, In: Caitlin E. Werrell and Francesco Femia, *Epicenters of Climate and Security: The New Geostrategic Landscape of the Anthropocene,* 52 (June 2017) www.climateandsecurity.org.

distribution.[28] This has been a concern for the US military,[29] as well as the general public, as insect-transmitted diseases like Zika move into Texas, Florida, and the Gulf Coast region.[30] Further, while Zika and other vector-borne diseases have been a concern for the southeast of the United States, in particular, recent research on the long-range internal migration of people from the coasts to the interior United States suggests a broader national concern regarding "climate augmentation" of disease.[31] Other infectious agents, such as malaria, dengue, tuberculosis, Hepatitis B, and intestinal parasites, can move with migrants and displaced populations, increasing risk for host communities.[32]

Climate change will also impact chronic health conditions such as cancer, hypertension, coronary heart disease, and diabetes, further straining public health infrastructure as people move into new communities. Changes to the background burden of chronic or latent diseases, as well as patterns of immunity will likely occur. Further, disease prevention, health promotion, and use of healthcare services will likely decrease. Among both rural-to-urban and international migrant populations, incidence of chronic diseases has increased relative to that of source populations. Researchers attribute this increase to changes in diet, acculturative stress, physical inactivity, isolation, and increased health risk behaviors such as smoking and hazardous use of alcohol.[33]

MENTAL HEALTH IMPACTS Impacts to mental health resulting from climate change and migration are particularly striking. Research on the mental health of refugees and displaced people reveals elevated rates of negative mental health outcomes.[34] These impacts are felt at the individual and communal levels during the pre- and post-displacement periods. Pre-displacement experiences of violence and trauma are associated with poor mental health. Even uncertainty can cause stress and is a risk factor for psychological distress.[35] Mental health risks also result from post-displacement stressors, such as

[28] Hawa, *Health and Climate Security, supra* note 27, at 53. See also McMichael et al., *An Ill Wind?, supra* note 3, at 647 (citing preliminary evidence of climate- related changes in the geographic and seasonal patterns of several infectious diseases over recent decades, including malaria, dengue fever, and tick-borne borreliosis and encephalitis).

[29] See generally, Hawa, *Health and Climate Security, supra* note 27.

[30] Hawa, at 52.

[31] Mathew E. Hauer, *Migration Induced by Sea-Level Rise Could Reshape the US Population Landscape*, Nature Climate Change, April 17, 2017.

[32] L. A. Palinkas et al., *The Journey to Wellness: Stages of Refugee Health Promotion and Disease Prevention,* J. of Immigrant Health (2003) 5: 19. https://doi.org/10.1023/A:1021048112073.

[33] Bermingham et al. 1999; Burns 2004; Montgomery et al. 2003; Palinkas 1995; Westermeyer 1993.

[34] McMichael et al., *An Ill Wind?, supra* note 3, at 649.

[35] Susan Clayton Whitmore-Williams, et al., *Mental Health and our Changing Climate: Impacts, Implications, and Guidance*, Am. Psychol. Ass'n and ecoAmerica (Mar. 2017).

fragmented social networks and separation from family, loss of familiar social contexts, and poor social connections or diminished sense of belonging.[36] Individuals have experienced acute impacts as a result of disasters, which climate change is forecasted to exacerbate. Acute mental health impacts include increases in trauma and shock, post-traumatic stress disorder (PTSD), anxiety, substance abuse, and depression. Indeed, acute traumatic stress is the most common mental health problem after a disaster.[37] These may have spillover effects, such as strained social relationships and impacts on physical health, including memory loss, sleep disorders, and immune suppression.[38]

The aftermaths of Hurricane Katrina and Superstorm Sandy, natural disasters that induced widespread displacement, are instructive. Among a sample population living in Katrina-affected areas, suicide and suicidal ideation more than doubled, one in six people met the diagnostic criteria for PTSD, and 49 percent of people living in Katrina's path developed an anxiety or mood disorder such as depression.[39] Researchers observed similar impacts immediately after Superstorm Sandy, with 14.5 percent showing symptoms of PTSD.[40] These impacts can persist. Spillover effects were also evident post-Katrina, where domestic violence and child abuse, for example, were found to increase among families that had experienced physical damage to their homes.[41]

The loss of place, itself, triggers negative mental health effects. Researchers have introduced the concept "solastalgia" in response to displacement related to environmental change.[42] Solastalgia describes the feeling that someone is losing a place that is important to them.[43] Researchers note that climate change irrevocably changes people's lived landscapes. A sense of desolation and loss similar to that experienced by people displaced from their home characterizes solastalgia. This impact tends to occur gradually, due to the slow-onset nature of underlying changes to the local environment.

Actual displacement entails a discrete and significant mental health risk of its own.[44] Researchers note that "[i]mmigrants are vulnerable to mental health problems, probably due to the accumulated stressors associated with the move,

36 McMichael et al., *An Ill Wind?*, *supra* note 3at 649.

37 Whitmore-Williams et al., *Mental Health and Our Changing Climate*, *supra* note 35 at 14

38 Fritze et al., *Hope, Despair and Transformation: Climate Change and the Promotion of Mental Health and Wellbeing* (2008).

39 Whitmore-Williams et al., *Mental Health and Our Changing Climate*, *supra* note 35 at 22.

40 Ibid.

41 Ibid. at 24.

42 G. Albrecht et al., *Solastalgia: The Distress Caused by Environmental Change*, 15(Suppl 1) Australas Psychiatry S95–98. (2007) www.ncbi.nlm.nih.gov/pubmed/18027145.

43 Whitmore-Williams et al., *Mental Health and Our Changing Climate*, *supra* note 35 at 25.

44 Ibid.

as well as with the condition of being in exile."[45] Forced displacement can threaten an individual's sense of continuity and belonging and can rupture the sense of connection to place that is integral to personal identity. This can result in feelings of alienation, diminished sense of self, and increased vulnerability to stress.[46] Generally speaking, climate and health researchers have identified loss of autonomy, loss of personal and occupational identity, and loss of control – and the attendant mental health consequences, such as stress and anxiety – as the products of slow or fast and irrevocable climate change impacts.[47] One can reasonably extrapolate from this research and view these health consequences as likely byproducts of climate-induced displacement, migration, and relocation as well. Indeed, the public health community notes that "ecomigration," resulting from the uninhabitability of local conditions, erodes social networks.[48] Further, because strong social networks are associated with health and well-being, the loss of networks constitutes a risk to an individual and community's sense of continuity and belonging – initiating and fueling the cycle described here.

Communities as a whole can also experience negative mental health impacts. This is especially relevant to planned relocation scenarios. Recent climate-related relocations – notably among indigenous communities in the United States[49] – provide clear evidence of how compounded stress from climate change can affect whole communities.[50] In particular, the pressures to move experienced by the Arctic Inuit demonstrate the grave impacts of the degradation and loss of ancestral homes, the impact on their cultural identity, and the prospect of ensuing relocation. The Inuit's altered interactions with the land have resulted in increases in food insecurity, sadness, anger, family stress, levels of conflict, drug and alcohol use, and significant stress associated with contemplation of relocation and actual resettlement.[51] Inuit community members reported that the loss of sea ice would make them feel like they "have

45 Ibid. at 25 (citing Kirmayer et al., *Common Mental Health Problems in Immigrants and Refugees: General Approach in Primary Care,* (2011)).

46 Ibid. (citing Adger et al, *Culutral Dimensions of Climate Change Impacts and Adaptation* (2013)).

47 Ibid. at 25, 27. Moser 2013.

48 Whitmore-Williams et al., *Mental Health and Our Changing Climate, supra* note 35 at 29.

49 These include the Inuit, Biloxi-Chitimacha-Choctaw and Quinault indigenous groups. See Kathy Lynn, Katherine MacKendrick, and Ellen M. Donoghue, Social Vulnerability and Climate Change Synthesis of Literature (2011) USDA 38-48 www.fs.fed.us/pnw/pubs/pnw_gtr838.pdf accessed March 28, 2017. See also, Cunsolo Willox et al. 2012, 2013, 2014; Durkalec et al. 2015 in Whitmore-Williams et al., *Mental Health and Our Changing Climate, supra* note 35 at 32 (noting that the "loss of any community is tragic, but the impact on native communities is particularly notable because it diminishes the cultural heritage").

50 Whitmore-Williams et al., *Mental Health and Our Changing Climate, supra* note 35 at 29.

51 Ibid. at 32.

no health" or they would be "lost" or "go crazy."[52] Chief Albert Naquin of the Biloxi-Chitimacha-Choctaw in Louisiana,[53] who is leading the relocation of his community, has said that climate change has meant a "loss of all our heritage, all our culture."[54] Climate change impacts, and particularly migration, further hamper the community's ability to maintain their health or to heal post-impact. These impacts undermine the social cohesion, cultural strength, food security, and autonomy that protect communities from threats to their well-being. Importantly, however, researchers have also found that social cohesion and social capital can improve communities' resilience to negative mental and physical impacts resulting from climate-related disaster.[55] Irrespective of socio-economic or cultural background, "communities with high levels of social capital and community leadership experience the quickest recoveries after a disaster and the highest satisfaction with community rebuilding."[56]

ADDITIONAL PUBLIC HEALTH IMPACTS Additional threats include the negative impacts climate change has on intergroup attitudes. A 2016 study showed that when temperatures were high, survey respondents "displayed more negative attitudes toward policies to support minorities and immigrants."[57] Another experimental study showed that "people who were thinking about climate change became more hostile to individuals outside their social group (that is, people they consider to be unlike them) and more likely to support the status quo and its accompanying social inequities."[58] Further, that hostility toward outsiders affirmed an individual's "group identity" in

[52] Durkalec et al. (2015); Whitmore-Williams et al., *Mental Health and Our Changing Climate, supra* note 35 at 32.

[53] Discussed further *infra*.

[54] Coral Davenport & Campbell Robertson, *Resettling the First American Climate Refugees*, N.Y. Times 3 (May 2016) http://www.nytimes.com/2016/05/03/us/resettling-the-first-american-climate-refugees.html, Whitmore-Williams et al., *Mental Health and Our Changing Climate, supra* note 35 at 32.

[55] Whitmore-Williams et al., *Mental Health and Our Changing Climate, supra* note 35, at 29; Daniel P. Aldrich and Michelle A. Meyer, *Social Capital and Community Resilience*, American Behavioral Scientist Vol 59, Issue 2, 254–269 (2014).

[56] Whitmore-Williams et al., *Mental Health and Our Changing Climate, supra* note 35, at 29; Daniel P. Aldrich and Michelle A. Meyer, *Social Capital and Community Resilience*, *American Behavioral Scientist* Vol 59, Issue 2, 254–269 (2014).

[57] Alexander H. Cohen & James S. Krueger, *Rising Mercury, Rising Hostility: How Heat Affects Survey Response*, Field Methods 28(2), 133–152 (2016) doi:10.1177/1525822X15627974 in Whitmore-Williams et al., *Mental Health and Our Changing Climate, supra* note 35 at 30.

[58] Immo Fritsche et al., *Global Warming is Breeding Social Conflict: The Subtle Impact of Climate Change Threat on Authoritarian Tendencies*, J. of Envtl. Psychol., 32(1), 1–10 (2012) in Whitmore-Williams et al., *Mental Health and Our Changing Climate, supra* note 35 at 30.

the face of a perceived threat.[59] "In a vicious cycle," explain authors of a comprehensive policy paper on climate change and health, "lower levels of social cohesion and connectedness, greater social inequalities, lack of trust between community members and for institutions, and other factors that inhibit community members from working together are associated with intergroup aggression."[60] This kind of intergroup aggression may have and will likely continue to hamper or upend sound policy responses to the public health effects of climate migration.

This kind of xenophobia represents one of a few dangerous distractions potentially resulting in misplaced resources and priorities vis-à-vis current public health needs. Climate change presents other physical impacts that divert resources from much needed existing services that face further stress. Government resources diverted to repair disaster-damaged infrastructure, for example, are no longer available for mental health agencies, educational institutions, and other services that provide social support.

Vulnerable populations are of particular concern. Prior research shows that an individual's identity or a community's composition can heighten adverse physical and mental health impacts. Race, gender, age (children and elderly), disability, disease burden, and poverty or historical oppression can increase vulnerability. During Hurricane Katrina, for example, fatalities were concentrated among elderly people, representing 49 percent of total deaths. African Americans had mortality rates 1.7 to four times higher than those of their white counterparts.[61] Women in refugee camps face additional health risks, related to reproduction and gender-based violence, among other things.[62] Children are at increased risk resulting from disruptions to the education system, such as occurred after Hurricane Katrina,[63] and losses related to social support networks.[64] Those burdened by disease, whether undernutrition or a chronic disease like diabetes, are substantially affected during the "emergency phase" of displacement.[65] Displacement also disproportionately affects the poor, with a range of health implications such as leaving PTSD untreated. Again, Katrina illustrates the consequences of lack of preparedness and insufficient response strategies coupled with

59 Whitmore-Williams et al., *Mental Health and Our Changing Climate*, *supra* note 35 at 30.

60 Fran H. Norris et al., *Community Resilience as a Metaphor, Theory, Set of Capacities, and Strategy for Disaster Readiness*, Am. J. of Community Psychol., 41, 127–150 (2008) doi:10.1007/s10464-007-9156-6

61 Whitmee et al., *Safeguarding Human Health in the Anthropocene Epoch*, *surpa* note 24, at 1995.

62 Ibid.

63 Whitmore-Williams et al., *Mental Health and Our Changing Climate*, *supra* note 35 at 35.

64 Ibid.

65 Whitmee et al., *Safeguarding Human Health in the Anthropocene Epoch*, *surpa* note 24, at 1995.

poor communication and inadequate access to services that support public health, particularly needed by those in low-income communities.[66]

### 11.1.2.2 Public Health and Climate Migration Scenarios

Climate change will influence population movement in distinct ways, each with their own temporal and geographical scales. Looking more closely at three scenarios reveals the public health impacts at stake in each. Developing country examples tend to predominate, as displacement, migration, and refugee scenarios – particularly those coupled with environmental degradation or disaster – are more prevalent in those regions. Researchers note that many places that will receive climate migrants are in developing regions where climate change will further strain public health resources that are already "lacking or inadequate."[67] This section will first look at instances of forced displacement globally and in the United States, then discuss planned relocation in the US Gulf Coast region, and then consider the circumstances of Pacific Islander migration and the potential relocation of individuals to host locations, like Hawai'i.

FORCED DISPLACEMENT – AFTER THE STORM McMichael et al. note that environmental changes and extreme climate events, which compromise livelihoods and introduce extreme health risks, will also drive people from their places of residence, spurring forced displacement.[68] That displacement will be, for the most part, internal and over short distances rather than cross-border. Forced displacement disrupts an individual's access to basic necessities for good health – such as food and water – as well as healthcare infrastructure and social networks.[69] The disruption to social networks produces mental health impacts and high-risk situations for disease transmission, including the spread of sexually transmitted diseases.[70] For refugees and those forcibly displaced under suddenly changed circumstances, infectious diseases have been a major cause of morbidity and mortality.[71] Diseases, such as measles, meningitis, and malaria, thrive in camps or settlements notable for, among other things, their overcrowding, poor ventilation, and inadequate water, sanitation, and access to immunization and healthcare

66 Ibid.
67 McMichael et al., *An Ill Wind?*, *supra* note 3, at 648 Carballo M. Smith and K. Pettersson, *Climate Change and Displacement*, *supra* note 26.
68 Ibid. at 646.
69 McMichael et al., *An Ill Wind?*, *supra* note 3, at 648.
70 Ibid. at 649.
71 Ibid. at 648.

facilities, spurring outbreaks.[72] As stated previously, vulnerable groups such as children, women, the elderly, and those with an existing disease burden have increased risks of adverse health outcomes resulting from forced displacement.[73]

Superstorm Sandy underscores the dangers of disruption to services and damage to healthcare facilities even in a developed country context. Coastal flooding during Sandy damaged and rendered inoperable critical healthcare infrastructure, including hospitals, nursing homes, adult-care facilities, primary and mental healthcare facilities, and pharmacies.[74] Many of these facilities were inoperable for extended periods. Generally speaking, for those displaced by disaster and evacuated from flood-prone neighborhoods, living in shelters for extended periods is associated with increased risk of negative health outcomes, including communicable diseases and complication from chronic health conditions with the interruption to medical care, including loss of medical record information and medications.[75]

PLANNED RELOCATION – A LOOK AT THE GULF COAST Planned relocation serves as an affirmative response to degraded conditions resulting from climate change. Ferris et al. define planned relocation as a "process in which persons or groups of persons move or are assisted to move away from their homes or places of temporary residence, are settled in a new location, *and* provided with the conditions for rebuilding their lives."[76] The process is carried out under the authority of the State, occurs internally, and is "undertaken to protect people from risks and impacts" of climate change. Relocations may be carried out at the individual, household, and/or community levels.[77]

These relocations are likely to involve large populations, at the community scale, that resettle within their own country. Relocation is meant to "reduce disaster risk" or "adapt to climate change," though it also risks the disruption of livelihoods and the loss of income, social networks, and cultural heritage.[78] Indeed, one of the more publicized climate-related relocations, the planned

72 Bart De Bruijn , *The Living Conditions and Well-Being of Refugees*, U.N. Development Program (2009); Murray et al., *The Australasian Society for Infectious Diseases Guidelines for the Diagnosis, Management and Prevention of Infections in Recently Arrived Refugees: An Abridged Outline*, Med. J. Australia 190(8):421–425 (2009); Rajabali et al., *Communicable Disease Among Displaced Afghans: Refuge Without Shelter* Nat. Rev. Microbiol. 7(8):609–614 (2009)

73 McMichael et al., at 648 (Toole 2005).

74 Patrick L. Kinney et al., *New York City Panel on Climate Change 2015 Report, Chapter 5: Public Health Impacts and Resiliency*, Ann. N.Y. Acad. Sci. 67–88, 72 (2015).

75 Kinney et al., *Chapter 5: Public Health Impacts and Resiliency, supra* note 74, at 72.

76 Elizabeth Ferris et al., A *Toolbox: Planning Relocations to Protect People from Disasters and Environmental Change*, https://isim.georgetown.edu/Planned-Relocations.

77 Ibid.

78 Ibid.

relocation of the Biloxi-Chitimacha-Choctaw tribe of Isle de Jean Charles in Louisiana,[79] has highlighted the surprisingly near-term impacts of climate change on island and coastal communities and concomitant damage to cultural heritage. Researchers note that few resettlement schemes have improved the lives and well-being of those who have resettled.[80] Indeed, some argue that for reasons of health alone, communities should avoid planned resettlement to reduce vulnerability because resettlements entail substantial risks to health and well-being.[81]

Nonetheless, planned relocation is occurring throughout the globe and, given current climate forecasts, may occur still more even if it is a scheme of last resort.[82] Pacific Island relocation planning, in places like Kiribati, has most often captured the public imagination. US domestic relocation scenarios are equally gripping. From the Arctic to the Pacific Northwest to the Mississippi Delta region, the US federal government, with state and local partners, has considered and is facilitating movement from high-risk areas.[83] As introduced in Section 11.1.1, Louisiana's Isle de Jean Charles is among the first set of communities undergoing planned relocation.[84] Recognizing the significant loss of land – a result of rising seas colliding with decades of channelization and oil and gas development – and the increased risk of staying in place as each hurricane season approaches, the Department of Housing and Urban Development funded the relocation of island residents.[85] This is a first for the US government, and a number of questions will arise if the government assists more communities' moves to higher ground. Paramount among these questions is, how do people relocate in a way that affords the opportunity to thrive, particularly in light of the myriad, documented health impacts that have arisen in comparable scenarios?

To answer this question researchers have drawn from resettlement schemes for development initiatives and infrastructure investments to mimic climate-

79 Davenport and Robertson, *Resettling the First American 'Climate Refugees'*, N.Y. Times, www.nytimes.com/2016/05/03/us/resettling-the-first-american-climate-refugees.html.

80 Roli Asthana, *Involuntary Resettlement: Survey of International Experience*, Econ. Polit. Wkly. 31:1468–1475 (1996); Michael Cernea, *The Risks and Reconstruction Model for Resettling Displaced Populations*, World Dev. 25(10):1569–1587 (1997); H. Kloos, *Health Aspects of Resettlement in Ethiopia*, Soc. Sci. Med. 30(6):643–656 (1990).

81 Ibid. at 649.

82 Michael Cernea, *The Risks and Reconstruction Model for Resettling Displaced Populations*, World Dev. 25(10):1569–1587 (1997).

83 See Burkett, *Justice and Climate Migration*, *supra* note 5.

84 Carolyn Van Houten, *The First Climate Refugees in the U.S. Race Against Time*, National Geographic May 25, 2016, http://news.nationalgeographic.com/2016/05/160525-isle-de-jean-charles-louisiana-sinking-climate-change-refugees/.

85 Ibid.; United States Department of Housing and Urban Development, *National Disaster Resilience Competition Grantee Profiles* (January 2016) 7–10 https://portal.hud.gov/hudportal/documents/huddoc?id=NDRCGrantProf.pdf.

triggered relocation. These situations involve resettlement of people who had no control over the destination and process of movement.[86] While the hope is that law and policy frameworks will ensure deep community engagement in decisions regarding relocation destinations and process, the need to move because of climate change is often not the community's choice or the byproduct of the community's conduct. Indeed, like many other injustices of climate change, those least responsible for the events forcing last-resort relocation scenarios are those who must avail themselves of a suboptimal adaptation measure relocation.[87] Food insecurity and unsafe water supplies are associated with planned resettlement.[88] Further, relocated individuals and communities encounter disease vectors "for which they have not developed behavioral controls or population level immunity."[89] Regarding mental health, resettlement schemes have produced multigenerational maladies.[90] Research documents, for example, the significant cultural disruption, depression, alcoholism, and dietary disorders that indigenous populations in North America and Australia suffer as a result of forced resettlement.[91]

The experience of the Isle de Jean Charles could be a harbinger of things to come for other Gulf Coast communities.[92] Petkova et al. have found that: "Through myriad pathways, climate change is likely to make the Gulf Coast less hospitable and more dangerous for its residents, and may prompt substantial migration from and into the region."[93] Severe erosion, subsidence, and the "ever-present potential for large-scale industrial accidents" collide with seasonal flooding, hurricanes, and tropical storms, inherent threats to the already low-lying region.[94] Researchers note that disease outbreaks inevitably follow large-scale disasters that impact public health systems, as was the case with a substantial increase in reported West Nile cases after Hurricane Katrina and a norovirus outbreak among evacuees in Texas.[95]

86 McMichael et al., *An Ill Wind?*, *supra* note 3, at 649.

87 See Burkett, *Justice and Climate Migration*, *supra* note 5.

88 Michael Cernea, *The Risks and Reconstruction Model for Resettling Displaced Populations*, World Dev. 25(10):1569–1587 (1997).

89 Ibid.

90 Ibid.

91 Ibid.

92 US Gulf Coast states include Texas, Louisiana, Mississippi, Alabama, and Florida. Petkova et al., *Climate Change and Health on the U.S. Gulf Coast: Public Health Adaptation is Needed to Address Future Risks*, at 9342.

93 Elisaveta P. Petkova et al., *Climate change and Health on the U.S. Gulf Coast*, *supra* note 92, at 9342, 9348.

94 Ibid. at 9343.

95 Ibid. at 9346.

For those who must migrate, severe and lasting mental health effects are anticipated because of "place attachment."[96] While an instance more akin to forced displacement, the experience of Gulf Coast residents after Katrina is instructive here. The loss of the bond to place had a negative effect. Roughly 1.5 million Gulf Coast residents were evacuated, and, as of 2015, thousands were still displaced.[97] Petkova et al. urge consideration of these kinds of psychological stressors and impacts associated with climate change and migration. They also note that permanent relocation will further stress "vital health and social services, including affordable housing, on the destination communities, where, for instance, access to health care might be reduced for all sectors of the population."[98]

MIGRATION – A LOOK AT THE PACIFIC For the purposes of this discussion, migration describes the movement of peoples, whether that movement is proactive, in anticipation of a significant stressor, or reactive, in response to a climate-related impact. Similar to the other scenarios, it will more likely involve internal, within country, migration and, in many cases, contribute to urbanization.[99] This is already occurring, particularly in the developing world.[100] The health outcomes, however, are not unidirectional under this scenario. They depend on the health of the migrant. People who migrate to poor urban settlements may experience "environmental, physical, and psychosocial health threats."[101] The "healthy migrant" effect, however, will in part determine the outcome of an individual migrant. The effect, described in demographic studies, shows that those in better health "elect to and are positively selected for migration compared with those in poor health."[102] Indeed, as noted by McMichael et al., "[i]n response to the health and physical safety risks posed by climatic environmental disasters, both acute and sustained, migration can be a survival strategy."[103] Rural to urban migrants, for example, are more likely to be young, employed, and better educated and, consequently, healthier than

96 Ibid. at 9348. "Place attachment is a bond between people and their natural environment, based on cognition and affect."

97 The population of the New Orleans region in 2010 declined by 25.4 percent and migrants remained dispersed throughout the United States, with as many as 250,000 in neighboring Texas. Petkova et al., *Climate Change and Health on the U.S. Gulf Coast, supra* note 92, at 9348.

98 Petkova et al., *Climate Change and Health on the U.S. Gulf Coast, supra* note 92, at 9348.

99 McMichael et al., *An Ill Wind?, supra* note 3, at 648.

100 Ibid.

101 Ibid. at 650 (citing Jon Sward, *Migration and Climate Change: How Will Climate Shifts Affect Migration Trends? Briefing Paper No. 12. Brighton,* Development Research Centre on Migration, Globalization and Poverty (2008)).

102 Ibid.

103 Ibid. at 647–648.

nonmigrants.[104] This does not, however, describe contemporary migration from the Pacific to the United States, discussed further in this section. Further, the effect is "attenuated" the longer the migrant resides in the host community.

Migrants, generally, are at higher risk of disease – both during transition and in residence in the new location. Rural to urban and international migrants are at higher risk of developing chronic diseases compared with those in their home community.[105] Among Mexican-born migrants, for example, "length of U.S. residency is directly associated with increasing risk of obesity."[106] Researchers attribute poor outcomes to changes in diet, stress related to cultural assimilation, physical inactivity, and isolation, among other things.[107] Further, poor chronic disease outcomes result from lack of access to healthcare services, including preventative care and early diagnosis.[108] During transition, rates of infection with diseases like tuberculosis, Hepatitis B, and malaria might increase as infectious agents can move with migrants, thus presenting greater risk to host populations.[109] Current patterns of migration to Hawai'i, for example, have introduced non-endemic infectious diseases like tuberculosis.[110] As discussed in this section, however, screening for infectious diseases is most robust at country borders. For internal migration, screening will be more difficult.[111]

The Pacific provides examples of potential planned relocation, discussed earlier, as atoll communities and entire nations contemplate moving to higher ground. The region also sheds light on individual migration from island nations to population centers in the United States, relative metropolises compared to their home communities. Pacific Islands, generally, are among the most vulnerable to the health impacts of climate change, compounding their limited capacity to manage and adapt to the risks it presents.[112] Their relative vulnerability results from their "unique geographic, demographic, and socioeconomic characteristics combined with their exposure to changing weather patterns associated with climate change," shared with many small island developing states. Climate and health impacts include "trauma from extreme weather events, heat-related illnesses, compromised safety and

104 Ibid. at 650.

105 Ibid.

106 Ibid. (citing Carlos H. Barcenas et al., *Birthplace, Years of Residence in the United States, and Obesity Among Mexican-American Adults*, Obesity 15(4):1043–1052 (2007)).

107 Ibid.

108 Ibid.

109 Palinkas et al., *The Journey to Wellness*, *supra* note 32.

110 State of Hawaii, Department of Health, *Tuberculosis Control Program, Data & Statistics* (Dec. 1, 2016) http://health.hawaii.gov/tb/data-statistics/.

111 McMichael et al., *An Ill Wind?*, *supra* note 3, at 650.

112 McIver et al., *Health Impacts of Climate Change in Pacific Island Countries*, *supra* note 2, at 1707.

security of water and food, vector-borne diseases, non-communicable diseases, population pressures, and health system deficiencies."[113] The climate-related health risks of noncommunicable diseases, disorders of mental and psychosocial health, and food-borne illnesses, such as ciguatera, are a concern in the Pacific "to an extent not documented elsewhere in the world."[114] The US Affiliated Pacific Islands are no exception.

Because of the unique history of the US Affiliated Pacific Islands, the health of some migrants is poor and, in fact, induces migration to the United States. As mentioned in this section, Pacific Island countries already experience the highest rates of noncommunicable diseases in the world and climate change may act as an additional driver.[115] Further, there is evidence that climate-induced relocation is already taking place in the region, internally and cross-border,[116] introducing further physical and psychosocial health impacts.[117] "In the case of the Marshallese, [for example,] the history of nuclear testing and displacement within the Marshall Islands has contributed to the prevalence of diabetes in particular. By impairing immunity, diabetes itself contributes to infectious diseases such as tuberculosis."[118]

Marshallese have been migrating to the United States in increasing numbers since 2000. Under the Compacts of Free Association (COFA), the citizens of the Marshall Islands can enter into the United States with the right to employment, without a visa and without requirements for health screening.[119] Despite increased in-migration, however, support for the Marshallese community is inconsistent at best and generally not commensurate with the resource needs of migrants and host communities. In 1996, Congress passed broad welfare reform (Personal Responsibility and Work

113 Ibid. at 1707.

114 Ibid. at 1711.

115 Ibid. at 1712.

116 Thomas Birk & Kjeld Rasmussen, *Migration from Atolls as Climate Change Adaptation: Current Practices, Barriers and Options in Solomon Islands*, Nat. Recourses Forum 38:1–13 (2014) doi:10.1111/14778947.12038; J.T. Locke, *Climate Change-Induced Migration in the Pacific Region: Sudden Crisis and Long-Term Developments*, Geogr. J. 175:171–180 (2009) doi:10.1111/j.1475–4959.2008.00317.x.

117 Colin D. Butler et al., Mental health, cognition and the challenge of climate change, In: Colin D. Butler, *Climate Change and Global Health* (Wallingford, 2014) CABI, 251–259; Rafael Reuveny, *Climate Change-Induced Migration and Violent Conflict*, Polit. Geogr. 26:656–673 (2007) doi:10.1016/j.polgeo.2007.05.001.

118 Yamada et al., *Sea-Level Rise and the Marshallese Diaspora*, *supra* note 25, at 3.

119 Compact of Free Association, *Agreement between the United States of America and the Marshall Islands*, U.S.-Marsh. Is., April 30, 2003, 80 Stat. 271 ("A person admitted to the United States under the Compact, or the Compact, as amended, shall be considered to have the permission of the Government of the United States to accept employment in the United States.") Palau and the Federated States of Micronesia as well.

Opportunity Reconciliation Act), which limited certain federal benefits, such as Medicaid, previously available to COFA migrants. This left destination states like the State of Hawai'i with the continued responsibility to support COFA migrants under its state-funded Medicaid program.[120] "Compact Impact" funding from the federal government to the State of Hawai'i to offset health and educational costs has since been inadequate.[121]

Migration from the Marshall Islands highlights how climate, migration, and health intersect.[122] Because of the wide range of health maladies from which Marshall Islanders suffer, the need to access healthcare already drives out-migration from the Marshalls to Hawai'i and the continental United States. The State of Hawai'i continues to struggle to meet the healthcare needs of Marshallese migrants. Further, the lack of sufficient language services by healthcare providers and literacy and cultural concerns compound these health challenges to migrants as well as the care systems of the receiving community.[123] Climate change will likely impact similarly situated sending communities and, potentially, further increase the costs of service to at-risk populations in receiving communities.

## 11.2 RELEVANT LAW AND POLICY FRAMEWORKS

The law and policy that addresses the synergistic effects of climate change and ill-prepared public health infrastructure is quite insubstantial. Like many areas of climate law, resilience will require augmenting existing infrastructure with greater capacity to anticipate and respond to climate crises as well as climate "surprises," those events that are unprecedented but are probable and/or severe. New law and policy scenarios to address the public health consequences of increased internal migration from cities or territories that are no longer viable may be necessary. This section looks at the laws that are relevant to the concerns climate change raises regarding public health and both internal and cross-border population movements.

120 Dina Shek and Seiji Yamada *Health Care for Micronesians and Constitutional Rights*, 70(11) Ha. Medical J. (2011), at 4. In the face of budget constraints, the State removed these people from its Medicaid rolls, though legal action later forced the state to resume funding.

121 Ibid. at 4–5.

122 See generally, Yamada et al., *Sea-Level Rise and the Marshallese Diaspora*, *supra* note 25, at 3–4.

123 Ibid. at 3.

### 11.2.1 *Public Health Law*

Public health infrastructure supports end-to-end health services, from prevention to treatment. As such, public health as it relates to climate change and migration seems to fall within the ambit of existing infrastructure, and additional adaptation schemes can buttress that infrastructure. The public health framework comprises surveillance, including authority to mandate medical examination or testing, medical countermeasures such as vaccines, antibiotics, and quarantine, as well as nonmedical responses. The latter encompasses combatting public nuisances that are sources of infection, such as mosquitoes, and public education about Lyme disease prevention or risks of exposure to toxic algae. Nonmedical responses become especially pertinent and urgent during disaster recovery.

The framework for addressing climate-induced health problems aims to prevent morbidity and mortality through three tiers of activities.[124] Primary preventions consist of monitoring, surveillance, intra and interagency communication, climate literacy education, and improving land use planning and water and waste management.[125] Secondary preventions concern climate change mitigation, on balance, with a focus on behavioral adaptation such as utilizing alternative forms of transportation, decreasing meat consumption, and reversion to traditional farming practices. Tertiary preventions are the treatment of diseases that climate change exacerbates.[126]

Looking ahead, efforts to strengthen essential services of public health most relevant to climate change are multiple and promising.[127] A few examples include the development of policies and plans that support individual and community health initiatives, such as emergency preparedness plans and municipal heatwave preparedness plans. Enforcement of laws and regulations that protect health and ensure safety, modification of these laws as needed, as well as the drafting and implementation of climate change and health laws and regulations can bolster these disaster response plans.

#### 11.2.1.1 Jurisdiction

Public health governance engages multiple jurisdictions, with relevant laws and actors at the federal, state, and local level. At the federal level the Centers

124 See e.g., *Climate Change-Induced Public Health Vulnerabilities for the State of Hawaii*, at 10 (on file with author).

125 Ibid.

126 Ibid.

127 See generally, Howard Frumkin et al., *Climate Change: The Public Health Response*, 98(3) Am. J. Public Health. 435–445 (2008 March) doi: 10.2105/AJPH.2007.119362.

for Disease Control (CDC) conducts research and provides guidance to the states to help manage public health risks.[128] The Department of Health and Human Services, under which the CDC sits, is also a key agency at the federal level. Climate change, and certainly the population movements it triggers, is of particular concerns to additional federal players, including the Department of State and the Department of Defense (DoD). With respect to health, the DoD has noted that the spread of pathogens and infectious disease is a major security concern, particularly as climate may "drive conflict and refugee migratory dynamics, increasing health risks for U.S. humanitarian assistance and disaster relief personnel."[129] All states have departments of health[130] that coordinate environmental health and administration of state and federal environmental laws alongside conventional healthcare services. Yet, public health is primarily controlled at the local level.[131] And, the state- and local-level initiatives to guide public health as it relates to climate change impacts are proliferating.[132]

#### 11.2.1.2 Healthcare Providers

Doctors, nurses, medical practices, hospitals, and hospital systems are particularly important to the success of long-term public health management, particularly as climate change introduces greater threats. Healthcare providers are some of the most trusted figures in society and, as such, have a unique opportunity to convey the seriousness of various climate change impacts on health to the public generally and their patients in particular. Providers can take a more active role in the well-being of individuals and communities when they are in transition and resettle in a new location. This is relevant to service providers in high-risk and sending communities as well as receiving areas. Providers can coordinate actions to change the health service jurisdiction by, for example, transferring information to new providers when an individual or group migrates.[133] In the relocation area, they can add or expand health

128 Center for Disease Control, *Climate and Health Policy*, www.cdc.gov/climateandhealth/policy.htm.

129 Hawa, *Health and Climate Security*, *supra* note 27, at 51–52.

130 Environmental Protection Agency, *US Agencies*, www.epa.gov/home/health-and-environmental-agencies-us-states-and-territories.

131 *New York State Public Health Manual*, 6 (2011), http://bit.ly/2mDg4aM; see also *Public Health Bench Books*, www.cdc.gov/phlp/publications/type/benchbooks.html for other examples of local public health laws and initiatives.

132 See Chapter 3.

133 See Ferris et al., *Planning Relocations to Protect People from Disasters and Environmental Change*, *supra* note 76, at 39.

centers in the host community and provide emergency assistance during disasters.[134]

Failure to provide adequate or comprehensive services can be as detrimental and determinative of outcomes as competently administered services can effect positive outcomes. As noted, for example, the lack of appropriate language services by healthcare providers coupled with literacy and cultural issues may affect a migrant's ability to efficiently navigate the healthcare system in Hawai'i.[135] Insufficient federal support, consistent with COFA obligations, regularly compounds that failure.[136]

Some jurisdictions are currently, and will be, better prepared than others to address climate migration and health-related challenges. Currently, however, gaps predominate, such as the lack of integration and coordination between relevant agencies and other stakeholders, and information gaps in disease tracking.[137] The localized nature of governance in public health may foster these gaps. In particular, poor, urban, or otherwise underserved areas may not have the resources to dedicate to proactive planning.

### 11.2.2 *Immigration and Migration Laws*

There is no federal or international law that covers the full scope of climate-induced displacement, migration, or relocation scenarios. It is important to note that most population movement will occur internally, possibly presenting additional burdens to the existing public health infrastructure.[138] For cross-border movement, there are no recognized legal frameworks that address environmental migration.[139] Even a patchwork approach that is just and

[134] Ibid.

[135] See Shek and Yamada, *Health Care for Micronesians and Constituional Rights*, *supra* note 120, at 6.

[136] See Shek and Yamada, *Health Care for Micronesians and Constituional Rights*, *supra* note 120, at 5–6. It is estimated that in 2007 Hawaii spent over $100 million annually in uncompensated education and health costs attributed to COFA migrants. The State of Hawaii receives ten million dollars in federal aid annually to offset these costs. See ibid. at 5.

[137] Centers for Disease Control and Prevention, CDC's *Climate-Ready States & Cities Initiative*, www.cdc.gov/climateandhealth/climate_ready.htm ("CDC's Climate-Ready States and Cities Initiative is helping 16 states and 2 cities develop ways to anticipate these health effects by applying climate science, predicting health impacts, and preparing flexible programs. CDC will help states and cities partner with local and national climate scientists to understand the potential climate changes in their areas. CDC will assist states and cities in developing and using models to predict health impacts, to monitor health effects, and to identify the area's most vulnerable to these effects.")

[138] A more in-depth discussion of increased impacts to the public health infrastructure and necessary amendments, particularly at the state and local level, is beyond the scope of this chapter.

[139] McMichael et al., *An Ill Wind?*, *supra* note 3, at 649.

timely remains elusive.[140] To the extent that the issue is of concern to the United States, it will impact national interests abroad or military or humanitarian engagements in so-called hotspots. There will likely be some cross-border migration into the United States; however, by all indications, these numbers will likely be a small proportion of the non–climate-related cross-border migration and, of course, the US population *en toto*.[141] In other words, the migration that would implicate domestic and international migration and refugee laws are manageable, *if managed.*

The rest of this section discusses relevant domestic and international laws in turn. Those discussions highlight health-related policies and procedures governing migration into the United States, with a focus on the visa and admission processes, and relevant international and multilateral migration and refugee laws.

#### 11.2.2.1 Domestic Immigration Laws and Health

Unlike cross-border migration, internal migration can occur freely and without incident. US citizens [142] can move freely between states under the privileges and immunities clause.[143] The absence of national or international legal frameworks squarely relevant to climate-induced international migration renders the migration status of those who cross borders uncertain and without recourse if faced with hostile reception.[144] This uncertainty coupled with the risk of poor reception can exacerbate mental health impacts.[145]

Migration into the United States and the visa process is within the domain of the federal government. Congress determines immigration laws, generally, and can place restrictions on immigration as long as those restrictions are constitutional. The State Department (DOS) and the Department of Homeland Security (DHS) are the central agencies that administer the laws and policies on the admission of foreign nationals.[146] Foreign nationals who wish to

140 *See* Burkett, *Justice and Climate Migration*, *supra* note 5.

141 See, e.g., Jaya Ramji-Nogales, *Migration Emergencies*, 68 Hastings L.J. 609, 618 (2017) ("While it is true that existing institutions have been overwhelmed by the absolute numbers of migrants, these migrant flows are small in relative terms: The migrant flows described above constituted approximately 0.2% of the European Union's population, 0.02% of the U.S. population, and 0.2% of Australia's population.")

142 Art. IV, § 2, Cl. 1.

143 US Const. Art. IV, § 2, Cl. 1.

144 McMichael et al., *An Ill Wind?*, *supra* note 3, at 649.

145 Ibid.

146 General authority for DOS is found in 8 U.S.C. § 1105 (Current through P.L. 115-43) and generally, 8 C.F.R. § 301 et seq; General authority for DHS to admit foreign nationals is found in 8 U.S.C. § 1103 (Current through P.L. 115-43), 8 C.F.R. § 2.1 (Current through Aug. 3,

immigrate must obtain a visa and submit to an inspection for admission. Health is one of a few reasons federal screeners can deem a foreign national inadmissible.[147]

Health-grounds for possible exclusion include having a communicable disease, lacking required vaccinations,[148] presenting a physical or mental disorder, and evincing drug abuse or addiction.[149] Regarding communicable "diseases,"[150] the Immigration and Nationality Act (INA) states that the Secretary of Health and Human Services (HHS) will determine which communicable diseases of public health significance will trigger inadmissibility.[151] The CDC, under HHS, coordinates protection against communicable diseases among foreign nationals and provides technical instructions to civil surgeons and panel physicians who conduct medical examinations for immigration.[152] Presence of mental disorder as grounds for possible exclusion is also notable in light of prior research emphasizing the mental health impacts on migrants and refugees leaving their crippled home and transiting to a new destination.

Presence of one of the health-related grounds for inadmissibility is not dispositive. Further analysis of the virulence of the disease is possible. In addition, the INA grants the Secretary of DHS the discretionary authority to waive some of the health-related grounds for exclusion.[153] For example, the Secretary by waiver might admit a close relative of a US citizen, permanent resident, or an alien issued an immigrant visa.[154] Further, the Secretary might waive inadmissibility for

2017), 8 C.F.R. § 100.1. DOS's Bureau of Consular Affairs (Consular Affairs) is the agency responsible for issuing visas, DHS's US Citizenship and Immigration Services (USCIS) is charged with approving immigrant petitions, and DHS's Customs and Border Protection (CBP) is tasked with inspecting all people who enter the United States. Ibid.

147 8 U.S.C. § 1182(a)(1)(A). Other grounds for exclusion include criminal history; security and terrorist concerns; public charge (e.g. indigence); seeking to work without proper labor certification; illegal entry and immigration law violations; ineligible for citizenship; and aliens previously removed. 8 U.S.C. § 1182.

148 8 U.S.C. § 1182(a)(1)(A). Vaccinations are statutorily required for mumps, measles, rubella, polio, tetanus, diphtheria, pertussis, influenza type B and hepatitis B.

149 8 U.S.C. § 1182(a)(1)(A).

150 The regulation includes categories of diseases in the list, including those determined to pose a threat to the U.S. 42 C.F.R. § 34.2(b).

151 8 U.S.C.§ 1182(a)(1)(A)(i); 42 U.S.C. § 264. 42 C.F.R. § 34.2(b); USCIS, *Communicable Diseases of Public Significance*, www.uscis.gov/policymanual/HTML/PolicyManual-Volume8-PartB-Chapter6.html (last visited Aug. 7, 2017). The four diseases are gonorrhea, infectious leprosy, infectious syphilis, and active tuberculosis.

152 42 C.F.R. § 70.2. The CDC, in conjunction with Customs and Border Protection (CBP) in DHS, operates twenty quarantine stations and has health officials on call for all ports of entry. There are 328 official ports of entry (POE) in the United States, including sixteen preclearance offices in Canada, Ireland, the Caribbean, and the United Arab Emirates. Wasem, at 8.

153 8 U.S.C.§ 1182(g).

154 8 U.S.C.§ 1182(g); Wasem, *supra* note 152, at 6.

those lacking proper vaccination and those who suffer from physical or mental disorders, under certain circumstances.[155] Finally, if he or she finds it in "the national interest," the Secretary may waive the application of any health-related grounds for inadmissibility.[156] It is plausible that the climate-triggered movement of peoples could rise to this level of "national interest."

### 11.2.2.2 International Migration Law

Relevant international, and multilateral, migration law consists primarily of the UN Convention and Protocol on the Status of Refugees, and more recent elements of the UN Framework Convention on Climate Change (UNFCCC).[157] The UN Convention and Protocol on the Status of Refugees defines a refugee as any person who, "owing to well-founded fear of being persecuted for reasons of race, religion, nationality, membership of a particular social group or political opinion, is outside the country of his nationality and is unable or, owing to such fear, is unwilling to avail himself of the protection of that country."[158] Because environmental migrants generally do not migrate urgently to flee persecution from within their origin state,[159] and because they generally cannot establish the legally enumerated grounds for seeking asylum from their

155 Ibid.

156 8 U.S.C.§ 1182(d).

157 Other international frameworks similarly are not particularly suitable for providing rights and protections to environmental migrants. The UN Convention on the Status of Stateless Persons, which provides rights to education, employment, housing, and nationality to stateless persons, may be relevant as distinct territories disappear as a result of climate change. However, this area of law is untested as to this situation, and is also poorly ratified. Article 15 of the UN Declaration on Human Rights articulates an international desire to preserve nationality, but lacks corresponding obligations on states. See Maxine Burkett, *Research Handbook on Climate Change, Migration and the Law* (Edward Elgar, 2017). Also of note is the International Organization for Migration (IOM), which is not a legal framework, but a UN-affiliated intergovernmental organization that now provides research and guidance on environmental and climate change migration issues. The IOM was initially founded to aid with migration in Europe following the Second World War. In 2007, member states requested the organization to work on environmental and climate change migration, most likely in response to the growing awareness of the problem. Environmental Migration Portal, *About*, http://environmentalmigration.iom.int/iom-and-migration-environment-and-climate-change-mecc (Last visited Aug. 7, 2017).

158 U.N. Convention Relating to the Status of Refugees, July 28, 1951, 189 U.N.T.S. 137; See Protocol Relating to the Status of Refugees, Jan. 31, 1967, 606 U.N.T.S. 267; U.N. High Comm'r. for Refugees, *Guiding Principles on Internal Displacement*, U.N. Doc. PR00/98/109 (July 22 1998).

159 However, it is possible that subsequent effects of climate change will cause violence and persecution of subgroups of people, possibly triggering the applicability of refugee status to those specific groups of people. See Burkett, *Research Handbook on Climate Change*.

origin state, they are not deemed refugees and do not qualify for protections afforded by the UN Convention and Protocol on the Status of Refugees.[160]

The UNFCCC first formally recognized the importance of the issue of climate-induced migration, displacement, and relocation in 2010,[161] under the Cancun Adaptation Framework. At the 21st Conference of Parties in 2015, there was further, though bounded, action on the issue. Paragraph 50 of the Paris Decision requested the establishment of a taskforce[162] "to develop recommendations for integrated approaches to avert, minimize and address displacement related to the adverse impacts of climate change."[163] Because the taskforce is still in its infancy, it is unclear what solutions will emerge, although there is hope that the UNFCCC will appropriately address the specific problem of climate migration in the future.

In the United States, the Refugee Act of 1980[164] governs the admissions of refugees. The purpose of the Act was to "provide a permanent and systematic procedure" for admission and resettlement of refugees admitted to the United States. Similar to the UN definition, the American definition of "refugee" does not incorporate "climate refugees."

In addition to the non-inclusion of climate refugees in the Act, the application process, roughly 18 to 24 months long, may not satisfy the needs of climate migrants fleeing urgent crises. In fact, the field of international migration law,

160 Regional conventions on refugees such as the Organization of African Unity Convention Governing the Specific Aspects of Refugee Problems in Africa, and the Cartagena Declaration on Refugees use broader definitions of refugees that could be interpreted to include environmental migrants, but would only apply in those regional areas. See Burkett, *Research Handbook on Climate Change.*

161 Paragraph 14(f) of the 2010 Cancun Agreement invited coordination and cooperation in regards to migration, displacement, and relocation caused by climate change in response to pleas from states most vulnerable to sea-level rise, but did not prescribe any response or binding obligations on states. Conference of the Parties ("COP") to the UNFCCC Dec. 1/CP.16, Cancun Agreement, 16th Sess., Nov. 29-Dec.10, 2010, U.N. Doc. FCCC/CP/2010/7/Add.1 ¶ 14(f) (Dec. 11, 2010). The 2012 Doha Decision encouraged further understanding of the impacts of climate change on migration, displacement, and human mobility, including noneconomic loss and damage, and mandated an institutional arrangement (later known as the Warsaw International Mechanism, in 2014) to conduct the research and assessment. COP to the UNFCCC Dec. 3/CP.18, Doha Amendment, 18th Sess., Nov. 26-Dec.8, 2012, U.N. Doc. FCCC/CP/2012/8/Add.1 ¶ 7(a)(2)(vi) (Dec 7, 2012); COP to the UNFCCC, Dec. 2/CP.19, Warsaw International Mechanism for Loss and Damage associated with Climate Change Impacts, U.N. Doc.FCCC/CP/2013/10/Add.1 (Nov. 23, 2013).

162 The taskforce is under the Warsaw International Mechanism. COP to the UNFCCC, Dec. 2/CP.19, Warsaw International Mechanism for Loss and Damage associated with Climate Change Impacts, U.N. Doc.FCCC/CP/2013/10/Add.1 (Nov. 23, 2013).

163 COP to the UNFCCC, Dec. 1/CP.21, Adoption of the Paris Agreement, U.N. Doc. FCCC/CP/2015/L.9./Rev.1 ¶ 50 (Dec. 12, 2015).

164 The Refugee Act of 1980, Pub. L. No. 96-212, 94 Stat. 102 (1980).

generally, buckles under existing migration and refugee scenarios – a discomfiting harbinger of how it will respond to the unique and emergent migration scenarios climate change might present. In a recent and trenchant critique of international migration law, Professor Jaya Ramji-Nogales suggests that the field of law is itself a chief contributor – if not the progenitor – of many contemporary migration and refugee crises.[165] A number of migration scholars deem the legal framework for migration to be inadequate to address these crises. Ramji-Nogales suggests it is also the cause of them. Taking international refugee law as one example, Ramji-Nogales notes that it benefits only a select group of migrants who fit into its narrow definition. Further, it encourages risk-taking activity, often resulting in exacerbated crises.

International migration law is dominated by the principle of *nonrefoulement*, which discourages the practice of returning refugees or asylum seekers to a country in which existing leaders are liable to subject those individuals to persecution. In order to avail oneself of the principle, however, a migrant must access state territory by, likely, embarking on a risky and perhaps extralegal journey. Once accessing the preferred nation-state, the refugee or asylum seeker must articulate a valid claim to obtain lawful immigration status.[166] The current system, Ramji-Nogales argues, "creates the conditions that encourage mass flows of migrants to show up at the borders of Australia, Europe, and the United States."[167] Given the inherent uncertainty, disorder, and ad hoc nature of this kind of process, Ramji-Nogales asks whether migration emergencies – or "crises" – are surprising events or "the logical and foreseeable outcomes of the structural failures of the global migration system."[168]

There are at least two immediate concerns raised here, both relevant to climate migration and health outcomes. First, an arguably systemic flaw produces migration events labeled crises, a term that connotes a temporary or isolated incident. Second, the crisis lens also "obscures the complex and long term structural causes of international migration," namely poverty and underdevelopment, food insecurity, and natural disasters.[169] Migration "crises" related to climate change and health will be neither temporary nor simple.

165 Ramji-Nogales, *Migration Emergencies*, 68 Hastings L.J. 609, *supra* note 141, at 609.
166 Ramji-Nogales, at 614.
167 Ibid. at 614.
168 See generally, Ibid. at 609. Further, Ramji-Nogales persuasively contends, "International law has constructed a deeply path-dependent approach to international migration that not only obscures systemic inequality but also consumes alternate conceptions of morality."
169 Ibid. at 624. Ramji-Nogales also notes the influence of cycles and structures of violence and, significantly, global power relations.

This is incredibly instructive for the climate-related migration scenarios. First, of most concern regarding this kind of migration is the absence of law to govern these scenarios. Given that established international migration law suffers from arguably fatal flaws, lawmaking relevant to climate-related population movements will be especially important to conduct carefully. Cognizant of potential path dependence, any lawmaking will need to consider if it unwittingly guarantees disorder by virtue of its own internal logic. Further, climate change presents unique and unprecedented crises, and the trigger itself is changing at an increasing rate. The balance of the global community, however, understands the severity of climate-related risks, even if the specifics are elusive. Ramji-Nogales wistfully writes: "One can imagine a legal regime that anticipated migration flows and enabled safe and lawful movement for migrants of all types."[170] Indeed, the global community could make such a regime part of its belated response to climate displacement and climate-related cross-border migration.[171]

In approaching the probable increase in climate-triggered movements and attendant health impacts, devoting resources to anticipate and process migration flows might galvanize the creation of institutions that can help ensure that these flows do not overwhelm countries and communities.[172] During the most recent migration crisis, migrant flows constituted 0.2 percent of each of the populations of the European Union (EU), United States, and Australia.[173] Given the relative size of cross-border migrant flows, better infrastructure choices can strengthen capacity. And, to the extent that climate introduces even greater numbers, which is plausible, planning or expanding capacity and the ability to manage a dynamic and increasing phenomenon is optimal. Finally, and perhaps most important, this kind of proactive planning is best able to incorporate the perspectives of the migrants themselves and ensure that states protect migrants' rights throughout the process of migration.

## 11.3 ADDRESSING EXISTING GAPS AND CRAFTING JUST INTERVENTIONS

In their comprehensive study on how to safeguard human health in the "Anthropocene," the Rockefeller Foundation-Lancet Commission on

[170] Ibid. at 614.

[171] Paris Agreement, para. 50.

[172] See Ramji-Nogales, *Migration Emergencies*, 68 Hastings L. J. 609, *supra* note 141, at 618 (arguing that had more resources been devoted to anticipating and processing migration flows, [the EU, United States, and Australia] might have created institutions capable of managing those flows without becoming overwhelmed. "In other words, this is arguably a crisis created by infrastructure choices, not a crisis of absolute capacity.").

[173] Ibid. at 618.

Planetary Health[174] note an evident lack of imagination, information, and implementation. In current efforts to safeguard public health in the so-called Anthropocene, "imagination challenges" result from conceptual and empathy failures that discount future harms for present gains and underestimate the disproportionate effect of harms to the poor and developing nations. Research and "information challenges" result from failures to address social and environmental drivers of ill health or a historical scarcity of transdisciplinary research and funding. Particularly relevant to climate change-induced migration and its likely health impacts, "an unwillingness or inability to deal with uncertainty within decision making frameworks" compound knowledge failures.[175] Finally, implementation failures, or governance challenges, result from governments or institutions delaying recognition and responses to threats, "especially when faced with uncertainties, pooled common resources, and time lags between action and effect."[176]

This comprehensive problem definition is consistent with similar analyses, which have emphasized the need to recognize the complexity and heterogeneity of migration to minimize any health inequities and to ensure access to services and the health rights of migrants, among other things.[177] As Ferris et al. note, however, compared to the construction of new housing, for example, support for livelihoods and social services are both "less visible and more complex."[178] Further, in the United States and other countries, strategies to protect infrastructure continue to develop and advance, while plans to ensure physical and psychological well-being remain elusive.

The lack of comprehensive and coherent policy tools adds to the urgency of identifying viable frameworks to address the suboptimal health outcomes resulting from climate-induced migration. More focused and interdisciplinary

174 Whitmee et al., *Safeguarding Human Health in the Anthropocene Epoch*, *supra* note 24, at 1973.

175 Whitmee et al., *Safeguarding Human Health in the Anthropocene Epoch*, *supra* note 24, at 1973.

176 Ibid.

177 McMichael et al., *An Ill Wind?*, *supra* note 3, at 650. See also, World Health Organization, *Health of Migrants: The Way Forward – Report of a Global Consultation*, Madrid, Spain March 2010. McMichael et al. find that "[o]f fundamental importance are the accessibility and quality of health care, health education, and prevention measures; establishment of adequate disease surveillance systems; control of pathogenic agents; targeting 'hidden' populations; monitoring and evaluating health interventions such as vaccination coverage; participation of affected populations in health care planning; coordination between services and administrative bodies and target populations; and consideration of the health needs of host populations."

178 Ferris et al., *Planning Relocations to Protect People from Disasters and Environmental Change*, *supra* note 76, at 55. Further, livelihoods and social services may not receive adequate attention.

research, education, and outreach are necessary – as are efforts at intensive coordination and planning over time. Given the complexity of the interventions moving forward, a fully elaborated and coherent framework is necessary but beyond the scope of this chapter. This section, however, attempts to identify a few approaches that may aid in easing negative public health outcomes in the context of displacement, migration, and relocation. It first identifies existing gaps and potential responses related to identifying and servicing at-risk populations. It then explores how communities and decision-makers might build resilience, support relocation, and provide access to healthcare throughout. The section concludes with a discussion of additional interventions that are prudent to employ now, and others that are more appropriate over the long-term.

### *11.3.1 Identifying At-Risk Populations and Circumstances*

There are numerous factors that might exacerbate one's sensitivity to physical and mental health impacts, generally, that are relevant to and may be heightened by the advent of climate change – and which the law must address with vigilance. These factors, noted earlier, include geographic location, presence of preexisting disabilities or chronic illnesses, and socioeconomic and demographic inequalities in education level, income, and age.[179] Children, for example, may experience changes in behavior, development, memory, executive function, decision-making, and scholastic achievement because of the stress from climate impacts and accompanying dislocation.[180]

As evinced in the disaster research literature, the poor and powerless are likely to suffer disproportionately, and in some cases fatally, from climate change.[181] For communities like those in the US Gulf Coast, public health adaptation can improve the capacity of the region to address present and future climate-related health hazards.[182] So, too, can addressing the underlying causes of vulnerability along lines of age, race, and socioeconomic status. For example, communities can invest in affordable housing, expanded employment opportunities, and improved public transportation in order to reduce extant disparity.[183] Petkova et al. cite specific state measures that are particularly effective at supporting vulnerable populations, such as Florida's support of the elderly. Specifically, the Florida Department of Elder Affairs works closely with the state's emergency

179 Whitmore-Williams et al.

180 Ibid.

181 Petkova, *Climate Change and Health on the U.S. Gulf Coast*, *supra* note 92, at 9351.

182 Ibid.

183 See Whitmore-Williams et al., at 48 (citing Baussan and Kelly [2016]).

operations during disasters to determine if area agencies on aging need special assistance. Further, the state uses census data and geographic maps to efficiently allocate resources in areas with greater concentration of older adults.[184] This kind of vulnerability mapping is frequently recommended and might benefit other vulnerable populations who are more susceptible, for example, to the health effects of heat extremes.[185] Maintaining and buttressing measures, like the Affordable Care Act,[186] will also aid those with pre-existing medical conditions who are also likely to suffer disproportionately from climate impacts.[187]

Finally, ensuring the rights and dignity of migrants and those displaced has inherent value and will also be key to maintaining physical and mental health through dynamic changes over the long-term. It will be important to know what mechanisms are in place, if any, to ensure that migrants can reestablish cultural norms, traditions, and heritage.[188] The presence of laws and policies that respect and protect civil and human rights, particularly for women, children, the elderly, disabled peoples, and the poor and powerless will help determine the value and success of proposed frameworks.

### 11.3.2 *Building Resilience for Displaced Individuals and Communities and Strategies for Planned Relocation*

Consistent across communities is the desire to remain in place, if conditions allow. To reduce negative impacts post disaster, certain public health interventions can help. It is critical for decision-makers and relevant partners to include an individual's or community's resilience capacity in the development of preparedness plans.[189] An inventory of the current capacity of and access to health infrastructure is an indispensable first step.[190] If a community's mental health infrastructure has a plan to serve people in need after a disaster, it will be important to gauge the infrastructure's capacity to serve potential increased needs as climate impacts grow and the number of recipients – migrants and hosts – increase. It is also important to note that in the context

184 Petkova et al., *Climate Change and health on the U.S. Gulf Coast*, *supra* note 92, at 9349–9350.

185 See, e.g. Kinney et al., *Chapter 5: Public Health Impacts and Resiliency*, *supra* note 74, at 77.

186 Patient Protection and Affordable Care Act, 42 U.S.C. 18001 (2010).

187 Petkova et al., *Climate Change and health on the U.S. Gulf Coast*, *supra* note 92, at 9349.

188 See generally, Ferris et al., *Planning Relocations to Protect People from Disasters and Environmental Change*, *supra* note 76, at 26, 41.

189 See generally Whitmore-Williams et al.

190 See generally, Ferris et al., questions for assessing a community's mental health infrastructure. Ferris, et al., *Planning Relocations to Protect People from Disasters and Environmental Change*, *supra* note 76, at 46.

of mental health resilience most people come through adversity "with positive adjustment and without psychopathology."[191] Some experience "post-traumatic growth," or the feeling of having gained something positive through a significant disruption.[192]

Community social support networks help to build such resilience. Accordingly, if efforts to prevent future migration are not successful, assuring timely relocation so as to maximally maintain social structures when forced displacement occurs is optimal.[193] So, too, are robust infrastructure and energy systems across geographic scales. They might help relieve burdens on potential host communities that might bear the brunt of forced displacement. For example, more than 11,000 households migrated to Harris County, Texas, from New Orleans following Hurricane Katrina, because Harris County was relatively close, out of harm's way, and had good infrastructure and emergency systems.[194] A greater number of well-prepared communities will allow for effective burden sharing, benefiting both the displaced and receiving communities.

Prior to entertaining important law and policy improvements to facilitate planned relocation, it is important to recall that, because of the substantial risks to health and well-being, researchers contend that the poor health outcomes resulting from planned relocation suffice as reasons to avoid planned resettlement altogether.[195] This is true, on balance, for most forced displacement and migration scenarios as well.[196] When planned relocation is unavoidable, researchers have identified indispensable elements of the relocation and resettlement process,[197] including consultation and participation of affected populations as well as undertaking monitoring and evaluation and ensuring accountability. These elements are relevant and crucial at all stages of planned relocation: from decision-making to undertaking relocation, to planning for

191 Whitmore-Williams et al., at 40.

192 Ibid.

193 Petkova et al., *Climate Change and health on the U.S. Gulf Coast*, *supra* note 92, at 9350.

194 *Mapping Migration After Hurricane Katrina*, National Geographic, 2015 https://blog.education.nationalgeographic.com/2015/08/27/mapping-migration-after-hurricane-katrina/.

195 See, e.g., McMichael et al., *An Ill Wind?*, *supra* note 3.

196 See, e.g., McMichael et al., *An Ill Wind?*, *supra* note 3, at 652 ("Although the range and extent of health risks associated with future climate related population movements cannot be clearly foreseen, the evidence of health outcomes of analogous movements of people indicates that health risks will predominate over health benefits. This, then, is an issue of considerable geopolitical, ethical, and economic importance."); Whitmore-Williams et al., at 44 ("Research in affected communities indicates that people do not wish to leave their homes, despite the changes they may experience (e.g. Cunsolo Willox et al., 2012).").

197 See Ferris et al., *Planning Relocations to Protect People from Disasters and Environmental Change*, *supra* note 76, at 3.

movement, and, finally, implementing the plan pending, during, and following relocation.[198]

Notably, and perhaps unsurprisingly, establishing and complying with an appropriate legal framework for planned relocation is also indispensable. In fact, legal preparedness is increasingly identified as an "essential component" of disaster risk reduction, evinced by its presence in the Sendai Framework for Disaster Risk Reduction,[199] among other instruments.[200] Proactive and preemptive legal frameworks help to avoid "shortcuts and oversights," which might result in further harm and rights violations arising if decision-makers try to establish these frameworks under intense time pressure and using reactive measures in response to disasters.[201] Proactive legal frameworks have other ancillary benefits, as they can also minimize conflict between domestic laws and customary laws, for example.

The risks that planned relocation introduces are present over long timescales. They are present during departure and transition to the new location. They are present at the settlement phases as well; during which time relocated individuals and communities are acclimatizing to new locations and circumstances and concurrently grappling with the dislocation from their previous livelihoods, with often attendant loss of culture or heritage.[202] They may need targeted assistance to weather the adjustment. Further, new and additional challenges may emerge. The long-term impacts of planned relocation require ongoing monitoring and response.[203]

To effectively understand the adequacy of existing legal frameworks and the most effective strategies to establish proactively an effective and rights-supporting framework, Ferris et al. suggest a number of useful approaches. A conceptual mapping exercise, if not already completed, can help determine if and how existing laws, policies, or regulations are relevant to planned relocation.[204] Additional considerations concern whether or not there are domestic

198 Ibid.

199 United Nations International Strategy for Disaster Reduction, *Sendai Framework for Ddren* (2015) www.unisdr.org/files/43291_sendaiframeworkfordrren.pdf. See 27(f).

200 See, e.g., The United Nations Framework Convention on Climate Change.

201 Ferris et al., *Planning Relocations to Protect People from Disasters and Environmental Change*, *supra* note 76, at 11.

202 McMichael et al., *An Ill Wind?*, *supra* note 3; Ferris et al., *Planning Relocations to Protect People from Disasters and Environmental Change*, *supra* note 76.

203 Ferris et al., *Planning Relocations to Protect People from Disasters and Environmental Change*, *supra* note 76, at 51.

204 Ibid. at 11.

laws that provide a legal basis and authority for the state to carry out planned relocation and whether or not these laws and regulations are consistent.[205]

Particularly for the most vulnerable, maintaining access to health-supporting services will be critical. This can be quite difficult given the myriad cultural and socioeconomic factors at play. For example, for Marshallese immigrants, patterns of healthcare-seeking behavior are distinct from many in their receiving communities. For cultural reasons, Marshallese migrants generally do not seek healthcare until they perceive a health crisis.[206] Marshallese migrants are also highly reliant on parochial networks for healthcare, given the strong mutual trust and value assigned to interdependency within their culture.[207] Marshallese migrants in Hawai'i, however, rely more on an allopathic healthcare system than traditional healthcare practices, making the governing frameworks in the United States relevant and important.[208]

Understanding the existence of patterns will be similarly important to ensure access across age groups, abilities, and socioeconomic statuses. For children, for example, provisions to ensure regular vaccinations and other preventive care during and after displacement will be critical for both well-being and to meet existing requirements that allow for continued access to other critical services such as public education.[209]

### 11.3.3 *Near- and Long-term Interventions*

#### 11.3.3.1 Prudent Now

The single most significant and urgent policy measure to address adverse health outcomes related to climate migration is to embark upon dramatic cuts in greenhouse gases immediately. As with all climate impacts, the degree to which we need to adapt to or suffer from unavoidable impacts is contingent upon the amount we arrest those impacts' root causes. Given that some impacts are at this point inevitable, adequate responsive measures will be necessary for the foreseeable future. In the near-term there are a number of specific measures decision-makers at all scales might deploy.

205 Ibid.

206 Jin Young Choi, *Seeking healthcare: Marshallese immigrants in Hawaii*, 39(1) Ethnicity and Health 73–92 (2008) www.shsu.edu/~jyc002/documents/Seekinghealthcare_MarshallesemigrantsinHawaii.pdf.

207 Ibid.

208 Ibid.

209 Ferris et al., *Planning Relocations to Protect People from Disasters and Environmental Change, supra* note 76, at 44.

Continued aid to developing countries could effect numerous, positive adaptations to facilitate better health outcomes. In the Pacific, for example, the health sector is already under significant strain because of shortfalls in data, information systems, human resources, technical capacity, infrastructure, and finance.[210] Further, and similar to global trends, in the Pacific region women and children will likely experience climate change and health impacts disproportionately.[211] To respond to cross-border migration in the Pacific, for example, while acknowledging and meeting the general preference to stay in their ancestral home countries, uninterrupted and expanded aid to the Affiliated Pacific Islands could head off morbidity at its source. Assisting with existing shortfalls as well as shoring up and improving water, sanitation, and hygiene systems, and vector controls would have myriad co-benefits and limit poor health outcomes.[212]

In the United States, governments can minimize the social and health costs of displacement, migration, and relocation by providing support for housing, health services, mental health services, employment, and education, paying compensation at a level equal to the standard of housing and materials in the host community, employing the people being moved wherever labor is required, and facilitating adequate time for community consultation and planning, among other things.[213]

Cross-sector and interdisciplinary coordination is critical. Similar to the coordination needed in the international context, which requires coordination between nongovernmental organizations, United Nations agencies, and governments, multiscalar and multisectoral engagement will be necessary in the domestic context. In the United States, states' participation in the CDC's national environmental tracking network will facilitate knowledge sharing. The State of Hawai'i Climate and Health Working Group, for example, found that a comprehensive and coordinated public health adaptation strategy will be most effective if it engages scientists and researchers, planners, and policymakers[214] to, for example, enhance predictive capability of chronic climate-associated disease threats through interdisciplinary and cross-agency

210 McIver et al., *Health Impacts of Climate Change in Pacific Island Countries, supra* note 2, at 1712.

211 Ibid.

212 Ibid. This holds for other developing country contexts. See McMichael et al., *An Ill Wind?, supra* note 3, at 651–652 ("A key component, particularly in lower income countries, will be to support public health capacity and action, health surveillance, and data collection.").

213 McMichael et al., *An Ill Wind?, supra* note 3, at 651.

214 Government of Hawaii, *Climate Change and Health*, http://climateadaptation.hawaii.gov/climate-change-and-health/.

teaming.[215] Dedicating resources to these efforts is a key action decision-makers can work to implement today.

In the meantime, concerted efforts to learn from scenarios that are similar to those that climate change presents can enhance preparedness. In particular, efforts that address health issues and access to migrant-focused health services, such as the European "Migrant-Friendly Hospitals," or coordination mechanisms, such as the United States-Mexico Border Health Commission, are instructive.[216] And, because most displacement and migration will occur internally, domestic-oriented efforts at the national, state, and local levels can apply similar approaches and underlying principles to US-based population movements.

### 11.3.3.2 Future Possibilities

Rethinking current legal frameworks will be important long-term work. Regarding cross-border migration, Ramji-Nogales argues convincingly that a more appropriate and modern legal system would anticipate migration flows as a part of contemporary reality.[217] She suggests that an appropriately designed legal system "might reasonably include finding solutions in home countries to minimize migrant outflows, but must also enable safe and lawful journeys as part of a coherent and comprehensive approach to global migration."[218] For those who must move to receiving communities a forward-looking and just framework might offer, for example, labor migration routes sufficient to meet labor market needs with a focus on protecting the most vulnerable. A temporary regime to transition from the existing policy approach to a more progressive one capable of handling the novel characteristics of climate change would be an initial step to engage players that are often loathe to sign on to a permanent treaty regime.[219] Regional migration governance might also succeed given the nature of climate migration and the potential for experimentation that regional agreements facilitate.

Healthcare providers and legal actors can also work together more intentionally, further to the benefit of migrants. In their article on healthcare for Micronesians and US constitutional rights, Shek and Yamada note that during the loss of healthcare funding for Marshallese healthcare in the United States, medical and legal advocates needed the tools of the other discipline to succeed.

215 Ibid.

216 McMichael et al., *An Ill Wind?*, *supra* note 3, at 651.

217 Ramji-Nogales, *Migration Emergencies*, 68 Hastings L. J. 609, *supra* note 141, at 647–648.

218 Ibid.

219 Ibid. at 649.

These tools were important as powerful additions to the voices of those most impacted, voices that are also indispensable. Shek and Yamada propose "that in order to better address the issue of health care access for Micronesian peoples, we must work together, as health and legal advocates, to define a more robust vision of both systems that includes reconciliation and community engagement."[220] Indeed, legal advocates might first engage providers to determine what existing law and policy frameworks hinder or advance work done at the intersection of climate and health. These will be critical engagements at the state and local levels, in particular. More assertive engagement might also be beneficial. Indeed, Whitmore-Williams et al. encourage mental health officials to, among other things, collaborate with colleagues, community, and climate advocates to influence future policy design and outcomes.[221]

Finally, in addition to the desperate need for greater research – transdisciplinary and otherwise[222] – understanding, coping with, and embracing the inherent uncertainty in the climate, health, and migration interplay will be vital. While the occurrence of climate displacement is established, the scope, in terms of numbers and timescales, will be difficult to capture with the precision decision-makers typically demand.[223] The health impacts of forced displacement, migration, and relocation are better documented now. How these phenomena will act synergistically will be hard to predict conclusively. The risks are much easier to convey. Improving risk communication to decision-makers and the public while employing strategies to facilitate decision-making under uncertainty[224] can help to avoid planning paralysis.

## 11.4 CONCLUSION

Concerns regarding impacts to health transcend political affiliation. In addition to the public's relative trust in health officials, the impacts of climate migration on health and the impacts of health on the well-being of migrants and hosts may spur concerted and compassionate action. The urgency and potential severity of those impacts militate in favor of swift action. The disproportionate impacts of the phenomenon further militate in favor of placing and maintaining the rights

220 Shek and Yamada, *Health Care for Micronesians and Constituional Rights, supra* note 120.

221 Whitmore-Williams et al., at 55.

222 Whitmee et al., *Safeguarding Human Health in the Anthropocene Epoch, supra* note 24, at 2018; see also McMichael et al., *An Ill Wind?, supra* note 3, at 652.

223 See generally, Ibid.

224 Ibid. at 2013 (identifying possible tools including the Thresholds Approach, Scenario Planning, and Resilience Thinking).

of those most impacted at the center of existing, revised, or new law and policy frameworks decision-makers erect in response.

The current political environment is, in some respect, an equally challenging phenomenon as the changing natural environment. Indeed, it might be more formidable. Reasoned and fact-based conversations about immigration and climate change are, arguably, unprecedented and are not guaranteed prospectively. Problem-solving at the intersection of climate and migration have not yet resulted in swift and coordinated efforts to craft just solutions. In fact, concerns about national security and public health are often deployed to incite fear and expedite regressive and exclusionary policies[225] by politicians, the media, and, sometimes unwittingly, researchers seeking to provide sound data and analysis to advance effective interventions. Considerations of ethical responses to the suffering of climate migrants and sound management to benefit those migrants and their hosts suggest, however, that swift action is critical.

225 See Ramji-Nogales, *Migration Emergencies*, 68 Hastings L. J. 609, *supra* note 141, at 648.

# PART IV

# INTERPLAY WITH INTERNATIONAL & DOMESTIC ENVIRONMENTAL LAW

# 12

# International Institutions and the Developing World

William Onzivu[1]

*This chapter considers public health impacts of climate change in developing countries and responsive interventions, with a focus on the United Nations Framework Convention on Climate Change(UNFCCC) governance frameworks, the World Health Organization (WHO), and its membership. Its scope encompasses the roles and mandates of international organizations and instruments relevant to climate change-related policy. It discusses the status of health in international climate law – particularly the UNFCCC, the Paris Climate Agreement, and related institutional decisions – and these instruments' implications for the prioritization of health in climate policy at the international and domestic levels. It then discusses the role of WHO policies to implement health requirements, especially in relation to the Paris Climate Agreement. The chapter concludes by proposing options to improve the status quo through adaptive governance at both the international and domestic levels and argues for several substantive, institutional, and procedural mechanisms of international climate law to promote and protect public health in developing countries.*

## 12.1 INTRODUCTION

This chapter's examination of international interventions responsive to the public health impacts of climate change in developing countries is divided into seven sections: Section 12.1 is the introduction. Section 12.2 discusses the impact of climate change on developing countries, focusing on the direct and indirect impacts. It does not, however, address the impacts of measures to tackle climate change. Section 12.3 examines the status of health in international climate law, with a focus on health in the UNFCCC, its COP Decisions and Resolutions, as well as the Paris Climate Agreement.

[1] World Health Organization Regional Office for Africa and Bradford University School of Law, UK.

Section 12.4 discusses recent international institutional responses in health and climate change by focusing on the role of WHO. It discusses both the strength and limits of WHO governance frameworks for health and climate change. Section 12.5 examines international climate change law and health in the contexts of strengths and shortcomings in the domestic implementation actions. Section 12.6 argues for tackling the international and domestic limitations in the progressive development and implementation of international climate law through the functional framework of adaptive governance. This offers a transformational pathway for reform of the implementation of the international climate law. Section 12.7 is the conclusion.

## 12.2 IMPACT OF CLIMATE CHANGE ON DEVELOPING COUNTRIES

The impacts of climate change on human health have been summarized in six categories of public health concern: heat-related health problems; respiratory problems; infectious diseases; waterborne diseases; food insecurity; and mental health problems.[2] The most marked increases in morbidity and mortality are predicted to occur within vulnerable populations such as the young, the poor, the medically frail, and the elderly.[3] According to the Intergovernmental Panel on Climate Change (IPCC), in recent decades, changes in climate have caused impacts on natural and human systems on all continents and across the oceans.[4] In many regions, changing precipitation or melting snow and ice are altering hydrological systems, affecting water resources in terms of quantity and quality. Many terrestrial, freshwater, and marine species have shifted their geographic ranges, seasonal activities, migration patterns, abundances, and species interactions in response to ongoing climate change. Based on many studies covering a wide range of regions and crops, negative impacts of climate change on crop yields have been more common than positive impacts. While climate change

[2] J. A. Patz et al., *Climate Change: Opportunities and Challenges for Global Health*, 312(15) JAMA 1565–1580 (2014).

[3] J. Krueger et al., *Human Health Impacts of Climate Change: Implications for the Practice and Law of Public Health*, 43 The J. of L., Medicine & Ethics 79–82 (2015) doi:10.1111/jlme.12223.

[4] Intergovernmental Panel on Climate Change, Summary for policymakers. In: C.B. Field et al., *Climate Change 2014: Impacts, Adaptation, and Vulnerability. Part A: Global and Sectoral Aspects. Contribution of Working Group II to the Fifth Assessment Report of the intergovernmental Panel on Climate Change*, 1–32 (Cambridge University Press, Cambridge and New York, 2014).

impacts are being felt in many places, differences in vulnerability and exposure arise from non-climatic factors and multidimensional inequalities often produced by uneven development processes. These differences shape differential risks from climate change. Impacts from recent climate-related extremes – such as heatwaves, droughts, floods, cyclones, and wildfires – reveal significant vulnerability of some ecosystems and many human systems to current climate variability. Climate-related hazards also exacerbate other stressors, often with negative outcomes for livelihoods, especially for people living in poverty. Notably, violent conflict increases vulnerability to climate change. To provide a sense of what this means: In Africa alone, between 75 and 250 million people are projected to be exposed to increased water stress due to climate change by 2020.[5] Ecosystems throughout the continent have already been impacted by climate change, and future impacts are projected to be substantial.[6] Although adaptation measures in Africa have begun to address some of the near-term risks to food production, such measures are inadequate to address the long-term risks posed by climate change.[7]

Climate change has increasingly become a global health challenge.[8] The IPCC and the World Health Organization (WHO) have concluded that projected climate change-related exposures are likely to affect the health of millions of people, increasing rates of death and disease – infectious and otherwise – by spurring heatwaves, floods, storms, wildfires, and droughts.[9] Indeed, it has long been understood that climate change threatens an increased burden of diarrheal diseases, increased frequency of cardiorespiratory

5 M. Boko et al., *Africa Climate Change 2007: Impacts, Adaptation and Vulnerability, Contribution of Working Group II to the Fourth Assessment Report of the Intergovernmental Panel on Climate Change* (Cambridge: Cambridge University Press, 2007).

6 I. Niang et al., Africa, In: *Climate Change*, 1199–1265 (2014).

7 Ibid.

8 See N. Watts et al., *Health and Climate Change: Policy Responses to Protect Public Health*, 386 Lancet 1861–1914 (2015); World Health Organizations, *Climate Change and Human Health: Publications*, www.who.int/globalchange/publications/en/ (listing publications addressing adverse health impacts of climate change and health-oriented policy responses).

9 K. R. Smith et al., Human health: Impacts, adaptation, and co-benefits. In: *Climate Change 2014*, at 709, 716; Simon Hales et al., *Quantitative Risk Assessment of the Effects of Climate Change on Selected Causes of Death, 2030s and 2050s* (World Health Organization 2014), www.who.int/globalchange/publications/quantitative-risk-assessment/en/; World Meteorological Organization, *Atlas of Health and Climate* (2012); Andrew K. Githeko and Alistair Woodward, International consensus on the science of climate and health: The IPCC third assessment report, In: World Health Organization et al., *Climate Change and Human Health, Risks and Responses* (Geneva, 2003), 47–50.

diseases,[10] and amplified transmission of infectious diseases such as malaria.[11] Climate change has already resulted in the increased incidence of malaria, dengue fever, yellow fever, onchocerciasis, and sleeping sickness, among other things.[12] Establishing the relationship between climate change and health faces challenges because climate change is just one among many contributors to health threats, and climate change often exacerbates existing conditions.[13] But there is no question that climate impacts will be most severe for those with low adaptive capacity, particularly those in the developing countries.[14] The IPCC characterizes the health impacts of climate change as being either direct or indirect.[15] Direct effects include temperature-related impacts, floods and storms, droughts, and increased ultraviolet radiation from ozone depletion. Indirect effects include ecosystem-mediated impacts such as vector-borne and other infectious diseases, food and waterborne infections, and air quality; and human institution-mediated impacts such as nutrition, occupational health, mental health, and violent conflict.

WHO and others have collated the direct and indirect effects of climate through its health and climate profiles.[16] The direct effects for health include lost work capacity and reduced labor productivity in vulnerable populations, increased numbers of warm days and nights; increase in frequency and intensity of heatwaves and increased fire risk in low rainfall conditions; excess heat-related mortality; increased incidence of heat exhaustion and heat stroke, particularly for outdoor laborers, athletes, and the elderly; exacerbated circulatory, cardiovascular, respiratory, and kidney diseases; increased premature

10 World Health Organization, *Health Guidelines for Episodic Vegetation Fire Events* (Geneva, 1999).

11 Willem J.M. Martens et al., *Potential Impact of Global Climate Change on Malaria Risk*, 103 Environmental Health Perspectives (1995).

12 Githecko and Woodward, 47–50.

13 Hans-Martin Fussel et al., Adaptation assessment for public health. In: Bettina Menne and Kristie L. Ebi *Climate Change and Adaptation Strategies for Human Health* 48 (Darmstadt: Springer, 2006); Jonathan A. Patz and R. Sari Kovats, *Hotspots in Climate Change and Human Health*.

14 K. R. Smith et al., Human health: impacts, adaptation, and co-benefits, In: C.B. Field et al., *Climate Change 2014: Impacts, Adaptation, and Vulnerability. Part A: Global and Sectoral Aspects. Contribution of Working Group II to the Fifth Assessment Report of the Intergovernmental Panel on Climate Change*, 709, 716 (2014).

15 Ibid.

16 WHO's health and climate country profiles, available here: www.who.int/globalchange/resources/country-profiles/en/ and from the Global Overview document that summarizes aggregated data, available here: http://apps.who.int/iris/bitstream/10665/208855/1/WHO_FWC_PHE_EPE_15.01_eng.pdf?ua=1., accessed August 3, 2017; See also L. Rudolph, S. Gould, and J. Berko, *Climate Change, Health, and Equity: Opportunities for Action* (Oakland, CA: Public Health Institute, 2015).

mortality related to ozone, and air pollution produced by fires, particularly during heatwaves; and greater risk of injury, disease, and death due to more intense heatwaves and fires.

The indirect effects comprise those mediated through natural systems, which include higher temperatures and humidity; changing and increasingly variable precipitation; higher sea surface and freshwater temperatures; accelerated microbial growth; survival, persistence, transmission, and virulence of pathogens; shifting geographic and seasonal distributions of cholera, schistosomiasis, and harmful algal blooms; lack of water for hygiene; flood damage to water and sanitation infrastructure, and contamination of water sources through overflow; and increased risks of food- and waterborne diseases. They also include accelerated parasite replication and increased biting rates; prolonged transmission seasons; re-emergence of formerly prevalent diseases; changing distribution and abundance of disease vectors; reduced effectiveness of vector control effectiveness, as well as increased risks of vector-borne diseases.

Effects heavily mediated by human systems include lower food production in tropics; lower access to food due to reduced supply and higher prices; combined effects of undernutrition and infectious diseases; and chronic effects of stunting and wasting in children. Outdoor and unprotected workers are obliged to work in physiologically unsafe conditions, or to lose income or livelihood opportunities. There are also consequences for health of lost work capacity and reduced labor productivity in vulnerable populations.

In addition to the direct and indirect effects of climate change, efforts to mitigate and adapt to climate change also have significant impacts on public health,[17] but while impacts of climate action are often left unexamined, they are beyond the scope of this chapter.

## 12.3 HEALTH IN INTERNATIONAL CLIMATE LAW

### 12.3.1 *The UNFCCC*

International climate change law is a progressively emerging major area of international law. Its main source is treaty law – principally consisting of the

[17] See Damilola S. Olawuyi, *The Human Rights-Based Approach to Carbon Finance* 14–15, 75–89 (2016) (discussing adverse public health impacts of projects credited with mitigating climate change); J. J. Cheng and P. Berry, *Health Co-Benefits and Risks of Public Health Adaptation Strategies to Climate Change: A Review of Current Literature*, 58(2) Int. J. Public Health 305–311 (2013).

United Nations Framework Convention on Climate Change (UNFCCC),[18] as further defined through the Kyoto Protocol (which lapses in 2020)[19] and other agreements and decisions, including the Paris Climate Agreement of 2015[20] – aimed directly at mitigating global warming, though it also encompasses aspects of the other international and domestic environmental laws, employed in new ways to respond to aspects of the climate change problem.[21] International climate law could significantly advance the promotion and protection of public health.[22] Foremost, the UNFCCC defines adverse effects of climate change as changes to the physical environment or biota that have significant deleterious effects on ecosystems, socioeconomic systems, or on *human health and welfare*.[23] It then calls upon the Parties, taking into account their common but differentiated responsibilities and their specific national and regional development priorities, objectives, and circumstances, to take climate change considerations into account in their social, economic, and environmental policies and actions.[24] In this context, Parties are encouraged to employ tools such as impact assessments to minimize adverse effects on the economy, on public health, and on the quality of the environment, of projects or measures undertaken by them to mitigate or adapt to climate change.[25] The UNFCCC obligation of Parties to promote and cooperate in scientific, technological, technical, socioeconomic, and other climate change research also implicates the need for a core role of public health research and evidence in the global climate legal regime.[26] As discussed further later, this cooperation has been made evident in proceedings of the Subsidiary Body for

[18] United Nations Framework Convention on Climate Change (UNFCCC), Rio de Janeiro, May 9, 1992, 31 Int' Legal Materials (1992), 849.

[19] Kyoto Protocol to the United Nations Framework Convention on Climate Change, Kyoto, December 10, 1997, in force February 16, 2005, 37 Int'l Legal Materials (1998), 22.

[20] Conference of the Parties, United Nations Framework Convention on Climate Change, Report of the Conference of the Parties on Its Twenty-First Session, Held in Paris from November 30, to December 13, 2015 – Addendum – Part Two: Action Taken by the Conference of the Parties at Its Twenty-First Session, Dec 1/CP.21, UN Doc FCCC/CP/2015/10/Add.1 (January 29, 2015) (Adoption of the Paris Agreement). The Agreement is included as an annex to this document (Paris Agreement).

[21] Jacqueline Peel, *Climate Change Law: The Emergence of a New Legal Discipline*, 32(3) Melbourne University L. Rev. 922 (2008).

[22] Yasmin Von Schirnding et al., *International Environmental Law and Global Public Health*, 80 Bulletin of the World Health Organization 970–971 (2002).

[23] United Nations Framework Convention on Climate Change (UNFCCC), Rio de Janeiro, May 9, 1992, 31 International Legal Materials (1992), 849, Article 1.

[24] Ibid. at Article 4(1)(f).

[25] Ibid.

[26] Ibid. at Article 4(1)(g); Article 9(1).

Scientific and Technological Advice to develop greater understanding of the health impacts of climate change under the Nairobi Work Program.[27]

Nonetheless, the link between mitigation and health is not thoroughly articulated in the UNFCCC, its Kyoto Protocol, or the recent Paris Climate Agreement, and while the UNFCCC has integrated considerations of health into its corpus, its primary objective is the stabilization of greenhouse gas concentrations in the atmosphere to prevent dangerous anthropogenic interference with the climate system.[28] Furthermore, the general commitments of the Convention such as the development of national inventories of anthropogenic sources of greenhouse gases, cooperation in developing climate-friendly technology, and removal of greenhouse gases from the atmosphere by the maintenance of sinks have limited direct relevance to health-related objectives or to the role of the health sector.[29] The lack of substantive references to health in the UNFCCC undermines what could otherwise be a clear and coherent legal basis for the promotion and protection of global and national public health under international climate law.[30] Consequently, as the WHO stated in 2011, "health is very poorly represented in the operating mechanisms for the climate change convention, including adaptation planning and finance, and within mitigation plans and finance streams."[31]

The UNFCCC international law and policy development agenda has primarily been carried out by the Conference of the Parties (COP), supported by a secretariat, subsidiary bodies, and a financial mechanism.[32] The COP, the supreme body of the Convention, has convened annually since 1995,[33] and implements the UNFCCC's agenda by performing several functions and adopting decisions.[34] COP decisions may be considered a "subsequent agreement

27 See Human health and adaptation: understanding climate impacts on health and opportunities for action UNFCCC document FCCC/SBSTA/2017/2, Subsidiary Body for Scientific and Technological Advice Forty-sixth session, Bonn, May 8–18, 2017, www.preventionweb.net/files/52825_02.pdf.

28 UNFCCC, *supra*, note 18, Article 2; Lisa F. Schipper, *Conceptual History of Adaptation in the UNFCCC Process*, 15 Rev. of European Community and Int'l Envtl. L. 82–92 (2006).

29 UNFCCC, Article 4(1)(a).

30 Ad Hoc Working Group on Long-Term Cooperative Action Under the Convention, Text to Facilitate Negotiations Among Parties, Note by the Chair, July 9, 2010, UN Doc. FCCC/AWGLCA/2010/8, available on the Internet at unfccc.int/resource/docs/2010/ awglca11/eng/08.pdf.

31 World Health Organization, *Improving Coherence of Climate Change, Health and Development Policy*, 1 (2011), www.who.int/globalchange/mediacentre/events/2011/WHO_Health_in_UNFCCC_COP17_.pdf.

32 UNFCCC, Articles 7 to 11.

33 UNFCCC, Article 7(4).

34 UNFCCC, Article 7(2).

between Parties regarding the interpretation or application of provisions of the treaty and are relevant in legal interpretation."[35] As described in this section, although COP decisions relating to health have been minimal, some offer guidance to developing country Parties on health policy and its implementation in an adaptation context.

### 12.3.2 *Health in UNFCCC COP Governance: Decisions and Resolutions*

The first COP to address health issues was the 1996 COP in Geneva, which declared that the "adverse effects" of climate change on human health were "potentially irreversible."[36] The Hague COP in 2000 made direct references to health in one decision and two resolutions, highlighting adaptation and monitoring for health in relation to diseases and disease control.[37] The 2001 Marrakesh COP affirmed health synergies in climate policy, the need to leverage climate funding for public health, the need for early warning systems, and prevention of disease.[38] Despite these nascent efforts, successive COPs through 2010 largely made only indirect references to health. More recently, health has again become more central. In Durban in 2011, the COP made extensive direct references to health in the Decisions, including in the criteria it adopted for carbon capture, which include not posing risks to human health.[39] In Lima in 2014, health is mentioned in one of the few technically substantive articles of the agreement, which "decides to continue the technical examination of opportunities with high mitigation potential, including

[35] Lavanya Rajamani, Addressing the *Post-Kyoto' Stress Disorder: Reflections on the Emerging Legal Architecture of the Climate Regime*, 58 Int'l and Comparative L. Quarterly 803–834 (2009); Article 31(3)(a) of the Vienna Convention on the Law of Treaties, May 22, 1969, in force January 27, 1980, 37 Int'l Legal Materials, 17 (1998); See also Robin Churchill and Geir Ulfstein, *Autonomous Institutional Arrangements in Multilateral Environmental Agreements: A Little Noticed Phenomenon in International Law*, 94 Am. J. of Int'l L. 623 (2000).

[36] UNFCCC, Report of the Conference of the Parties on its Second Session, Geneva, 8 to July 19, 1996, UN Doc. FCCC/CP/1996/ 15/Add.1, unfccc.int/resource/docs/ cop2/15a01 .pdf.

[37] UNFCCC, Report of the Conference of the Parties on its Sixth Session, The Hague, November 13–25, 2000, UN Doc. FCCC/CP/2000/5/Add.2., unfccc.int/ resource/docs/cop6/ 05a02.pdf (last accessed on July 13, 2017).

[38] UNFCCC, Report of the Conference of the Parties on its Seventh Session, Marrakesh, October 29 to November 10, 2001, UN Doc FCCC/CP/2001/13/Add.1., unfccc.int/resource/ docs/cop7/13a01.pdf (last accessed on July 15, 2017).

[39] See Appendix B of Annex to Decision 10/CMP (Modalities and procedures for carbon dioxide capture and storage in geological formations as clean development mechanism project activities), Additional requirements for carbon dioxide capture and storage project activities under the clean development mechanism, http://unfccc.int/meetings/durban_no v_2011/meeting/6245/php/view/decisions.php#c.

those with adaptation, health and sustainable development co-benefits, in the period 2015–2020,"[40] and specifies a series of actions to be undertaken by the UNFCCC Secretariat in support of this goal.[41] As discussed further in the following section, the Paris Climate Agreement placed a large emphasis on public health. In 2016, the Marrakech COP yielded the Ministerial Declaration on Health, Environment and Climate Change.[42]

The COP's treatment of public health forms an important basis for the progressive development of health-related norms in international climate law and policy that can contribute to global and national public health promotion and protection. COP decisions on health are of important operational significance and influence key legal outcomes in the climate change negotiations. The COP decisions do not impose new obligations on Parties, as they are not legally binding, but these instruments have enriched and expanded the normative content of international climate change law by increasing the visibility and centrality of public health concerns.

### 12.3.3 *Health in the Paris Climate Agreement 2015*

The Paris Climate Agreement of 2015 has made a significant contribution to integrating health in international climate law.[43] An adoption Decision of the Agreement has also highlighted the importance of health in the operationalization of the Agreement.[44] The Agreement and its Adoption Decision aim to achieve climate mitigation, adaptation, and finance through a series of cooperative frameworks and mechanisms, each of which establishes different legal rights and obligations for Parties, and explicitly makes provision for the needs of developing country Parties, especially the most vulnerable.[45]

40 Report of the Conference of the Parties on its twentieth session, held in Lima from December 1–14, 2014, Decisions adopted by the Conference of the Parties, 1/CP.20, Lima Call for Climate Action, Paragraph 19, Document FCCC/CP/2014/10/Add.1, at http://unfccc.int/resource/docs/2014/cop20/eng/10a01.pdf.

41 Ibid.

42 Ministerial Declaration on "Health, Environment and Climate Change", the UN Framework Convention on Climate Change (UNFCCC COP 22), Marrakech, 15 November 2016, www.who.int/globalchange/mediacentre/events/Ministerial-declaration-EN.pdf accessed August 5, 2017.

43 Paris Agreement (adopted 12 December 2015) UN Doc FCCC/CP/2015/L.9/Rev.1.

44 Conference on the Adoption of the Paris Agreement (December 12, 2015, opened for signature April 21, 2016) UN Doc FCCC/CP/2015/L.9/Rev.1.

45 *Advancing the Paris Agreement on Climate Change for Sustainable Development* (Legislative Comment), 5(2) Cambridge J. of Int'l and Comparative L. 202–237 at 207 (2016).

The Agreement's precise objective, the bottom up requirement on nationally determined contributions (NDCs), as well as the requirements on loss and damage demonstrate a progressive development of international climate law.[46] While health is not directly referred to in the substantive provisions of the Agreement, the Preamble acknowledges that climate change is a common concern of humankind and encourages Parties, when they take action to address climate change, to respect, promote, and consider their respective obligations on human rights, the right to health, the rights of indigenous peoples, local communities, migrants, children, persons with disabilities, and people in vulnerable situations and the right to development, as well as gender equality, empowerment of women, and intergenerational equity.[47] The express reference to a right to health opens a broader discourse of the important role health can play in international climate law.[48]

Besides the inclusion of health in the preambular paragraphs, the relevance of health can be inferred from the language of several other provisions. First, Article 2 of the Agreement establishes the global goal of limiting the increase in the global average temperature to well below 2 °C above pre-industrial levels and pursuing efforts to limit the temperature increase to 1.5 °C above pre-industrial levels. In establishing this goal, the Agreement recognizes that this would significantly reduce the risks and impacts of climate change – including risks to public health. Second, Article 2 also seeks to increase the ability to adapt to the adverse impacts of climate change and to foster climate resilience and development that

46 See H. M. Osofsky et al., *The 2015 Paris Agreement on Climate Change: Significance and Implications for the Future*, 46 Envtl. L. Rep. News & Analysis 10267, 10268 (2016); see Chitzi C. Ogbumgbada, *The Paris Agreement: An Imperfect but Progressive Document*, 8 Int'l Energy L. Rev. 320–323 (2016); Miguel Schloss, *Climate Change: Moving the Paris Agreement From Words to Deeds*, 4 Int'l Energy L. Rev., I.E.L.R. 146–150 (2016).

47 Preamble to the Paris Agreement, Paris Climate Agreement, adopted April 22, 2016, entered into force November 4, 2016, http://unfccc.int/files/essential_background/convention/application/pdf/english_paris_agreement.pdf access, Article 2; Conference of the Parties, United Nations Framework Convention on Climate Change, Report of the Conference of the Parties on Its Twenty-First Session, Held in Paris from November 30 to December 13, 2015 – Addendum – Part Two: Action Taken by the Conference of the Parties at Its Twenty-First Session, Dec 1/CP.21, UN Doc FCCC/CP/2015/10/Add.1 (January 29, 2015) ('Adoption of the Paris Agreement'). The Agreement is included as an annex to this document ('Paris Agreement'). See also Benoit Mayer, *Human rights in the Paris Agreement*, 6 Climate Law 109, 112 (2016); Annalisa Savaresi, *The Paris Agreement: A New Beginning*, University of Edinburgh Law School Research paper 2016/03.

48 See A. Dietzel, (2017), *The Paris Agreement – Protecting the Human Right to Health?*. Glob Policy. doi:10.1111/1758-5899.12421, assessing the right to health in the context of Paris Climate Agreement.

involves low levels of greenhouse gas emissions intensity.[49] Health is also arguably indirectly given priority through references to the United Nations' Sustainable Development Goals (SDGs), which treat health as a core concern, and integrate issues relating to climate change and public health.[50] The linkage of the SDGs to the Paris Climate Agreement can make health outcomes more relevant to international climate law and policy. To this end, the Adoption Decision of the Paris Climate Agreement also clarifies that the Subsidiary Body for Scientific and Technological Advice will undertake a work program under the framework for nonmarket approaches to sustainable development;[51] invites all UN agencies and financial institutions to provide information on how their development assistance and climate finance programs incorporate climate-proofing and climate resilience measures;[52] and, among others, recognizes the co-benefits of voluntary mitigation actions for adaptation, health, and sustainable development.[53]

The substantive and procedural requirements of the Paris Agreement have obvious consequences for public health.[54] For example, it provides that "adaptation action should follow a country-driven, gender-responsive, participatory, and fully transparent approach, taking into consideration vulnerable groups, communities, and ecosystems ... based on and guided by the best available science." [55] This resonates with the importance of evidence-based public health policy to address challenges of climate change.[56] Furthermore, WHO and others have recognized that there are public health co-benefits that will arise from providing financial

[49] Paris Climate Agreement, adopted 22 April 2016, entered into force 4 November 2016 and at http://unfccc.int/files/essential_background/convention/application/pdf/english_paris_agreement.pdf, Article 2 accessed 6 July 2017; Conference of the Parties, United Nations Framework Convention on Climate Change, Report of the Conference of the Parties on Its Twenty-First Session, Held in Paris from 30 November to 13 December 2015 – Addendum – Part Two: Action Taken by the Conference of the Parties at Its Twenty-First Session, Dec 1/CP.21, UN Doc FCCC/CP/2015/10/Add.1 (29 January 2015) ('Adoption of the Paris Agreement').The Agreement is included as an annex to this document ('Paris Agreement').

[50] See Health related Sustainable Development Goals, *World Health Statistics 2016*, http://who.int/gho/publications/world_health_statistics/2016/EN_WHS2016_Chapter6.pdf

[51] Conference on the Adoption of the Paris Agreement (December 12, 2015, opened for signature April 21, 2016) UN Doc FCCC/CP/2015/L.9/Rev.1., para 40.

[52] Ibid. at paragraph 44.

[53] Ibid. at paragraph 109.

[54] Margaretha Wewerinke-Singh and Curtis Doebbler, *The Paris Agreement: Some Critical Reflections on Process and Substance*, 39(4) University of New South Wales L. J.(2016), at 1486.

[55] See Paris Climate Agreement 2015, Article 7, paragraph 5.

[56] J. J. Hess et al., *An Evidence-Based Public Health Approach to Climate Change Adaptation*, 122 Environ. Health Persp. 1177–1186 (2014); http://dx.doi.org/10.1289/ehp.1307396;

resources and technology development and transfer for climate change mitigation and adaptation.[57] The importance of health as a co-benefit of mitigation can also bolster health adaptation efforts of some developing country Parties to the UN Framework Convention on Climate Change.[58] Indeed, Parties have recognized the integral role public health has to play in fulfillment of the Paris Agreement's goals. According to one study of 184 Intended Nationally Determined Contributions submitted to the UNFCCC in advance of COP21 in Paris, almost two-thirds included explicit mention of health.[59] Notably, however, the study also found significant regional variation: countries in Africa (88.9 percent), Asia (69.1 percent), and Latin America and the Caribbean (81.8 percent) were far more likely to integrate health into their INDCs than countries in Western Europe (13.8 percent) and Eastern Europe (13 percent). (The United States' INDC did not include express reference to health.)[60] Of the 121 INDCs that mention health, approximately 75 percent included health in the context of adaptation, while about 25 percent mentioned health in the context of mitigation.[61]

The growing importance of a public health lens on climate change is also already visible in some of the work being conducted under the Nairobi Work Program on impacts, vulnerability, and adaptation to climate change (NWP), which continues to carry out its activities consistent with and in relation to new processes established under the Paris Agreement. In particular, recognizing the risks climate change poses to human health, Parties to the UNFCCC have sought "to inform adaptation planning and actions at the regional, national and subnational levels" through a sequence of activities that has included a survey of Parties and expert organizations, designing and hosting an event, the 10th Focal Point Forum, convening Parties and health experts to discuss climate change action and public health; production of a synthesis paper on the subject; and further discussion at the Subsidiary Body on Scientific

R. C. Brownson et al., *Evidence-Based Public Health: A Fundamental Concept for Public Health Practice*, 30 Annu. Rev. Public Health 175–201 (2009); D. Campbell-Lendrum et al., *Global Climate Change: Implications for International Public Health policy*, 85(3) Bull. World Health Org. 235–237 (2007); K. L. Ebi et al., *Integration of Public Health with Adaptation to Climate Change: Lessons Learned and New Directions* (Leiden:Taylor & Francis, 2005).

57 See Paris Climate Agreement 2015, Articles 9, 10.

58 D. Campbell-Lendrum and C. Corvalán, *Climate Change and Developing-Country Cities: Implications For Environmental Health and Equity*, J. Urban Health 84(Suppl 1): 109 (2007), https://doi.org/10.1007/s11524-007-9170-x.

59 Elizabeth Wiley et al., *Health in Intended Nationally Determined Contributions*, www.mcgill.ca/epi-biostat-occh/files/epi-biostat-occh/analysis_of_indcs_1.pdf.

60 Ibid.

61 Ibid.

and Technical Advice.[62] The synthesis paper concludes with a number of proposals for collaboration and action, including enhancing research and health information systems; adopting a comprehensive approach to integrating health into climate change adaptation plans, projects, and programs and to combining climate change with other determinants and drivers of healthcare systems; developing the capacity of the healthcare workforce and educational institutions in order to develop climate-resilient healthcare systems; strengthening intersectoral action and multilevel governance; promoting climate-resilient and sustainable health infrastructure and technologies; and scaling up financial investments/flows toward adaptation plans and actions addressing health.[63]

The NWP's synthesis report on climate change, public health, and adaptation reflects the same collaborative imperative adopted by the Paris Agreement, which invites the participation (as observers) of the United Nations and its specialized agencies – such as the WHO – at sessions of the UNFCCC Conference of the Parties,[64] and encourages them to support the efforts of Parties to implement actions under the Convention. The implication of the report is that health can be included as a component of the global stock take, with a potential for repositioning health at the core of international climate law and policy. The role of the WHO is discussed further in the next section.

The Paris Climate Agreement has provided a clear linkage between public health and international climate law in its preambular text, and its goals and mechanisms are properly understood to incorporate public health protection. References to sustainable development, the role of scientific advice and institutions, and the invitation to international cooperation, including among UN specialized agencies, can all be read to imply a heightened focus on public health policies and outcomes. Many countries, especially countries in the developing world, have recognized these realities, and have expressly incorporated public health into their plans for climate action under the Paris Agreement. However, the effectiveness of the Agreement to advance health objectives will depend on even bolder participation by the health sector within the existing institutional mechanisms for the progressive development and implementation of the Agreement.[65]

62 See *Human Health and Adaptation: Understanding Climate Impacts on Health and Opportunities for Action*, UNFCCC document FCCC/SBSTA/2017/2, Subsidiary Body for Scientific and Technological Advice Forty-sixth session, Bonn, May 8–18, 2017.

63 Ibid. at paras. 11, 47–54.

64 Article 16.1 of the Paris Agreement.

65 William Onzivu, *Health in Global Climate Change Law: The Long Road to an Effective Legal Regime Protecting Both Public Health and the Climate*, 2010(4) Carbon and Climate L. Rev. 364 (2010).

## 12.4 RECENT INTERNATIONAL INSTITUTIONAL RESPONSES IN HEALTH AND CLIMATE CHANGE: ROLE OF WHO

WHO is the international community's most prominent global public health agency.[66] The Charter of the UN enshrines WHO with a mandate to promote and protect health within the UN system,[67] and its constitution empowers it to provide leadership in global public health by promoting health cooperation.[68] WHO is a key actor with a major interest in the protection of the global environment.[69] WHO has adopted a number of resolutions in the area of environment and health,[70] and WHO and the United Nations Environment Program (UNEP) have long promoted environmental protection in developing countries.[71]

WHO has shown important leadership on climate change.[72] Together with UNEP and the World Meteorological Organization (WMO), it has been active on climate change and public health issues.[73] Indeed, WHO member states since 2008 have recognized that climate change is a health issue.[74] In this context, WHO's World Health Assembly passed a key resolution, "Climate Change and Health," that continues to inform WHO's technical work to date.[75] The resolution requests WHO to further strengthen its existing

66 Patricia W. Birnie and Alan E. Boyle, *International Law and the Environment* 9–31 (1st edn. 1992) 61 (stating that the WHO makes significant contributions to setting international standards, assists in drafting regulations, and provides consultants on relevant regulations).

67 U.N. Charter art. 57 (declaring that various specialized agencies established by intergovernmental agreement shall be brought into relationship with the UN in accordance with the provisions of Article 63).

68 World Health Organization Const., art. 19, 62 Stat. at 2681–82, 14 U.N.T.S. at 187–189.

69 Birnie and Boyle, *International Law and the Environment*, 61.

70 See, e.g., Protection of the Marine Environment, WHA Res. 50.14, World Health Assembly, 50th Association, 8th plenary meeting (May 12, 1997) (urging States to implement the Global Program of Action for the Protection of the Marine Environment from Land-based Activities at national, regional, and international levels); Promotion of Chemical Safety, with Special Attention to Persistent Organic Pollutants, WHA Res. 50.13, World Health Assembly, 50th Ass., 8th plenary meeting (May 12, 1997) (Calling upon States to, inter alia, involve national health efforts in implementing UNCED and WHO decisions relating to currently identified persistent organic pollutants).

71 The WHO/FAO/UNEP joint Panel of Experts on Environmental Management for Vector Control and the WHO/ILO/UNEP International Program on Chemical Safety are important joint projects.

72 F. Godlee, *Climate Change*, 349 British Medical J. 5945 (2014).

73 Githeko and Woodward, *International Consensus on the Science of Climate and Health*, at 47–50.

74 World Health Organization, Sixty-First World Health Assembly, *Resolution* WHA61.19, May 24, 2008.

75 Ibid.

program of support to countries, and to ensure that health is fully represented in the international climate change debate.[76] Similar resolutions followed, adopted by governing bodies of the WHO Regions of Africa, Southeast Asia, the Americas, the Western Pacific, and the Eastern Mediterranean.[77]

In 2009, WHO reported progress on the implementation of the resolution to its Executive Board and this progress was acknowledged by the World Health Assembly.[78] The report stated that WHO's health leadership in climate change focused on supporting Member States through advocacy and awareness raising, partnerships with other organizations of the United Nations system and other sectors, promotion and support of the generation of scientific evidence, and strengthening health systems to protect populations from the threats posed by climate change.

WHO established a health and climate change program due in part to its impact on human lives and health and because the responsibility for protecting lives and well-being ultimately falls on the health sector. Furthermore, investing in health protection and adaptation can save lives now and increase resilience to climate change and mitigation efforts with co-benefits for public health. Health sector activities and infrastructure are energy intensive and can contribute to reducing greenhouse emissions.

In 2008, WHO adopted a work plan that comprises, first, advocating and raising awareness to better understanding the links between climate change and health to motivate behavioral change, and public support, for mitigation and adaptation; second, strengthening partnerships by representing health in the UN Framework Convention on Climate Change and the Intergovernmental Panel on Climate Change, coordinating networks of climate negotiators, public health actors, and scientists, and convening intersectoral policy mechanisms at the national level; third, enhancing scientific evidence by mobilizing leading expertise and institutions worldwide to improve the evidence base of the linkages of health and climate, the disease burden and health costs attributable to climate change, and effective interventions to protect and promote health through both adaptation and mitigation; and finally strengthening health systems by leveraging its country presence, policy, and technical capacity to support member states to strengthen the climate resilience of health systems and communities, including through capacity building, monitoring health risks from climate change, and providing technical assistance to health adaptation projects in all WHO regions.

76 Ibid.

77 Onzivu, *Heath in Global Climate Change Law*, *supra* note 65 at 382.

78 WHO Executive Board Resolution EB124.R5, Climate change and health, 124th Session of the WHO Executive Board, January 22, 2009, http://apps.who.int/gb/ebwha/pdf_files/EB124-REC1/B124_REC1-en.pdf.

The Program has made significant achievements, which include the mobilization of commitment by 193 Member States, resulting in 2008 World Health Assembly Resolution on climate change and health, backed by Regional Resolutions for action, providing leadership to the health sector in the UN system and the IPCC, extensive climate and health research providing evidence on the interactions between both, technical assistance to health vulnerability assessments, workshops on awareness raising, and supporting the implementation of adaptation projects in Member States. WHO has also established Climate Change and Health Focal Points in the WHO Regional Offices. Finally, WHO has continued to work with other specialized agencies and programs, such as the World Meteorological Organization (WMO), the United Nations Environment Program (UNEP), and the United Nations Development Program (UNDP) on climate change-related initiatives, including the establishment of a joint specialized WHO/WMO climate and health office to strengthen contribution of meteorological services to international health, promote technical assistance to pilot activities, and generally strengthen collaboration between WHO and WMO.[79]

WHO also represents the interests of the health sector in global climate negotiations and provides a link to operational health programs in the field. In recent years, WHO has participated in the UNFCCC institutional regime in several ways: First, WHO has provided scientific expertise, input, and advice to the Subsidiary Body of Scientific and Technological Advice and the Nairobi Work Program.[80] Under the Nairobi Work Program, WHO committed to strengthen its scientific, normative, and policy development functions, enhancing operational programs such as combating infectious disease, improving water and sanitation services and hygiene practices, providing health support in emergencies, and supporting health Ministries and the health sector globally.[81] Second, WHO participated in the Ad Hoc Working Group on Long Term Cooperative Action under the Convention (AWG-LCA) established in Bali in 2007. Finally, in recent years, WHO has also supported several side meetings to the Conference of Parties to highlight the health aspects of climate change policies and initiatives.[82]

79 See WHO Press Release, *WMO and WHO Establish Joint Office for Climate and Health* www.who.int/globalchange/mediacentre/news/joint-office/en/.

80 See Onzivu, *Health in Global Climate Change Law, supra* note 65, at 364.

81 55 Decision 12/CP.15, Program budget for the biennium 2010–2011, UN Doc. FCCC/CP/2009/11/Add.1, December 18–19, 2009.

82 Wold Health Organization, *Global Conference Sets Health Action Agenda for the Implementation of the Paris Agreement*, WHO Conference Conclusions, July 8, 2016, www.who.int/globalchange/conferences/second-global/conclusions/en/.

Perhaps the most visible of these events took place immediately after COP21 in Paris, where WHO convened the Second Global Conference on Health and Climate.[83] The conference, hosted by WHO and the Government of France and in close collaboration with the Government of Morocco, brought together more than 300 government ministers, health experts and practitioners, nongovernmental organizations, and experts in climate change and sustainable development. The conclusions called for countries to adopt a new approach to link health economics assessment and climate change, for example, by calculating the avoided healthcare costs when countries invest in mitigation of climate emissions and protection from climate risks. In support of this recommendation, WHO announced a working group to articulate a new coherent approach to health economics and climate change. Countries were also encouraged to spend more on protecting health from risks linked to climate change, such as extreme weather events and outbreaks of infectious disease, and in cleaner energy sources, more sustainable transport systems, and urban planning that also reduces major health risks, for example, air pollution in cities and in the home. Recommendations also focused on scaling up financing on climate change and health through additional resources and mechanisms and existing resources earmarked for adaptation. They urged different stakeholder groups to work together to highlight ways to address climate change and improve health, for example, through a broad health and environment climate coalition.

One important outcome of the conference was a commitment to measure the progress that countries are making in protecting health from climate change, and to report through the WHO UNFCCC Climate and Health Country Profiles project[84] and Sustainable Development Goal indicators. At the time of writing, the Climate and Health Country Profiles project had published profiles on forty countries.

Ultimately, WHO is a technical agency that faces financial and jurisdictional limitations, among others. Yet, it remains a critical international organization that is spearheading international cooperation to promote international climate law and policy that also advances global public health goals, particularly in the more vulnerable developing countries. The challenges WHO faces are both normative and operational, in significant part due to the transition from international to global health. This shift is underscored by the increasing vulnerability of geopolitical boundaries to communicable and noncommunicable diseases,

83 Second Global Conference, *Health and Climate*, July 7–8, 2015, Paris, France, www.who.int/globalchange/mediacentre/events/climate-health-conference/en/.

84 See World Health Organization, *WHO Health and Climate Country Profiles* www.who.int/globalchange/resources/countries/en/.

obsolesce of the distinction between national and international health, and the emergence of a multiplicity of international health actors.[85] These challenges are compounded by WHO's antipathy to more robust utilization of the full range of international legal instruments to promote global health and a focus on medico-technical approaches to health, even in the face of increasing determinants of health, such as the global environment including climate change, international trade, and investments largely shaped by international law.[86] However, WHO also faces the challenge of delicately balancing the requests of Member States and unilaterally undertaking normative or institutional reforms.[87] This adversely impacts on WHO's efforts to integrate substantive normative work into the body of its global health agenda. In the face of these global health challenges, repositioning health in international climate law requires a rethink through a functional framework that is argued in the concluding section of this chapter.

To sum up, WHO is the leading UN specialized agency working to address climate-driven public health impacts. The absence of discussion of other international organizations in this chapter owes to space limitations in addressing a sprawling topic rather than a deliberate decision to omit the importance of the work of other UN agencies and programs such as the World Meteorological Organization (WMO), United Nations Development Program (UNDP), United Nations Environment Program (UNEP), the Food and Agricultural Organization(FAO), the World Bank, the African Development Bank, or the Asian Development Bank whose climate change mandates often intersect *inter alia* with public health.

## 12.5 INTERNATIONAL CLIMATE CHANGE LAW AND HEALTH: DOMESTIC IMPLEMENTATION ACTIONS

Global climate change law imposes obligations on countries to implement measures that serve health objectives. Under the UNFCCC, at both the international and domestic levels, parties are called upon to cooperate on

85 See Obijiofor Aginam, WHO as a norm entrepreneur in global health governance, In: Michael Freeman et al., *Law and Global Health: Current Legal Issues*, 560–567 (Oxford: Oxford University Press, 2014).

86 D. P Fidler, *International Law and Infectious Diseases* (Oxford: Clarendon, 1999); Obijiofor Aginam, *Global Health Governance: International Law and Public Health in a Divided World* (Toronto: University of Toronto Press, 2005); Jennifer Prah Ruger and Derek Yach, *The Global Role of WHO, Global Health Governance*, Vol II, No.2 (Fall 2008/ Spring 2009); Derek Yach, *World Health Organization Reform – A Normative or an Operational Organization?* 106(11) Am. J. Public Health 1904–1906 (November, 2016).

87 Gian Luca Burci and Claude-Henri Vignes, *World Health Organization* (London: Kluwer Law International, 2004).

the development and elaboration of plans for coastal zone management, water resources, and agriculture; and to tackle desertification, drought, and floods.[88] Parties are required to consider public health implications of all measures or projects undertaken to mitigate or adapt to climate change. This is a key health-related commitment for parties, one that offers a legal pathway for accountability and litigation in case public health implications are not being considered by countries.[89] Parties are also called upon to undertake a variety of efforts within their own borders, including climate change-related education, information exchanges, and training that involves civil society organizations. In conformity with these obligations, a number of developing countries have adopted climate change legislation,[90] integrated climate change into existing national development plans,[91] developed specific national plans of climate action,[92] strengthened capacities for international reporting of their domestic climate change obligations,[93] reoriented their existing environmental laws,[94] or adopted entirely new legislation to address some requirements of the UNFCCC and or the Kyoto Protocol.[95] The 2015 Globe Report[96] reported that since 1997, the number of climate change laws and policies has doubled every five years. However, the status of health protection in these laws has not been reported or discussed. By 2014, several developing countries had adopted framework environmental laws to tackle climate change. These include Indonesia, Democratic Republic of Congo, Senegal, Tanzania, and Zimbabwe.[97] Other developing countries have modified legislation that promotes environmental management to also implement global climate

88 UNFCCC, Article 4(1)(e).

89 UNFCCC, Article 4(1)(f).

90 See COP22, *The Global Climate Legislation Study 2016 – Key Trends, Globe* (2016) www.lse.ac.uk/GranthamInstitute/wp-content/uploads/2016/11/The-Global-Climate-Legislation-Study_2016-update.pdf.

91 W. Onzivu, *Tackling the Public Health Impact of Climate Change: The Role of Domestic Environmental Health Governance Mechanisms in Developing Countries*, 43(3) The Int'l Lawyer 131 (2009), discussing National Action Plans of Bangladesh, Maldives, and Vanuatu.

92 International Climate Change Law, *Health Promotion and Evolving Domestic Climate Law in Developing Countries*, 18(1) African Yearbook of Int'l L. 257–285 (2012), discussing domestic climate laws and policies in Brazil, Philippines, and South Africa.

93 Ibid.

94 Ibid. citing framework environmental laws of South Africa.

95 See the Kenya Climate Change Act. No. 11 of 2016. Date of Assent: May 6, 2016. Date of Commencement: May 27, 2016, file:///C:/Users/onzivuw/Downloads/ClimateChangeAct110f2016.pdf.

96 Michal Nachmany et al., *The 2015 Global Climate Legislation Study*, Grantham Research Institute on Climate Change and the Environment, GLOBE and Inter-Parliamentary Union, 2015, www.lse.ac.uk/GranthamInstitute/wp-content/uploads/2015/05/Global_climate_legislation_study_20151.pdf.

97 Ibid.

change legal requirements. For example, the Uganda National Environmental Management Act has been converted to a framework law to help implement that country's climate commitments.[98] Similarly, the National Environmental Management Act of South Africa[99] has been utilized as an enabling act aiming at improving environmental management while facilitating sustainable development and improving coordination and governance of environmental issues on matters of climate change, including the reduction of emissions. While several developing countries have some soft policy basis for climate change actions,[100] most developing countries have not adopted comprehensive climate change laws or modified older framework laws, and their piecemeal legislation that pertains indirectly to climate change mitigation or adaptation tends to ignore issues such as health, which are often considered secondary in the implementation of a climate change regime. Notably, most do not have climate and health focal points in the Ministries of Health.[101] Even if more developing countries had adopted climate-specific domestic laws, or adapted existing laws and related mechanisms to implement their climate change obligations, there are limitations relating to how much health impacts of climate change can be addressed by framework environmental legislation in developing countries.[102] Even the adoption of climate-specific laws such as the Kenya Climate Change Act has not been the panacea to the normative integration of health into climate law and policy. For example, the Act does not mention health in the substantive body and establishes a National Climate Change Council chaired by the President of Kenya, whose role is to provide leadership to domestic efforts to tackle climate change.[103] Yet, the council's membership includes environment, energy, finance, local government, and economic planning ministries and excludes the Health Ministry.[104] The case of Kenya's Climate Change Law demonstrates other limitations of laws that regulate climate change in

98 Uganda, National Environment Statute, 7–8, LXXXVIII Uganda Gazette 21 Supp. (1995).

99 South Africa, National Environmental Management Act No 62 of 2008 (G-31789), (www.environment.co.za/documents/legislation/, accessed July 4, 2017.

100 See *WHO Health and Climate Country Profiles* 2015 www.who.int/globalchange/resources/country-profiles/en/.

101 Ibid.

102 W. Onzivu, *International Environmental Law, the Public's Health and Domestic Environmental Governance in Developing Countries*, Am. University Int'l L. Rev. (2006), 597.

103 The Kenya Climate Change Act. No. 11 of 2016. Date of Assent: 6th May, 2016. Date of Commencement: 27th May, 2016, file:///C:/Users/onzivuw/Downloads/ClimateChangeAct11of2016.pdf.

104 Ibid. at Section 7.

developing countries. The challenges to the implementation of health goals include weak compliance with and enforcement of adopted laws, weak administrative and institutional coordination to achieve set legislative goals, inconsistent policies and legal requirements across health and other sectors, weak technical capacity, and a lack of law implementation resources.[105] Furthermore, efforts to address health issues in the context of tackling climate change must also build on existing public health laws, which are, in most developing countries, obsolete or archaic, having been adopted in the first half of the twentieth century. Therefore, efforts to integrate health into the corpus of domestic climate law and policy require a rethinking of frameworks for the progressive development and implementation of legal regimes that are flexible, adaptable, and effective. In the face of these challenges, the role of WHO in supporting Member States on developing and implementing climate legislation that leverages health promotion is pivotal. However, WHO has not yet built its own capacity in the provision of legal technical assistance to countries to promote health-friendly climate legislation. For example, WHO's work plan on climate and health (2014–2019), while highlighting policy support in climate change and health, does not specify how support for the development of climate change legislation will be implemented in the program.[106]

## 12.6 TOWARD EMBEDDING GLOBAL HEALTH IN THE PROGRESSIVE DEVELOPMENT AND IMPLEMENTATION OF INTERNATIONAL CLIMATE LAW AND POLICY: THE ROLE OF ADAPTIVE GOVERNANCE

This chapter has highlighted challenges facing health in international climate legal regimes at three levels: first, despite the progress, its substantive, procedural, and institutional frameworks have not fully embraced global health needs; second, international health actors require renewed pathways to better engage with international climate regimes; and third, the domestic frameworks for the implementation of both international health and climate law are inadequate and require significant reforms to ensure the effective integration of health in domestic climate change governance. The author proposes three areas of reform: first, at the UNFCCC COP, the creation of a subsidiary technical body on health to advise the UNFCCC COP on health in the

[105] Onzivu, *Health in Global Climate Change Law*, *supra* note 65, at 364.

[106] See WHO work plan on climate change and health, 2014–2019, endorsed by the Executive Board of the World Health Organization at its 136th Session, February 2015, www.who.int/globalchange/health_policy/climate-change-and-health-workplan-2014-2019.pdf.

progressive development and implementation of international climate law; second, the creation of an advisory Committee at the World Health Organization to advise WHO on its health and climate change leadership within the jurisdiction and mandate of the organization and its work as an international actor in the progressive development and implementation of international climate law; and third, a call on Parties to the UNFCCC who are also Members of the World Health Organization to undertake deliberate legal reforms to embed health and climate change in domestic law and policy. Finally, these reforms can only be underpinned by a framework to effect these reforms, and the author advances the functional theory of adaptive governance to realize these reforms.[107]

*First*, the establishment of a UNFCCC technical body on health and climate change could provide an important forum to contribute to the optimization of achievement of health goals in international climate law. It could also help identify areas for further work by the COP on the intersections of health and climate change. The UNFCCC establishes a Conference of Parties to keep under regular review the implementation of the Convention and any related legal instruments that it adopts and take decisions within its mandate for the effective implementation of the Convention.[108] The UNFCCC COP is also empowered to establish such subsidiary bodies as deemed necessary for the implementation of the Convention.[109] Furthermore, under the Paris Climate Agreement, the COP to the UNFCCC is empowered to keep under regular review the implementation of this Agreement, take decisions necessary to promote its effective implementation, and establish such subsidiary bodies as deemed necessary for the implementation of this Agreement.[110] Since the adoption of the UNFCCC, several bodies have been established to support the work of the COP. These include among others the Subsidiary Body for Scientific and Technological Advice (SBSTA), Subsidiary Body for Implementation (SBI), Adaptation Committee, The Paris Committee on Capacity-building (PCCB), and the Ad Hoc Working Group on Long-term Cooperative Action under the Convention (AWG-LCA), among others.[111] Establishing a subsidiary body on health would be an innovative action toward addressing sectoral challenges to health and providing a pathway to engagement of other sectors to achieve an integrated functional international climate legal regime. It could commence

107 UNFCCC, Article 4(1)(i), Article 6.
108 UNFCCC, Article, 7(1)(2).
109 UNFCCC, Article 7(2)(i).
110 Paris Agreement, Article 16 (1, 4.).
111 See UNFCCC, *Organogram of its Bodies* http://unfccc.int/bodies/items/6241.php.

as an ad hoc body and evolve into a standing committee reporting under the Paris Agreement or fully under the UNFCCC. Its membership could be open ended, and inhabited by governments, experts in global health and climate change, and intergovernmental and nongovernmental organizations. The body could report to the COP through, for example, the Subsidiary Body for Scientific and Technological Advice (SBSTA).

*Second*, the creation of a WHO Advisory Committee on Health and Climate Change can be a critical institutional change that will catalyze WHO's normative, operational, and institutional leadership in global health and climate change. The Constitution of WHO provides it a wide mandate on matters of global health. The Constitution establishes the World Health Assembly, the Executive Board, and the Secretariat and empowers them to support the work of the organization through decision-making.[112] Among the functions of the World Health Assembly is the establishment of committees deemed necessary for the work of the Organization.[113] In this context, WHO has established expert panels, committees, or study groups. By December 2016, WHO had established a total of forty-four expert panels and advisory groups.[114] These included those meeting regularly, such as WHO Expert Committee on Biological Standardization (ECBS), Joint FAO/WHO Expert Committee on Food Additives (JECFA), WHO Expert Committee on Specifications for Pharmaceutical Preparations, WHO Expert Committee on Selection and Use of Essential Medicines, and WHO Expert Committee on Drug Dependence. Others include those on ageing and health, health promotion, health, science and technology policy, and injury and violence prevention.[115] Due to the high prioritization of climate change globally, the advisory Body on climate change and health should be funded, empowered with a clear and broad mandate, and shall meet regularly to set and implement an advisory agenda for WHO's leadership in climate change and health.

*Third*, systematic domestic legal reforms are required to fully implement international climate law to leverage health protection benefits in developing countries. WHO has developed key indicators to develop domestic climate and health policies.[116] These elements include requirements for a national

112 WHO Constitution, Article 9.

113 WHO Constitution, Article 18(e).

114 WHO Executive Board, Reports of advisory bodies, Document EB140/50 Add.1 140th session, December 12, 2016, http://apps.who.int/gb/ebwha/pdf_files/EB140/B140_50Add1-en.pdf?ua=1.

115 Ibid. Annex

116 WHO, *Climate and Health Country Profiles 2015, a Global Overview* 20, http://apps.who.int/iris/bitstream/10665/208855/1/WHO_FWC_PHE_EPE_15.01_eng.pdf?ua=1; See also WHO *Operational Framework for Building Climate Resilient Health Systems*, http://apps.who.int/iris/bitstream/10665/189951/1/9789241565073_eng.pdf?ua=1.

focal point for climate change in the Health Ministries, a national health adaptation strategy approved by the relevant Government entity, and reporting to the UNFCCC that includes health implications of climate change mitigation policies. The domestic policy must also include implementation of projects or programs on health adaptation to climate change, and entail actions to build institutional and technical capacities to work on climate change and health, conduct of a national assessment of climate change impacts, vulnerability and adaptation for health, inclusion of climate information in Integrated Disease Surveillance and Response (IDSR) systems, including development of early warning and response systems for climate-sensitive health risks, and implementation of activities to increase climate resilience of health infrastructure. It also includes the need for budgetary allocation and valuation of health co-benefits in climate change mitigation. However, a significant aspect of these indicators are operational, and even if some countries may have all or some of these in their policy and operational plans, the lack of a coherent law to embed these indicators can fall short of integrating health into domestic implementation of international climate law. This is due to the importance of including all substantive, institutional, procedural, and compliance aspects of a domestic climate law that often are not sectoral in approach yet must meet the needs and objectives of health. Therefore, there is a need for developing countries to adopt comprehensive climate laws that incorporate key sectoral challenges to health.

The establishment of advisory bodies at UNFCCC COP and WHO, as well as effective domestic legal reforms, cannot be undertaken in a structural and systemic vacuum. Promoting global health in international climate law and policy requires functional frameworks of action in which legal and policy interventions can be implemented, reviewed, and improved. Indeed, international climate law has provided some space to address global public health challenges. WHO and partners have increased efforts to integrate health concerns in the progressive development and implementation of international climate law. However, these efforts have had significant limits. First, international climate laws continue to focus on mitigation, energy issues, or forestry issues and have not integrated health perspectives to climate change mitigation. Second, adaptation, where health is posited, does not receive the priority it deserves in developing countries. Third, generally, the body of international climate law poses an ambivalence for health protection with limited clarity on the scope of health obligations. Fourth, in developing countries the implementation of international climate law is piecemeal and has limits in addressing sectoral concerns such as health and its interactions with climate change. For example, intersectoral coordinating mechanisms are weak among the various key sectors included in climate change legislation, such as forestry, agriculture, energy, water, environment generally, and land use and health. Yet, climate change threats and their impact on health are transcendent in nature, requiring new tools and legal requirements. It is in view of these challenges that the need for renewed thinking on pathways to better integrate health in international climate law becomes critical.

Efforts to counteract the health impacts of climate change through international climate law and policy and through the work of international actors such as WHO have limitations. Climate change laws and policies evolve based on emerging evidence, political commitment of Parties, and the impact of laws and policies in promoting mitigation and adaptation. Implementation of commitments, including those in relation to health at the global level, has been slow. These challenges reflect the governance problem in international climate law. Climate change is a problem to which solutions cannot readily be found in existing mechanisms and which persists even after the application of the best-available practices, driving the emergence of experimentalist governance due to its high complexity, low visibility, and global dimension.[117] These characteristics point toward the experimentalist idea of strategic uncertainty, where traditional problem-solving strategies fail.[118] On the other hand, climate change certainly constitutes a collective action problem, one that cannot be solved by a single entity but requires the cooperation of others who often have disparate interests and incentives, raising the costs of transacting or negotiating a cooperative solution.[119] Therefore, experimentalist experiences can make a valuable contribution to climate change governance, providing a new framework for dealing with the failure of traditional problem-solving strategies and multilevel cooperation in decision-making.[120]

In this context, promoting global public health in international climate law requires a concrete functional framework to improve policy and practice in health and climate law at the domestic level. A persuasive functional framework to strengthen health in both international and domestic climate law and

[117] R. J. Lazarus, *Super Wicked Problems and Climate Change: Restraining the Present to Liberate the Future*, 94 Cornell L. Rev. 1153 (2009).

[118] G. de Búrca et al., *Global Experimentalist Governance*, 44 British J. of Pol. Sci. 477 (2014).

[119] D. H. Cole, *Climate and Collective Action*, 61(1) Current Legal Problems 229, 232 (2008).

[120] Chiara Armeni, *Global Experimentalist Governance, International Law and Climate Change Technologies*, 64(4) Int'l and Comparative L. Quarterly 875–904, at 885 (2015).

policy is adaptive governance. Adaptive governance institutions are those "capable of generating long-term, sustainable policy solutions" to complex and dynamic natural resource problems through collaboration among diverse resource users and governmental agencies.[121] An ideal adaptive governance structure would react to surprises not only from the ecological system but also from human institutions and ecological sustainability.[122] Adaptive management focuses on the interaction of resource management and science, and recognizes that because science is constantly evolving our understanding of natural systems or the effect of human interactions on these systems is rarely, if ever, complete.[123] Adaptive management monitors outcomes and maintains flexibility so that policies can be altered should predictions prove inaccurate or scientific understanding advances. This theory would ensure that the international and domestic climate change laws and policies in developing countries are dynamic and can respond to evolving public health challenges of climate change.

Adaptive governance rooted in physical and biological sciences has its origins in complexity theory.[124] At the core of complexity theory is not a cluster of unrelated systems but a complex adaptive system.[125] The relationships between the interrelated parts of this system – the combination of their autonomy and interaction and the capacity of the system for self-adaptation and self-reorganization – and the potential for new properties emerge.[126]

Three common themes have evolved in the social sciences: the complexity and uncertainty of social life; the need for intensified social learning to counteract this uncertainty; and the role of governance systems in facilitating such learning. This has necessitated research into organizational processes and the creation of reflexive institutional frameworks to facilitate learning and innovation across institutions. Moreover, international relations scholarship on states in complex interdependence has focused on developing flexible and adaptive institutions that respond to global change. Furthermore, work in public law, on learning-centered alternatives to traditional command, and

121 J. T. Scholz and B. Stiftel, *Adaptive Governance and Water Conflict: New Institutions for Collaborative Planning*, 5 (Washington, DC: Resources for the Future Press, 2005).

122 Ibid. at 2.

123 J. B. Ruhl, *Thinking of Environmental Law as a Complex Adaptive System: How to Clean Up the Environment by Making a Mess of Environmental Law*, 34 Houston L. Rev. 933 (1997).

124 P. Cilliers, *Complexity and Post Modernism*, 112–140 (London: Routledge, 1998).

125 J. N. Rosenau, *Distant Proximities: Dynamics Beyond Globalization*, 212–213 (Princeton, NJ: Princeton University Press, 2003).

126 J. N. Rosenau, NGOs and fragmented authority in globalizing space. In: Yale Ferguson and Barry R. J. Jones, *Political Space: Frontiers of Change and governance in a Globalizing World*, 261–280 (New York: State of New York Press, 2002).

on regulatory frameworks has spurred experimentalist new governance approaches.[127] The integration of the social sciences enriches adaptive governance, transcending natural, ecological, human, and social processes and institutions. Adaptive policy making has been viewed as cautious, macroscopic, incremental, experimental, well-grounded, context dependent, flexible, provisional, accountable, and sustainable.[128] Adaptive governance can better inform health and climate change law and policy if the following pillars are integrated into this theory. The importance of institutional learning in adaptive governance resonates with the above proposals to establish subsidiary/advisory forums at the UNFCCC COP and WHO, respectively, to promote health in international climate law.

*First*, environmental justice is a key pillar to strengthen adaptive reforms. The participation of the health sector in climate law and policy development is crucial to the effective optimization of health into the corpus of the progressive development and implementation of international climate law. An environmental justice framework offers an effective support pillar to embed public health in international climate law and policy by promoting the effective participation of the health and other sectors in international climate change regulation.[129] The drafting, negotiation, adoption, and application of environmental laws (such as climate law instruments) induce justice considerations such as the distributive and corrective effects of laws and decisions pertaining to health, the environment, and natural resources.[130] Well-established concepts in environmental law, whether based on custom or statutes, appear neutral on their face; however, a closer study may reveal disproportionate burdening, or restricting effects for certain groups and/or categories once the concepts are applied. An environmental justice framework ensures full participation of all actors and provides mechanisms for the full recognition of health actors as legitimate inhabitants of international and domestic

[127] P. Hill, *Understanding Global Health Governance as a Complex Adaptive System*, 6(6) Global Pub. Health 59 (2011).

[128] R. S. Whitt,'*Adaptive Policymaking: Evolving and Applying Emergent Solutions for U.S. Communications Policy*, 61(3) Federal Communications L. J. 483, 500–504 (2009).

[129] See C. Okereke and P. Coventry, *Climate Justice and the International Regime: Before, During, and After Paris,* 7 WIREs Clim. Change 834–851 (2016), doi:10.1002/wcc.419 on the importance of a justice paradigm in assessing international climate law.

[130] J. Ebbesson, Introduction. In: J. Ebbesson and P. N. Okowa, *Dimensions of Justice in Environmental Law in Environmental Law and Justice in Context*, 1 (Cambridge: Cambridge University Press, 2009).

climate regulation.[131] For communities that already bear the brunt of deleterious public health impacts, such as the resurgence of malaria, heatwaves, and vector-borne diseases in Africa, an environmental justice paradigm would ensure that equity, recognition, and participation are integrated. This enables the embedment of health in international and domestic climate change law. Recognition ensures that public health is a core inhabitant of international climate change legal regime.[132] This provides a concrete framework for optimizing health protection in climate change legal regimes, and helps to tackle the public health burdens of climate change that unequally fall on the third world.[133]

*Second*, strengthening health vis-à-vis climate change requires an environmental ethics framework because it will enable a balance between the protection of the natural climate and human health. Environmental ethics provide an important theoretical basis to reinforce adaptive governance because it requires human stewardship for the natural world and managing it sustainably for future generations.[134] This entails responsibility on Parties to the UNFCCC and related agreements to support health capacity building in tackling climate change. This perspective of environmental ethics reflects the basis of sustainable development and its three pillars.[135] To sum up, an environmental ethic provides effective bedrock to strengthen adaptive governance as a functional tool to optimize health protection in international climate law.

*Third*, introducing reforms in international climate governance such as new bodies under the UNFCCC COP, at WHO, or in domestic climate

[131] *United Nations Millennium Declaration*. New York: United Nations, 2000. United Nations General Assembly Resolution 55/2, www.un.org/milleennium/declaration/ares552e.pdf, para.6. The United Nations Millennium Declaration states that, "global challenges must be managed in a way that distributes the costs and burdens fairly in accordance with basic principles of equity and social justice. Those who suffer, or benefit least, such as those from the third world, deserve assistance from those who benefit most."

[132] WHO, *Closing the Gap: Policy into Practice on the Social Determinants of Health: Discussion Paper*, World Conference on Social Determinants of Health, Rio De Janeiro, Brazil, October 19–21, 2011, 3, 17–21, 33, 42.

[133] J. Prah Ruger, *Global Health Justice*, 2 Public Health Ethics 261–275, 261–263 (2009).

[134] M. Stenmark, *Environmental Ethics and Policy Making* (Aldershot: Ashgate Publishing Company, 2002); M. Malden et al., *Environmental Ethics: An Anthology* (Oxford: Blackwell, 2003); A. Brennan and Y.-S. Lo, *Understanding Environmental Philosiphy*.

[135] World Commission on Environment and Development, *Our Common Future* (Oxford: Oxford University Press, 2007); Experts Group on Environmental Law of the WCED, *Environmental Protection and Sustainable Development*, 46 (London: Graham & Trotman, 1986); Rio Declaration, 31 ILM 874 (1992), Principle 1, endorsed by the UNGA Res. 47/190 (1992), para. 2.; M. C. Segger and A. Khalfan, *Sustainable Development Law: Principles, Practices and Prospects* (Oxford: Oxford University Press, 2004).

governance requires evaluation. The evidentiary basis and the impact of international climate law and policy in achieving its objectives are fundamental principles of the legal regime. It is in this context that placing public health at the core of climate and health discourse requires an evaluation of the degree of health promotion and protection in international climate change law. Evaluation refers to a careful retrospective assessment of the merit, worth, and value of administration, output, and outcome of government interventions, which is intended to play a role in future practical action situations.[136] Across jurisdictions in the developing world, there is no mandatory requirement for a systematic and consistent evaluation of legislation. Law Reform Commissions in developing countries are often underfunded and unable to highlight obsolete or weak legislation, including those that fail to come up with ways of tackling climate change and its impacts. While evaluation has been institutionalized in some parts of the developed world, this is largely not the case in the developing world. This is because it is a costly exercise that requires enormous financial and technical capital that the developing world may not prioritize. Despite the weak legal regimes in existence today, evaluation is a promising tool to integrate health protection in the progressive development and implementation of international climate law.

*Finally*, embedding health in the progressive development and implementation of international climate law and policy will also require multisectoral coordination at both the global and domestic levels. The membership of key institutional forums of the UNFCCC, the Paris Climate Agreement, and the domestic levels will need to be expanded to address human health dimensions of climate change. Significant membership from the health sector, as well as expanding its mandate will help it to integrate public health issues within its remit. This will require a collaborative approach to international climate change and health governance to ensure harmonized coordinated actions and legal strategies. The constituents of environmental law and policy often involve the environmental, health, agriculture, and other sectors that may be involved in the objectives of a specific environmental treaty or other law. Collaboration needs to occur among the inhabitants of an environmental regulatory space. Both the UNFCCC and the Paris Climate Agreement have created subsidiary mechanisms to spearhead implementation actions. While health sector membership of such mechanisms is significant, the creation of

136 E. Vedung, *Public Policy and Program Evaluation*, 2–3 (New Brunswick, NJ: Transaction Publishers, 2000).

additional sectoral bodies or committees in the context of multisectoral action as a key pillar of adaptive governance can catalyze actions at the intersection of human health and climate change both on the international and domestic planes.

## 12.7 CONCLUSION

While the progressive development and implementation of international climate law has recognized the importance of integrating health dimensions within its corpus, there are still fundamental gaps and challenges impeding a fuller integration of health into the functioning of the regime. The place of health in the UNFCCC structures and improved pathways in the adopted Paris Climate Agreement demonstrate the potential promise for harmonizing health into international climate law to effectively attain both health and climate goals. In this context, the catalytic role of the World Health Organization and domestic legal reforms in developing countries can greatly improve the status quo. To address the gaps in health climate regime integration, the creation of international advisory bodies at both the UNFCCC and WHO, coupled with domestic legal reforms, will improve the legal regime. However, these reforms need to be underpinned by a systematic functional framework of adaptive governance. This framework also needs to be reinforced by an environmental justice, ethical, law, and policy evaluation and multisectoral coordination to ensure its efficacy in improving legal and policy pathways to better position health in international climate law and policy.

# 13

# How Existing Environmental Laws Respond to Climate Change and Its Mitigation

Justin Gundlach

*Existing environmental laws interact with public health priorities and with aspects of the changing climate in numerous and varied ways. This chapter does not attempt to catalogue those interactions, but instead focuses on two that are especially important and illustrative of the operation and limitations of existing environmental laws vis-à-vis climate change-driven challenges. The first interaction is between pollution levels boosted by climate change and pollution control laws that employ health-based standards to determine pollution limits. The second is between a wider array of existing laws and the effects of climate change mitigation measures on public health. Examining these interactions reveals the inadequacy of existing laws to the tasks of (1) tracking the public health impacts of – much less adapting to – climate change, and (2) ensuring that climate change mitigation efforts reflect a rational accounting of impacts on public health, whether from foregoing mitigation or undertaking it.*

## 13.1 INTRODUCTION

Existing environmental laws interact with public health priorities and with aspects of the changing climate in numerous and varied ways. This chapter does not attempt to catalogue those interactions, but instead focuses on two that are especially important and illustrative of the operation and limitations of existing environmental laws vis-à-vis climate change-driven challenges. The first interaction is between pollution levels boosted by climate change and pollution control laws that employ health-based standards to determine pollution limits. The second is between a wider array of existing laws and the effects of climate change mitigation measures on public health. Examining these interactions reveals the inadequacy of existing laws to the tasks of (1) tracking the public health impacts of – much less adapting to – climate change, and (2) ensuring that climate change mitigation efforts reflect a rational accounting of impacts on public health, whether from foregoing mitigation or undertaking it.

## 13.2 POLLUTION CONTROL STATUTES: CLIMATE CHANGE AND THE AMPLIFICATION OF PUBLIC HEALTH IMPACTS

As climate change alters the pathways through which pollution arrives at points of contact with people, it often causes the same volume of released pollution to result in a higher degree of human exposure. For instance, the same volume of agricultural runoff will generate more – and more virulent – algae in warmer water.[1] The case is similar for ozone, which forms more readily from the same quantity of precursors amid higher ambient air temperatures. Thus, as the US Global Change Research Program's assessment of climate change and health concludes: "Climate change will make it harder for any given regulatory approach to reduce ground-level ozone pollution in the future as meteorological conditions become increasingly conducive to forming ozone over most of the United States."[2] The report also points out the clear and unavoidable implication of this trend: "*Unless offset by additional emissions reductions*, these climate-driven increases in ozone will cause premature deaths, hospital visits, lost school days, and acute respiratory symptoms."[3] But pollution control statutes will not automatically effectuate those offsetting reductions, whether in relation to ozone formation or other forms of pollution. Indeed, it is unlikely that they will even signal clearly that such reductions are necessary to maintain the current public health status quo.[4]

With the exception of the Endangered Species Act, which focuses on the protection of other species, environmental protection statutes – and certainly all pollution control statutes – make their basic objective the protection of public health and welfare. Pollution control statutes, however, apply different

1 See Rob Herman, *Toxic Algae Blooms Are on the Rise*, Scientific American (Blog), Sept. 7, 2016, https://perma.cc/5D3U-FMF2; see also Robin Kundis Craig, *Climate Change Comes to the Clean Water Act: Now What?* 9 Wash. & Lee J. Energy, Climate, & Env't 28–31 (2009) ("if waters are warming because of both climate change impacts and discharges of heated effluent, reducing the point source discharges may forestall the worst effects of increasing temperatures and hence increase the resilience of the species and ecosystems in those waters").

2 U.S. Global Change Research Program, *Climate and Health Assessment, Key Finding 1: Exacerbated Ozone Health Impacts*, https://health2016.globalchange.gov/air-quality-impacts#finding-42 (emphasis added).

3 Ibid.

4 At least one commentator goes a step farther, arguing that the environmental impacts of climate change will both exacerbate pollution and dampen public support for effective environmental regulation. Matthew D. Zinn, *Adapting to Climate Change: Environmental Law in a Warmer World*, 34 Ecology L.Q. 61, 65 (2007).

standards to the protections they impose.[5] Under the Clean Air Act, for instance, standards fall into three categories. Some are set using a cost-benefit analysis that balances public health risks against polluters' costs. Others are set based on "feasibility," meaning that polluters must use whatever pollution control technologies have been shown to be both maximally effective and not so expensive as to drive an industry into bankruptcy. Still others – of particular interest in this chapter – are "health-based," meaning that they are to reflect the level of pollution that research has identified as safe for humans to encounter in the ambient environment.

A limit prescribed by this third category starts with a fixed value, often derived from dose-response or concentration-response curves that measure the correlation between the risk of a particular adverse health outcome and increasing levels of exposure to a particular pollutant.[6] Levels of risk on a dose-response curve are sensitive to the concentration of a given pollutant and the status of the human exposed to it (e.g. child, immuno-compromised, pregnant, exercising[7]), but are generally abstracted from circumstances at the point of exposure, like the ambient air temperature. In addition to this fixed, abstract value, pollution limits also incorporate factors that reflect how the pollution at issue enters, travels through, and persists in the ambient environment. While climate change does not affect the fixed value, it does alter the dynamic ones.[8]

### 13.2.1 *Two Examples of Health-based Standards amid a Changing Climate*

The following examples – the Clean Air Act's regulation of ground-level ozone and the Resource Conservation and Recovery Act's regulation of coastal hazardous waste transfer and storage facilities – illustrate how health-based standards *could* make pollution control laws a spur to adaptation.[9]

5 Congressional Research Service, *Cost and Benefit Considerations in Clean Air Act Regulations* 3–7 (May 2017), https://perma.cc/YMH9-DVF6 (listing statutory provisions that instruct, permit, or prohibit the Administrator from taking cost into account when setting a particular standard).

6 See, e.g., EPA, *Integrated Science Assessment of Ozone and Related Photochemical Oxidants* 2–32 to 2–34 (2013).

7 Ibid. at 5–18 to 5–19 (discussing effects of activity); 5–64 to 5–67 (discussing effects of life stage).

8 See P. C. D. Milly et al., *Stationarity is Dead: Whither Water Management?* 5863 Science 573–574 (Feb. 2008) (discussing disruptive effects of climate change on models that inform water policy).

9 See Daniel A. Farber, *Climate Adaptation and Federalism: Mapping the Issues*, 1 San Diego J. Climate & Energy L. 259, 266 (2009) ("climate impacts might be managed in part through current federal law.").

### 13.2.1.1 Ground-Level Ozone under the Clean Air Act

Ground-level ozone does not emerge from smoke stacks or tailpipes, but rather forms when nitrogen oxide (NOx) and volatile organic compounds (VOCs) – which do emerge from smoke stacks or tailpipes – comingle in the presence of sunlight or another heat source.[10] Thus, even if levels of NOx and VOCs remained flat in the coming decades, ozone's prevalence in the ambient air would likely grow as average temperatures rise across the United States.[11] Notably, over the past several decades, levels of ozone in the ambient air have fallen far less than other pollutants subject to regulation under the Clean Air Act.[12]

The Clean Air Act directs EPA and the various states to take a series of regulatory steps with respect to ground-level ozone.[13] The first is adoption by EPA of a National Ambient Air Quality Standard (NAAQS) – a health-based limit above which concentrations of ground-level ozone in the ambient air have been shown to do harm to public health and welfare.[14] To determine that limit, EPA seeks the advice of the Clean Air Scientific Advisory Committee (CASAC), which evaluates available scientific evidence and issues recommendations about where the limit should fall and how to measure compliance with it.[15] The Clean Air Act, as interpreted by the Supreme Court, directs EPA to treat ozone as a "non-threshold contaminant," meaning that the NAAQS for ozone can theoretically be set to zero.[16] Once the NAAQS is set, other

[10] EPA, *Ozone Pollution: Ozone Basics*, https://perma.cc/3EEW-2ZKZ (updated Apr. 5, 2017).

[11] See, e.g., Perry E. Sheffield et al., *Modeling of Regional Climate Change Effects on Ground-Level Ozone and Childhood Asthma*, 41 Am. J. Preventive Med. 251, 255 (2011), https://perma.cc/93RP-W2YP (estimating increases in emergency room visits to treat acute episodes of pediatric asthma due to effects of climate change on ozone formation).

[12] EPA, *Air Trends: Air Quality-Trends and Summary*, https://perma.cc/SA4C-QNCR (last updated July 26, 2017).

[13] EPA, *Process of Reviewing the National Ambient Air Quality Standards*, https://perma.cc/8ZRK-FW7F (last updated Dec. 1, 2016).

[14] See Clean Air Act §§ 108, 109; 40 C.F.R. §§ 50.09 (primary ozone NAAQS), 50.10 (secondary).

[15] Clean Air Act § 109(d)(2) (establishing independent scientific review committee to recommend new or updated NAAQS to EPA Administrator); EPA Clean Air Scientific Advisory Committee (CASAC): *Charter*, https://perma.cc/E5UA-UTFL;7 see also *Integrated Science Assessment (ISA) of Ozone and Related Photochemical Oxidants* (Final Report, Feb 2013); History/Chronology (Apr. 4, 2017), https://perma.cc/4JZN-V6JY; EPA, 78 Fed. Reg. 11,172, 11,172 (Feb. 15, 2013) (describing steps of ozone NAAQS development involving CASAC). But see also Michael A. Livermore and Richard L. Revesz, *Rethinking Health-Based Environmental Standards*, 89 N.Y.U. L. Rev. 1184 (2014) (criticizing EPA's approach to deriving nominally health-based NAAQS in a way that manifestly but secretly takes non-health factors into account).

[16] See *American Trucking Ass'ns* v. *EPA*, 175 F.3d 1027, 1034 (D.C. Cir. 1999), rev'd on other grounds sub nom Whitman v. *American Trucking Ass'ns* (2001) (citing 62 Fed. Reg. 38,863 (July 18, 1997) and 61 Fed. Reg. 65,637, 65,651 (1996)). Livermore and Revesz have pointed out

components of the Clean Air Act's machinery make it a national yardstick: Areas where ozone concentrations do not exceed it are in "attainment," and those few areas where ozone concentrations do exceed it are in "nonattainment."[17] (See Figure 13.1.)

As Figure 13.1 shows, there are gradations of nonattainment, ranging from marginal to extreme. Different rules apply to sources of pollutants that generate ozone precursors in attainment and nonattainment areas; the rules governing nonattainment areas reflect the extremity of each area's noncompliance.[18] Perhaps most importantly for this chapter's purposes, in nonattainment areas, major new or modified emitting facilities may only receive a permit to pollute if they demonstrate that they will achieve the Lowest Achievable Emission Rate, which the Clean Air Act defines as either the most stringent emission limitation prescribed by any state for that source category, or the most stringent limitation actually achieved by that source category.[19]

Actually dealing with sources of ozone precursors is a task for the state agencies that update and enforce compliance with State Implementation Plans (SIPs), and for the state attorneys general who seek to enforce compliance by other states with the Cross State Air Pollution Rule's restrictions on emissions that make downwind states' compliance with their own SIPs more difficult.[20] While states operate with guidance and subject to EPA oversight,[21] their approaches are not mechanistic, even with respect to measuring pollution levels, which has been shown to be strategic: In marginal areas, the strategy aims to avoid thresholds for a higher category of classification.[22] Similarly, state-level action to address pollution does not follow uniformly or

that EPA does not actually adhere to the logic of a "non-threshold" health-based standard with respect to all non-threshold pollutants and often adopts limits that reflect other considerations as well. Livermore and Revesz, *Rethinking Health-Based Environmental Standards*, *supra* note 15, at 1184, 1188–89, 1246.

17 40 C.F.R. § 51.491 (defining "attainment area" and "nonattainment area").

18 42 U.S.C. §§ 7501–7515 (prescribing state implementation plan requirements for nonattainment areas); *Id.* § 7511d (imposing fines on exceedance of prescribed limits in severe or extreme nonattainment areas).

19 42 U.S.C. § 7501(3).

20 See, e.g., Complaint for Injunctive Relief, *Maryland* v. *Pruitt*, 1:17-cv-02873-JKB (D. Md. Sept. 27, 2017) (alleging that NOx emissions from at least 36 power plants in upwind states constitute a violation of the Clean Air Act's "good neighbor" provision).

21 See EPA, *Technology Transfer Network / NAAQS Ozone Implementation: Related Documents and Data*, https://archive.epa.gov/ttn/ozone/web/html/related-3.html (last updated Feb. 21, 2016) (collecting federal guidance and state-submitted documents).

22 Corbett Grainger et al., *How States Comply with Federal Regulations: Strategic Ambient Pollution Monitoring* (Oct. 2016) https://perma.cc/A4WP-JH3L (finding that state and local authorities place monitoring stations strategically to affect results).

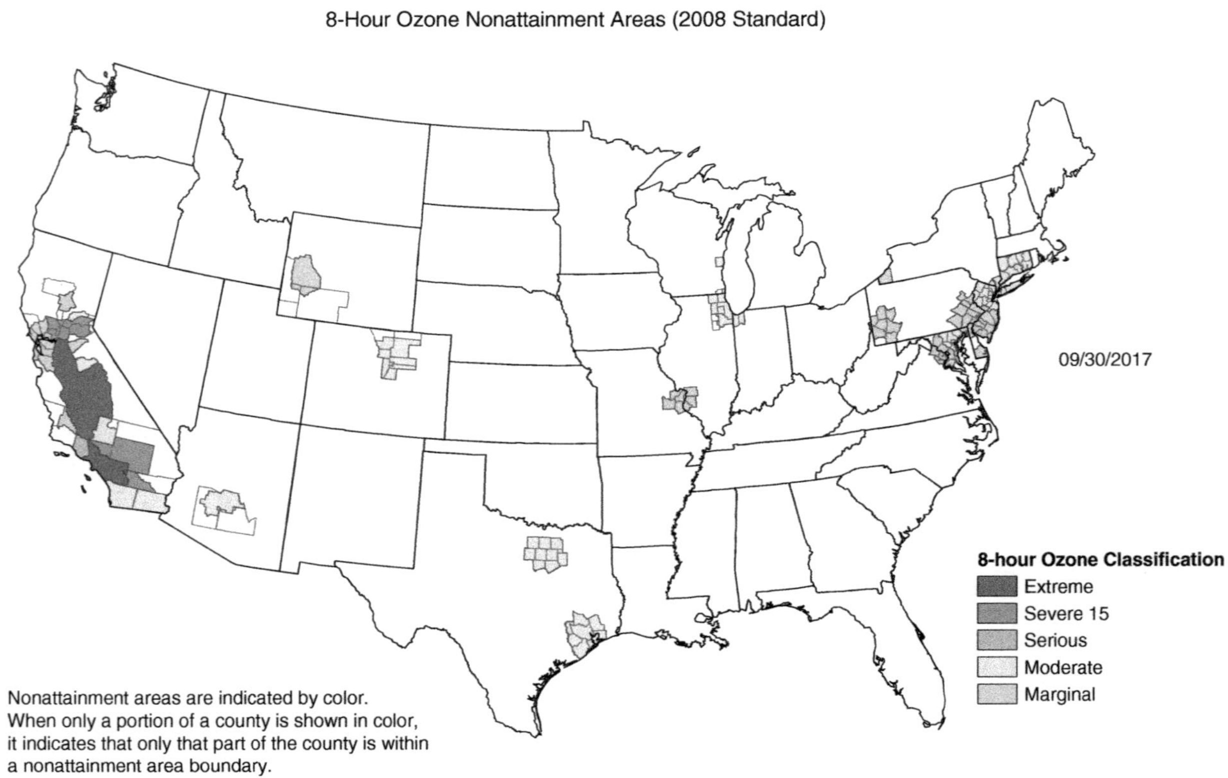


FIGURE 13.1 US counties, by attainment status under 2008 8-hour ozone standard

straightforwardly from setting a NAAQS, but often reflects diverse approaches of variable rigor.[23]

#### 13.2.1.2 The Resource Conservation and Recovery Act and Coastal Hazardous Waste Facilities

"[The Resource Conservation and Recovery Act (RCRA)]'s primary purpose... is to reduce the generation of hazardous waste and to ensure the proper treatment, storage, and disposal of that waste which is nonetheless generated, 'so as to minimize the present and future threat to human health and the environment.'"[24] To accomplish this goal, which expressly contemplates future circumstances but says nothing about climate change, RCRA imposes various requirements on the entities involved in managing solid or hazardous wastes.[25] Some of those requirements relate to the detailed tracking of wastes, which in turn facilitates identification of instances of "release," i.e. failures to prevent those wastes from migrating out of the containers, tanks, or impoundments used to transport, store, or dispose of them. Other requirements relate to facility design and management; as New Jersey's Department of Environmental Protection summarizes, a RCRA permit "outlines facility design and operation, lays out safety standards, and describes activities that the facility must perform, such as monitoring and reporting. Permits typically require facilities to develop emergency plans, find insurance and financial backing, and train employees to handle hazards."[26]

Much as rising temperatures promise to generate more ground-level ozone from the same volume of NOx and VOCs, rising sea levels promise to heighten the prospective risk of release at coastal hazardous waste facilities due to various forms of flooding – risks illustrated dramatically by Hurricane

[23] Jinghui Lim, *The Impact of Monitoring and Enforcement on Air Pollutant Emissions*, 49 J. Reg'y Econ. 203, 220 (Apr. 2016), https://perma.cc/7LQ3-VC7U (identifying deterrent effect from different levels of penalties imposed by state agencies on NOx emitters); *EPA, Office of Inspector General, EPA and States Not Making Sufficient Progress in Reducing Ozone Precursor Emissions in Some Major Metropolitan Areas*, Report No. 2004-P-00033 (Sept. 29, 2004).

[24] *Meghrig v. KFC W., Inc.*, 516 U.S. 479, 483 (1996) (quoting 42 U.S.C. § 6902(b)); *see also* 42 U.S.C. § 6901 et seq.

[25] RCRA § 3019, 42 U.S.C. § 6939a(a) (requiring applicant seeking a permit for new landfill or surface impoundment facility that will contain hazardous waste to submit an assessment of public exposure from "reasonably foreseeable potential releases" that are "related to the unit").

[26] New Jersey Department of Environmental Protection, *Waterfront Permitting Made Simple (or at Least Simpler): The Guide to New York City and New Jersey Waterfront Permitting*, https://perma.cc/7F62-GNNV.

Harvey's inundation of multiple facilities near Houston.[27] And, much as the Clean Air Act sets forth fixed pollutant concentration standards based on human health impacts, RCRA establishes a standard that is also based on human health and would seem to transcend particular environmental circumstances. However, the differences between the two statutes' health-related standards are significant as well. Whereas the Clean Air Act bases source-specific pollution limits on the numbers and testing protocols specified in a NAAQS, RCRA articulates its health-related basic standard not as a numeric threshold, but in its definition of entities potentially subject to a citizen suit:

> any past or present generator ... transporter, or ... owner or operator of a treatment, storage, or disposal facility, who has contributed or who is contributing to the past or present handling, storage, treatment, transportation, or disposal of any solid or hazardous waste which may present an *imminent* and *substantial endangerment* to health or the environment.[28]

Because various courts' applications of this standard give it meaning, the "threshold" it provides for separating compliant from noncompliant conduct is sometimes fuzzy and frequently contested – especially the terms "imminent and substantial endangerment."

In the context of rising sea levels and encroaching shorelines, does this definition effectively require coastally sited permit holders to adapt or else be found to be the source of imminent and substantial endangerments to the health of nearby individuals and communities? The case, *Conservation Law Foundation* v. *ExxonMobil*, which has survived a motion to dismiss as of September 2017,[29] addresses this question in the context of the Everett Terminal, a coastal petroleum transfer and storage facility located near Boston. CLF asserts that downscaled climate models make clear that the facility is already vulnerable to flooding in ways that ExxonMobil has not addressed.[30] ExxonMobil responds that it complies with the requirements of the facility's permit – which does not expressly contemplate climate change impacts – and therefore cannot be in violation of RCRA.[31] Because the basic RCRA standard at issue is not a pollution concentration threshold against which compliance can

[27] Ralph Vartabedian, *Harvey Pounded the Nation's Chemical Epicenter. What's In the foulsmelling floodwater left behind?* L.A. Times, Aug. 31, 2017, https://perma.cc/MY5G-8R24.

[28] RCRA § 7002(a)(1)(B), *codified at* 42 U.S.C. § 6972(a)(1)(B) (emphasis added).

[29] Order, *Conservation Law Foundation, Inc.* v. *ExxonMobil Corp. et al.*, C.A. No. 1:16-cv-11950-MLW (D. Mass. Sept. 13, 2017).

[30] Complaint for Declaratory and Injunctive Relief and Civil Penalties, *Conservation Law Foundation, Inc.* v. *ExxonMobil Corp.*, C.A. No. 1:16-cv-11950-MLW (D. Mass. Sept. 29, 2016).

[31] Memorandum of Law in Support of Defendants' Motion to Dismiss, *Conservation Law Foundation, Inc.* v. *ExxonMobil Corp.*, C.A. No. 1:16-cv-11950-MLW (D. Mass. Dec. 6, 2016).

be measured with precision, resolution by the courts will be based on a legal analysis as well as – and possibly more so than – empirical measurements.

### 13.2.2 *Pollution Control Laws and Stationarity*

Climate change is replacing stable historical environmental baselines with dynamic ones. This point is almost a tautology, but it does not guarantee that those responsible for enforcing environmental laws will abandon the premise of environmental stationarity that tacitly informs most pollution control laws.[32] As the examples discussed in the previous section show, mere evidence of stationarity's obsolescence is not enough. Strong signals that existing pollution control will become increasingly ineffective without more stringent and aggressive enforcement have not (yet) altered the approaches taken by federal and state agencies to applying existing environmental laws to pollution in a changing environmental context. This is understandable in so far as law enforcement takes direction from policy makers; if policy makers in agencies ignore – or are prevented from acting on – those signals, a course correction will only come from accumulated courtroom defeats or legislation.

The still-unresolved questions of whether and how GHG emissions are to be regulated under the Clean Air Act hint at what we can expect of the coming fights – like *CLF* v. *ExxonMobil* – over applying environmental laws in ways that respond to changing environmental conditions. The Supreme Court's 2007 decision in *Massachusetts* v. *EPA* started the wheels of the Clean Air Act turning,[33] but now, more than ten years later, US political leaders dispute the scientific consensus underlying that decision and have debated internally how best to apply the Act so as to avoid regulating GHG emissions with any material effect.[34] This persistent political ambivalence over climate change mitigation policy does not augur well for the application of existing environmental laws – directly or indirectly – to effectuate adaptation goals. This is not only because it demonstrates political leaders' and policy makers' willingness to ignore evidence but also, and no less importantly, because causation in the adaptation context is less straightforward than in the context of mitigation: Dozens of noisy factors accompany the signal of a warmer climate amplifying the contribution of pollution precursors to human exposure to pollution.

32 See P. C. D. Milly et al., *Stationarity is Dead, supra* note 1483.

33 *Ma. v. EPA*, 594 U.S. 497 (2007).

34 Andrew Resuccia and Alex Guillen, *Pruitt Takes Fire From Conservatives in Climate Showdown*, Politico, Mar. 28, 2017, https://perma.cc/K53Q-ZHBR (describing debate among climate skeptics within EPA over tactical efficacy of trying to rescind the 2009 endangerment finding on which mitigation under the Clean Air Act is based).

Thus, while environmental laws can be expected to be a source of partial information as the climate changes, environmental enforcement agencies' understanding and application of those laws should not be expected to evolve without conscious intervention at least by regulatory policy makers and, likely, eventually, by legislators as well.

## 13.3 PUBLIC HEALTH CO-BENEFITS AND CO-HARMS OF CLIMATE CHANGE MITIGATION MEASURES

Positive and negative public health impacts can be expected to result from mitigation efforts,[35] but the scales are not balanced: Mitigation is likely to yield far more positive public health impacts than negative ones. Current law credits only some of the positive impacts and fails to anticipate or avoid many of the negative ones. This section considers several likely public health impacts of mitigation – co-benefits first, then co-harms – and describes how the law takes them into account.

Before considering examples of co-benefits and co-harms from mitigation, it is important to first note that this chapter borrows from the Deep Decarbonization Pathways Project for its vision of the potential scale and scope of mitigation efforts in the United States over the coming decades.[36] That vision entails large investments in energy efficiency, the electrification of the transportation and building sectors, and the near-total replacement of fossil-fueled electricity generation with renewable and non-emitting sources.[37] Putting this borrowed vision in the background of this chapter clarifies the degree of mitigation currently thought possible without basic disruptions to the economy and society. This in turn anticipates (and refutes) the suggestion that because fossil fuels have played an important role historically in improving public health it is not possible to decouple future economic and societal development – and concomitant improvements to public health – from reliance on fossil fuels.

A further preliminary note relates to this chapter's geographic focus on the United States, which has important implications for the types of climate

35 Kirk R. Smith et al., *Energy and Human Health*, 34 Ann. Rev. Pub. Health 159 (Mar. 2013) DOI: 10.1146/annurev-publhealth-031912-114404.

36 See J. H. Williams et al. *Pathways to Deep Decarbonization in the United States: The U.S. Report of the Deep Decarbonization Pathways; Project of the Sustainable Development Solutions Network and the Institute for Sustainable Development and International Relations* (Nov. 2015) https://perma.cc/92LD-G6VN (describing pathways by which the United States could reduce its annual GHG emissions by 80 percent below 1990 levels by 2050).

37 Ibid. at vii.

mitigation co-benefits and co-harms considered herein.[38] Some forms of mitigation have very different public health implications in the United States as compared to the developing world. An especially clear example is the climate and public health tradeoff related to air conditioning (AC). As a changing climate leads to warmer average temperatures and more frequent and intense heatwaves, access to AC will become increasingly important to public health outcomes (see Chapter 7). Most AC units currently rely on the compression of fluorinated gases (F-gases) to generate cool air.[39] Unfortunately, f-gases' extremely large global warming potential (GWP) coefficients make them a threat to climate stability.[40] AC is widely available in those parts of the United States that experience hot weather, but not in the developing world. Recent examples from Pakistan and India,[41] and elsewhere,[42] illustrate the acute health-related dangers of heat without recourse to artificial means of cooling.[43] In the coming decades, greater spending power combined with growing numbers of hot days in developing countries is expected to drive huge volumes of AC sales.[44] This will, on the one hand, provide greater protection for public health vis-à-vis heat hazards. On the other hand, however, it will potentially drive a large increase in the release of F-gases into the atmosphere. Thus, whereas an aggressive program of

38 See, e.g., Damilola Olawuyi, *Climate Justice and Corporate Responsibility: Taking Human Rights Seriously in Climate Actions and Projects*, 34 J. Energy & Nat. Resources L. 1–18 (2016) (describing adverse impacts of climate mitigation projects); Damilola Olawuyi, Fostering accountability in large scale environmental projects: Lessons from CDM and REDD+ projects. In: J. Wouters et al., *Improving Delivery in Development: The Role of Voice, Social Contract, and Accountability*, 127–147 (Washington, DC: World Bank 2015).

39 O. Abdelaziz et al., *The Future of Air Conditioning for Buildings* (July 2016), https://perma.cc/959T-JJKD.

40 See Pallav Purohit and Lena Höglund-Isaksson, *Global Emissions of Fluorinated Greenhouse Gases 2005–2050 with Abatement Potentials and Costs*, 17 Atmos. Chem. Phys. 2795, 2804 (2017) (estimating changes in f-gas emissions by region under different scenarios).

41 Michael Wehner et al., *The Deadly Combination of Heat and Humidity in India and Pakistan in Summer*, 97(12) (2015); Stephanie C. Herring et al., *Explaining Extreme Events of 2015 from a Climate Perspective, Bulletin of the American Meteorological Society*, s81–s86 (Dec. 2016), https://perma.cc/E4KX-AAYT

42 Tom Kosatsky, *The 2003 European Heat Waves*, Eurosurveillance (noting estimate of 22,080 excess deaths amid 2003 heatwaves in Europe); Stephen Whitman et al., *Mortality in Chicago Attributed to the July 1995 Heat Wave*, 87(9) Am. J. of Pub. Health 1515–1518 (1997), https://perma.cc/R6SD-VBGE.

43 Yan Wang et al., *Heat stroke Admissions During Heat Waves in 1,916 US Counties for the Period From 1999 to 2010 and Their Effect Modifiers*, 15 Envtl. Health 83 (Aug. 2016) (identifying correlation between access to central AC and lower rates of hospitalization during heat waves).

44 Abdelaziz, Goetzler et al., The Future of Air Conditioning for Buildings, https://perma.cc/959T-JJKD.

climate mitigation might call for restrictions on the use of AC reliant on F-gases in the United States and elsewhere, such restrictions would impose significant co-harms on public health in large swathes of the developing world.

### 13.3.1 *Co-benefits of Mitigation*

This section briefly describes two examples of climate change mitigation efforts that would yield significant public health co-benefits. It follows that description with a discussion of the limited and inconsistent ways in which the law recognizes such co-benefits.

#### 13.3.1.1 Reducing Fossil Fuel Use

Material threats to public health have been identified in relation to the air and water pollution resulting from nearly all aspects of getting at and using fossil fuels: coal mining,[45] the "conventional" drilling and exploitation of land-based and off-shore oil deposits,[46] and wells using "unconventional" high volume hydrofracture

45 Mountaintop removal mining (MTR) is notoriously polluting of the air and water around the mountains that it transforms into flattened moonscapes; however, a National Academies panel has been convened to resolve ongoing disputes about the public health impacts of that pollution by assessing the quality of existing studies and to identify what research questions remain open. Stephen Lee, *Evidence Thin That Mountaintop Mining Harms Health: Coal Groups*, BloombergBNA Daily Environment Report, Jul 19, 2017 ("On July 11, a National Academy of Sciences panel heard evidence from researchers broadly connecting mountaintop removal mining to higher rates of heart disease, cardio-vascular disease, stroke, and death . . . However, the researchers who presented their work also repeatedly stressed that the data are still too thin and that more study is needed."); see also National Toxicology Program, *Mountaintop Removal Mining: Impacts on Health in the Surrounding Community* (last updated Nov. 16, 2016), http://bit.ly/2tdDESl (for list of studies on adverse public health impacts of MTR in Central Appalachia, follow link to "Included and Excluded Studies (Updated Nov. 16, 2016)" spreadsheet and select "Included (n=37)" tab).

46 Impacts are especially severe in less-developed countries with weaker regulatory regimes. See Cristina O'Callaghan-Gordo et al., *Health Effects of Non-Occupational Exposure to Oil Extraction*, 15 Envtl. Health 56 (2016), https://perma.cc/WW8Z-U4VR. Impacts of off-shore drilling arise chiefly from spills or occupational exposures. L. C. Peres et al., *The Deepwater Horizon Oil Spill and Physical Health Among Adult Women in Southern Louisiana: The Women and Their Children's Health (WaTCH) Study*, 124 Envtl. Health Persp. 1208–1213 (2016), https://perma.cc/X65W-WKUB; Blanca Laffon et al., *Effects of Exposure to Oil Spills on Human Health: Updated Review*, Part B 19 J. of Toxicology and Envtl. Health, 105–128 (2016), http://dx.doi.org/10.1080/10937404.2016.1168730; Jo S. Stenehjem et al., *Self-Reported Occupational Exposures Relevant for Cancer Among 28,000 Offshore Oil Industry Workers Employed Between 1965 and 1999*, 12 J/ of Occupational and Envtl. Hygiene 458–168 (2015) http://dx.doi.org/10.1080/15459624.2014.989358.

drilling;[47] gas and oil refining;[48] the transportation of coal via train,[49] and of gas and oil via train,[50] tanker,[51] and pipeline;[52] gas and oil storage;[53] gasoline

47 Shaina L. Stacy, *A Review of the Human Health Impacts of Unconventional Natural Gas Development*, 4 Current Epidemiology Reports 38–45, (Mar. 2017); Nancy E. Lauer et al., *Brine Spills Associated With Unconventional Oil Development in North Dakota*, 50 *Environ. Sci. Technol.* 5389–5397 (2016); New York State Department of Health, *A Public Health Review of High Volume Hydraulic Fracturing for Shale Gas Development* (Dec. 2014), https://perma.cc/X2KB-XYXQ (describing identified adverse air and water quality impacts, and concluding that there is too much uncertainty about fracking's potential public health impacts to allow it in New York); Maryland Institute for Applied Environmental Health, *Potential Public Health Impacts of Natural Gas Development and Production in the Marcellus Shale in Western Maryland*, 26–83 (July 2014) https://perma.cc/X2KB-XYXQ (listing "community impacts" and "occupational impacts").

48 Emmanuelle Lavaine and Matthew Neidell, *Energy Production and Health Externalities: Evidence From Oil Refinery Strikes in France*, 4 J. of the Association of Envtl. and Resource Economists 447–477 (Apr. 2017) https://perma.cc/74R6-78B5; Michael T. Kleinman et al., *Emissions From Oil and Gas Operations in the United States and Their Air Quality Implications*, 66 J. of the Air & Waste Management Association 1165–70 (2016) https://perma.cc/HRW7-5FGP.

49 Multnomah County Health Department, *The Human Health Effects of Rail Transport of Coal through Multnomah County, Oregon: A Health Analysis and Recommendations for Further Action* (Feb. 2013), https://perma.cc/3JAC-Q3AB.

50 Melissa Genereux et al., *Two Years After the Train Derailment: Lac-Megantic (Quebec, Canada) Residents are Still Suffering*, 26 Eur. J. Public Health 207–208 (2016) (suppl_1); Ann Hayward Walker et al., *Consensus Ecological Risk Assessment of Potential Transportation-Related Bakken and Dilbit Crude Oil Spills in the Delaware Bay Watershed, USA*, 4 J. of Marine Sci. and Engineering 23, tbl. 1 (2016), https://perma.cc/29MG-9GN8 (listing incidents of spills of Bakken or Dilbit crude from railways between 2010 and 2014).

51 Heon Lee Cheol et al., *Acute Health Effects of the Hebei Oil Spill on the Residents of Taean, Korea*, 43(2) J. Prev. Med. Public Health 166–173 (2010), https://perma.cc/GZ3H-8K3U; Gina M. Solomon and Sarah Janssen, *Health Effects of the Gulf Oil Spill*, 304(10) JAMA 1118–1119 (Sept. 2010), https://perma.cc/SWN2-W6LM.

52 Romany M. Webb, *Safety First, Environment Last: Improving Regulation of Gas Pipeline Leaks*, KBH Center for Energy, Law & Business, Research Paper No. 2015-14 (Sept. 2015), http://bit.ly/2tjmjU8 (noting prevalence of leaks and explosions in aging US gas transmission and distribution infrastructure); Elizabeth Douglass, *Two Years After Exxon's Mayflower Spill, Will Tougher Pipeline Rules Go Beyond Talk?*, InsideClimate News, Mar. 30, 2015, https://perma.cc/SE76-N63S; Ayana R. Anderson, *Health Effects of Cut Gas Lines and Other Petroleum Product Release Incidents – Seven States, 2010–2012*, 64(22) CDC Morbidity and Mortality Weekly Report (MMWR) 601–605 (June 12, 2015), https://perma.cc/2LK5-KDRJ.

53 Nathaniel Rich, *The Invisible Catastrophe*, N.Y.T. Mag., March 31, 2016, https://perma.cc/GG9U-Y22W ("In November, readings taken by SoCalGas near its facility found benzene concentrations fluctuating wildly between 0.3 p.p.b. and a nightmarish 30.6; readings taken by the company in Porter Ranch shot as high as 5.5 p.p.b."); Drew R. Michanowicz et al., *A National Assessment of Underground Natural Gas Storage: Identifying Wells with Designs Likely Vulnerable to a Single-Point-of-Failure*, 12 Envtl. Research Letters 064004 (May 2017), https://perma.cc/7G23-NA73.

distribution;[54] and, of course, the combustion of fossil fuels, whether by industrial facilities, for electricity generation, or in vehicles.[55] The mechanisms through which these activities expose people to pollutants are numerous and varied, but generally fall into three categories: pollutants that reduce local or regional air quality;[56] toxic pollutants initially emitted by smokestacks that are deposited in watersheds and from there enter drinking water supplies and food chains;[57] and pollutants introduced directly into surface or subsurface waters.[58]

54 Markus Hilpert et al., *Hydrocarbon Release During Fuel Storage and Transfer at Gas Stations: Environmental and Health Effects*, 2(4) Current Envtl. Health Reports 412–422 (Dec. 2015) https://perma.cc/MG59-YCR5; Vicky Huppé et al., *Residential Proximity to Gasoline Service Stations and Preterm Birth*, 20 Envtl. Sci. & Pollution Res. 7186 (Oct. 2013), doi:10.1007/s11356-013-1677-y.

55 Heidi Vreeland et al., *Oxidative Potential of PM2.5 During Atlanta Rush Hour: Measurements of In-Vehicle Dithiothreitol (DTT) Activity*, 165 Atmospheric Env't 169–178 (2017) (in-vehicle PM2.5 samples collected during rush hour exceeded concentrations of particulates measured on roadsides); Frederica P. Perera, *Multiple Threats to Child Health From Fossil Fuel Combustion: Impacts of Air Pollution and Climate Change*, Envtl. Health Perspectives (June 2016), http://dx.doi.org/10.1289/EHP299 ("The data summarized here show that by sharply reducing our dependence on fossil fuels we would achieve highly significant health and economic benefits for our children and their future."); Ryan Wise et al., *The Environmental and Public Health Benefits of Achieving High Penetration of Solar Energy in the United States* (May 2016), https://perma.cc/94PW-ZSG2; L. Noel and R. McCormack, *A Cost Benefit Analysis of a V2G-Capable Electric School Bus Compared to a Traditional Diesel School Bus*, 126 Appl. Energy 246–255 (2014) (estimating public health benefits from lower levels of air pollution).

56 American Lung Association, *State of the Air 2017*, 32-40 (2017) (providing overview of adverse health effects of exposure to particulate matter and ozone); G. F. Nemet et al., *Implications of Incorporating Air-Quality Cobenefits Into Climate Change Policymaking*, 5 Envtl. Research Letters 19 (2010).

57 Eric Roston, *Coal Plants Might be Even More Toxic Than We Thought*, Bloomberg (2017) www.bloomberg.com/news/articles/2017-08-08/coal-plants-might-be-even-more-toxic-than-we-thought; Adeline R. Lopez et al., *Coal Ash Constituents at the Base of Aquatic Food Webs: Processes Affecting Bioaccumulation and Trophic Transfer*, UNC Water Resources Research Inst. Rep. No. 465 (May 2016) (tracing arsenic from coal-fired power plants through aquatic food chains); EPA, *National Emission Standards for Hazardous Air Pollutants From Coal- and Oil-Fired Electric Utility Steam Generating Units and Standards of Performance for Fossil-Fuel-Fired Electric Utility, Industrial-Commercial- Institutional, and Small Industrial-Commercial-Institutional Steam Generating Units*, 77 Fed. Reg. 9304 (Feb. 16, 2012) ("up to 29 percent of modeled watersheds have populations potentially at-risk from exposure to [mercury] from U.S. EGUs.").

58 William D. Burgos et al., *Watershed-Scale Impacts From Surface Water Disposal of Oil and Gas Wastewater in Western Pennsylvania*, 51 Environ. Sci. Technol. 8851 (2017), (reporting on pollutants introduced into Pennsylvania surface waters after treatment of hydrofracture back-flow by treatment facilities); Jennifer S. Harkness et al., *Evidence for Coal Ash Ponds Leaking in the Southeastern United States*, 50 Environ. Sci. Technol. 6583–6592 (2016) https://perma.cc/MM9R-2TCW; Amrika Deonarine et al., *Environmental Impacts of the Tennessee Valley Authority Kingston Coal Ash Spill, 2. Effect of Coal Ash on Methylmercury in Historically Contaminated River Sediments*, 47 Environ. Sci. Technol. 2100–2108 (2013), https://perma.cc/WVS7-Z873.

The practical task of estimating these co-benefits is facilitated by several factors. Legally mandated pollution controls have called into being sophisticated techniques and technologies for tracking pollution levels, and the measurable volumes of – and predictable relationships among – GHG and non-GHG pollutants make it relatively easy to extrapolate the ultimate effects of non-GHG emissions reductions that would follow from GHG emissions reductions.[59] Similarly, the pathways through which non-GHG air pollution burdens public health have been studied extensively.[60] Furthermore, EPA generally estimates the costs and benefits of proposed air quality standards, even when the Clean Air Act directs that such estimates should not inform the agency's decision about a particular standard.[61] Of course, debate persists over the nature and extent of adverse public health impacts attributable to some of the activities listed in this section.[62] Nonetheless, it cannot be gainsaid that public health co-benefits would accrue from the reduced exposures to conventional and toxic pollutants that would follow from reducing the extraction, transportation, and consumption of fossil fuels.[63]

#### 13.3.1.2 Improving Agricultural Nitrogen Use Efficiency

To grow, crops need reactive nitrogen, which is naturally scarce (unreactive nitrogen is abundant, but not useful to plants).[64] Historically, the availability

59 See Justin V. Remais et al., *Estimating the Health Effects of Greenhouse Gas Mitigation Strategies: Addressing Parametric, Model, and Valuation Challenges*, 122 Environ. Health Persp. 447–455 (May 2015).

60 See, e.g., Philip J. Landrigan, *Air Pollution and Health*, 2(1) The Lancet: Public Health e4–e5 (Jan. 2017) ("Air pollution is one of the great killers of our age."); Frank J. Kelly and Julia C. Fussell, *Air Pollution and Public Health: Emerging Hazards and Improved Understanding of Risk*, 37(4) Envtl. Geochemistry & Health 631–649 (2015).

61 See Livermore and Revesz, *Rethinking Health-Based Environmental Standards*, *supra* note 15, at 1188 n.13.

62 See, e.g., Madelon L. Finkel and Jake Hays, *Environmental and Health Impacts of 'Fracking': Why Epidemiological Studies are Necessary*, 70 J. of Epidemiology and Community Health 221–222 (2016) https://perma.cc/98NF-A82G (observing that existing studies of public health impacts are inconclusive, partly due to the long latency periods of adverse medical outcomes from exposures likely to result from proximity to drilling).

63 Andy Haines, *Health Co-Benefits of Climate Action*, 1(1) The Lancet Planetary Health e4–e5 (Apr. 2017) (collecting and summarizing studies); Fernando Garcia-Menendez et al., *U.S. Air Quality and Health Benefits From Avoided Climate Change Under Greenhouse Gas Mitigation*, 49 *Environ.* Sci. Technol. 7580–7588, 7586 (2015) https://perma.cc/4RE8-BJVB (identifying significant benefits to air quality from the mitigation of climate change even ignoring accompanying reductions in ozone and PM pollution levels). This chapter does not attempt to answer the larger question of whether and under what conditions net public health benefits would follow from the reduction and substitution of fossil fuel use. See notes 20 and 21, *supra*, and accompanying text.

64 Unreactive nitrogen ($N_2$) makes up much of the Earth's atmosphere. Reactive nitrogen takes several forms, including nitrate ($NO_3$), ammonia ($NH_3$), and nitrous oxide ($N_2O$).

of naturally occurring reactive nitrogen was a limiting factor for crop growth;[65] the synthesis of nitrogen fertilizers thus marked a crucial advance in agricultural technology.[66] Demand for synthetic nitrogen fertilizers is now endemic to agriculture,[67] as is the excessive provision of those fertilizers on land under cultivation.[68] Excess reactive nitrogen that is spread on croplands often becomes a pollutant. As nitrate ($NO_3$), it does the following: contaminates drinking water aquifers and causes acidification, hypoxia (oxygen depletion), and the eutrophication that feeds harmful algae blooms in coastal and lake ecosystems.[69] As nitrous oxide ($N_2O$), it depletes stratospheric ozone[70] and is a potent greenhouse gas – 298 times more potent than $CO_2$.[71] These harmful effects are not de minimis. Excessive use of nitrogen fertilizer is a key cause of the dead zone in the Gulf of Mexico at the mouth of the Mississippi River,[72] and is responsible for roughly $1.7 billion of the $4.8 billion spent annually on removing nitrates from drinking water supplies.[73] "Agricultural soils management," which includes fertilization and manure management, generates 75 percent of the anthropogenic $N_2$ O emitted by US sources.[74] Furthermore, the greater intensity of precipitation attributable to climate change exacerbates the transmission of nitrogen

65 Vaclav Smil, *Enriching the Earth* (2004).

66 Ibid.

67 See UN Food and Agriculture Organization, *World Fertilizer Trends and Outlook to 2019: Summary Report*, 7 (2016), www.fao.org/3/a-i5627e.pdf (tabulating world demand for nitrogen fertilizer by region).

68 Allen G. Good and Perrin H. Beatty, *Fertilizing Nature: A Tragedy of Excess in the Commons*, 9 PLoS Biology e1001124 (Aug. 2011).

69 Marc Ribaudo et al., USDA Economic Research Service, *Nitrogen in Agricultural Systems: Implications for Conservation Policy* 1 (Sept. 2011), https://perma.cc/TYD8-U5NS; see also E. Sinha et al., *Eutrophication Will Increase During the 21st Century as a Result of Precipitation Changes*, 357 Sci. 405–408 (July 2017) (describing how climate change will amplify the effects of reactive nitrogen pollution).

70 See generally, United Nations Environment Program (UNEP), *Drawing down Nitrous Oxide to Protect the Ozone Layer* (2013).

71 EPA, *Inventory of U.S. Greenhouse Gas Emissions and Sinks, 1990–2015*, ES-3 tbl. ES-1 (2017) (listing GHGs and indicating global warming potentials of each relative to $CO_2$); see also A. R. Ravishankara et al., *Nitrous oxide (N2O): The Dominant Ozone-Depleting Substance Emitted in the 21st Century*, 326 Sci. 123 (2009). Notably, $N_2O$ is recognized as an ozone-layer-depleting gas under the Vienna Convention, but was not listed in the Montreal Protocol among pollutants to be phased out of use. Vienna Convention for the Protection of the Ozone Layer, Annex I, ¶4(b)(i), Mar. 22, 1985; Montreal Protocol on Substances that Deplete the Ozone Layer, Sept. 16, 1987.

72 EPA Science Advisory Board, *Reactive Nitrogen in the United States: An Analysis of Inputs, Flows, Consequences, and Management Options*, 34–35 (2011) https://perma.cc/7399-63ME.

73 Ibid. at 4.

74 EPA, *Inventory of U.S. Greenhouse Gas Emissions and Sinks, 1990–2015*, at ES-6 tbl. ES-2 (2017).

from agricultural soils.[75] Thus, as reactive nitrogen is a source of multiple forms of harmful pollution, improving its use efficiency (i.e. reducing instances of excessive application) on croplands would yield significant climate change mitigation benefits as well as public health co-benefits, such as drinking water provision.[76]

### 13.3.1.3 When Co-benefits are Legally Cognizable in the United States

Though co-benefits are hugely significant to the economics and public health implications of climate change policy,[77] in the United States they are only relevant legally if they appear in the following narrow set of circumstances: a legal challenge to a federal regulation adopted in part because of its co-benefits (or despite its co-harms).[78] This relevance stands on two feet, one administrative, the other legal. The administrative foot is the Executive Order that directs agencies to only adopt a proposed regulation after analyzing its benefits and its costs.[79] The Office of Management and Budget (OMB), which

75 Terrance D. Loecke et al., *Weather Whiplash in Agricultural Regions Drives Deterioration of Water Quality*, 133 Biogeochemistry 7–15 (Mar. 2017) https://link.springer.com/article/10.1007/s10533-017-0315-z (measuring increases in N resulting from rainfall made more severe by climate change).

76 David R. Kanter et al., *Managing a Forgotten Greenhouse Gas Under Existing U.S. Law: An Interdisciplinary Analysis*, 67 Envtl. Sci. and Pol'y 44, 45 (2017); see also USDA ERS (noting that among best management practices available to reduce harmful effects of reactive nitrogen pollution, only improved nitrogen use efficiency reliably reduces all adverse effects).

77 See Ian Parry et al., *How Much Carbon Pricing is in Countries' Own Interests? The Critical Role of Co-Benefits*, 6 Climate Change Econ. 1550019 (Nov. 2015); San Francisco Department of Public Health, *Assessing the Health Co-benefits of San Francisco's Climate Action Plan* (Aug. 2013); J. Jason West et al., *Co-Benefits of Mitigating Global Greenhouse Gas Emissions for Future Air Quality and Human Health*, 3 Nature Climate Change 885 (2013); Nemet et al., *Implications of Incorporating Air-Quality Cobenefits Into Climate Change Policymaking*, *supra* note 56, at 2–3 (noting that air quality co-benefits in particular "are large" but are, as of 2010, largely ignored by climate policy decision-makers).

78 See Carolyn Cecot and W. Kip Viscusi, *Judicial Review of Agency Benefit-Cost Analysis*, 22 Geo. Mason L. Rev. 575, 576 (2015) (discussing history of agency cost-benefit analysis and noting that "there must be some legal challenge involving the regulation and its BCA to trigger a judicial review"). The sole exception to this statement as of this writing appears to be California's law, AB 1532, adopted in 2012. California Global Warming Solutions Act of 2006: Greenhouse Gas Reduction Fund (2012). That law requires the agencies responsible for making California Climate Investments (i.e. spending the revenue from the state's greenhouse gas emissions cap-and-trade program) to maximize the co-benefits resulting from those expenditures. California Air Resources Board, *Cap-and-Trade Auction Proceeds: Co-benefit Assessment Methods*, https://perma.cc/KY8B-WRSU.

79 Executive Order 12,866 instructs agencies to conduct a cost-benefit analysis of any significant proposed regulation and to adopt regulations whose societal benefits exceed their costs. Exec. Order No. 12866, *Regulatory Planning and Review*, 58 Fed. Reg. 51735 (Oct. 4, 1993); see also Exec. Order No. 13,563, 76 Fed. Reg. 3821 (Jan. 21, 2011) (affirming Exec. Order No. 12,866).

has more fully prescribed how agencies should conduct a cost-benefit analysis, defines a co-benefit (which it calls an "ancillary benefit") as "a favorable impact of the alternative under consideration [i.e. the rule,] that is typically unrelated or secondary to the purpose of the [rule]."[80] OMB also makes plain that co-benefits are relevant to cost-benefit analysis: agencies are to "look beyond the direct benefits and direct costs of your rulemaking and consider any important ancillary benefits and countervailing risks," and to apply to them "[t]he same standards of information and analysis quality that apply to direct benefits and costs."[81]

The basic legal footing for co-benefits' relevance is the Administrative Procedure Act, as interpreted by federal courts. The Act directs that agency rulemakings should reflect consideration of the factors contemplated by the relevant statute, and courts have read this direction to inform the scope of costs and benefits agencies should consider in the rulemaking process.[82] Several courts have confirmed the relevance of co-benefits to rules' administrative validity and legality.[83] In *U.S. Sugar Corp.* v. *EPA*, for instance, the D.C. Circuit upheld an EPA regulation after agreeing with EPA that:

> its consideration of these co-benefits [from lowering different Hazardous Air Pollutants (HAPs) than those primarily targeted by the proposed standard] was not a regulation of other pollutants; rather, it was simply choosing not to ignore the purpose of the CAA – to reduce the negative health and environmental effects of HAP emissions – when exercising its discretionary authority under the Act.[84]

See also generally Maeve P. Carey, *Congressional Research Service, Cost-Benefit and Other Analysis Requirements in the Rulemaking Process* (Dec. 2014), https://perma.cc/23YS-JXSN.

80 OIRA, OMB, EOB, *Regulatory Impact Analysis: A Primer*, 7 (Aug. 15, 2011) https://perma.cc/5Z6D-FJJP, OMB Circular A-4, which OMB issued in 2003 to "standardiz[e] the way benefits and costs of Federal regulatory actions are measured," OMB Circular A-4, at 1 (2003), 68 Fed. Reg. 58366 (Oct. 9, 2003), offers an example of a valid co-benefit: "e.g., reduced refinery emissions due to more stringent fuel economy standards for light trucks." Ibid. at 26.

81 OMB Circular A-4, at 26.

82 *Michigan* v. *EPA*, 135 S. Ct. 2699, 2706 (2015) ("agency action is lawful only if it rests 'on a consideration of the relevant factors.' *Motor Vehicle Mfrs. Assn. of United States, Inc. v. State Farm Mut. Automobile Ins. Co.*, 463 U.S. 29, 43").

83 See, e.g. *Ctr. for Biological Diversity* v. *Nat'l Highway Traffic Safety Admin.*, 538 F.3d 1172, 1198–1203 (9th Cir. 2008) (reversing agency for quantifying only ancillary costs but not ancillary benefits); *Am. Trucking Ass'ns* v. *EPA*, 175 F.3d 1027, 1051–52 (D.C. Cir. 1999) (holding that EPA was wrong to disregard some indirect effects of proposed rule), *rev'd on other grounds sub nom. Whitman* v. *Am. Trucking Ass'ns*, 531 U.S. 457 (2001); *Competitive Enter. Inst.* v. *Nat'l Highway Traffic Safety Admin.*, 956 F.2d 321, 327 (D.C. Cir. 1992) (remanding fuel economy standard on grounds that agency failed to weigh ancillary costs of lives lost because more people would buy and drive smaller vehicles, which are less protective in accidents).

84 *United States Sugar Corp.* v. *EPA*, 830 F.3d 579, 625 (D.C. Cir.), *reh'g en banc*, 671 F. App'x 822 (D.C. Cir. 2016), *reh'g en banc in part*, 671 F. App'x 824 (D.C. Cir. 2016), *cert. denied sub nom. Am. Mun. Power* v. *EPA*, No. 16-1168, 2017 WL 1134103 (US June 26, 2017).

Consistent with this reasoning, EPA has also counted as co-benefits criteria pollutants reduced by the regulation of other criteria pollutants,[85] and criteria pollutants reduced by the regulation of GHGs.[86]

Though they have confirmed co-benefits' administrative and legal relevance, courts have yet to explore the limits of that relevance.[87] Much debate over the judicial interpretation of co-benefits has lately focused on EPA's decision to adopt the Mercury and Air Toxics Standard, which the Supreme Court remanded but did not vacate in *Michigan* v. *EPA*.[88] EPA estimated $4 to $6 *million* in benefits from the rule's reduction of the harmful effects of mercury on the mental capacity of people exposed to it ("IQ points lost"), $9.6 *billion* in annual costs, and $37 to $90 *billion* in benefits from indirect or ancillary reductions in fine particulate matter ($PM_{2.5}$).[89] Thus direct benefits did not outweigh costs, and roughly 90 percent of the benefits tallied were co-benefits arising from effects incidental to the emissions targeted by the rule. However, because EPA did not rely on this calculation to justify the rule,[90] the court did not resolve several questions related to it, such as:

85 EPA, *Final Ozone NAAQS Regulatory Impact Analysis* 6–30 (Mar. 2008) https://perma.cc/443A-5M3L.

86 EPA, *Carbon Pollution Emissions Guidelines for Existing Stationary Sources: Electric Utility Generating Units*, 80 Fed. Reg. 64662, 64679 (Oct. 23, 2015); EPA, *Regulatory Impact Analysis for the Clean Power Plan Final Rule*, 4-11 to 4-42 (Aug. 2015), https://perma.cc/4ETS-BFU4, (describing calculation of monetized benefits from $PM_{2.5}$ reduction and non-monetized benefits from reductions of other criteria pollutants).

87 See Cecot and Viscusi, *Judicial Review of Agency Benefit-Cost Analysis*, *supra* note 78, at 576–577, 592–603 (surveying cases that have addressed validity of agencies' different approaches to cost-benefit analyses, e.g. choice of discount rate to be applied to future costs or benefits).

88 *Michigan* v. *EPA*, 135 S. Ct. 2699 (2015) (remanding but not vacating the rule). EPA announced that the rule's requirements would remain in force while it revised the explanation of its legal basis pursuant to the Supreme Court's instructions. The D.C. Circuit rejected industry challenges to this step, *White Stallion Energy Center, LLC* v. *EPA*, No. 12-1100 (D.C. Cir. Dec. 15, 2015), and EPA issued that revised explanation in due course. EPA, Supplemental Finding that it is Appropriate and Necessary to Regulate Hazardous Air Pollutants from Coal- and Oil-Fired Electric Utility Steam Generating Units, 81 Fed. Reg. 24419 (Apr. 25, 2016). EPA's Supplemental Finding was immediately challenged by the same industry coalition, and the D.C. Circuit then granted a request by Trump Administration's EPA to hold the case in abeyance and to stay the rule while EPA reviewed the 2016 Supplemental Finding. *Murray Energy Corporation* v. *EPA*, No. 16-1127 (D.C. Cir. Apr. 27, 2017).

89 *Michigan v. EPA*, 135 S. Ct. at 2706, *supra* note 88.

90 Ibid. at 2711. Indeed, the gravamen of the case was whether the Clean Air Act directed EPA to only consider the rule's effects on health or to also consider its cost implications.

* To what extent should courts defer to agency decisions about the significance of co-benefits to a rule's adoption?[91]
* Should an agency weigh co-benefits that would accrue from reductions of $PM_{2.5}$ below the level embodied in the relevant NAAQS, i.e. the level determined by EPA to be safe?[92]
* If costs outweigh direct benefits but are outweighed by co-benefits, should the rule be considered beneficial?

The mismatch between the practical and legal relevance of public health co-benefits to climate policy highlights an important gap in regulatory accounting, but one that is difficult to bridge. Perhaps, like the development and implementation of the Social Cost of Carbon, the highly technical task of articulating how best to integrate measures of co-benefits into regulatory decision-making is one for a multiagency working group or the National Academies of Sciences.[93]

### 13.3.2 *Co-harms of Mitigation*

Researchers who have examined the co-harms of mitigation in the developed world, including the United States, have identified rising costs of energy and food, and employment impacts of the transition away from fossil-fuels as the most likely to be significant.[94] Notably, these are all indirect sources of co-harms – in marked contrast to the more direct sources of co-benefits described in the previous section. The three examples of co-harms discussed in this section highlight how existing laws address – or fail to address – them.

91 See Michael Abramowicz, *Toward a Jurisprudence of Cost-Benefit Analysis*, 100 Mich. L. Rev. 1708 (2002) (proposing that courts develop a "common law of cost-benefit analysis" rather than ceding the whole of that analytic field to agencies).

92 EPA did so in the case of the MATS rule. *See* EPA, *National Emission Standards for Hazardous Air Pollutants from Coal and Oil Fired Electric Utility Generating Units*, 77 Fed. Reg. 9304, 9305 (Feb. 16, 2012); EPA, *Regulatory Impact Analysis for the Final Mercury and Air Toxics Standards 5-1 to 5-3* (Dec. 2011) https://perma.cc/6TD2-RDXJ (summarizing health co-benefits calculation).

93 National Academies of Sciences, *Valuing Health for Regulatory Cost-Effectiveness Analysis* (2006).

94 *Modeling Toronto's Low Carbon Future – Technical Paper #4: Considerations of Co-benefits and Co-harms Associated with Low Carbon Actions for TransformTO*, 39 tbl.5 (Jan. 2017) http://taf.ca/wp-content/uploads/2017/02/170127_FINAL_Tech-Paper4_Cobenefits.pdf (higher housing costs due to density; employment impacts from displacement).

### 13.3.2.1 Biofuels Production and Consumption

Biofuels are derived from plants and function as substitutes for liquid fossil fuels, chiefly motor fuels. In the United States, the federal Renewable Fuel Standard – a purchasing mandate for US motor fuel distributors – drives their production.[95] Congress adopted the mandate to reduce fossil fuel imports and use.[96] These goals have been achieved to some degree,[97] but biofuels production also has a host of adverse public health impacts, including air and water pollution.[98] It also boosts food prices, with implications for nutrition and food security both in the United States and abroad.[99] In addition, some biofuels' consumption has been found to generate higher levels of ozone than the fossil fuels they replace.[100]

It is important to note that biofuels derived from different feed stocks – corn starch, corn stover (i.e. stalks and husks), sugar cane, and switch grass – differ from one another in several important respects, including their lifecycle GHG emissions, production costs and volumes, effects on markets for foodstuffs and livestock feed, and the pollution they generate.[101] Thus different biofuels mitigate climate change better or worse (or not at all) and also yield greater or lesser public health co-harms.[102] Ethanols derived from cornstarch

95 EPA, *Renewable Fuel Standard Program: Overview for Renewable Fuel Standard*, https://perma.cc/EE5P-ZDJD (last updated June 7, 2017).

96 Energy Independence and Security Act of 2007, Pub. L. No. 110-140 (2007) (establishing Renewable Fuel Standard 2 (RFS2) for purposes that include energy security and promotion of renewable fuel sources); 40 C.F.R. pt. 80 (codifying RFS2, which distinguishes among fuels based on their lifecycle greenhouse gas emissions levels); see also Energy Policy Act of 2005, Pub. L. No. 109-58 (2005) (establishing the initial standard, RFS1, for purpose of enhancing energy security).

97 *See* Congressional Budget Office, *The Renewable Fuel Standard: Issues for 2014 and Beyond* 5–8 (2015) (noting volume of biofuels produced in the United States but also reporting challenges to achieving production volumes at intended scale).

98 S. Kent Hoekman et al., *Environmental Implications of Higher Ethanol Production and Use in the U.S.: A Literature Review, Part I – Impacts on Water, Soil, and Air Quality, Renewable and Sustainable Energy Reviews* (2017), http://dx.doi.org/10.1016/j.rser.2017.05.050.

99 Congressional Budget Office, *The Impact of Ethanol Use on Food Prices and Greenhouse-Gas Emissions* 8–12 (2009) (describing effects of RFS on food prices, and the knock-on effect on federal food assistance programs).

100 Noah Scovronick et al., *Air Quality and Health Impacts of Future Ethanol Production and Use in São Paulo State, Brazil*, 13 Int'l J. Envtl. Res. & Pub. Health 695 (2016); Alberto Salvo and Franz M. Geiger, *Reduction in Local Ozone Levels in Urban São Paulo Due to a Shift From Ethanol to Gasoline Use*, 7 Nature Geoscience 450 (2014), doi:10.1038/ngeo2144.

101 EPA, *Biofuels and the Environment: First Triennial Report to Congress*, Report no. EPA/600/R-10/183F (2011); National Research Council, *Renewable Fuel Standard: Potential Economic and Environmental Effects of U.S. Biofuel Policy* (2011).

102 EPA, *Renewable Fuel Standard Program (RFS2) Regulatory Impact Analysis* 497 (Feb. 2010) (describing program impacts and noting key differences between fuel types).

generally perform worst in all respects, apart from being the easiest to produce cost-effectively in large volumes in the United States.[103]

### 13.3.2.2 Disuse of Fossil Fuel Extraction, Processing, Storage, and Waste Sites and Facilities

Large-scale climate change mitigation means the cessation of at least some fossil fuel extraction, transport, refining, storage, and consumption, which in turn means the disuse of a host of facilities where toxic materials are processed and stored. In this way, decarbonization will resemble the historical shift away from other disused energy technologies, such as manufactured gas, which have left toxic legacies behind them.[104] As the fact of Congress' passage of the Surface Mining Reclamation and Control Act (SMCRA), Resource Conservation and Recovery Act (RCRA), and Comprehensive Environmental Response, Compensation, and Liability Act (CERCLA) make plain, without mandated decommissioning and remediation, disused sites and facilities would in most if not all cases present significant dangers to public health. Thus the shuttering of facilities that currently play important roles in emitting GHGs, if not followed by effective decommissioning and site remediation, can generate co-harms by damaging the surrounding environment on an ongoing basis.

For reasons described more fully in subsection 13.3.2.4, the track records for decommissioning and remediation by the coal mining sector (particularly in Appalachia) and segments of the oil and gas sector suggest that co-harms following mitigation-driven facility closures are a likelihood rather than a mere possibility.

### 13.3.2.3 Impacts Transmitted by Local Economies and Tax Bases

Mitigation policies raise the cost of doing business for firms whose profitability relies on it being cheap or costless to emit GHGs.[105] If effective, such policies

103 For a thorough discussion of the impacts of corn starch ethanol production and the implications of increasing that production, see Hoekman et al., *Environmental implications of Higher Ethanol Production and Use in the U.S.*, http://dx.doi.org/10.1016/j.rser.2017.05.050; and S. Kent Hoekman and Amber Broch, *Environmental implications of Higher Ethanol Production and Use in the U.S.: A Literature Review. Part II – Biodiversity, Land Use Change, GHG Emissions, and Sustainability, Renewable and Sustainable Energy Reviews* (2017), https://doi.org/10.1016/j.rser.2017.05.052.

104 See generally, Allen W. Hatheway, *Remediation of Former Manufactured Gas Plants and Other Coal-Tar Sites* (2012).

105 See, e.g., Akio Yamazaki, *Jobs and Climate Policy: Evidence from British Columbia's Revenue-Neutral Carbon Tax* (Dec. 2015) (finding that emissions-intensive industries saw employment declines after imposition of carbon tax).

have net positive effects on public health,[106] but could have net-positive, neutral, or net-negative effects on employment and the economy.[107] Whatever their valence in the aggregate, positive and negative economic effects will not be distributed evenly; some localities will lose.[108] It follows that the public health impacts of raising costs – in some cases prohibitively – on firms with significant exposure to mitigation policies in a given locality will be mixed and complex. On the one hand, extractive industries, power plants, and high-emitting industrial facilities pollute their surroundings, and some of the jobs they support are unhealthy for the workers doing them.[109] On the other hand, unemployment in the United States has been shown to harm public health by leading to higher rates of smoking and drinking,[110] prescription drug use,[111] heart disease,[112] suicide,[113] and all-cause mortality.[114]

106 See generally, *Health and climate change*, The Lancet (2009), www.thelancet.com/series/health-and-climate-change; see also John M. Balbus et al., *A Wedge-Based Approach to Estimating Health Co-Benefits of Climate Change Mitigation Activities in the United States*, 127 Climatic Change 119–210 (Nov. 2014).

107 See Mita Bhattacharya et al., *The Effect of Renewable Energy Consumption on Economic Growth: Evidence From top 38 Countries*, 162 Applied Energy 733 (Jan. 2016) (identifying a positive relationship between renewable energy development and economic growth in numerous countries, and a slightly negative relationship in some others, including the United States).

108 Appalachian Regional Commission, *Appalachian Coal Industry, Power Generation and Supply Chain*, 12 (Mar. 2016), http://bit.ly/2wkZbXJ (showing concentration of job losses in particular counties); see also Headwaters Economics, *Communities at Risk from Closing Coal Plants* (Mar. 2017), https://perma.cc/PVZ9-LNJP. see also *Transition Stories (2017)*: http://delta-institute.org/delta/wp-content/uploads/In-Transition-Stories-From-Coal-Plant-Communities-Delta-Institute-Aug-2017.pdf (summarizing statistics and narrative descriptions for numerous US communities in which a coal-fired power plant has closed or is slated for closure).

109 United States Department of Labor, Bureau of Labor Statistics, *Injuries, Illnesses, and Fatalities in the Coal Mining Industry*, www.bls.gov/iif/oshwc/osh/os/osar0012.htm.

110 T. Falba et al., *The Effect of Involuntary Job Loss on Smoking Intensity and Relapse*, 100 Addiction 1330–1339 (Sept. 2005); S. M. Montgomery et al., *Unemployment, Cigarette Smoking, Alcohol Consumption and Body Weight in Young British Men*, 8 Eur. J. Pub. Health 21 (1998).

111 Daniel Kozman et al., *Association Between Unemployment Rates and Prescription Drug Utilization in the United States, 2007–2010*, 12 BMC Health Serv. Res. 435 (2012).

112 Stefan Walter et al., *The Health Effects of US Unemployment Insurance Policy: Does Income From Unemployment Benefits Prevent Cardiovascular Disease?*, 9 PLoS One e101193 (July 2014).

113 A. Reeves et al., *Increase in State Suicide Rates in the USA During Economic Recession*, 380 Lancet 1813–1814 (2012); Timothy J. Classen and Richard A. Dunn, *The Effect of Job Loss and Unemployment Duration on Suicide Risk in the United States: A New Look Using Mass-Layoffs and Unemployment Duration*, 21 Health Econ. 338–350 (Mar. 2012) ("mass layoffs may be powerful localized events where suicide risk increases shortly afterward").

114 David J. Roelfs et al., *Losing Life and Livelihood: A Systematic Review and Meta-Analysis of Unemployment and All-Cause Mortality*, 72 Social Sci. Med. 840–854 (Mar. 2011); see also Christopher B. McLeod et al., *Unemployment and Mortality: A Comparative Study of Germany and the United States*, 102 Am. J. Pub. Health 1542–1550 (Aug. 2012) (finding that unemployment leads to mortality increases among low- and medium-skilled Americans, but not Germans of similar job skill levels).

Importantly, the mechanisms that would link these results to mitigation policies are complex, difficult to parse, and often involve not just downsizing employers but also community institutions confronted with falling local spending and taxes (severance, sales, and property).[115]

### 13.3.2.4 Existing Legal Responses to Co-harms

As explained in this section, existing legal responses to the co-harms described in the previous section suffer from important shortcomings.

**Biofuels.** Various laws address some – but only some – of the co-harms of biofuel production and consumption. Air and water pollution control laws treat the farming and processing of biofuels much like other forms of farming and feedstock processing, requiring that qualifying facilities receive permits to operate and abide by those permits' requirements.[116] Of course, these laws' treatment of agricultural activities is notoriously permissive,[117] and biofuels production and processing fits within many of the same carve-outs that account for that permissiveness. Similarly, the 2017–2025 Corporate Average Fuel Economy standards that govern emissions from passenger vehicles do not make any special provision to address harms arising from ethanol as opposed to standard gasoline.[118] The legislation imposing the Renewable Fuel Standard on fuel distributors does not expressly contemplate co-harms, much less address them,

[115] Brian Dabson et al., *Case Studies of Wealth Creation and Rural-Urban Linkages* 41–42 (Apr. 2012), https://perma.cc/H69Y-JDHN (noting how transition away from extractive industry crunches funding for local government services, including those supportive of public health); see also Saqib Rahim, *N.Y.'s steely coal towns brace for transition*, E&E News, Apr. 18, 2017, https://perma.cc/CS4Y-HK4H (reporting impending loss of $6 million in annual taxes paid to the local government).

[116] See EPA Region 7, *Environmental Laws Applicable to Construction and Operation of Ethanol Plants* (Nov. 2007), https://perma.cc/LL6D-G9K8.

[117] J. B. Ruhl, *Farms, Their Environmental Harms, and Environmental Law*, 27 Ecology L.Q. 263, 263–264 (2000) ("Farms are one of the last uncharted frontiers of environmental regulation in the United States. Despite the substantial environmental harms they cause . . . environmental law has given them a virtual license to do so.").

[118] Incentives for ethanol use embedding in the 2012–2016 CAFE standards were discontinued in the 2017–2025 standards. The only special treatment of note under the new standards is extra credits toward CAFE compliance for manufacturers of flex-fuel vehicles (i.e. those capable of running on mixtures of 85% ethanol and 15% gasoline) *if* those manufacturers can provide evidence of actual vehicle miles travelled using ethanol. EPA and NHTSA, 2017 and Later Model Year Light-Duty Vehicle Greenhouse Gas Emissions and Corporate Average Fuel Economy Standards; Final Rule, 77 Fed. Reg. 62,624, 62,823–24 (Oct. 15, 2012).

whether they result from pollution, land use changes, or upward pressure on food prices.[119]

**Remediation.** Several laws – chiefly SMCRA, RCRA, and CERCLA – direct firms to conduct decommissioning and remediation of fossil fuel-related facilities and sites, provide agencies and citizens with causes of action against instances of harmful nonfeasance, and make state and federal agencies a partial backstop to nonfeasance. Thus coal mines, gas and oil wells, pipelines, oil refineries, and even gas stations, among other things, cannot simply be abandoned, but must be dealt with in a way that protects public health by preventing the release of any toxic substances onsite and in some instances restoring their location to something like its original state.[120] However, even though these laws provide for minimal amounts of bonding (i.e. the setting aside of funds) or retroactive liability to incentivize thorough cleanup of a site, underinvestment in decommissioning and site remediation is endemic in segments of the energy sector.[121] In relation to coal, this seems to owe to financial pressure on mining operations in the eastern United States,[122] and to the unwillingness of state governments and the federal Office of Surface Mining and Reclamation Enforcement to compel more aggressive saving for

[119] EPA, in compliance with the administrative requirements discussed in Section 13.2, estimates price changes expected to result from federal Renewable Fuel Standard's purchasing mandate and production targets. See EPA, *Regulatory Impact Analysis: Renewable Fuel Standard Program* 325–32 (Apr. 2007), www.epa.gov/sites/production/files/2015-08/documents/420r070 04.pdf. It does nothing to address those price changes, however.

[120] See, e.g., SMCRA §§ 508 (requiring all permit applications to include reclamation plan for restoration of mined land to its "pre-mining capability") 509(c) (requiring purchase of reclamation bonds); 42 U.S.C. § 6991b(d) (codifying RCRA provisions relating to underground storage tank remediation financing), CERCLA § 121(d)(1) (defining standard for site cleanup as "at a minimum which assures protection of human health and the environment").

[121] See, e.g., Zachary C. M. Arnold, *Preventing Industrial Disasters in a time of Climate Change: A Call for Financial Assurance Mandates*, 41 Harv. Envtl. L. Rev. 243, 268–270 (2017) (describing weaknesses in RCRA financial assurance mechanism for underground storage tanks as applied in some but not all states); Jayni Foley Hein et al., *Self-Bonding in an Era of Coal Bankruptcy: Recommendations for Reform* (Aug. 2016) ("outstanding self-bond obligations currently total $3.86 billion nationwide, of which $2.4 billion is held by coal companies currently in bankruptcy"); Lucas Davis, *Modernizing Bonding Requirements for Natural Gas Producers* (June 2012) ("Minimum bond amounts were set in 1960 and have never been updated for inflation.").

[122] That pressure was the result of several factors, Susan F. Tierney, *The U.S. Coal Industry: Challenging Transitions in the 21st Century*, ES-1 & -2 (Sept. 2016) (listing factors, competition from natural gas in the electricity sector high among them). DOE, *Staff Report to the Secretary on Electricity Markets and Reliability* 13 (Aug. 2017), https://perma.cc/3J2H-S6LT ("The biggest contributor to coal and nuclear plant retirements has been the advantaged economics of natural gas-fired generation.").

remediation.[123] In the production and distribution portions of the oil and gas sector, the story is similar in some ways,[124] but different in others. While adequate maintenance of a facility or drilling site with healthy financial prospects is a sound investment for a well-capitalized company, such companies often sell when facilities' profitability begins to fade – and generally sell to less well-capitalized buyers with incentives to wring out whatever profitability remains. This wringing out involves investing less in maintenance and set-asides for eventual decommissioning and environmental remediation.[125]

Mitigation can be expected to increase the number of facilities implicated by these laws, which provide fragmentary coverage[126] and are often too weak or too easily avoided to ensure adequate investment or action on the part of private parties.[127] Furthermore, the agencies that are meant to provide a backstop are often overwhelmed and unable to make up for private nonfeasance.[128] Thus the environmental and public health consequences of

123 Tripp Baltz et al., *No Collateral Needed for Cleanup in Some States Despite Mine Bankruptcies*, Bloomberg BNA, Mar. 30, 2017, https://perma.cc/6VT9-ZYG4 ("Several states are still willing to let mines operate without putting up collateral for land cleanup even though three of the country's biggest coal companies only recently emerged from bankruptcy."); see also Eric L. Dixon and Kendall Bilbrey, *Abandoned Mine Land Program: A Policy Analysis for Central Appalachia and the Nation* 73 .5.2 (July 2015), https://perma.cc/63MK-5UQ5 (showing a steep and secular decline of fees collected for remediating abandoned coal mines in Appalachian states; adjusting for inflation, in 2013, collected fees were less than 25 percent of those collected in 1979).

124 Jacqueline Ho et al., *Resources for the Future, Plugging the Gaps in Inactive Well Policy* (May 2016), https://perma.cc/3J7T-54SZ (noting that decommissioning costs generally far exceed bonding requirements in surveyed states).

125 See Keith Goldberg, *5 Legal Battles to Watch as Oil, Gas Bankruptcies Surge*, Law360, Mar. 17, 2016, https://perma.cc/N35Y-EUBQ (describing fights over environmental liabilities incurred or held by bankrupt drilling firms).

126 See, e.g., Stephen Lee, *Mountains of Slag Spur Unlikely Alliance to Clean Up Waste*, Bloomberg BNA Daily Environment Report, July 25, 2017 ("'If the nation's 18 remaining refuse plants also shut down, the 800 coal refuse piles scattered throughout Pennsylvania – to say nothing of the hundreds more in West Virginia, Kentucky, and other Appalachian states – will remain in place indefinitely, because neither the federal government nor the states have the money to clean them up,' said John Stefanko, deputy secretary for the Office of Active and Abandoned Mine Operations at the Pennsylvania DEP.").

127 See, e.g., Sylvia Carignan, *Superfund Skepticism Spurs States to Seek Toxic Cleanup Alternatives*, Bloomberg BNA Daily Environment Report, Sept. 18, 2017.

128 See EPA, *Hazardous Waste: Baselines for Resource Conservation and Recovery Act (RCRA) Corrective Action Sites*, https://perma.cc/Y2MK-3JP4 (last updated Dec. 28, 2016) (providing links to document listing 3,779 sites prioritized by EPA for cleanup pursuant to RCRA); Rachael Bale, *The Suprising Reason Abandoned US Mines Haven't Been Cleaned Up*, Reveal-The Center for Investigative Reporting, Nov. 4, 2014, https://perma.cc/7TYB-2ETC (reporting that at least 500,000 mines remain unremediated in the United States).

underfunded decommissioning and remediation represent externalities that existing law and policy have failed to price into the cost of doing business in the fossil fuel industry.

**Local economic impacts.** Existing legal responses to this category of coharms all fit into the category of programs devised to support a "just transition"[129] – a term coined by unions in the 1980s and applied since to various policy-driven displacements of industry, whether rooted in trade, environmental protection, or something else.[130] These responses tend to be ad hoc, temporary, and reflective of the political circumstances that prevailed at the time of their formulation.

One fairly recent example of a federal response to regional economic impacts that were not driven by mitigation policies, but by policies that resemble mitigation policies, is the Secure Rural Schools and Community Self-Determination Act of 2000 (SRS Act).[131] The SRS Act addressed the economic impacts of the timber industry's decline in localities that, because of their location on or near federal lands, were prohibited from using taxation to raise revenues.[132] Since 1908, they relied instead on a portion of the receipts from federal timber leases – revenues that fell alongside timber sales in the 1990s when the industry shrank and in some places collapsed.[133] The SRS Act originally scheduled its last payments for FY 2006, but that end date was extended a further seven years through several reauthorizations.[134] An earlier example, the Clean Air Employment Transition Assistance Program, was enacted as part of the Clean Air Act Amendments of 1990 with the express purpose of ameliorating those amendments' impacts on employment in affected sectors.[135] Both of these were straightforward in their logic and

[129] See, e.g., Robert Pollin and Brian Callaci, *A Just Transition for U.S. Fossil Fuel Industry Workers*, American Prospect, July 6, 2016, https://perma.cc/6V2J-YRU6 (proposing national program to accompany decarbonization driven by federal policy).

[130] Samantha Smith, *Just Transition: A Report for the OECD* (May 2017), https://perma.cc/Q6RJ-JFXR (summarizing history of transition policy category and providing examples of transition policies from several international contexts).

[131] Pub. L. No. 106-393, *codified at* 16 U.S.C. §§7101–7153.

[132] Katie Hoover, *Reauthorizing the Secure Rural Schools and Community Self-Determination Act of 2000* (Mar. 2015), https://perma.cc/U5GV-W65B.

[133] See Bruce Sorte et al., *Economic Impacts on Oregon Counties of the Termination of the Secure Rural Schools and Community Self-Determination Act* (P.L. 106-393) (Dec. 2008), https://perma.cc/9V78-6QJJ (estimating impacts of Act's impending sunset in 2008).

[134] Ibid.

[135] 29 U.S.C. § 1662e ("For purposes of this section, the term 'eligible individual' means an individual who – (A) is an eligible dislocated worker . . . and (B) has been terminated or laid off, or has received a notice of termination or lay off, as a consequence of compliance with the Clean Air Act.").

focused in their aims: Federal money would, for a limited period, offset local economic impacts that federal policy had created in whole or in part.

A more recent example of a federal transition program included two components: the POWER (Partnerships for Opportunity and Workforce and Economic Revitalization) Initiative, launched in 2015, and the Assistance for Coal Communities (ACC) 2017 program (collectively, the Initiative). Legal authority for the Initiative roots primarily in the 1965 law, amended in 1975, 1998, and 2002, that established the Appalachian Development Commission,[136] but it draws from a variety of sources with diverse statutory bases for funding,[137] and directs that funding to the support of diverse activities.[138] In contrast to the SRS Act and Clean Air Employment Transition Program, the Initiative was not conceived of or codified by Congress, but by the Obama Administration to address existing consequences of coal communities' decline.[139] It was intended to address *anticipated* consequences – actual or perceived – of the Clean Power Plan,[140] whose opponents framed it as the latest attack in the "war on coal," and one that could cause the Appalachian coal supply chain to collapse.[141] Thus the Initiative was, arguably, the first example of a deliberate (if not explicit) federal response to co-harms of climate change mitigation policy in the United States.

136 Appalachian Regional Development Act (ARDA) of 1965, *as amended* (codified at 40 U.S.C. § 14101–704.

137 See White House, *Fact Sheet: The Partnerships for Opportunity and Workforce and Economic Revitalization (POWER) Initiative* (Mar. 27, 2015) https://perma.cc/H86P-CFPC ("The POWER initiative will coordinate use of appropriated FY 2015 funds from a range of federal programs, while following the relevant statutory and regulatory requirements for each program."); Consolidated Appropriations Act of 2017, Pub. L. No. 115-31 (including funds for ACC 2017).

138 Appalachian Regional Commission, *POWER Award Summaries by State* (Jan. 2017), www.arc.gov/images/grantsandfunding/POWER2017/ARCPOWERAwardSummariesbyState1-25-2017.pdf.

139 Appalachian Citizens' Law Center, *POWER+ For the People*, https://perma.cc/HQJ2-UWX4 ("The 'POWER Initiative' is a federal initiative that provides funding to assist communities struggling with the decline of the coal industry in strengthening and diversifying their economies.").

140 80 Fed. Reg. 64,661, 64,676 (Oct. 23, 2015) ("In the final rule, the EPA encourages states, in designing their state plans, to consider the effects of their plans on employment and overall economic development to assure that the opportunities for economic growth and jobs that the plans offer are manifest. We also identify federal programs, including the multi-agency Partnerships for Opportunity and Workforce and Economic Revitalization (POWER) Initiative.").

141 See Lydia Wheeler, *Mining Group Asks EPA to Stay Its Climate Rule*, *The Hill*, Aug. 3, 2015, https://perma.cc/L484-JDZ3 ("Since more than 90 percent of coal sold in the U.S. is used for power generation, the NMA said the EPA rule would also cripple domestic coal production and eliminate thousands of mining jobs.").

## 13.4 RECOMMENDATIONS

The following recommendations are ambitious. That ambition should not mark them as "academic" (i.e. logically sound but actually unfeasible). Rather, it should highlight the large gap that separates our current legal regime from one that would translate current scientific understanding of climate change and public health outcomes into limitations on polluting activity and the prudent selection and calibration of climate change mitigation measures.

### 13.4.1 *Adopt Legislation Clarifying the Relevance of Climate Change to Permits Issued under Environmental Laws*

Debates over whether and how Congress should address climate change have focused almost exclusively on mitigation measures that would address GHG emissions; adaptation has largely been ignored. As described in the previous section, patterns of enforcement of environmental laws will signal that the climate is changing and that adaptation is necessary to protect public health, but that signal is likely to be compromised and also accompanied by a great deal of noise. Congress should ensure that the signal is unmistakably clear. It can do so by amending key provisions of several statutes – at a minimum, RCRA, CERCLA, and the Clean Water Act – to condition permitting, whether for facilities or whole states, on the conduct of a climate change vulnerability assessment within the past five to ten years and demonstrated efforts to address any significant vulnerabilities it identifies. For coastal facilities, key vulnerabilities would presumably arise from storms and flooding; for, say, statewide water pollution permits under the Clean Water Act, they could arise equally from risks of drought or extreme precipitation.

### 13.4.2 *Fund a Study to Develop Recommendations for the Analysis of Ancillary Effects of Regulations with an Eye to Potential Judicial Review*

Much divides the disparate views of those who dispute how agencies should assess co-benefits and co-harms of a proposed regulation – and how courts should review an agency's assessment. Because these issues are both important and contested, it would be valuable to develop basic points of agreement and to articulate points of disagreement with precision so that further research is more likely to resolve them. The National Academies of Sciences and the US Judicial Conference are both potential sources of neutral expertise that could contribute authors and editors to such an effort.

### 13.4.3 *Compel Insurance Coverage for Community Public Health Impacts As Well As Bonding for Site Remediation*

Planning and prudent investment can avoid many of the public health co-harms that arise from economic transition. Many of those co-harms are avoidable through anticipatory bonding or investment. Firms that are responsible for facilities/sites that will require decommissioning/remediation should prevent local communities from enduring the public health impacts that would arise directly from their abandonment. In addition to bonding requirements, therefore, firms should also be required to carry an insurance policy against direct adverse public health impacts of potential abandonment (i.e. not including indirect effects like unemployment or economic straits). The premiums charged for the policy would reflect the insurer's assessment of the adequacy of money set aside for bonding – adequate bonding would mean a nominal premium, inadequate bonding a potentially enormous premium.

## 13.5 CONCLUSION

Existing environmental laws are not well suited to addressing the public health impacts of climate change or the impacts of efforts to mitigate it. As this chapter has explained, even laws that make public health their lodestar cannot be expected to reveal fully the costs of climate change or to compel action that effectively addresses or avoids those costs. With respect to mitigation, existing law has thus far failed to take account of the public health impacts of several sources of GHG emissions, and is unlikely to anticipate or avert the impacts of mitigating those emissions. This bleak assessment is the basis for several ambitious recommendations that would not provide complete solutions to the problems identified herein, but would provide important foundations on which to build toward effective solutions.

# 14

# Incorporating Public Health Assessments into Climate Change Action

Jessica Wentz

*As policy makers weigh cost-benefit and risk-risk tradeoffs in their responses to climate change, they should factor public health impacts into their calculations. This chapter discusses how and where public health can be made a prominent criterion for decision making about climate change mitigation and adaptation. It describes the types of health-related considerations that would be relevant in the context of actions that are expressly intended as a response to climate change as well as actions that have unintended consequences related to climate change. It then identifies specific laws that require consideration of health impacts in the planning and decision-making processes for these types of actions. Finally, it identifies key gaps in the legal mandates for and institutional capacity to conduct climate change-related health assessments and presents recommendations on how to fill those gaps.*

## 14.1 INTRODUCTION

As discussed throughout this book, climate change will have wide-ranging impacts on human health. Increases in severe weather such as coastal storms, heavy precipitation events, and heatwaves will threaten the lives, safety, and well-being of millions of people. More gradual changes such as sea-level rise, increases in average temperatures, and changes in average precipitation will affect public health in other ways – for example, by displacing people from their homes, reducing the availability of freshwater, adversely impacting agricultural production and food security, and contributing to conflicts driven by resource scarcity. Scientists also predict that warmer temperatures will increase the prevalence of diseases like malaria, exacerbate local air quality problems, and lead to longer, more severe allergy seasons.

Responses to climate change can also generate secondary health effects, both positive and negative. For example, the closure of fossil fuel-fired power plants will reduce not only greenhouse gas emissions but also emissions of conventional air pollutants such as nitrogen oxides ($NO_x$) and sulfur dioxide ($SO_2$). At the same time, policies aimed at increasing the proportion of renewable energy in the energy mix may have the effect of increasing energy prices, potentially making it more difficult for low-income households to cover heating and cooling costs, and this, too, can have an adverse impact on public health.[1] Similarly, while retreat from encroaching coastlines avoids the adverse effects that follow from coastal storms and flooding, the devaluing of homes and displacement of communities can also generate adverse public health effects.

These public health considerations need to be accounted for when deciding how and whether to proceed with actions aimed at mitigating or adapting to climate change. They should also be accounted for when deciding how to proceed with actions that have unintentional consequences related to climate change, such as actions that generate substantial quantities of greenhouse gas emissions and actions that may compound the health risks associated with climate change. Generally speaking, the relevant health considerations will differ depending on whether the action deals with climate change mitigation (i.e. the action will generate or reduce greenhouse gas emissions), adaptation (i.e. the project has the potential to compound or mitigate public health risks associated with climate change), or both.

This chapter discusses how public health can be made a prominent criterion for decision-making about climate change mitigation and adaptation. Section 14.2 describes the types of public health considerations that are relevant for climate change-related actions, including actions that are specifically intended as climate change response measures, as well as actions that have unintended consequences related to climate change. Section 14.3 identifies specific laws and policies that direct decision-makers to account for public health impacts and could be used to incorporate climate change-related health considerations into public decision-making processes. While there are only a few laws that require a comprehensive and independent health impact assessment (HIA) for a narrow subset of actions, there are many laws that require an assessment of public health outcomes as part of a broader impact assessment or planning process, and these laws apply to a wide range of actions entailing mitigation and adaptation considerations. Finally, Section 14.4 discusses gaps in legal mandates and technical support for assessing the health impacts of climate change-related actions and presents recommendations for filling those gaps.

[1] Chapter 13 contains a more detailed discussion of the public health effects of climate change mitigation efforts.

## 14.2 PUBLIC HEALTH CONSIDERATIONS FOR CLIMATE CHANGE-RELATED ACTIONS

The types of public health considerations that are relevant for climate change-related actions will differ depending on whether the action has implications for climate change mitigation, adaptation, or both.[2] This section thus outlines two alternative frameworks for assessing health outcomes in the context of climate change mitigation and adaptation.

### 14.2.1 *Public Health Considerations in the Context of Climate Change Mitigation*

The public health considerations outlined in this section would be relevant to any actions that have a meaningful impact on greenhouse gas emissions,[3] either by reducing or increasing those emissions. Such actions would include measures specifically intended to mitigate climate change, such as emission standards for stationary and mobile sources of greenhouse gases, renewable energy portfolio standards, and energy efficiency standards. They would also include measures that have the co-benefit of reducing greenhouse gas emissions, such as emission standards for conventional air pollutants. Finally, they would include measures that would contribute to increases in greenhouse gas emissions, such as fossil fuel subsidies and leasing programs that allow additional fossil fuels to reach the market.

When considering how and whether to proceed with these types of actions, decision-makers should account for two types of public health considerations: (1) the health impacts of the greenhouse gas emissions that are either generated or avoided as a result of the action under consideration, and (2) any secondary public health impacts, such as the effect of rising energy costs on low-income households and the air quality benefits of reducing fossil fuel use.

[2] Examples of actions spanning both mitigation and adaptation considerations include energy efficiency programs and comprehensive land and resource management plans.

[3] This chapter recognizes that it is extremely difficult to define what constitutes a "meaningful impact" on greenhouse gas emissions. Metrics have been developed to approximate significance in particular contexts – for example, the US Environmental Protection Agency recognizes that 25,000 tons per year of carbon dioxide equivalent ($CO_2e$) is meaningful enough to impose a greenhouse gas reporting obligation on facilities that meet or surpass this threshold. Mandatory Reporting of Greenhouse Gases; Final Rule, 74 Fed. Reg. 56260 (Oct. 30, 2009) (codified at 40 CFR Parts 86, 87, 89 et al.). Ultimately, the determination of significance will depend on the context of the action, and can best be understood as part of a spectrum (more or less significant) rather than binary (significant vs. not significant), with the depth of the impact analysis tailored to the level of significance.

#### 14.2.1.1 Health Impacts of Greenhouse Gas Emissions

The first step to understanding the health impacts of a project's greenhouse gas emissions footprint is to quantify the amount of greenhouse gases that will be generated or abated as a result of the project. Such quantification has become relatively commonplace in environmental reviews and other forms of project assessment.[4] Quantification is also frequently treated as the final step in the greenhouse gas assessment – the justification being that the greenhouse gas footprint of a particular project cannot be traced to any specific on-the-ground impacts of climate change. But any suggestion that there is not a causal connection between a project's greenhouse gas footprint and on-the-ground impacts is inaccurate. The reality is that the emissions generated or abated by a particular action will either contribute to or mitigate global climate change, and this effect will have implications – however modest – for all of the on-the-ground impacts of climate change. As noted by Justice John Paul Stevens, writing for the Supreme Court in *Massachusetts* v. *EPA*, "[a] reduction in domestic emissions would slow the pace of global emissions increases, no matter what happens elsewhere."[5]

Recognizing this, the next step in the assessment of public health impacts stemming from an action's greenhouse gas footprint would be to provide a general and qualitative summary of how the impacts of climate change will affect public health. There are many authoritative reports and studies to draw on when preparing such a summary,[6] and this book contains a wealth of information on linkages between climate change impacts and public health outcomes. The summary should describe, as a starting point, the global public health risks of climate change as well as those risks which are most relevant to the United States and the region and locality where the action is located.

The assessment should also consider the scale and magnitude of the emissions impact on these public health outcomes. One option for doing this is to assign a cost value to emissions, using a tool such as the social cost of carbon

4 See Saloni Jain et al., *How Did Federal Environmental Impact Statements Address Climate Change in 2016?* (New York: Sabin Center for Climate Change Law, 2017); Jessica Wentz, *Survey of Climate Change Considerations in Federal Environmental Impact Statements, 2012–2014* (New York: Sabin Center for Climate Change Law, 2016).

5 *Massachusetts* v. *E.P.A.*, 549 U.S. 497, 526 (2007).

6 See, e.g., George Luber et al., Ch.9: Human health, In: Jerry Melillo et al., *U.S. Global Change Research Program, Climate Change Impacts in the United States: The Third National Climate Assessment* (U.S. Global Change Research Program, 2014); Interagency Working Group on Climate Change and Health, *A Human Health Perspective on Climate Change* (National Institute of Environmental Health Sciences & Environmental Health Perspectives, 2010); World Health Organization, *Fact Sheet: Climate Change and Health*, https://perma.cc/VF6Z-DW7R.

($SC\text{-}CO_2$), relied upon by the US federal government until recently.[7] Scientists have also developed estimates for the social cost of methane ($SC\text{-}CH_4$) and nitrous oxide ($SC\text{-}NO_2$). The challenge of using these estimates to approximate health outcomes is that they are under-inclusive, insofar as they only cover a subset of costs related to public health outcomes (specifically, costs stemming from increases in the prevalence of climate-related diseases, such as malaria and dengue fever, and pollution), and they cover a wide range of other costs beyond public health outcomes (such as economic damages from net changes in agricultural productivity, property damage from increased flood risk, and changes in energy system costs).[8] Notably, these estimates do not reflect the effects of heat stress, extreme weather events, wildfires, infrastructure damage, increasing food prices, food and water shortages, scarcity-driven conflicts, and human migration on human health.[9] This situation may change in the near future, as the $SC\text{-}CO_2$ is currently being reviewed and may

7 The $SC\text{-}CO_2$ was used in federal rulemakings involving greenhouse gas reductions prior to President Trump's issuance of Executive Order 13783 on Promoting Energy Independence and Economic Growth (Mar. 28, 2017), which stated that the technical support documents that established and updated the $SC\text{-}CO_2$ were "no longer representative of governmental policy." This order creates uncertainty about how agencies should monetize greenhouse gas impacts in rulemakings. For examples of regulatory impact analyses that used the social cost of carbon, methane, and nitrous oxide, see: EPA, *Regulatory Impact Analysis for the Clean Power Plan Final Rule*, EPA-452/R-15-003 (2015); EPA, *Regulatory Impact Analysis of the Final Oil and Natural Gas Sector: Emission Standards for New, Reconstructed, and Modified Sources*, EPA-452/R-16-002 (2016); EPA and NHTSA, *Joint Technical Support Document: Final Rulemaking for 2017–2025 Light-Duty Vehicle Greenhouse Gas Emission Standards and Corporate Average Fuel Economy Standards*, EPA-420-R-12-091 (2012). State and local governments have also begun to use the federal estimates for the social cost of carbon, methane, and nitrous oxide to support policymaking, regulation, and planning, most often in the context of utility planning. See, e.g., Colorado Public Utilities Commission Decision No. C17-0316, issued April 28, 2017, Proceeding No. 16A-0396E, at paragraph 88 (requiring Colorado utility resource planners to use the $SC\text{-}CO_2$ to account for the costs of emissions in planning decisions); New York Public Service Commission Decision No. 15-01168/15-E-0302, decided Aug. 1, 2016, at 20 (using the $SC\text{-}CO_2$ as the basis for zero emission credit pricing in utility contracts); 20 Ill. Comp. Stat. Ann. 3855/1–75 (same).

8 Interagency Working Group on Social Cost of Carbon, *Technical Support Document – Social Cost of Carbon for Regulatory Impact Analysis Under Executive Order 12866* (2010); Interagency Working Group on the Social Cost of Greenhouse Gases, *Addendum to the Technical Support Document on Social Cost of Carbon for Regulatory Impact Analysis Under Executive Order 12866: Application of the Methodology to Estimate the Social Cost of Methane and the Social Cost of Nitrous Oxide* (2016); Interagency Working Group on the Social Cost of Greenhouse Gases, *Technical Update of the Social Cost of Carbon for Regulatory Impact Analysis Under Executive Order 12866* (2016).

9 The National Academies of Sciences, Engineering, and Medicine, Committee on Assessing Approaches to Updating the Social Cost of Carbon et al., *Valuing Climate Damages: Updating Estimation of the Social Cost of Carbon Dioxide* (Washington, DC: National Academies Press, 2017).

be updated to cover a broader range of impacts, including health impacts (and similar changes may be made to the SC-$CH_4$ and SC-$N_2O$).[10] But even with the anticipated changes, the estimates will likely remain somewhat overinclusive and underinclusive with respect to health impacts.

This does not mean that the SC-$CO_2$, SC-$CH_4$, and SC-$N_2O$ should not be used in a health assessment, but the limitations of these estimates should be explicitly recognized in assessment documents. Additionally, as discussed in Section 14.4, government agencies and public health advocates should also consider whether it would be helpful and feasible to develop a separate metric for specifically valuating the health impacts of emissions.[11]

Assigning a monetary value to the health impacts of greenhouse gas reductions or increases associated with a proposed action is particularly important when comparing the costs and benefits of that action. It would be irrational not to monetize these impacts when other costs and benefits are monetized, and such an omission may very well render the decision unlawful. As held by the Ninth Circuit Court of Appeals in *Center for Biological Diversity* v. *National Highway Traffic Safety Administration (NHTSA)*, an agency "cannot put a thumb on the scale" of its cost-benefit analysis by ignoring the cost of greenhouse gas emissions or the value of emission reductions.[12]

The scope and depth of the health impact analysis should correspond with the magnitude of the greenhouse gas impact: for smaller actions, such as the approval (or reauthorization) of a single coal-fired power plant, it may be sufficient to provide a qualitative summary of public health risks associated with climate change and a quantification of the social cost of the emissions generated by the project. For proposals with a large greenhouse gas impact, such as the promulgation of national emission standards for power plants or motor vehicles, there should be a more in-depth review of public health impacts from emissions generated or avoided.

As climate change attribution science progresses, it may be possible to link the greenhouse gas emissions from such major actions to quantifiable health impacts – for example, avoided deaths, avoided instances of displacement, or decreases in the incidence of asthma and infectious diseases.[13] These types of

10 National Academies of Science, *Project Information: Assessing Approaches to Updating the Social Cost of Carbon*, https://perma.cc/Q7MT-37YB.

11 There are existing studies that provide monetized estimates of the health co-benefits of greenhouse gas emission reduction policies.

12 *Center for Biological Diversity* v. *National Highway Traffic Safety Administration*, 538 F.3d 1172 (9th Cir. 2008).

13 Part III of this book (Chapters 7–12) reviews the types of health risks that are associated with climate change, the extent to which these risks have been quantified, and the challenges of

estimates could be developed by taking overall (global or national) projections of climate change impacts on public health and then scaling those projections based on the proportional contribution of the project to global or national emissions.[14] This sort of quantitative data could also feed into the development of public health impact cost estimates for greenhouse gas emissions.

#### 14.2.1.2 Health Impacts of Secondary Effects

The assessment of health impacts for actions that generate or abate greenhouse gas emissions should also encompass any secondary public health effects of those actions. This ensures a balanced and holistic approach to the health assessment. Such effects may include beneficial impacts on public health ("co-benefits") as well as adverse impacts ("co-harms"), and will vary depending on the action that is being reviewed. Table 14.1 outlines some examples of secondary effects from actions that entail mitigation considerations. Note that this is an illustrative rather than comprehensive list.

A variety of studies and tools can be used to quantify some of these public health co-benefits and co-costs associated with actions that generate or reduce greenhouse gas emissions. One key tool is the US Environmental Protection Agency (EPA)'s Co-Benefits Risk Assessment (COBRA) Screening Model, which can be used to estimate the effect of clean energy policies on conventional air pollutants such as $SO_2$ and $NO_x$ as well as the corresponding public health benefits.[15] There are also many other studies of the health benefits

quantification. See also Maryam Navi et al., *Developing Health-Related Indicators of Climate Change: Australian Stakeholder Perspectives*, 14 Int'l J. of Envtl. *Research and Public Health* 552 (2017); World Health Organization, *Quantitative Risk Assessment of the Effects of Climate Change on Selected Causes of death, 2030s and 2050s* (2014); June J. Cheng and Peter Berry, *Development of key indicators to quantify the health impacts of climate change on Canadians*, 58(5) Int'l J. of Pub. Health 767 (2013).

14 Some quantified estimates of public health impacts at the global and national levels are already available. The US National Climate Assessment (NCA), for example, provides general protections of health impacts such as increases in temperature-related deaths. However, the uncertainty and range of these estimates is still quite high, and quantitative data is not available for all health impacts. See Luber et al., *Ch:9: Human Health, supra* note 6.

15 EPA, *Co-Benefits Risk Assessment (COBRA) Screening Model*, www.epa.gov/statelocalclimate/co-benefits-risk-assessment-cobra-screening-model (last visited Mar. 9, 2017). For additional guidance and other tools recommended by EPA to estimate the air quality co-benefits of a clean energy program see EPA, *Assessing Air Quality, Greenhouse Gas, and Public Health Benefits*, https://perma.cc/9EG9-F7YY (last visited Mar. 9, 2017).

TABLE 14.1 *Secondary health effects of actions that generate or reduce greenhouse gases*

| Action | Co-Benefits<br>*Positive health outcomes stemming from:* | Co-Harms<br>*Negative health outcomes stemming from:* |
|---|---|---|
| $CO_2$ emission cap for power plants | • Reductions in air pollutants (other than greenhouse gases)<br>• Reductions in water pollutants<br>• Reductions in harmful waste products, such as coal ash<br>• Increase in employment in green energy sectors | • Possible increase in electricity prices<br>• Decrease in employment in fossil fuel sectors |
| Transportation plan aimed at decreasing motor vehicle use, increasing use of other options (walking, biking, public transit) | • Reductions in air pollutants (other than greenhouse gases) from vehicles<br>• Reductions in traffic accidents<br>• Reductions in traffic noise<br>• Increase in physical activity | • Increased exposure to pathogens for public transit users[16] |
| Authorization of coal-fired power plant (new or continued operation) | • Local employment at power plant<br>• Local economic benefit | • Increase in air pollutants (other than greenhouse gases)<br>• Increase in water pollutants<br>• Increase in harmful wastes<br>• Possible decrease in employment in green energy sectors |

16 See Joy Troko et al., *Is Public Transport a Risk Factor for Acute Respiratory Infection?* 11(16) BMC Infectious Diseases 1 (2011) (finding that recent bus or tram use was associated with an almost six-fold increase in the occurrence of acute respiratory infection).

TABLE 14.1 *(continued)*

| Action | Co-Benefits *Positive health outcomes stemming from:* | Co-Harms *Negative health outcomes stemming from:* |
|---|---|---|
| Approval of natural gas pipeline | • Employment opportunities<br>• Possible lower power prices | • Increase in air pollutants (other than greenhouse gases) from construction and operation<br>• Risk of water pollution / soil contamination from leaks, spills<br>• Risk of harm caused by gas explosion |

stemming from reductions in co-pollutants associated with greenhouse gas mitigation programs that can be referred to when conducting a health assessment.[17] Notably, while the EPA tool and these studies focus on co-benefits of greenhouse gas mitigation actions from decreases in conventional air pollutants, they can also be used to assess co-*harms* of actions that increase greenhouse gas emissions from *increases* in conventional air pollutants.

Chapter 13 contains a more detailed discussion of the potential secondary effects of greenhouse gas mitigation measures, including the potential for adverse health impacts.

### 14.2.2 *Public Health Considerations in the Context of Climate Change Adaptation*

The public health considerations outlined in this section are relevant to actions that are expressly intended as a response to the impacts of climate change (typically referred to as "adaptation" or "resilience" measures) as well as actions that may exacerbate or mitigate public health risks associated with

[17] See, e.g., John Balbus et al., *A Wedge-Based Approach to Estimating Health Co-Benefits of Climate Change Mitigation Activities in the United States*, 127 Climatic Change 199 (2014); Britt Grossman et al., *The Ancillary Benefits from Climate Policy in the United States*, 50(4) Envtl. and Resource Economics 585 (2011); Darby Jack and Patrick Kinney, *Health Co-Benefits of Climate Mitigation in Urban Areas*, 2(3) Current Opinion in Envtl. Sustainability 172 (2010); Michelle Bell et al., *Ancillary Human Health Benefits of Improved Air Quality Resulting From Climate Change Mitigation*, 7 Envtl. Health 41 (2008).

climate change. This second class of actions is quite broad: It encompasses physical actions located in areas that will be adversely affected by climate change, such as coastlines and floodplains exposed to sea-level rise and/or extreme precipitation events, forest areas at increased risk of wildfire and biological transformation, and urban areas where the heat island effect will intensify as average and extreme temperatures increase, as well as policies and programs that affect the vulnerability of people to the impacts of climate change. Programs and policies pertaining to the provision of affordable housing and shelters are excellent examples of the sorts of policies that, while not explicitly geared toward climate change adaptation, can either exacerbate or mitigate public health risks related to climate change.

Health impacts need to be accounted for when evaluating the relative costs and benefits of, or allocating scarce resources to, actions expressly intended to improve adaptation or resilience. Take, for example, a coastal municipality that is developing a sea-level rise adaptation plan. Municipal decisions about whether to allow sea wall construction should reflect assessments of public health consequences. Such assessments should encompass secondary or unintended impacts on public health – for example, installing a seawall might protect one property from sea-level rise and storm surge while simultaneously causing displaced wave action to destroy an adjacent living shoreline that provides an ecosystem supportive of mosquito predators. By eliminating a natural source of mosquito control, sea walls could protect some properties while also undermining a tool for reducing the transmission of vector-borne diseases.

For actions that may inadvertently exacerbate or mitigate risks associated with climate change, the purpose of the health assessment would be to better understand how the effects of the action and of climate change interact, and then use this information in deciding whether and how to proceed with the action. This assessment could be used to provide a more compelling justification for actions that would mitigate climate-related risks, such as programs aimed at increasing access to public shelters and emergency services. Conversely, this type of assessment might lead a decision-maker to reconsider an action that would compound climate-related risks, such as a development project that would intensify the urban heat island effect in a city that already experiences dangerously high temperatures during the summer.

The first step to assessing health outcomes in this context would be to conduct a vulnerability assessment that evaluates how climate change will affect the area where the action would be implemented and identifies the corresponding public health risks. The second step would be to evaluate how the proposed action would affect existing climate-related risks, and whether climate change may affect the proposed action in a way that would generate new risks. For

TABLE 14.2 *Examples of public health considerations for projects that may mitigate or exacerbate risks associated with climate change*

| Project | Relevant climate impacts | Public health considerations |
|---|---|---|
| Seawall | • Sea-level rise<br>• More severe storms | • Reduced risk of death or harm caused by flooding and inundation<br>• Improved psychological health (due to risk reduction)<br>• Possible direct and indirect harms to adjacent properties or communities if built |
| Coastal wetlands restoration | • Sea-level rise<br>• More severe storms<br>• Rising temperatures | • Reduced risk of death or harm caused by flooding and inundation<br>• Improved psychological health<br>• Improved water quality<br>• Possible alleviation of urban heat island effect (if located in urban area)<br>• Reduction in vector-borne diseases (by providing habitat for predators of mosquitoes) |
| Floodplains development | • Sea-level rise<br>• More severe storms<br>• Changing precipitation patterns<br>• Rising temperatures | • Health risks associated with increased exposure to storms, flooding, erosion, and inundation<br>• Possible reduction in capacity of natural habitat to act as storm buffer<br>• Possible compounding of heat effects from urban development and climate change<br>• Possible compounding of risk of saltwater intrusion into groundwater resources due to increased water consumption |

TABLE 14.2 *(continued)*

| Project | Relevant climate impacts | Public health considerations |
|---|---|---|
| Managed retreat from vulnerable areas | • Sea-level rise<br>• More severe storms<br>• Inland flooding | • Reduced risk of death or harm caused by flooding and inundation<br>• Possible psychological and physical health harms caused by relocation of people and communities |

example, when deciding whether to permit the construction of a facility where hazardous chemicals will be stored, the assessment should consider whether sea-level rise or more severe floods may breach the facility and cause the release of chemicals into the ambient environment.[18] Finally, the assessment should evaluate whether the proposal can be sited, designed, or implemented in a way to minimize public health risks associated with climate change.

There are existing resources that can help with this analysis. In particular, there is already a substantial body of literature on how to assess the public health effects of climate change for planning and adaptation purposes,[19] as well as literature on assessing specific health outcomes of adaptation measures (including co-benefits or co-harms).[20]

[18] The effects of Hurricane Harvey on Houston illustrate the perils of failure to plan for climate change and enhanced flood risk, particularly in the context of permitting decisions for facilities that might release hazardous substances during a flood. See Dena Adler, *Hurricanes' Contaminated Floodwaters Might Crest Next Wave of Climate Change Litigation*, Climate Law Blog, Sept. 19, 2017 (citing reports detailing the massive quantities of toxic chemicals and pollutants that were released as a result of the flooding caused by Hurricane Harvey).

[19] See, e.g., Adele Houghton and Paul English, *An Approach to Developing Local Climate Change Environmental Public Health Indicators, Vulnerability Assessments, and Projections of Future Impacts*, J. of Envtl. and Pub. Health, Article ID 132057 (2014); Helen Brown et al., *A Health Impact Assessment Framework for Assessing Vulnerability and Adaptation Planning for Climate Change*, 11 Int'l J. of Envtl. Research and Pub. Health 12896 (2014); Paul English et al., *Environmental Health Indicators of Climate Change in the United States: Findings from the State Environmental Health Indicator Collaborative*, 117 Envtl. Health Persp. 1673 (2009); Centers for Disease Control and Prevention (CDC), *Assessing Health Vulnerability to Climate Change: A Guide for Health Departments* (2015).

[20] See, e.g., June J. Cheng and Peter Berry, *Health Co-Benefits and Risks of Public Health Adaptation Strategies to Climate Change: A Review of Current Literature*, 58 Int'l J. of Pub. Health 305 (2013); United Nations Framework Convention on Climate Change, *Assessing the Costs and Benefits of Adaptation Options: An Overview of Approaches* (2011) (while not focusing exclusively on health impacts, this document contains useful information on how to quantify and qualitatively assess health benefits from adaptation measures).

## 14.3 LEGAL FRAMEWORK FOR INCORPORATING PUBLIC HEALTH ASSESSMENTS INTO CLIMATE CHANGE-RELATED ACTIONS

Various laws require government actors to consider health impacts as part of the public planning and decision-making process for issuing regulations and other major actions. However, this requirement is typically part of a broader legally required impact assessment process in which health impacts do not always feature prominently. Only a handful of laws require an independent and comprehensive health impact assessment (HIA) prior to the approval of certain types of actions. As a result, comprehensive HIAs are typically conducted on an ad hoc basis at the discretion of government agencies or nongovernmental public health advocates.

This section reviews the types of federal, state, and local laws that require consideration of health impacts and could be used to incorporate climate change-related health considerations into public decision-making processes. Two key findings are that: (i) the analysis of public health outcomes in an HIA tends to be far more comprehensive than that which is conducted in the context of a broader impact assessment,[21] but (ii) because an independent HIA is rarely required by law, other impact assessment laws, particularly those pertaining to regulatory and environmental impact assessments, will continue to be an important vehicle for ensuring that health impacts are assessed in the context of climate change-related actions.

### 14.3.1 *Health Impact Assessment (HIA)*

A health impact assessment (HIA) is a "systematic process that uses an array of data sources and analytic methods and considers input from stakeholders to determine the potential effects of a proposed policy, plan, program, or project on the health of a population and the distribution of those effects within the population."[22] The goals of HIA are both procedural (to ensure that health is a key factor in decision-making) and substantive (to improve health outcomes). HIA is also used to demonstrate the value (both positive and negative) of a proposed action to decision-makers and the public. This is certainly the case for climate change mitigation and adaptation actions, where public health benefits are an important justification for any costs incurred as a result of these actions.

[21] Rajiv Bhatia and Aaron Wernham, *Integrating Public Health Into Environmental Impact Assessment: An Unrealized Opportunity for Environmental Health and Justice*, 116(8) Envtl. Health Persp. 991 (2008).

[22] National Research Council Committee on Health Impact Assessment, *Improving Health in the United States: The Role of Health Impact Assessment*, 5 (Washington, DC: The National Academies Press, 2011).

The components of an HIA typically include: (i) screening to determine the need and value of HIA; (ii) scoping to determine what sorts of health impacts should be considered; (iii) the assessment of health impacts; and (iv) the development of recommendations based on that assessment, for example, recommendations on how to mitigate adverse health impacts.[23] In some instances, the HIA may also include provisions related to monitoring and evaluation of health impacts and measures aimed at addressing adverse health impacts.[24] The types of health impacts considered in an HIA can be divided into two categories: first, the effect of the action on *health determinants* – such as air pollution – and second, the corresponding effect on specific *health outcomes* – such as the incidence of respiratory disease.[25] HIAs typically employ a holistic definition of health,[26] consistent with the World Health Organization's approach: "Health is a state of complete physical, mental and social well-being and not merely the absence of disease or infirmity."[27]

There are very few US laws that require HIA. At the federal level, there are no general mandates to conduct HIA, and thus requirements for environmental reviews, discussed in this section, have been the primary vehicle for considering health impacts in the context of federal projects, programs, and regulations.[28] There are, however, a number of environmental statutes that require some form of health assessment to inform the establishment of environmental standards. The Clean Air Act, for example, requires the Environmental Protection Agency (EPA) to establish National Ambient Air Quality Standards (NAAQS) for certain air pollutants[29] at a level sufficient to

23 Centers for Disease Control and Prevention, *Health Impact Assessment*, https://perma.cc/M2LW-66E2 (last visited Mar. 10, 2017).

24 Ibid.

25 Health determinants encompass the range of personal, social, economic, and environmental factors that determine the health status of individuals or populations. These include, for example, air pollution, water pollution, outdoor temperatures, extreme weather events, and access to goods and services such as food, water, shelter, air conditioning, and heating. Health outcomes encompass any changes in the health status of an individual or group when attributable to the proposed action, including both intended and unintended outcomes. These include, for example, any changes in the incidence of death, injury, physical disease, mental illness, and emotional stress.

26 See, e.g., California Department of Health, *A Guide for Health Impact Assessment* (2010), 6.

27 Constitution of the World Health Organization (1948), *available at* https://perma.cc/9DTU-MTV3.

28 James G. Hodge Jr. et al., *Integration of Health & Health Impact Assessments Via Environmental Policy Acts* (Baltimore, MD: The Network for Public Health Law, 2016).

29 EPA has not attempted to establish a NAAQS for greenhouse gas emissions, thus the NAAQS process has not become a vehicle for assessing the public health impacts caused by those emissions. This could change in the future, but many commenters believe that the NAAQS process is a poor fit for greenhouse gas regulation, because it regulates ambient air pollution

protect public health and welfare,[30] which means that EPA must conduct a health assessment to determine the proper level for the NAAQS. But these types of provisions only apply to specific forms of regulatory action and do not encompass the broad scope of federal actions that implicate climate change adaptation and mitigation.

At the state and local level, there have been a number of legislative proposals to require or support HIA in certain contexts, but only a handful of these proposals have been enacted.[31] Table 14.3 lists the state laws that have been enacted that either require or support HIA.

TABLE 14.3 *State laws requiring or supporting HIA*

| State | Summary of HIA law |
|---|---|
| California | Executive Order S-04-10 (2010) called for the establishment of a Health in All Policies (HiAP) Task Force to identify priority programs, polices, and strategies to improve the health of Californians while also meeting other social and environmental goals within the state.[32] The HiAP Task Force has generated and contributed to a number of guidebooks and collaborative public health programs,[33] including a 2013 guide for state and local governments on how to incorporate health considerations into policy and planning decisions.[34] |
| Massachusetts | S.B. 2087 (2009) requires HIAs to "determine the effect of transportation projects on public health and vulnerable populations."[35] |

rather than emissions. See, e.g., The Center for Clean Air Policy, A *Pragmatic Approach to Regulating Greenhouse Gases Under the Clean Air Act* (2009).

30 42 U.S.C. § 7409.

31 Health Impact Project, *Health Impact Assessment Legislation in the States* (2015) (see Table 1, p. 5–6, for a list of proposed and enacted HIA legislation); Doug Farquhar, *An Analysis of State Health Impact Assessment Legislation* (Washington, DC: National Conference of State Legislatures, 2014) (finding that fifty-six bills in seventeen states have been introduced since 2009, including thirteen bills that supported the use of full-fledged HIAs as defined by current guidance and practice standards in the field).

32 California Executive Order S-04-10 (2010), https://perma.cc/6SBG-UAZX.

33 For a list of the HiAP Task Force's accomplishments, see *California's Health in All Policies Task Force Fact Sheet*, https://perma.cc/W3LP-7CSN.

34 Linda Rudolph et al., *Health in All Policies: A Guide for State and Local Governments* (Public Health Institute and American Public Health Association 2013), https://perma.cc/E99G-C47L.

35 Codified at: General Laws of Massachusetts, Chapter 6C, § 33.

TABLE 14.3 (*continued*)

| State | Summary of HIA law |
|---|---|
| Vermont | H. 202 (2011), enacted as Act 48, requires the Department of Health to recommend a plan to implement a "public health impact assessment process" to evaluate the health effects of local, municipal, and state policy and planning decisions.[36] |
| Washington | S.B. 6099 (2007) required an HIA to examine the impact of a bridge replacement project on "air quality, carbon emissions, and other public health issues."<br>Wash. Rev. Code 43.20.270 /43.20.285 (2006) calls for the creation of an interagency council on health disparities and authorizes health impact reviews as one mechanism to address those disparities. |

Due to the lack of legal mandates, HIA is usually conducted on an ad hoc and voluntary basis,[37] often with funds or other forms of support from nongovernmental actors.[38] In many instances, an HIA is conducted entirely by a nongovernmental actor and then submitted to a government agency as an input to the decision-making process.[39] When HIA is undertaken voluntarily by a government agency, it is often due to public pressure to evaluate health impacts in the context of a particular project and/or the availability of external support for the assessment (such as when other parties offer to partner in the HIA process).[40]

Some jurisdictions also have provided technical and financial support for HIA. Oregon, for example, does not have legislation mandating HIA, but the Oregon Health Authority nonetheless developed a HIA program that provided technical and financial support for the development of HIAs (or measures "aimed at developing state and local capacity to conduct HIAs").[41] As part of this program, the Health Authority developed guidance on how to conduct

36 The HIAs conducted by the Vermont Department of Health are available at: https://perma.cc/M7RG-K2B9.

37 Aaron Wernham, *Health Impact Assessments are Needed in Decision Making About Environmental and Land-Use Policy*, 30(5) Health Affairs 947 (2011).

38 Emily Bourcier et al., *An evaluation of health impact assessments in the United States, 2011–2014*, 12 Preventing Chronic Disease: Pub. Health Research, Practice, and Pol'y 1 (2015) (Table 1 lists the characteristics of 23 HIAs prepared from 2011 through 2014); Health Impact Project, *HIA in the United States*, https://perma.cc/XLZ9-TMDJ.

39 Ibid.

40 Bhatia and Wernham (2008), *Integrating Public Health Into Environmental Impact Assessment, supra* note 21.

41 Oregon Health Authority, *Health Impact Assessment*, https://perma.cc/7PGD-PLD5.

HIA in the context of local climate change policies.[42] However, due to the lack of any legislative mandate or dedicated allocation of funds for HIA activities, the funding for the Oregon Health Authority's HIA program has been "unstable, primarily grant funded, with a greater demand than there are funds."[43] California's HiAP Task Force, noted in Table 14.3, performs a similar function as the Oregon Health Authority's HIA program.

### 14.3.2 *Regulatory Impact Analysis*

The federal government and some states have enacted laws requiring government agencies to conduct a regulatory impact analysis (RIA) prior to the issuance of major regulations.[44] The primary goal of these laws is to ensure that decision-makers fully weigh the social costs and benefits of these regulations, and as such, RIAs frequently contain some sort of formal cost-benefit analysis (CBA). While public health considerations are frequently accounted for in RIAs, the scope, depth, and quality of the analysis varies. For example, the full range of public health impacts associated with a proposal are not always assigned a monetary value for the purposes of conducting CBA, which may result in these impacts receiving less attention by decision-makers than monetized impacts.

The federal requirements for RIA originate from several different sources, including the Unfunded Mandates Reform Act,[45] several executive orders,[46] and the Office of Management and Budget (OMB) Circular A-4.[47] All of these legal authorities direct federal agencies to evaluate costs and benefits of regulations prior to their issuance. Health impacts are explicitly cited as a factor to consider when conducting CBA, but there is no explicit guidance on how to evaluate health impacts. In the absence of guidance, the extent to

[42] Oregon Health Authority, *Climate and Health Impact Assessment*, https://perma.cc/4Y2P-XDJX.

[43] Tia Henderson, *Funding Health Impact Assessment in Oregon and Washington: An Analysis*, 3 (Portland, OR: Upstream Public Health 2015).

[44] The federal requirements are discussed in this section. For state law examples, see Cal. Gov't Code § 11346.3; N.Y. A.P.A. Law § 202-a; Mich. Comp. Laws Ann. § 24.245; Del. Code Ann. tit. 29, § 10404A; Ky. Rev. Stat. Ann. § 13A.240; N.C. Gen. Stat. Ann. § 150B-21.4;.Tex. Gov't Code Ann. § 2001.0225 (only requiring RIA for major environmental rules).

[45] Pub. L. 104–4 (1995).

[46] Key examples include Executive Orders No. 12291 (1981), 12866 (1993), 13563 (2011), and 13579 (2011).

[47] For a detailed discussion of these legal sources and how they interact, see Maeve P. Carey, *Cost-Benefit and Other Analysis Requirements in the Rulemaking Process* (Washington, DC: Congressional Research Service, 2014). See also OMB, *Regulatory Impact Analysis: A Primer* (2011).

which federal RIAs encompass an evaluation of health outcomes depends on the nature of the regulation being reviewed. Unsurprisingly, the analysis of health impacts is typically a major component of RIA for regulations aimed at environmental protection.

There are some federal precedents for evaluating climate change-related health impacts in RIA. During the Obama administration, EPA accounted for the effects of climate change on public health in RIAs for regulations establishing greenhouse gas emission standards for vehicles and stationary sources.[48] EPA took a two-part approach to evaluating health benefits in the CBA for these rules: First, it used the SC-$CO_2$ to determine the monetary value of $CO_2$ emission reductions and then it modeled the effect of the rule on co-pollutants (conventional air pollutants) and calculated the economic benefit of those emission reductions. EPA also provided a fairly detailed discussion of the health benefits that were captured in the RIA but could not be quantified.[49] The assessment of health outcomes has played a fundamentally important role in justifying these rules to the public. For example, when reviewing the impacts of the Clean Power Plan (which established $CO_2$ emission standards for existing power plants), EPA found that the total benefits of the rule would range from $32 to $54 billion in 2030 (taking into account public health and other impacts) whereas the compliance costs would have only been $5.1 to $8.4 billion in 2030.[50]

Various states also use some form of RIA to evaluate the impact of the regulations they promulgate. Examples of such jurisdictions include California, New York State, and Michigan.[51] The level of detail and extent to which public health assessments are included varies quite a bit depending on the jurisdiction and the type of regulatory proposal. Many of these RIA laws require consideration of "benefits" and "costs" but there is no explicit reference to "public health impacts."[52] Some only require a calculation of costs and

[48] EPA, *Regulatory Impact Analysis for the Clean Power Plan Final Rule*, EPA-452/R-15-003 (2015); EPA, *Regulatory Impact Analysis of the Final Oil and Natural Gas Sector: Emission Standards for New, Reconstructed, and Modified Sources*, EPA-452/R-16-002 (2016); EPA and NHTSA, *Joint Technical Support Document: Final Rulemaking for 2017–2025 Light-Duty Vehicle Greenhouse Gas Emission Standards and Corporate Average Fuel Economy Standards*, EPA-420-R-12-091 (2012).

[49] See, e.g., *Regulatory Impact Analysis for the Clean Power Plan Final Rule*, Ibid. at 4–46 to 4–48.

[50] Ibid., at ES-9, ES-20-21 (these figures cover estimates for both mass- and rate-based implementation approaches and reflect a 3 percent discount rate applied to climate and health benefits).

[51] Cal. Gov't Code § 11346.3; N.Y. A.P.A. Law § 202-a;).

[52] See, e.g., N.Y. Comp. Codes R. & Regs. tit. 9, § 3.100 (I)(b)(v); Mich. Comp. Laws Ann. § 24.245(3).

not benefits.[53] While adverse public health outcomes can and should be considered as costs and beneficial health impacts should be considered as benefits, this is not always the case – the focus tends to be on the direct economic impacts of the proposal. Other constraints, such as word limits[54] and limited agency resources, can also prevent state and local actors from fully evaluating health impacts in their RIAs.

### 14.3.3 *Environmental Impact Assessment*

Environmental impact assessment (EIA) provides another opportunity for incorporating consideration of health impacts into public planning and decision-making processes. EIA laws passed at the federal, state, and local levels generally require some consideration of public health impacts. However, as detailed in this section, the evaluation of health impacts in EIA documents is typically far less comprehensive than what would be found in a standalone HIA.

At the federal level, the National Environmental Policy Act (NEPA) requires federal agencies to evaluate the environmental impacts of major proposals and to prepare a comprehensive environmental impact statement (EIS) for proposals "significantly affecting the human environment."[55] As noted by Bhatia and Wernham (2008), the use of the term "human environment" demonstrated the "intent of congress for NEPA to promote public policy attentive to the interrelated needs of human well-being and environmental integrity."[56] Reinforcing this idea, the protection of public health features prominently in the stated objectives of the statute.[57] The implementing regulations also specify that the environmental "effects" that must be considered under NEPA include health effects,[58] and that the "degree to which the proposed action affects public health or safety" is one of the criteria that must be applied when determining whether a proposal will have a significant effect on the environment.[59] Based on these

[53] See, e.g., Del. Code Ann. tit. 29, § 10404A.

[54] See, e.g., New York State, *Regulatory Impact Statement* (SAPA § 202-a) Instructions (revised Jan. 2017), https://perma.cc/UAN8-42VZ (no word limit on description of "costs" but 500-word limit on description of "needs and benefits" – which is the only category where public health impacts might be discussed).

[55] 42 U.S.C. § 4331(C).

[56] Bhatia and Wernham (2008), *Integrating Public Health Into Environmental Impact Assessment*, *supra* note 21, at 992.

[57] See 42 U.S.C. § 4321(b)(2)-(3), (c).

[58] 40 C.F.R. § 1508.8.

[59] 40 C.F.R. § 1508.27.

provisions, the National Research Council has emphasized that assessing health effects in NEPA reviews is "a matter of law and not discretion."[60]

Seventeen states and several cities have enacted statutes requiring some form of EIA.[61] The scope and depth of the EIA requirements varies by jurisdiction, but generally speaking, there is a good deal of overlap with the requirements of NEPA.[62] In some cases, the mandate to consider public health outcomes may be even more explicit. The New York City Environmental Quality Review (CEQR) Technical Manual contains some of the most detailed guidance on how to consider public health impacts in an EIA process. Its chapter on public health defines relevant terms, provides instructions for determining whether a full public health assessment is appropriate, and outlines a step-by-step process for conducting a public health assessment in the context of a CEQR review.[63]

Consistent with these statutory and regulatory directives, EIA documents typically examine a range of issues that can affect human health, such as whether the proposed action will generate air pollution, water pollution, and other hazardous substances. However, as noted by the National Research Council Committee on Health Impact Assessment, EIA "has traditionally included at most only a cursory analysis of health effects."[64] Other studies have similarly found that the assessment of actual public health outcomes through EIA tends to be "rare and narrowly focused on toxic exposures" and that a "comprehensive and systematic approach to human health impacts in EIA practice has not evolved."[65] In many cases, EIA

60 National Research Council *Improving Health in the United States, supra* note 22, at 12.

61 Council on Environmental Quality (CEQ), *State NEPA Contacts* (last updated June 2013) https://perma.cc/8GXS-EA4D.

62 See, e.g., California Environmental Quality Act (CEQA), California Environmental Quality Act, 13 Cal. Pub. Res. Code § 21000(d) ("it is the intent of the Legislature that the government of the state take immediate steps to identify any critical thresholds for the health and safety of the people of the state and take all coordinated actions necessary to prevent such thresholds from being reached"); CEQA regulations, CCR § 15126.2(a) (environmental analysis should include "health and safety problems caused by the physical changes" associated with the project); New York State Environmental Quality Review Act (SEQRA), NYS Envtl. Conservation Law § 8–0103 ("The capacity of the environment is limited, and it is the intent of the legislature that the government of the state take immediate steps to identify any critical thresholds for the health and safety of the people of the state and take all coordinated actions necessary to prevent such thresholds from being reached"); SEQRA regulations, 6 CRR-NY 617.2(l) ("Environment means the physical conditions that will be affected by a proposed action, including land, air, water, minerals, flora, fauna, noise, resources of agricultural, archeological, historic or aesthetic significance, existing patterns of population concentration, distribution or growth, existing community or neighborhood character, and human health").

63 NYC Mayor's Office of Environmental Coordination, *CEQR Technical Manual*, Chapter 20 (March 2014), https://perma.cc/VEY2-5SY5.

64 National Research Council (2012), *Improving Health in the United States, supra* note 22, at 12.

65 Bhatia and Wernham *Integrating Public Health Into Environmental Impact Assessment, supra* note 21, at 991 (citing National Research Council, *Cumulative Environmental Effects of Oil and Gas Activities on Alaska's North Slope* (Washington, DC: National Academies Press, 2003); A. Steinemann, *Rethinking Human Health Impact Assessment*, 20 Environ. Impact

preparers will address public health by simply characterizing the pollutants and wastes that will be generated by the project, without linking this to specific health outcomes, such as increases in the occurrence of death, disease, and disability.[66] For example, a study of forty-two federal EISs published from 1979 to 1996 found that 62 percent of them contained no mention of specific health impacts accruing from exposure to pollutants and other aspects of the proposed actions.[67] EIA documents also tend to ignore how changes in the built environment, social determinants, and economic impacts can affect public health.[68]

There are several factors that explain why health impacts do not feature more prominently in EIA documents. The first is that most of the EIA laws do not provide explicit guidance on the types of public health impacts to include in an EIA or the process by which to identify, evaluate, and mitigate those impacts. The New York City CEQR Technical Manual is one exception to this rule. The second factor is that public health agencies are often not involved in the EIA process and most jurisdictions lack a formal mechanism for their engagement.[69] The third factor is that there has not been much litigation defining the scope of agency obligations to evaluate health outcomes in EIA documents.[70]

The situation is changing, however, due to both public pressure and, to a lesser extent, litigation.[71] In recent years, there have been several examples of

Assess. Rev. 627 (2000); K. Davies and B. Sadler, *Environmental Assessment and Human Health: Perspectives, Approaches, and Future Direction* (Ottawa: Health Canada, 1997).

66 Bhatia and Wernham, *Integrating Public Health Into Environmental Impact Assessment, supra* note 21, at 991, 993.

67 Steinemann, *Rethinking Human Health Impact Assessment*, at 627. It is possible that health impacts have become a more prominent consideration in more recent federal EISs, but the author is not aware of any study confirming this.

68 Bhatia and Wernham, *Integrating Public Health Into Environmental Impact Assessment, supra* note 21; Davies and Sadler, *Environmental Assessment and Human Health, supra* note 65.

69 National Research Council, *Improving Health in the United States, supra* note 22, at 17; Bhatia and Wernham, *Integrating Public Health Into Environmental Impact Assessment, supra* note 21, at 991. The only federal agency that is consistently involved in the federal EIA process is the Environmental Protection Agency, which reviews EISs and provides recommendations on both the framing of the environmental analysis as well as comments on the acceptability of the environmental impacts of the proposed action. EPA is granted this authority under Section 309 of the Clean Air Act, 42 U.S.C. § 7609.

70 National Research Council, *Improving Health in the United States, supra* note 22, at 17.

71 For example, in California, there has been some litigation on how health impacts should be considered in CEQA reviews, and courts have emphasized the need to connect environmental outcomes with specific health outcomes. In one case, the court held that an environmental impact report (EIR) was inadequate because it did not analyze the respiratory health impacts that corresponded with anticipated air quality impacts from construction of two shopping centers. The court explained that: "[a]fter reading the EIRs, the public would have no idea of the health consequences that result when more pollutants are added to a non-attainment basin." *Bakersfield Citizens for Local Control* v. *City of Bakersfield*, 124 Cal. App. 4th 1184, 1220 (2004). In another case, the court found that an EIR was inadequate because it did not fully consider the health impacts of authorizing pesticide use. *Californians for Alternatives to Toxics* v. *Dep't of Food & Agric.*, 136 Cal. App. 4th 1, 17 (2005).

integrated HIA and EIA for federal and state projects. For example, the 2007 North Slope Development HIA, which was conducted as part of a broader EIS, assessed the health impacts of oil development on Alaskan Native Communities in response to public pressure from those communities for a full HIA.[72] A similar HIA was integrated into the EIS for the 2012 Thomson Oil and Gas Leasing Decision in Alaska.[73] EPA is also taking steps to develop more formal frameworks for integrating HIA and EIA. In 2010, CDC and EPA Region 4 signed a pilot agreement to increase consideration of public health in NEPA assessments.[74] There is very little public information about the implementation of this agreement, but it appears to have been the impetus for an EPA Region 4 HIA training and an HIA that was recently conducted for a green infrastructure project in Atlanta, Georgia.[75] More recently, in 2015, the national EPA office issued a memorandum to its regional NEPA directors highlighting the utility of using HIA as a decision-support tool for NEPA reviews.[76] The memorandum also stated that EPA intended to: (i) develop screening and scoping tools that EPA reviewers could use to determine whether a proposal undergoing NEPA review would benefit from an HIA or HIA elements, (ii) develop web-based training to educate regional NEPA staff on HIA, and (iii) partner with other federal agencies to conduct a pilot project integrating HIA into an EIS or environmental assessment (EA).[77] EPA has yet to issue a progress update on the implementation of these actions.

Turning to the issue of climate change: EIA preparers now routinely account for greenhouse gases in EISs and other NEPA documents.[78] However, the analysis typically focuses on the amount of emissions that will be generated as a result of the proposed action, without discussing the social costs of those emissions and the

72 A. Wernham, *Inupiat Health and Proposed Alaskan Oil Development: Results of the First Integrated Health Impact Assessment / Environmental Impact Statement for Proposed Oil Development on Alaska's North Slope*, 4(4) EcoHealth 500 (2007). For other examples of how HIA can be integrated into EIA, see Hodge et al., *Integration of Health & Health Impact Assessments Via Environmental Policy Acts, supra* note 28.

73 U.S Army Corps of Engineers, Point Thomson Project FEIS (2012), Appendix R.

74 S. H. Kennedy et al., *Health Impact Assessment in the National Environmental Policy Act: Leveraging Resources Across Federal Agencies for Environmental Health and Informed Decisionmaking* (2010).

75 See EPA Region 4, Proctor Creek's Boone Boulevard Green Street Project Health Impact Assessment (HIA), EPA/60/R-14/400 (2015).

76 EPA, *Memorandum on Promoting the Use of Health Impact Assessment to Address Human Health in Reviews Conducted Pursuant to the National Environmental Policy Act and Section 309 of the Clean Air Act* (Nov. 10, 2015), https://perma.cc/4AY4-727Z.

77 Ibid.

78 See Jain et al., *How Did Federal Environmental Impact Statements Address Climate Change in 2016?*; Wentz et al., *Survey of Environmental Considerations in Federal EISs, 2012–2014*.

implications for public health. There are two court decisions holding that federal agencies must assign a monetary value to greenhouse gas emissions when other costs and benefits are also monetized,[79] but there is no independent requirement to use the SC-$CO_2$, SC-$CH_4$, or SC-$N_2O$ in NEPA documents. Some EIA preparers have also begun to account for the effects of climate change in the area where the proposal will be implemented and the corresponding implications for the environmental outcomes of the proposal.[80] But again, this analysis typically does not entail an in-depth assessment of health risks and outcomes.

Some state governments have also begun to develop procedures for integrating HIA and climate considerations into EIA processes. The Minnesota Department of Health, for example, has developed an informal HIA screening tool for projects undergoing state environmental assessment (the screening phase before a full review) as well as a training module for incorporating health and climate change into environmental review, and has performed a desktop HIA for a mixed-use development project during the assessment phase.[81] The department is now working to better incorporate health into environmental review by adding additional sections and questions within the review procedures on this topic and by developing better guidance documents. Other states, including California and Massachusetts, are also working to integrate EIA, HIA, and climate change through new rules and guidance for project proponents.[82]

### 14.3.4 *Local Land Use and Transportation Planning*

The requirements for local land use and transportation planning vary widely depending on the jurisdiction, but some form of impact assessment is often required as a component of the planning process (this may take the form of EIA or a separate land use/transportation impact assessment). There are also

79 *Ctr. for Biological Diversity* v. *Nat'l Highway Traffic Safety Admin.*, 538 F.3d 1172, 1198 (9th Cir. 2008);
*High Country Conservation Advocates* v. *United States Forest Serv.*, 52 F. Supp. 3d 1174, 1191–92 (D. Colo. 2014).

80 See Jain et al., *How Did Federal Environmental Impact Statements Address Climate Change in 2016?*, *supra* note 78; Wentz et al., *Survey of Environmental considerations in Federal EISs*, *supra* note 78.

81 Minnesota Department of Health & Minnesota Climate and Health Program, *Incorporating Health and Climate Change into the Minnesota Environmental Assessment Worksheet (2012): Minnesota Department of Health, What's In Your EAW? Incorporating Health and Climate Change into the Environmental Assessment Worksheet* (2013).

82 See, e.g. 2011 California Environmental Quality Act (CEQA) Handbook, § 15064.4: Determining the Significance of Impacts from Greenhouse Gas Emissions; Appendix G: Environmental Checklist Form (2016); Massachusetts Executive Office of Energy and Environmental Affairs, GHG Emission Policy and Protocol (2010).

substantive mandates in many jurisdictions that indicate that health should be a criterion in the development of land use and transportation plans and accompanying requirements. In particular, there are many state and local laws that authorize local governments to enact plans and accompanying requirements as may be necessary to protect or promote the health, safety, and general welfare of the local community.[83]

As with other planning processes, HIAs are sometimes integrated into local land use and transportation planning on an ad hoc basis, typically when there is public pressure to assess health impacts or a public health advocate that volunteers to conduct the analysis. According to one study, approximately one third of all HIAs completed in the United States between 2004 and 2014 involved an assessment of planning decisions and their effect on local health outcomes.[84] Preliminary research suggests that these HIAs conducted in this context have improved health outcomes without stymying local development.[85] Even where a formal HIA is not conducted, public health considerations often feature prominently in local land use and transportation planning processes. However, many jurisdictions lack formal legal requirements pertaining to the evaluation of public health outcomes in the context of local planning processes. As a result, health outcomes may be ignored or may receive a cursory analysis, as is the case with many EIAs.

Some states have contemplated adopting legislation that would integrate HIAs into land use and transportation planning, but most of these proposals have not been enacted. For example, in 2007, both California and Maryland considered integrating HIAs into land use planning but neither measure passed.[86] Massachusetts is the one state that has enacted a law expressly requiring HIAs for transportation planning projects.[87] In other states and at the federal level, EIA laws continue to be an important tool for ensuring that governments and project proponents evaluate the public health effects of land use and transportation planning decisions.

This is not to say that public health outcomes are ignored in the absence of a legally mandated framework for evaluating those outcomes. To the contrary, many states and local governments have made a concerted effort to integrate public health assessments into planning processes, primarily through guidance and technical support. For example, the Minnesota Department of

83 See Joseph Schilling, *The Public Health Roots of Zoning – In Search of Active Living's Legal Genealogy*, 96 Am. J. Prev. Med. 28 (2005).

84 Health Impact Project & American Planning Association, *Health Impact Assessment Can Inform Planning to Promote Public Health* (2016).

85 Wernham, *Inupiat Health and Proposed Alaskan Oil Development*, *supra* note 72, at 950–951.

86 Health Impact Project (2015), *Health Impact Assessment Legislation in the States*, *supra* note 31, at 4.

87 General Laws of Massachusetts, Chapter 6C, § 33.

Health provides guidance and training on health, planning, and climate change, and the state also has a Metro Healthy Comprehensive Plan Workgroup that serves as a resource for city and county planners in the Minneapolis metropolitan area.[88] Other states, such as California, as well as nongovernmental groups, such as the American Planning Association, have also published guidance on this topic.[89]

### 14.3.5 *Facility-Level Approvals and Assessments*

Health impacts can also be assessed when decision-makers are deciding whether to issue a permit or other facility-level approval. In some cases, some sort of health risk assessment (HRA) is legally required for individual facilities in order to commence and continue operations. Such HRAs often focus on a particular health concern, rather than an overarching assessment of public health outcomes like that found in an HIA. For example, California's Air Toxics "Hot Spots" Information and Assessment Act (AB 2588), requires sources of toxic air pollutants that are designated as "high priority" to submit a HRA to air pollution control and air quality management districts within 150 days after they are designated as such.[90] The HRA must describe how toxic emissions are released from the facility, how they disperse throughout the community, and the potential for those toxic pollutants to impact human health.[91] Facility-level approvals may also be subject to the EIA requirements discussed in the previous section.

These types of facility-level assessment could provide a basis for incorporating climate change-related health considerations into approvals required for the facilities. Where a facility is subject to EIA, the assessment of such considerations could be quite broad, encompassing both the contribution of

[88] Minnesota Department of Health, *Comprehensive Plans*, www.health.state.mn.us/topics/places/plans.html (see resources on public health, such as "Healthy Planning Training" and the "Healthy Planning 'How-To' Guide"); Hennepin County, *Active Living: Metro Healthy Comprehensive Plans Workgroup*, www.hennepin.us/your-government/projects-initiatives/active-living. See also Minnesota Department of Health & Minnesota Department of Transportation, *Connecting Health and Transportation: A Health Impact Assessment of the Minnesota Statewide Multimodal Transportation Plan* (2016);

[89] See, e.g., California Department of Health, *A Guide for Health Impact Assessment, supra* note 26; Anna Ricklin and Nick Kushner, *Healthy Plan Making: Integrating Health Into the Comprehensive Planning Process: An Analysis of Seven Case Studies and Recommendations for Change* (Chicago and Washington, DC: American Planning Association 2013).

[90] California Health and Safety Code, Division 26, Part 6, Chapter 4, § 44360.

[91] Ibid.; California Air Resources Board and California Air Pollution Control Officers Association, *Risk Management Guidance for Stationary Sources of Air Toxics*, (July 23, 2015), https://perma.cc/ZG26-F4S4.

the facility to climate change and the impacts of climate change on the facility (as may be relevant). There are also ways to integrate climate change-related health considerations into more targeted HRAs for facilities. For example, future HRAs conducted under the California air toxics program could incorporate an analysis of whether and how warmer temperatures could contribute to air toxic exposures. Similarly, HRAs conducted for hazardous waste and waste treatment facilities could be updated to account for climate-related risks such as flooding and sea-level rise.[92] Guidance or regulations may be needed to induce facilities to conduct such an analysis.[93]

## 14.4 LEGAL GAPS, CAPACITY CONSTRAINTS, AND RECOMMENDATIONS

The lack of a comprehensive or systematic legal framework for conducting HIA means that health-related considerations may not be a prominent criterion in decisions that have implications for climate change mitigation and adaptation. Granted, many of the major actions undertaken as a response to climate change – such as federal controls on $CO_2$ emissions – have been accompanied by relatively detailed assessments of public health benefits, as these assessments provide an important justification for the actions. However, in other contexts – such as actions that may *contribute* to greenhouse gas emissions, or land use-based approaches to climate change mitigation, or in a wide range of adaptation actions related to infrastructure or human settlements – public health considerations may very well be ignored or downplayed in the absence of legal requirements compelling decision-makers to thoroughly consider these considerations. The same is true for actions that may compound or mitigate health risks related to climate change.

Institutional capacity and resource constraints could also pose a significant barrier to incorporating climate change-related health considerations into planning and decision-making processes. Many of the government agencies and private actors that engage in projects with significant implications for

92 The federal Resource Conservation and Recovery Act (RCRA) requires each application for a landfill or surface impoundment containing hazardous waste to be accompanied by an assessment of the "potential for the public to be exposed to hazardous wastes or hazardous constitutes through releases related to the unit." RCRA § 3019, 42 U.S.C. 6939a. A health assessment is required for facilities that pose a "substantial potential risk to human health." Ibid. Sea-level rise and extreme weather events could increase the risk of exposure for facilities located on coastlines or in floodplains, and therefore should be considered in this context.

93 *See* EPA Office of Solid Waste and Emergency Response, *Climate Change Adaptation Implementation Plan*, Appendix A (2014) (recognizing that hazardous waste permitting requirements may need to be updated in order to account for the potential impacts of climate change on these facilities).

climate change mitigation and adaptation do not have specific expertise in public health analysis, and as noted previously, they do not always coordinate with public health agencies when assessing costs, benefits, and risk-risk trade-offs of their actions.

This section will discuss how federal and state regulators could overcome some of the legal gaps and implementation challenges so as to ensure that public health considerations are accounted for at all levels of decision-making on climate change mitigation and adaptation.

### 14.4.1 *Addressing the Legislative Gap*

While some form of regulatory impact analysis is frequently required for regulatory actions, this is not the case for legislative actions: the federal government and most states do not have laws which require comprehensive impact evaluations for legislative proposals.[94] There are thirteen states that require the preparation of *fiscal* impact statements for legislative proposals, but these statements do not encompass public health considerations (or a wide range of other considerations) and in most states the requirement is limited to voter initiatives.[95] Legislative impact analyses are occasionally conducted on an ad hoc basis, typically by nongovernmental entities as a means of building support for or opposition to the proposed law. But without any legal requirements or systematic guidelines for these assessments, there is no way of ensuring their accuracy, completeness, or consistency.

This legal gap is particularly problematic from the perspective of integrating public health considerations into responses to climate change-related actions, because many of these actions – particularly those that establish new legal requirements – will need to be legislative in nature, owing to the current lack of climate change legislation at the federal level and in many states. New laws aimed at addressing climate change would benefit from an evaluation of how the law would promote better public health outcomes as compared with other legislative approaches.

Legislative bodies could address this gap by enacting legislation or establishing guidelines on how and when legislative impact assessments should be conducted. Such laws or guidelines could specify the manner in which public

94 The one exception in federal law is the National Environmental Policy Act (NEPA), which requires an environmental review for legislative proposals developed by administrative agencies and submitted to Congress. 42 U.S.C. § 4332(C). This provision does not apply to the vast majority of legislative proposals because they are not drafted by federal agencies.

95 National Council of State Legislators, *Preparation of Fiscal Analysis*, https://perma.cc/A52E-W7BS.

health outcomes should be addressed in the overarching impact assessment. One such proposal was recently introduced in Minnesota – the proposed bill (SF 742) would direct the state Legislative Audit Commission, with input from legislators and legislative committees, to select up to five legislative topics each year that would then undergo HIA pursuant to the terms of the bill.[96]

### 14.4.2 *Avoiding Cursory Analysis of Health Impacts under Existing Impact Assessment Regimes*

The other fundamental legal gap, discussed at length in Section 14.3, is the absence of any requirement in most jurisdictions for independent HIAs, whether on their own or as part of other impact assessment regimes. This absence means that most legislators and regulators need not account for the health impacts of their decisions in a meaningful way.

There are two ways to address this gap. The first would be to expand the legal requirements for conducting independent HIAs. However, political support for expanding HIA requirements appears to be limited: The federal government has not expressed any intention to consider or enact such requirements, and although many states have contemplated such legislation, only a handful have actually enacted it. There is also a pragmatic rationale for opposing this approach: If independent HIA requirements are imposed on top of other planning and impact assessment requirements, this could become burdensome for government agencies and project proposers. New HIA requirements could increase the cost and time required to implement an action and create new opportunities for litigation challenging the action. That said, the time and resources invested in the assessment process (as well as the prospect and outcome of litigation) can result in more deliberative planning and better informed decisions.[97]

The second approach would be to develop laws and guidelines for existing planning and impact assessment processes that contain clear directives on how to account for public health outcomes in these regimes. Such an integrated approach may be more politically feasible and less burdensome than establishing new HIA requirements, as it would build on existing legal and institutional frameworks. Regulatory impact analysis and environmental impact analysis present the most promising legal frameworks for incorporating health-related

[96] Minnesota SB 742, 90th Legislature (2017–2018).

[97] Another advantage to requiring comprehensive and independent HIA is that it would be possible to reach actions not covered by EIA or RIA. For example, in some EU countries such as England and Sweden, HIA has evolved independently of EIA and is applied in a wide range of policy decisions not subject to EIA. Bhatia and Wernham, *Integrating Public Health Into Environmental Impact Assessment*, *supra* note 21, at 993.

considerations. The federal and state entities responsible for promulgating standards for RIA and EIA could draw on existing HIA guidance from other jurisdictions – for example, Canada and Australia have adopted formal guidance for incorporating HIA into EIA.[98] There are state guidance documents that could also be referred to, including the New York City CEQR Technical Manual, the California Department of Public Health's HIA Guidebook,[99] and a variety of guidebooks published by nongovernmental public health advocates.[100] Further complementing these resources are the examples of integrated EIA-HIA that have already been conducted by federal and state agencies,[101] and the public health assessments in some RIAs.[102]

### 14.4.3 *Addressing Capacity and Resource Constraints and Promoting a Uniform Approach to Health Assessments*

The two main legal gaps identified previously are not the only barriers to effective consideration of health impacts in the context of climate change-related actions. There are also capacity and resource constraints within the agencies and other stakeholders that would be conducting these impact assessments. Additional guidance and tools that make it possible to streamline the climate change and health assessment would help to resolve some of these resource and capacity constraints while also promoting greater uniformity in assessments.

*First*, guidance on how to account for climate change-related public health outcomes in the context of different types of projects would be helpful. As noted in Section 14.3, the Oregon Health Authority has developed a toolkit for conducting HIA in the context of actions intended as a response to climate change.[103] This toolkit is a valuable resource, and provides some

98 EnHealth, *Health Impact Assessment Guidelines, Canberra, Australia (2001)*; Health Canada, *Canadian Handbook on Health Impact Assessment (2004)*. For additional examples of international guidance, see World Health Organization, *Health Impact Assessment.*

99 Rajiv Bhatia, A *Guide for Health Impact Assessment* (Sacramento: California Department of Public Health 2010), https://perma.cc/JZ8E-REZN.

100 See, e.g., Rudolph et al. (2013), *Health in All Policies, supra* note 34; Rajiv Bhatia et al., *A Health Impact Assessment Toolkit: A Handbook to Conducting HIA, 3rd Edition* (Oakland, CA: Health Impact Partners, 2011), https://perma.cc/2D8U-2YH5.

101 See Section 14.3.3. See also Janet Collins, *Health Impact Assessment: A Step Toward Health in All Policies*, 302(3) J. of the Am. Medical Assn 315 (2009) (discussing examples of integrated EIA and HIA from California and Alaska); Bhatia and Wernham *Integrating Public Health Into Environmental impact Assessment, supra* note 21; Hodge et al. *Integration of Health & Health Impact Assessments, supra* note 28.

102 See Section 14.3.2.

103 Oregon Health Authority, *Climate and Health Impact Assessment,* https://perma.cc/4Y2P-XDJX.

examples of what factors would be considered in the HIA for different types of climate change-related actions. But the examples are not comprehensive – they do not cover all categories of action, nor do they cover all categories of health effects. Thus there is ample room for additional guidance in this area. This guidance could draw on experience from HIAs that have been conducted in the context of real-world actions involving climate change mitigation and adaptation considerations, including:

* *The California Department of Public Health's HIA for the AB 32 Cap and Trade Program*: This document outlines a number of pathways by which cap and trade programs can affect public health. It also contains recommendations on how to minimize adverse health outcomes and maximize beneficial health outcomes when designing such programs.[104]
* *Climate Smart Communities Scenarios HIA*: This document assesses the health impacts of low carbon transportation and land use planning scenarios in Portland Oregon.[105]
* *Southeast Florida Regional Climate Action Plan HIA*: This document assesses health outcomes under 110 climate change adaptation and mitigation recommendations set forth in the Southeast Florida Regional Climate Change Compact's Regional Climate Action Plan (RCAP) to determine how to best prioritize the RCAP recommendations and maximize positive health outcomes.[106]

Even in the absence of formal guidance, these documents can serve as models for conducting HIA for similar projects.

*Second*, tools to streamline and standardize health assessments of climate change would be valuable in this context. The Social Cost of Carbon, Social Cost of Methane, and Social Cost of Nitrous Oxide are great examples of streamlining and standardizing tools: The person preparing an impact assessment for a project that generates or abates greenhouse gas emissions can use these tools to quickly estimate the social impact of those emissions without conducting any complex research or modeling. Similar tools could be developed to better elucidate the public health outcomes of

104 California Department of Public Health (CDPH), *Health Impact Assessment of a Cap-and-Trade Framework* (2010), https://perma.cc/82YB-SD96. See also Maxwell Richardson et al., A Health Impact Assessment of California's Proposed Cap-And-Trade Regulations, 102(9) Am. J. of Pub. Health e52 (2012).

105 Mandy Green et al., Oregon Health Authority, *Climate Smart Scenarios: Health Impact Assessment* (2013), https://perma.cc/AH4G-DSBE.

106 Southeast Florida Regional Climate Change Compact, Regional Climate Action Plan Health Impact Assessment: Minimizing the Effects of Climate Change in Southern Florida (2014), https://perma.cc/7G8V-MBMM.

greenhouse gas emissions – for example, a "public health cost" estimate for each ton of greenhouse gas emissions (similar to the "social cost" but focusing exclusively on health costs), or a tool that generates estimates of specific public health outcomes per ton of emissions. Granted, it may be difficult to reach consensus on specific public health outcomes in the near future, but as climate change attribution science progresses, this would be a feasible task.

*Finally*, better coordination with public health agencies and nongovernmental public health organizations will also be critical for agencies that are assessing health impacts of proposed actions but lack expertise in the assessment of health impacts. There are many documented examples of how such coordination can improve not only the quality of assessments but also public health outcomes, while simultaneously reducing the burden on the lead reviewing agency.[107] While such coordination could occur on an ad hoc basis, developing a formalized procedure whereby public health agencies review and provide input on EIA and RIAs would be helpful in this context.

All of these actions could help to address capacity and resource constraints. However, it is important to keep in mind that additional guidance, technical support, and coordination can only go so far – dedicated financial resources and skilled personnel will be needed to perform these assessments and to act on the findings.

## 14.5 CONCLUSION

There are a number of different impact assessment laws that provide opportunities for integrating public health considerations into the overall analysis of whether and how to proceed with climate change-related actions. But these laws do not cover all types of actions, nor do they contain specific directives on how to assess public health outcomes. As a result, public health is not always a prominent criterion in planning and decision-making processes. Rather than developing new laws that would require HIA for actions with important implications for climate change, this chapter recommends improving the existing impact assessment frameworks to

[107] See California Department of Health, *A Guide for Health Impact Assessment*, *supra* note 26, at 18, 33, 38 (reviewing some examples of collaboration on HIA); U.C. Berkeley Health Impact Group, *HOPE VI to HOPE SF: San Francisco Public Housing Redevelopment – A Health Impact Assessment* (2009) (HIA conducted as an informal partnership between the San Francisco Department of Public Health, researchers at U.C. Berkeley, Human Impact Partners, and Telegraph Hill and Mission Neighborhood Community Centers).

provide for better consideration of health impacts. In addition, this chapter recommends expanding the scope of impact assessment requirements to cover legislative proposals. Finally, this chapter recommends that regulators develop tools aimed at streamlining and standardizing the assessment of health impacts and overcoming institutional capacity and resource constraints.

# Index